Maritime Governance and Policy-Making

Michael Roe

Maritime Governance
and Policy-Making

 Springer

Prof. Michael Roe
School of Management
Plymouth Business School
Plymouth University
Plymouth PL4 8AA
UK

ISBN 978-1-4471-4152-5   ISBN 978-1-4471-4153-2 (eBook)
DOI 10.1007/978-1-4471-4153-2
Springer London Heidelberg New York Dordrecht

Library of Congress Control Number: 2012938542

© Springer-Verlag London 2013

This work is subject to copyright. All rights are reserved by the Publisher, whether the whole or part of the material is concerned, specifically the rights of translation, reprinting, reuse of illustrations, recitation, broadcasting, reproduction on microfilms or in any other physical way, and transmission or information storage and retrieval, electronic adaptation, computer software, or by similar or dissimilar methodology now known or hereafter developed. Exempted from this legal reservation are brief excerpts in connection with reviews or scholarly analysis or material supplied specifically for the purpose of being entered and executed on a computer system, for exclusive use by the purchaser of the work. Duplication of this publication or parts thereof is permitted only under the provisions of the Copyright Law of the Publisher's location, in its current version, and permission for use must always be obtained from Springer. Permissions for use may be obtained through RightsLink at the Copyright Clearance Center. Violations are liable to prosecution under the respective Copyright Law.

The use of general descriptive names, registered names, trademarks, service marks, etc. in this publication does not imply, even in the absence of a specific statement, that such names are exempt from the relevant protective laws and regulations and therefore free for general use.

While the advice and information in this book are believed to be true and accurate at the date of publication, neither the authors nor the editors nor the publisher can accept any legal responsibility for any errors or omissions that may be made. The publisher makes no warranty, express or implied, with respect to the material contained herein.

Printed on acid-free paper

Springer is part of Springer Science+Business Media (www.springer.com)

"Postmodernism is the Swiss Army knife of critical concepts. It's definitely overloaded and it can do almost any job you need done."
Louis Menard, 2009

Preface and Acknowledgments

...two different opinions were in conflict: one recognised the Newts as a new working class, and sought to ensure that all social legislation should be extended to them as regards the working day, holidays with pay, health insurance and old age pensions, and so on; the other view, on the contrary, was that the Newts constituted a dangerous and growing menace to working men, and that Newt labour should be simply prohibited on the grounds that it was anti-social. Against this argument not only did the representatives of the workers raise objections, but the delegates of the workers pointed out that the Newts were not only a new labour force, but also a big and increasingly important outlet as consumers... The International Labour Office, of course, could not ignore these objections; and as a result of lengthy negotiations a compromise was arrived at so that 'employees mentioned above as belonging to Group A (amphibian) could only be employed under water, or in the water, or if on shore then only within a distance of ten yards from high–water mark; that they should not mine for coal or petroleum at the bottom of the sea; that they should not manufacture paper, textiles, or artificial leather from seaweed for consumption on land,' etc: these regulations imposed on the activities of the Newts were put together in a code of nineteen paragraphs, of which we give no details chiefly because as a matter of course nobody ever respected them; but as a solution to the Newt problem, on broad and truly international lines dealing with industrial and social questions, the code referred to above was a meritorious and imposing effort. Capek, *Válka s mloky* (*War With The Newts*) (1936), pp. 230, 231

Many public policy initiatives in the shipping and ports sectors have failed significantly in the past 25 years, with particularly notable examples from the European Union and its relationship with other jurisdictions. These range from the inconsistencies associated with liner shipping regulation, through repeated disasters in the bulk and ferry shipping sectors, to problematic relations among the European Union, the IMO, the OECD and major maritime nation-states and culminating in the recent inadequacies in EU ports policy initiatives and open dispute between European Commissioners over division of responsibilities in the maritime sector.

Maritime policy analysis presented in the following pages takes Harvey's (1982, 1989, 1990) concept of a 'spatial fix' with regards to capital accumulation that requires continuous space–time compression to sustain the capitalist system's desire for growth and development. Passing from a Post-Fordist/Modernist environment characterized in governance terms by a period of state-centred hierarchies whereby maritime policy-making focused upon state initiatives, state influence in supra national and international bodies such as the EU and IMO, and state control of regional and local finance, we have moved from around 1970–1980 into a period of Postmodernism characterized by accelerated and intense compression of space and time as globalization continues to progress which has made unviable the modernist state-centric governance hierarchy. Whilst the term Postmodern may have its detractors the existence of a new global scenario is undeniable. Meanwhile the archetypal state-centric, hierarchical political and governance structures of the maritime sector have yet to catch up contributing to the policy failures that have increasingly characterized the sector in recent years.

The focus is upon how governance in the maritime sector needs to adapt to meet the demands of a Postmodernist environment and the continuing pressure for accelerated space-time compression. It identifies the failures of governance, the processes which characterize maritime governance, and a selection of potential responses to the problems identified. It also questions the lack of enthusiasm shown by the shipping industry for revision of the governance framework within which it currently works. Is this merely the response of an industry characterized by regulatory anachronism or is it a reflection of the advantages that can be gained from a weak and inefficient governance structure for an industry grounded in an environment where capital accumulation is paramount?

There really are too many people to thank individually and without whose help and encouragement this project would never have been completed. However, it would be unforgivable not to mention the Onassis Foundation for their financial support and in particular Frederique Hadgiantoniou in Athens, the help of Venus Lun at Hong Kong Polytechnic University and of course colleagues at the University of Plymouth who were especially supportive in providing a research grant, a sabbatical from teaching and also while I recovered from an unexpected and unappreciated subarachnoid haemorrhage. Notable thanks also go to my ever faithful insulin pump—not everybody carries around a major organ in a pocket, after all.

Finally, enormous thanks as always go to Liz, Joe and Siân for making it all worthwhile.

Meanwhile in a season that at the moment is looking promising, hope springs eternal that Charlton Athletic will soon return to where they belong.

West Hoe, Plymouth, March 2012

http://www.nhs.uk/conditions/Subarachnoid-haemorrhage/Pages/Introduction.aspx
http://www.guysandstthomas.nhs.uk/services/specialistmedicine/diabetes/diabetes.aspx
http://www.londondiabetes.nhs.uk/content.aspx?pageid=100729
http://www.cafc.co.uk/page/Home

Contents

1 **Failure? What Failure?** ... 1
   1.1 Failure? What Failure? Are You Sure? ... 2
      1.1.1 Liner Shipping Regulation ... 2
      1.1.2 Tramp Shipping ... 4
      1.1.3 European Register of Shipping ... 4
      1.1.4 Double Hulled Tankers ... 6
      1.1.5 Tonnage Tax ... 7
      1.1.6 Ports ... 9
      1.1.7 The IMO, The European Union, and Nation-States ... 11
      1.1.8 UNCLOS ... 17
      1.1.9 Criminalization of Seafarers ... 18
      1.1.10 Places of Refuge ... 19
      1.1.11 Security ... 20
      1.1.12 Environmental Issues ... 20
      1.1.13 Regional Policy ... 27
      1.1.14 Logistics ... 28
      1.1.15 EU Policy Documents ... 28
      1.1.16 World Trade Organisation (WTO) ... 31
      1.1.17 Bulker and Tanker Construction ... 31
   References ... 33

2 **Governance** ... 41
   2.1 Governance Models ... 55
   2.2 Networks ... 61
   2.3 Other Governance Models ... 72
   2.4 Governance Failure ... 76
   2.5 Maritime Governance ... 82
   2.6 Conclusions ... 94
   References ... 94

3 Hierarchies 111
- 3.1 The Hierarchical Model 113
- 3.2 The Effectiveness of Hierarchies 116
- 3.3 The European Union, Governance and Hierarchies 125
- 3.4 Conclusions 128
- References 130

4 The Nation-State 137
- 4.1 The Decline of the Nation-State 153
- 4.2 State Survival 161
- 4.3 Time for a Review! 174
- 4.4 The Nation-State Under Global Pressure 175
- 4.5 Conclusions 182
- References 184

5 Globalization 197
- 5.1 A Definition of Globalization 199
- 5.2 Globalization and the Nation-State 203
- 5.3 Globalization and Space–Time Compression 206
- 5.4 The Relationship Between Disorganization, Globalization, and Capitalism 218
- 5.5 The Nation-State, Territory, Governance, and the Role of Globalization 232
- 5.6 The Consequences of Continued Globalization for Contemporary Maritime Governance 240
- 5.7 Conclusions 244
- References 245

6 Modernism 255
- 6.1 Modernism: Its Character and Characteristics 256
- 6.2 Modernism to Postmodernism 278
- 6.3 Conclusions 288
- References 288

7 Postmodernism 295
- 7.1 The Origins of Postmodernism 296
- 7.2 The Characteristics of Postmodernism 301
- 7.3 The Relationship of Postmodernism to Modernism and Post-Fordism 307
- 7.4 Postmodernism and Governance 326
- 7.5 Postmodernism: The Case Against 341
- References 343

8	**Maritime Postmodernism in Practice**		353
	8.1	Postmodernism and Organizations	354
	8.2	Postmodernism, Transport, and the Maritime Sector	358
		8.2.1 Gender, Postmodernism, Transport, and Shipping	362
		8.2.2 Postmodernism, the Third World, and Transport	363
		8.2.3 The Postmodern Waterfront	364
		8.2.4 Postmodernism and Maritime Architecture	367
		8.2.5 Maritime Postmodernity: Blogs, Pods, and E-mails	368
		8.2.6 Screen Trading, E-booking, and Internet Cargo Management	370
		8.2.7 The Media, Interest Groups, and NGOs	371
		8.2.8 Port Infrastructure, Organization, and Technology	372
		8.2.9 Shipping and Technology	375
		8.2.10 Logistics, JIT, and Lean Supply Chain Management (SCM)	376
		8.2.11 Marketing and the Postmodern Maritime Sector	377
		8.2.12 Transport, The Arts, and Postmodernity	378
		8.2.13 Naval Strategy and the Postmodern	381
		8.2.14 Labor	382
		8.2.15 Green Shipping	383
		8.2.16 Shipping Organization	384
		8.2.17 Postmodern Maritime Governance	385
	References		388
9	**So, What Next?**		395
	9.1	The Attitude of the Shipping Industry	396
	9.2	The Problems of Implementing Governance and Policy Change in the Maritime Sector	403
	9.3	So What Is to be Done?	406
		9.3.1 Institutional Change	407
		9.3.2 Stakeholders	408
		9.3.3 Speed	413
		9.3.4 Flows	418
	9.4	Cosmic Capitalism and the Outer Spatial Fix: The Face of Shipping Future?	422
	9.5	Some Final Thoughts	423
	References		424
Author Biography			433
Index			435

Abbreviations

AMSI	Association of Marine Scientific Industries
ANZDL	Australia New Zealand Direct Line
APL	American President Lines
ASEAN	The Association of Southeast Asian Nations
BIMCO	The Baltic and International Maritime Council
CEMT	Confederation of European Maritime Technology Societies
CESA	Community of European Shipyards Association
CGA	CGM Compagnie Maritime d'Affrètement Compagnie Générale Maritime
CLECAT	Liaison Committee of European Forwarders
COREPER	Committee of Permanent Representatives in the European Union
CRPM	Conference of Peripheral Maritime Regions
EAFPO	European Association of Fish Producer Organizations
EBA	European Boating Association
EBU	European Barge Union
ECASBA	EC Association of Shipbrokers and Agents
ECJ	European Court of Justice
ECSA	European Community Shipowners Association
EIA	European Intermodal Association
EEDI	Energy Efficiency Design Index
EEZ	Exclusive Economic Zone
EFIP	European Federation of Inland Ports
EHMC	European Harbourmasters Committee
ELAA	European Liner Affairs Association
EMEC	European Marine Equipment Council
EMF	European Metalworkers Federation
EMPA	European Maritime Pilots Association
EMSA	European Maritime Safety Agency
ESC	European Shippers Council
ESPO	European Seaport Organisation
ETA	European Tug Operators

ETF	Euopean Transport Workers Federation
ETTC	European Towing Tank Community
EU	European Union
EUDA	European Dredging Association
EURACS	European Association of Classificiation Societies
EURMIG	European Recreational Marine Industry Group
EUROGIF	European Oil and Gas Innovation Forum
EUROPECHE	Association of National Organizations of Fishing Enterprises
EUROPIA	European Petroleum Industry Association
EUROS	European Register of Shipping
FCL	Full Container Load
FEAP	Federation of European Aquaculture Producers
FEMAS	Federation of European Maritime Association of Surveyors and Consultants
FEPORT	Federation of European Ports Private Operators
FFA	Forward Freight Agreements
G8	Group of 8
GATS	General Agreement on Trade in Services
GATT	General Agreement on Tariffs and Trade
GSCC	Greek Shipping Co-Operation Committee
GTN	Global Transportation Network
GVC	Global Value Chains
HELCOM	Baltic Maritime Environment Protection Committee
IACS	International Association of Classification Societies
IAMI(EU)	Organization of Maritime Institutes of the EU
ICC	International Chamber of Commerce
ICP UN	Informal Consultative Process on Oceans and the Law of the Sea
ICS	International Chamber of Shipping
ICT	Inland Container Terminal
IGO	International Governmental Organization
IIMS(EG)	International Institute of Marine Surveyors
ILO	International Labour Organization
IMCO	Inter-Governmental Maritime Consultative Organization
IMarEST	Institute of Marine Engineering, Science and Technology
IMF	International Monetary Fund
INGO	International Non-Profit Non-Governmental Organization
IMO	International Maritime Organisation
INE	Inland Navigation Europe
INTTRA	Maritime e-Commerce Platform
ISU	International Salvage Union
JIT	Just In Time
KNVR	Koninklijke Vereniging van Nederlandse Reders (Netherlands Register of Shipping)
LR	Lloyd's Register

MARPOL	International Convention for the Prevention of Pollution from Ships
MBO	Management by Objectives
MEPC	Marine Environment Protection Committee
MSC	Mediterranean Shipping Company
NAFTA	The North American Free Trade Agreement
NASA	National Aeronautics and Space Administration
NGO	Non-Governmental Organization
NVOCC	Non-Vessel Operating Common Carrier
OCEAN	Organization of European Community Ship Suppliers
OMC	Open Method of Coordination
OECD	Organisation for Economic Cooperation and Development
PPP	Public-Private Partnerships
QUANGO	Quasi-Autonomous Non-Governmental Organisation
SCTW	Standards of Training, Certification and Watchkeeping Convention
SEMP	Ship Efficiency Management Plan
SOLAS	International Convention for the Safety of Life at Sea
SBM	Single Buoy Mooring
SCM	Supply Chain Management
SEMP	Ship Energy Management Plan
SPM	Single Point Mooring
STW	Standards of Training and Watchkeeping
TACA	Trans Atlantic Conference Agreement
TMM	Transportacion Maritima Mexicana
TMN	Traditional Maritime Nations
TQM	Total Quality Management
UN	United Nations
UNCLOS	United Nations Conference on the Law of the Sea
UNCTAD	United Nations Conference on Trade and Development
WCO	World Customs Organization
WHO	World Health Organisation
WSC	World Shipping Council
WTO	World Trade Organisation

Chapter 1
Failure? What Failure?

Van Loon (1932, pp. 3–4) gave us much that we need to consider:

> If everybody in this world of ours were six feet tall and a foot and a half wide and a foot thick (and that is making people a little bigger than they usually are), then the whole of the human race… could be packed into a box measuring half a mile in each direction….
>
> If we transported that box to the Grand Canyon of Arizona and balanced it neatly on the low stone wall that keeps people from breaking their necks when stunned by the incredible beauty of that silent witness of the forces of Eternity, and then called little Noodle the dachshund, and told him… to give the unwieldy contraption a slight push with his soft brown nose, there would be a moment of crunching and ripping as the wooden planks loosened stones and shrubs and trees on their downward path, and then a low and even softer bumpity–bumpity-bump and a sudden splash when the outer edges struck the banks of the Colorado River.
>
> Then silence and oblivion!
>
> The human sardines in their mortuary chest would soon be forgotten.
>
> The canyon would go on battling wind and air and sun and rain as it has done since it was created.
>
> The World would continue to run its even course through the uncharted heavens.
>
> The astronomers on distant and nearby planets would have noticed nothing out of the ordinary.
>
> A century from now, a little mound, densely covered with vegetable matter would perhaps indicate where humanity lay buried.
>
> And that would be all.

Meanwhile:

> Whilst there is a temptation to focus all reform attention on the flag states as the weakest link in the regulatory chain, it is unrealistic to hope that flag state supervision alone will be adequate to counteract the pervasive market forces that operate on shipowners. (Lloyd's List, 20th May, 2005)

With this quote from Lloyd's List a number of the issues that are central to current maritime policy-making are brought to the forefront. The relationship among disreputable flag states, globalization and markets, and the role of shipowners is the focus of many of the problems facing the maritime sector and the stimulus for many policies generating much debate (see for example Roe 2007a, b, c, d, e, 2008a, b,

2009a, b, c, d, 2010a, b; Roe and Selkou 2006; Selkou and Roe 2004). But beware—maritime policies will remain ineffective regardless of their laudable and appropriate aims if their derivation, reflection of stakeholders' interests, and the governance of how they come about and delivered, is substantially lacking.

The purpose of this book is quite simple—to outline the significance of the problems facing maritime policy and governance across all policy jurisdictions—international, supranational, national, regional, and local; to establish likely causes and in particular if what has been observed is a consequence of an underlying policy and governance inadequacy; and to suggest some moves that might make some difference. The devil is in the detail of course, for maritime policy and governance are ill-defined, misunderstood, and commonly deliberately misrepresented. Our story is one with roots long ago and it is necessary to accommodate much of its history if we are to begin to understand the causes of the failures which we shall examine and to make any progress in addressing their solution. Consequently, the reader will need to be patient as there are no quick fixes in the policy arena, certainly none which have substance or durability, and there is a requirement of those taking part for endurance, grit, and determination (as well as a little foresight and talent). However, our story will be lightened by tales of death and destruction, injury and corruption, disease and immorality, incarceration and violence—a case of a cornucopia of sex, drugs, rock and roll, and ships. A happy ending cannot be promised, in fact not even an ending, but it does come with a guarantee that it will make you think.

So where do we begin? Perhaps with some maritime policy failures that give teeth to the problems that are claimed to exist and which might guide us in beginning to understand why these problems occur.

1.1 Failure? What Failure? Are You Sure?

Much of the discussion on policy failure will concentrate upon the situation in the maritime sector in the European Union (EU). This is because there are many examples of policy debate and governance dispute that have emerged at this level of jurisdiction and over many years both generally (see for example Smith 1996) and related to the maritime sector (Bredimas and Tzoannos 1981). However, because of the globalized nature of shipping, all maritime policies tend to exhibit jurisdictional interlocking and as a result discussion crosses over to evidence emerging at international, national, regional, and local levels as well. There are many examples of policy failure that we can take. Our choice is not wholly unlimited but its range and scope at the same time is almost breathtaking.

1.1.1 Liner Shipping Regulation

We can start some years back with a policy anomaly that illustrates the scale and longevity of the degeneration of maritime policy-making and governance across all levels of jurisdiction. The 1986 package of measures for the maritime sector agreed by the EU included the infamous 4056/86 Regulation that provided exemption for liner conferences operating into or out of the EU (then Community) from the Treaty of Rome rules on competition (Paixaio and Marlow 2001; Selkou and Roe 2005). Liner conferences in their traditional form clearly breached the Treaty in many ways and other commercial and industrial sectors in the EU have always been prevented from collusion and the formation of cartels under any circumstances with severe fines applied to those stepping out of line. Not so liner shipping which was allowed to continue to enter into agreements on schedules, rates, and the like—albeit with some restrictions and with a degree of enforced transparency. This anomaly emerged from a complex and heated political debate between the European Commission and the Council of Transport Ministers, with some intense pressure from leading (generally larger) liner shipping companies such as Maersk, P & O, Nedlloyd, and Hapag Lloyd. Despite heavy criticism, the exemption continued to be applied from its inception in 1986 and was only finally eliminated in 2008 despite almost universal condemnation apart from a few shipowners that were the major beneficiaries from the protection offered.

Regulation 4056/86 was inconsistent, incoherent, and irrational in the context of the Treaty of Rome, the principles of the EU and their relatively rigid application to all other sectors. For 20 years, simply its existence was a clear policy failure as it violated the most significant principle that drives the EU in all that it does—that of free competition to achieve maximum economic benefits across the EU. It also represented an aberration of governance in its emergence from a political tiff between two authorities of the EU, rather than as a rational consequence of structured and open debate of the issues. Its ineffectiveness was apparent from its failure in application. Examples abound, but one that was especially notable featured fines imposed on operators in 1998 for collusion through the Trans-Atlantic Conference Agreement (TACA) which were substantial but subsequently quashed on appeal in 2003 (Lloyd's List 2006aa).

From 2006 onwards, there were major steps taken to abandon 4056/86 so that by 2008, the Treaty of Rome was fully applicable to the liner shipping industry and conferences became illegal for all operations into and out of the EU (Lloyd's List 2006w, dd, ee, kk, nn, yy; Tradewinds 2006a, b, c). While in itself this ended a major maritime policy anomaly, the process of removal has been fraught with argument and in particular accusations of vagueness by the European Commission with respect to what should come next and how much collusion between operators would continue to be acceptable (Tradewinds 2007c). The European Liner Affairs Association (ELAA), a Brussels-based lobbying group, commented that the guidelines issued by the European Commission in September 2007 were too 'vague and in some cases contradictory' and 'carriers will have to assess whether it is safe to gather and publish information' (Tradewinds 2007d). In addition, there was continued distrust between the Council of Ministers and the Commission with the former Council of Competition Ministers demanding the right to be consulted on each stage of the phasing out by the Commission (Lloyd's List 2006aa). Meanwhile in 2007, shippers continued to express fears about guidelines which the

EU were preparing for the liner sector which they considered too lenient to shipowners and later in 2009, continued to challenge the shipping industry over regulation of the liner trades by the EU (Lloyd's List 2007o, 2009ee). Even from the grave, 4056/86 will continue to haunt EU maritime policy-making.

1.1.2 Tramp Shipping

A related problem has featured only much more recently and concerns the position of tramp shipping in relation to EU competition policy. For many years tramp shipping was assumed to be exempt from the competition rules but suddenly, in late 2006, the Directorate General for Competition announced that it would enforce rules on operators which if violated could result in fines of up to 10 % of the group turnover (Lloyd's List 2006cc). The position has always been vague in that competition law did apply to tramp pools but its application was left in the hands of member states while the similar situation relating to liner conferences was a Commission responsibility. The sudden change of emphasis was unclear and unexpected. The Council of Ministers' Regulation shifting powers from member states to the Commission was agreed without notice to stakeholders who were expecting change but not for some months. In addition to the unexpected timing, the impact of the new Regulation was unknown on the industry and its derivation, timing, and impact had remained a close secret. Rather unconvincingly, the Commission was felt to prefer to test the effect of competition law in practice rather than spend time predicting and considering the effects before implementation.

1.1.3 European Register of Shipping

Meanwhile, the initial attempts at launching a European Register of Shipping (EUROS) under the 1989 package of maritime proposals presents another catalog of disasters in policy terms. Wrapped up with a variety of add-ons to make it more tempting, EUROS offered privileged access to aid cargoes and a variety of potential tax concessions in exchange for joining a second (European) register which came with restrictions on manning (Smith 1993). As an attempt to counter the growth of flags of convenience by offering a European registry with higher standards but with options to reduce the strict manning requirements of state flags, it failed miserably falling between the economic advantages of flags of convenience and the standards which were ensured (at some cost) by conventional registries.

In fact EUROS never really even began to get started. Fiercely resisted by shipowners and labor unions that saw it as a threat to profits and jobs respectively, the European Commission failed to understand that policy proposals of this type

1.1 Failure? What Failure? Are You Sure?

needed the full support and agreement of those most affected from their initial conception. This was never even considered. The policy emerged from high-level discussions and was expected to filter down the jurisdictional hierarchy from supranational through national member states to those at the operational level—the shipowners and labor unions—who would enthusiastically take up the idea despite the failure to incorporate their needs and fears in the process. This pattern of governance failure has been repeated in recent years over the development of EU ports policy—more on which later. EUROS was eventually abandoned after many years of sitting on the agenda of high-level groupings in the EU, a victim of a misconceived policy that had failed to incorporate the views of the various stakeholders that had central interests in the issues it raised.

Interestingly, the whole issue of a European register was raised again by the Commission in the debates on maritime policy through 2007 and into 2008, despite an intriguing combination of clear and widespread opposition and apathy. The public consultation process that led to the publication of the 2007, Blue Maritime Policy Paper, contained a strong theme that hinted at the creation of a new EUROS linked to the concept of a European Maritime Space and (as we shall see later) the concept of maritime cabotage. All three concepts generated fierce opposition amongst the maritime community with almost no exception—and yet remained at the forefront of Commission proposals throughout 2006 and 2007 (Lloyd's List 2006l).

The proposals had first emerged during February 2006 with suggestions that the Commission had plans for a European registry leaked to newspapers and led to a fierce denial of any such idea (Lloyd's List 2006a). Stemming from ideas of European Flag State control and a single line in notes accompanying the Erika 3 package of legislation referring to the 'development of a European flag', the Commission stated that they did 'not foresee the adoption of a European flag state'. However, the fact that the use of a single phrase in an obscure part of documentation can lead to extensive debate at the European Parliament and across the European media is indicative of a failure of trust across jurisdictions and between governance authorities.

Despite the Commission denials, the debate continued throughout 2006 with the Green Paper questioning whether 'an optional EU register be made available?' and Commission President Jose Barroso quoted as saying that 'he would not hesitate to propose the EU flag... if they (*sic*) were deemed necessary after the consultation phase' (Lloyd's List 2006l).

Clear indication of the intensity of debate over maritime policy throughout the Green Paper consultation phase came from the appointment of Fotis Karamitsos as Director for Maritime Policy by the Commission, who stressed in September 2006 that the Commission had not made up its mind on a European flag (Lloyd's List 2006bb). However, finally, at least for the time being, in October 2006 the plans for a European flag were put to rest when the Commission's Deputy Director for Energy and Transport, Zoltan Kazatsay commented at the Maritime Industries Forum that the policy was 'going to fail'. In response the secretary general of the European Community Shipowners Association compared the situation with that of EUROS in the early 1990s stating:

Why did EUROS fail at that time? The simple reason is that it intruded into the competence of member states on the company taxation and the taxation of seafarers and social security. (Lloyd's List 2006ff).

Although the issue was raised again by European Commissioner Joe Borg in October 2006, it was only in the context of encouraging free and open discussion of all potential policies (Lloyd's List 2006hh). In that sense the renewed discussion of European flags was healthy in that it was not a policy pursued by the Commission in opposition to member states and stakeholders. However at the same time, the interest and resentment generated reflects an ongoing distrust that exists between each group and a failure of communication itself implying difficulties within the maritime governance community.

1.1.4 Double Hulled Tankers

Maritime policy has continued to exhibit a catalog of failures since then. One area of policy that has given trouble for many years has stemmed from environmental concerns about shipping and in particular something that has raised much public debate—the safety of oil tankers and the infrequent but disastrous effects of a number of incidents in EU waters in recent years. These have included the most infamous of all with the sinking of the Prestige and Erika and the catalogue of political, policy, and governance failures that followed. These culminated in policy decisions in France and Spain at the national level concerning the compulsory introduction of double-hulled tankers in advance of similar legislative and policy proposals both at the international (IMO) and the supranational (EU) levels (De Vivero and Mateos 2004).

In a later section, we shall be considering the traditional hierarchical organization of maritime policy-making and governance that has been adopted and whether it is a cause of some of the policy problems that we can identify. A dominant principle of hierarchical policy-making is that policies emerge in broad, generic form from the higher jurisdictional levels, to cascade down the hierarchy to lower levels where they are applied to the sector with appropriate, locally relevant detail adopted. The adoption of new rules concerning double-hulled tankers by individual nation-states in the wake of the *Prestige* was in clear conflict of hierarchical policy-making with agreement descending through jurisdictions and implementation at the lowest level possible, closest to the stakeholders who have most to gain and lose. Instead, we had national member states taking the lead because of the delays in implementing policy at superior jurisdictions and consequently sending messages up the hierarchy to higher levels as well as down to those involved in the day-to-day operation of shipping. The failure of the key characteristics of the hierarchical model that underpins maritime sector governance generated loud objections to the unilateralism that followed. The issue that is important here is not the policy adopted—in fact those at the peak of the

1.1 Failure? What Failure? Are You Sure?

hierarchy had similar views toward tankers as those at the nation-state level—but that the model around which all policy is supposed to be derived and applied and which forms a fundamental framework for the maritime industry should be violated to such an extent—the consequence of which is the potential for confused and contradictory policies and significantly raised opportunities for policy abuse and manipulation by the shipping industry itself. A governance model based on hierarchy, but ignored at will by lower jurisdictions cannot do anything but generate random and widespread policy failure.

1.1.5 Tonnage Tax

One area that has been almost universally seen as a success for maritime policy in the EU has been the relentless growth of tonnage tax initiatives to the extent that very few member states now do not operate such a system (Brownrigg et al. 2001; Selkou and Roe 2004). Of course it is not only EU member states that have introduced tonnage taxation and examples exist in India, Korea, South Africa, Norway, and beyond (Lloyd's List 2006s). However, it is the success of these policies within the EU that is most dramatic and which has attracted substantial favorable comment since the relaxation of the EU Guidelines for State Aid to Shipping that took place from 1997 (Leggate and McConville 2005). However, in whose terms is the policy of tonnage tax a success and how much does it represent an anomaly or worse, an example of hypocrisy stimulated by the desire for financial success generated by those with most to gain?

Tonnage taxation varies member state by member state, but in all cases represents a mechanism that replaces some form of national corporation tax by a specific taxation that is aimed at the maritime sector. The characteristics of variation are considerable. In the UK only direct shipping activities can be included and if a company opts for tonnage taxation instead of corporation tax, then all such activities of the company must be included. This contrasts with the Irish version where virtually any maritime-related activity can be included. Taxation rates vary considerable as do requirements for membership of the national registry, links with training, and lengths of commitment. However, and the key point to attributing 'failure' status to the tonnage tax policy in the EU, all such taxation is a subsidy; a state-aid as defined in the EU in its guidelines, under which tonnage taxation is permitted as an exception by the Commission to the rules of the Treaty of Rome.

Two major measures of policy success are those of consistency and equitable application. Tonnage tax offers neither. As a shipowner, tonnage tax is a boon offering a range of subsidized taxation regimes across reputed flag states in the EU, which with the capital, legal, and physical mobility of shipping, can be played off one against another to achieve the maximum benefit for shipping companies, which threaten to move vessels between flags and capital regimes at will unless their taxation requirements are met. This is neither equitable nor consistent. It is a policy unavailable to any other commercial sector in the EU despite similar characteristics of financial, legal, or even physical mobility (for example the airline industry, international road freight industry, international telecommunications, satellite television etc.), while the benefits to be derived from tonnage taxation are hard to identify apart from to shareholders of the companies concerned and consequential spin-offs in related financial markets. In many cases a link to the respective EU flag or to stimulating interest in the maritime labor market is not even a requirement and any supposed benefits of safety and to the environment claimed by registries from established states remain unrealized. Inconsistencies in the tonnage tax policy were further emphasized in early 2007 by comments from the Managing Director of Carnival Cruising, David Dingle, who in commenting upon the development of a new maritime policy in one breath called upon the necessity to retain tonnage taxation in the EU while also demanding that the European Commission should not be protectionist (Lloyd's List 2007c). By definition, tonnage taxation is an EU protectionist measure, facilitating taxation privileges only for those allowed to receive them and particularly so if entry to the tonnage tax club is reserved for EU member state flags—as is the case for some member states.

Tonnage tax thus benefits very few, does so inequitably across the EU, fails to achieve broader benefits for public goods (safety, the environment, security), acts as an exception to the Treaty of Rome which guides the rest of the activities within the EU, and can be justified only in terms of sustaining a maritime base within the EU—something that is questionable at best and at worst simply a privilege. Its inconsistency is clear in the variety of forms it takes across the EU and in how it is applied in each case. In 2008, the UK Chamber of Shipping and EU Commission clashed on the definition of eligible shipping activity with the UK government applying the EU regulations while under severe pressure from maritime interests not to do so (Lloyd's List 2008d). In May 2008, issues relating to demands for back-tax in Norway raised their head generating claims that it would result in the end of the Norwegian shipping community (Lloyd's List 2008i). In January 2010, it was removed from the jurisdiction of the Commissioner for Transport and placed under the remit of the Commissioner for Competition suggesting that even the EU Commission was doubtful about its suitability for an organization committed to removing or at least lessening the interference of the state in the marketplace (Lloyd's List 2010a). This move generated comments from MEPs that the European Parliamentary Transport Committee was increasingly unsure of who was responsible for what, reflecting doubts about not only the policy but the policy-making institution itself. This was further emphasized when the Commission Directorate conducted its own investigation into the activities of the International Association for Classification Societies without even informing the Transport Directorate (Lloyd's List 2009a). But we divert and must return to tonnage taxation. In policy-making terms it has no place in the current hierarchical arrangement as it violates all other taxation standards at the supranational level and has no basis at the international level in achieving international standards for safety, security, and the environment. And as such it represents a glaring example of policy failure—not for the shipping companies that take every advantage of it but for the shipping (and wider) community as a whole.

1.1.6 Ports

A further troublesome area focuses upon EU Port Policy which quite publicly, has staggered forwards (and backwards) over the last 10 years, in protracted attempts (and failures) to introduce common competitive standards across ports in the EU (Farrell 2001). Heavily promoted by the European Commission as a necessary move toward achieving a European market for port services, the ports policy looked to enforce a competitive regime in larger ports across the EU. This included a requirement to facilitate at least two providers of services (for example for stevedoring, ro–ro ramp provision, container terminal services, and even pilotage) so that port users would have choice. The result would be increased competition, higher quality services, and reduced costs—or so at least the Commission believed. Services included would cover both traditionally commercial but also those traditionally state-run and monopolistic in character—pilotage being the most characteristic of these. In addition the proposals included allowing vessels to use their own labor to load/unload ships to provide further competition in the market.

Largely supported by ship-owner representatives, the ports policy was promoted twice in an attempt to push it through the EU legislative process—but each time it failed due to severe objections from labor unions, port representatives, the various service providers in ports, and a host of other discontented organizations. Few member states saw the need for a policy of this type—the UK for example could not see how it could be made to work in a country where all port facilities were privatized and some port service providers had contracted for long periods into a non-competitive situation. Others argued that competition existed between ports (sometimes between ports in different countries) and competition within them was wholly unnecessary.

Proponents of the EU ports policy geared up for a third attempt throughout 2006 and 2007 and in the process seemingly ignored the opposition that continued to exist at all levels (Lloyd's List 2006c, g, h) exemplified by the views of the European Seaports Organisation (ESPO) (Lloyd's List 2007f). The 2006 Transport Commissioner failed to rule out further attempts at a common ports policy despite continued fierce opposition from almost everyone affected. The situation was further confused by his claim that policy 'action' was not synonymous with legislation leaving the future open to much debate and little certainty (Lloyd's List 2006h). Meanwhile the Green Paper on maritime policy issued in 2006 for the new maritime policy public consultation phase raised the issue again with suggestions that an EU ports policy was a necessity. This generated little enthusiasm. The policy appears to be unwanted and largely (if not entirely) unneeded by all levels of the policy-making hierarchy, by the industry, by its representatives, and by the market in general. This is not to say that *a* ports policy is not needed but what is needed is one that reflects the full range of stakeholders involved in the sector and also one that is not developed and dictated to the industry with inadequate consultation.

Ports policy degenerated further throughout 2007 following continued and long-running clashes between the Commissioners for Transport and Competition and the Director for Maritime Affairs over their respective roles—each seeing themselves as central to policy making in this area (Lloyd's List 2006n, tt). The confusion was not helped when the Chairman of the European Commission Task Force John Richardson attempted to clarify part of the situation, claiming there had been a 'semantic misunderstanding' and that 'Mr Barrot's team would continue to handle 'maritime' issues, whereas Mr Borg's team would handle marine issues'. The respective rights of each are not for us to debate—but what is central is the difficulty of generating and interpreting coherent policies in such a context where there is an open and fractious policy dispute between the individuals responsible for them. The end of 2007 saw the publication of the EU Commission Maritime Blue Policy Paper and a new EU Ports Policy Consultation Paper both of which contained proposals for policy strategy. The Commission indicated that it was looking for progress on ports policy but using the 'soft law' option—a vague concept that implies Commission 'interpretations' rather than legislation which could be used by the European Court of Justice when required (Lloyd's List 2007g, i). The response to such suggestions has not been popular with port stakeholders who view such measures as policy by 'stealth', whereby unpopular policies can be implemented without debate or formal agreement by those outside the Commission. Is it any wonder that the EU ports policy lacks coherence?

The debacle that EU ports policy represents was made clear in 2009 when members of the European Parliament claimed that the European Commission was now not prepared to tackle market distortions in ports because of the repeated failure to introduce a ports policy and that further disagreements within the Commission between the Competition and Transport Directorates had delayed state-aid guidelines for over 10 years (Lloyd's List 2009ff, gg). The OECD (Organisation for Economic Cooperation and Development) even chimed in later that year suggesting there had been 'almost two decades of failed attempts' to open European ports to greater competition, that many ports were 'well behind the productivity frontier', and that port services liberalization remained a 'previously untouched sector' (Lloyd's List 2009hh).

To make matters worse, by June 2010 there were claims by delegates to the ESPO Annual Conference that the proposed policy would have been useful providing some sort of framework for the 'social dialogue' with which the EU has replaced policy proposals especially in the coming discussions upon deregulation, terminal concessions, and safety (Lloyd's List 2010b). Issues such as self-handling have been abandoned as far too contentious but other issues would be covered by Brussels 'soft law' which would emerge from this dialog but would lack structure, consistency, and detail reflected in the 3 years it had taken even to agree to some of the teams to be present at negotiations between the industry and the EU.

The issue of ports policy has not gone away and the Commission remains clearly tenacious in its desire to reform the sector for in September 2011, there were hints emerging from the EC that the issues of cutting red tape, improving financial transparency, and a return to self-handling were to form a significant part of a review of port services leading to a package of measures in 2013 (Lloyd's List 2011g). Predictably, the return to a revitalization of port policy in the EU was welcomed by the European Community Shipowners Association (ECSA) but the waters were muddied by a statement from the President contradicting the rumours suggesting that 'the issue of self-handling has already become a non-issue during the discussions on the two port directives that were rejected' (Lloyd's List 2011h).

1.1.7 The IMO, The European Union, and Nation-States

The issue of jurisdictional hierarchy remains central to other policy problems in the EU maritime sector and has been a feature of policy debate between the member states, other nations, and the Commission for many years (see for example Lloyd's List 2005h, 2006f, o, oo, 2008b). Arguments have raged over the respective roles of international, supranational and national jurisdictions in such fundamental areas as shipping safety with little signs of resolution possibly reflecting something seriously amiss in the maritime governance structure (see for example EurActiv 2009; Lloyd's List 2008j, 2009ii, jj, 2010gg; Tradewinds 2008d, e, f). In recent years a long-running dispute about maritime policy representation has been festering as the European Commission continues to attempt to gain a seat for itself (representing all the EU member states) at the International Maritime Organization (IMO). Supported by perhaps one member state in this approach, the EU continues to suggest that for the member states to have a balanced and strong say in IMO policy-making—in response to pressure from the USA in particular—it is necessary that all members speak with a coherent and single voice which only the EU as a single representative body can provide. The weight of the EU presence would match the existing member states' votes but would overcome arguments and disagreements between them which tend to fragment their views in opposition to other large states and voting blocs. Discussion on agreeing an EU maritime viewpoint would take place in Brussels rather than London so that a coherent and single stand could be taken and subsequently adhered to.

Very few member states support the position of the European Commission and in fact, along with much of the maritime industry, the majority fiercely oppose it (Lloyd's List 2007a). The IMO has been consistent in its opposition as it sees its role as a collective of nation-state representatives despite the fact that the EU does have representation (in lieu of the member states) on other UN and international policy-making bodies. In addition, industry representatives from the International Chamber of Shipping (ICS) and the International Shipping Federation (ISF) condemned any proposals to substitute the member states by an EU representative at the IMO, particularly because it would dilute the technical debates to which member states actively contribute local knowledge (Lloyd's List 2006ll).

Persistent attempts were continued by the EU through its Transport Commissioner to gain EU membership of the IMO and in 2008 it was announced that the forthcoming paper on maritime policy would contain a chapter on 'reinforcement of Europe's presence in international instances' (Lloyd's List 2008l). By January 2009 the Commission was emphasizing that it did not consider the need to gain member state permission to apply for IMO membership (Lloyd's List 2009ll). Constitutional lawyers had advised the Council of Ministers that approval was unnecessary although senior officials did comment that politically it perhaps would be best to achieve a consensus. The Maritime Strategy Document 2009–2018 included:

> For the EU Member states to act as an efficient team that can rely on strong individual players, requires enhancing the recognition and visibility of the EU within the IMO by formalising the EU co-ordination mechanisms and granting informal status, if not full membership, to the EU within this organization

Whereas prior to the IMO meetings, where the Commission currently has observer status, the position of the member states is informally coordinated by the Commission, membership by the EU at the IMO would formalize this process and require member states to abide by the EU majority view. The Commission would therefore have more political weight in the maritime area at the IMO. The Commission would continue to seek Council of Ministers' agreement to become an IMO member replacing the member states, but would only need a qualified majority to achieve this and those in disagreement would have to implement the decision.

This policy dispute is at the heart of the problems that besiege maritime policy-making and reflects the difficulties inherent in the current hierarchical, state-centred approach. The EU sees a need that is unsupported by its member states who, in theory, should be taking the policy lead from the hierarchy above—in this case the EU Commission and other EU bodies. The Commission continues to press ahead with its ambitions at the IMO despite objections from those above it in the policy hierarchy (the IMO) and those below (the member states). The issue is not whether the policy would work if imposed—but whether the policy-making process at present can accommodate such demands, reflects the needs and desires of those it represents, or whether it is outmoded, inappropriate, and as such, a liability. Awareness of this dispute between the IMO and EU and an increasing trend by the EU and its member states toward unilateralism is clear from the increasing number of statements made by Efthymios Mitropoulos, the then Secretary General of the IMO, through 2005, 2006, and 2007 which referred to these issues and which stressed how the IMO saw itself as an international organization increasingly forced to defend its mandate to set global standards of safety at sea, vessel-sourced pollution prevention, and maritime security:

1.1 Failure? What Failure? Are You Sure?

There is also what I consider to be a misguided trend toward seeking political solutions to shipping-related matters without even bringing them to the international body that was specifically set up to deal with them. (Lloyd's List 2005b)

Particular concern has been noted by the IMO over the EU in its consideration of safety issues, mirroring the US policy moves in 1990 which resulted in the US Oil Pollution Act, creating a unilateral safety regime for tankers without consultation with other states through the IMO. Further unilateralism can be anticipated especially with regard to issues that strike a chord with the public, media and politicians—pollution is one clear issue but others include recent difficulties over ship recycling where member state governments and pressure groups are driving policies rather than emerging from the established hierarchical governance framework. In Denmark and the Netherlands exports of scrap vessels to India have been stopped and the UK now insists that all government owned ships must be scrapped in OECD, environmentally sound facilities. These are responses to what are considered the inadequate 2003 voluntary scrapping policies of the IMO and are aimed to bridge the gap that exists until further international law can be agreed. Meanwhile, ship scrapping policy in the EU was placed in further disarray by internal squabbling that emerged between Commissions (Lloyd's List 2009ss, tt).

Clearly the opinions of both national governments and pressure groups representing public views are important to the policy-making process, but to reiterate the problem once again—if a hierarchical policy-making framework exists, and there is the need to coordinate policy to make it work through an international industry, then unilateralism produces chaos and inefficiencies. The answer may well not lie in the structures which exist at present and perhaps other governance approaches may need to be adopted. It is clear that 'the responsibility for resolving this complex web of environmental and human issues must, in the final analysis, be sorted by entities with an international brief like the IMO and the International Labour Organization (ILO)' (Tradewinds 2006a) but the current governance framework clearly also does not work and the apparent slow speed of the IMO, justifiable or otherwise, is generating policy failure through a process of insidious unilateralism in the context of hierarchically structured maritime governance that abhors such moves.

Meanwhile concern has been expressed about the deteriorating relationship among the IMO, the EU, and member states. For example Aart Korteland, Chairman of the Royal Association of Dutch Shipowners (KNVR):

It becomes more and more clear that the Commission is trying to dictate a common policy to the member states (Lloyd's List 2005e)

Pressure has also been placed on the IMO to revise policy-making procedures as member states view current geographical representation at the IMO Council (the main policy-making body) as unfair (Lloyd's List 2006ll). The countries that make up the council are felt by many new maritime powers to be unrepresentative—examples being that the USA is a member of the senior category of the council although its ship register is small, while Liberia, one of the world's biggest, is not represented at all. The result is undemocratic policy-making, limited stakeholder involvement, unilateralism, and policy failure. By 2009 the situation had changed little (Lloyd's List 2009k) when the established nations remained firmly in control. Indicative was that the Cook Islands, Iran, the Marshall Islands, and Pakistan all failed to gain places on the IMO Council as representatives of countries with a special interest in maritime transport. The Marshall Islands has one of the world's largest ship registries and yet has never held a position on the Council.

These inadequacies of representation manifest themselves in IMO policies that are unwanted or treated with suspicion by members. Take the voluntary member state audit scheme—viewed by a number of major registries as 'worthwhile and laudable' but also as generating accelerated flag hopping and with the potential to be abused as a marketing ploy. It would be further disadvantaged by the fact that a number of organizations to which flag states would have the opportunity to delegate certification would not be subject to any IMO audit—but only to nation-state monitoring. This confused state of affairs would allow individual member states to operate IMO-certificated registries without recourse to the IMO itself, creating a governance nightmare (Lloyd's List 2005c, g, 2006b, x). The US Coastguard for example, recognized 155 certification organizations, all of which were outside the ambit of the IMO. To quote the Director General of Belize Registry, Angelo Mouzouropoulos:

> the IMO's approach would only encourage unilateral legislation at Brussels level and elsewhere. (Lloyd's List 2006b)

The debate on the relationship between the EU and its member states has continued to rumble on without resolution and at times has even spread over into IMO's activities elsewhere around the world. In January 2009, the Commission reiterated its view that it did not need member state permission to apply for membership of the IMO (Lloyd's List 2009b) although it was admitted that it might be advisable to do so. The latest maritime policy document from the Commission (Commission of the European Communities 2009b) suggested that observer status was a necessity and full membership desirable obviating the need for informal coordination meetings with member states before IMO meetings to ensure that the views put forward were in line with those of the Commission where the latter had ultimate competence.

In February of the same year the European Parliament joined in the debate about IMO/EU relationships describing the maritime governance overseen by the IMO as weak and suggesting that it:

- Lacked the power to apply its own rules.
- Was too flexible with signatories.
- Could only come up with non-binding resolutions even if action was essential (Lloyd's List 2009c).

The following week it was reported that a collection of EU member states had blocked key measures at a recent IMO STW sub-committee, with Denmark acting to ensure that all member states voted as one (Maritime Global Net 2009). Claims followed from UK trade unions that it was becoming difficult for any member state

1.1 Failure? What Failure? Are You Sure?

to express an independent view and that Chairs of IMO Committees were assuming that all EU states supported the same view unless it was pointed out otherwise.

It was not the EU alone however, that was struggling with the issue of representation. The Asian Shipowners' Forum, representing shipowner groups from China, Hong Kong, Taiwan, SE Asia, Japan, and South Korea, made it clear in 2008 that they wished to obtain observer status at the IMO and were given indications that this was a distinct possibility by 2010 (Lloyd's List 2009d, x, mm). This might involve direct representation at the IMO or the development of an Asian IMO Centre to provide a conduit for policy-making pressure. However, by late May 2009 they had abandoned the plan deciding to work through national governments and with other groups that already had representation, a decision thought to stem from pressure from the Chinese, Japanese, and South Korean governments (Lloyd's List 2009e).

Meanwhile, shipowners from all parts of the world were urging the European Commission to drop plans to set up a regional cargo liability regime and instead to ratify the convention adopted by the UN in 2009 'in the interests of international uniformity' (Lloyd's List 2009f). In addition, the Nautical Institute announced plans to obtain consultative status at the IMO particularly in light of the increasing pressures on seafarers who had little direct voice internationally. This suggested that membership through national governments was not working as well as it should be (Lloyd's List 2009g). The IMO welcomed the application suggesting that even the IMO could see the benefit of wider consultation and one not necessarily allied to nation-state membership.

After a few months of comparative silence the EU reviewed its unofficial policy of pushing for full IMO membership following ratification of the Lisbon Treaty. Contained within the Treaty is the principle that the 'Union'—a combination of member states and institutions that make up the EU—should be represented at all international organizations including the IMO (Lloyd's List 2009h). The situation at that time was that only the Commission had membership and that was at observer level. While the IMO remained fairly relaxed about whatever EU membership would mean, member states were not with clear opposition from Greece, Malta, and Cyprus fearing loss of influence. The Commission continued to exert pressure, indicating in a communication to the next European Parliament Transport Committee meeting that:

> The EU should consistently seek membership in international organizations that are relevant for maritime affairs. The difficulty which traditional intergovernmental organizations face in trying to accommodate the specificity of the EU needs to be overcome.

However, even with EU member-state agreement, difficulties would remain in gaining full membership for the EU. The IMO Convention would need to be amended as only nation-states can be members at present—and this in turn would require member states to agree to this. Hardly likely in the current climate of opposition to the whole idea (Lloyd's List 2009i, j).

Problems have also emerged in the IMO's relationships across jurisdictions with other governmental and non-governmental bodies. For example in December 2009, the IMO announced plans to make audits of member countries mandatory by 2015 following the success of the STCW White List with the aim of enhancing safety and environmental protection (Lloyd's List 2009l). However, some member states expressed fears that it would be 'intrusive of individual governments' practices' as it violated traditional sensitivity surrounding sovereignty. Non-governmental ambitions to gain representation at the IMO included those from EMEC (European Marine Equipment Council) who saw an increasing role of technology created by IMO policies and wished to join, CESA (Community of European Shipyards Association) who already had representation, and the Nautical Institute with ambitions to raise the impact of the seafarer at IMO Committee level (Lloyd's List 2008k, 2009kk).

The IMO also debated their relationship with member state governments especially over the issue of piracy suggesting that states must act both individually and collectively through the IMO if unilateralism is to be avoided (Lloyd's List 2008m, 2009zz).

However, it is the IMO/EU relationship that remains at the heart of the debate and the clear failure to resolve membership and representation differences continues. By late summer 2010 'simmering tension between the IMO Secretariat and the European Commission may have reached an entente cordiale' but the issue of direct EU membership had not been resolved (Lloyd's List 2010c). In particular the continued disagreement of what was termed 'common but differentiated responsibilities' concerning climate change and the IMO, EU and UN Framework Convention on Climate Change (UNFCCC) remained a running sore although some progress toward a resolution was seen at last in July 2011. The IMO remained accused by other jurisdictions of being slow with the example of shipbreaking which first featured in the news in 2006 (Tradewinds 2006a). This was followed by the Hong Kong Convention on Shipbreaking which was expected to take at least 10 years to be ratified by the 63 nations that adopted the Convention (Lloyd's List 2010d). Clearly, like all arguments there are two sides to this issue— the IMO may well be an organization designed to be slow but steady (and in that it achieves anything with 170 or so member states is a minor miracle in itself), but the nation-state members have it in their own hands to move more quickly (Lloyd's List 2010e). The year 2010 also saw deterioration in the efficiency of fee payment by member states reflecting a scant disregard for the IMO's work and status. At least one large flag state missed payment day in January 2010, felt to be a deliberate protest against the large rise which had been imposed (Lloyd's List 2010f). Others had also failed to pay up by May and it was reported that the worse culprits were actually members of the governing council rather than the membership as a whole (Lloyd's List 2010g). Serious defaulters included Panama, Greece, Malta, Marshall Islands, Japan, China, the USA, Cyprus, and South Korea, some of which had debts from earlier years as well and many of which are major shipping flags. While some paid up soon after, others remained defiant. Concern was also expressed at the lack of financial transparency at the UN agency which releases almost no official financial data despite its role as a public institution (Lloyd's List 2010h), representing a 'veil of secrecy that remains the unacceptable face of the modern shipping industry'.

Only some of the myriad of policy problems that have characterized the IMO and its relations with the EU have been outlined here, including in December 2008 when it was reported that there were concerns in the IMO over the EU's plans for class society recognition (Lloyd's List 2008o); while March 2010 saw further reports speaking of leading flag states continuing to refuse to pay fees to the IMO as an act of conscious protest against a rise of 14.9 % (Lloyd's List 2010kk). Overall, it seems clear that the general jurisdictional arrangement of maritime policy-making could work better to improve the lot of the sector and it may well be that the current system and its institutional components, inherited from a pre-war design for international jurisdiction are far from the optimum.

1.1.8 UNCLOS

Further debate was generated following the EU Maritime Commissioner Joe Borg's call for EU full membership of the UN Informal Consultative Process on Oceans and the Law of the Sea (ICP) in a speech in September 2005 (Lloyd's List 2005d). The EU had only observer status while each of the individual member states of the EU were full members. Borg went on to say:

> We have made it clear that we attach great importance to resolving the discrepancy that currently exists between the European Community's observer status and *its competences—whether exclusive or mixed—with respect to many issues that are being discussed in the ICP.* (Lloyd's List 2005d) (*emphasis added*)

Now these comments by Borg go back to the heart of the EU governance and policy problems (Lloyd's List 2005a). The Commission sees an irrationality in policy competences and representation. Their view is that their role is to develop and impose maritime, legal policies but they are represented in fragmented form (it could be argued that they are not represented at all) and therefore their competencies are prejudiced. This is a governance issue which is made more severe by the fact that the member states who make up the EU are in total disagreement with their own representatives from the Commission over who should take part in such fora (and the IMO representation issue is much the same). From this sorry mess is supposed to emerge clear, agreed, and directed policy, implemented across jurisdictions that in fact, remain blurred by the debate and imposed on an industry that is subject to considerable complexity as a consequence of its globalized activity. What chance is there of anything clear, agreed, or directed happening?

Borg went on emphasizing the debate between the Commission, the International body (ICP) and the member states:

> As a contracting party in its own right of both UNCLOS… and of the UN Fish Stock Agreement, the European Community (*sic*) has accepted legal obligations with respect to

oceans and the law of the sea that are particularly relevant to the ICP agenda. We hope that the UN will grant the EU the status that fully reflects its rights and obligations under the international law within the ICP. (Lloyd's List 2005d)

This call for increased influence came as the Commission was developing its new maritime policy (more of which later) which was anticipated to call for more powers within Exclusive Economic Zones (EEZ) off EU coastal states, which are currently governed by UNCLOS. Fears were expressed that the Commission would be looking to regional measures rather than using international bodies to redesign UNCLOS particularly with regard to amending established intervention rights to provide enhanced environmental protection (Lloyd's List 2005d). Research conducted for the Commission suggested that a greater power of intervention by member states or the Commission, than is allowed for under UNCLOS would be a logical way forward to protect Europe's coasts against marine accidents and vessel - sourced pollution. Central to the concept (and to objections to it from international policy-making bodies) was the idea of legalizing intervention outside a member state's territorial waters but within their 200 nautical mile EEZ. This measure would be taken by the EU without recourse to the IMO or other international policy-making bodies, generating justifiable accusations of unilateralism.

The principle was further emphasized by Borg in July 2006, looking to enable EU actions in waters beyond national jurisdiction (Lloyd's List 2006u). Resistance to the EU effectively extending jurisdiction over commercial shipping from the existing 12 miles to the 200 miles EEZ limit was expressed by the Secretary-General of the European Community Shipowners Association who also insisted that any changes to UNCLOS had to be made through the appropriate international body and not regionally. Further criticism came in December 2006 with the International Chamber of Shipping and the ISF both suggesting that firm opposition would come from the industry and many member state governments to the EU proposed changes to UNCLOS (Lloyd's List 2006ww).

By now the difficulties that characterize EU maritime policy-making should be apparent and represent a catalog of dismal under-achievement if not outright calamity. However, there are more that can be documented and it is important to do so to gain an idea of the extent of the problem. While policy making will always engender its fair share of failure even when the frameworks for policy development are appropriate and the practise of policy implementation well tuned, maritime policy-making is characterized by a seemingly unprecedented poor profile which should alert the need for change.

1.1.9 Criminalization of Seafarers

The European Commission has generated a barrage of criticism over its policy toward ship-related pollution where the approach adopted to deter such events is to criminalize the seafarers involved (Lloyd's List 2006i, k, m, q, mm). The Directive came in force in October 2005 (and was implemented in the laws of the member

1.1 Failure? What Failure? Are You Sure?   19

states in March 2007) despite extensive lobbying from all sides of the shipping industry and from member state governments in Greece, Malta, and Cyprus and also extensive, if subtle, criticism from the Secretary-General of the IMO, Efthymios Mitropoulos during a 2005 meeting with the EU Commissioner for the Environment, questioning whether:

> imposing criminal sanctions in such cases, was the appropriate response. (IMO 2005)

The pollution incident might well be accidental but this would not prevent seafarers involved being prosecuted if the pollution took place anywhere within EU waters. Putting the rights and wrongs of this to one side at least for the moment, support for this measure was very limited from member states and also from the industry as a whole. While it is of course the case that some legislation and policy has to be introduced by authorities from time to time that is wholly unpopular, in most cases there is normally at least an understanding that the policy is necessary. Not so with criminalization which has been widely seen as far too extreme and quite probably targeting the wrong people.

Resistance to the policy has continued with particular pressure from the International Chamber of Shipping and BIMCO. However, in both cases the policy soup has been mired as they withdrew formal opposition, in the first case aiming to direct their efforts toward specific seafarers' rights and in the second as the ICS felt that their national membership made the situation over-complex and that their true role was one as a lobbying rather than a political organization. However, the industry continued to press against the Directive through the London High Court where representatives from Intercargo, the Greek Shipping Cooperation Committee (GSCC), Lloyd's Register (LR), and the International Salvage Union (ISU) sought a judicial review to have the case referred to the European Court of Justice. The key to the case was a governance dispute centering on conflict between the Directive and treaty obligations under the International Convention for the Prevention of Pollution from Ships (MARPOL) which had not been resolved.

Further criticism came in the form of accusations that the Directive was a unilateral measure designed to break away from international law. The whole debate was made even more unclear in that although it is an EU Directive, the international shipping industry was actually challenging the UK Department of Trade and Industry which had imposed the Directive in the UK. The heady mix of national, international, supranational and cross-national representation battling it out over a policy unwelcome by most of those for whom it was designed suggest policy, governance, and jurisdictional chaos at its worst.

1.1.10 Places of Refuge

Meanwhile, the European Commission is also finding difficulties in implementing its plans for places of refuge for vessels at times of poor weather or when technically troubled by accident or mechanical failure. While the aim of the policy is reasonably well understood by all those involved including member states' governments, its implementation is proving extremely difficult suggesting a policy process that can deliver legislation but not necessarily enforce it—certainly within a reasonable timescale (BIMCO 2003). Directive 2002/59/EC, which includes the requirements for member states to designate places of refuge, has proved to be extremely difficult for some to implement to the extent that the EU Commission had to resolve to a campaign to encourage implementation at member state level—for something that actually is a legal requirement (EMSA 2004).

1.1.11 Security

Security has been a rising star in policy terms for some years now with major concerns over shipping's role in the movement of terrorist materials, terrorists, and illegal immigrants. Two issues have emerged where confusion and deliberate moves to avoid the established policy-making framework have taken place. The first in 2005 was when the USA lobbied national governments in the EU rather than EU institutions in an attempt to reach agreement over two projects considered key to security strategy. Despite the existence of established policy routes which lay through the EU before moving on to national government discussions, the USA government decided to deal directly with member states concerning the Container Security Initiative and the Galileo Satellite project thus resorting to bilateralism in the face of unwanted multi-lateral procedures. Commenting on such events, former US Ambassador to the EU, Rockwell Schnabel claimed that it was a tactic 'commonly employed' (Lloyd's List 2005f).

The EU spent much time in 2005–2006 deriving its own 24-h rule for container screening based on the USA approach (which we have just noted was itself developed avoiding discussion with the EU). This was adopted in 2006 and employed from 2009. However, it has been heavily criticized by those most involved—the liner shipping sector and more specifically, the World Shipping Council (WSC)—as having no security value because of flaws in its drafting (Lloyd's List 2006z). The flaw in Commission Regulation 1250 is that it does not oblige non-vessel operating common carriers (NVOCC) or freight forwarders to file advance manifest declarations with European customs and in so doing goes against World Customs Organization and US/Canadian guidelines. Claims by the WSC are that the Commission received but ignored advice to this effect from ECSA, the European Shippers Council (ESC) and interests representing freight forwarders, aviation, and road transport.

1.1.12 Environmental Issues

Environmental policy-making both within the EU and elsewhere, has shown rather more coherence than other sectors although it is still easy to find examples of failure to collaborate, coordinate, and cooperate between the various jurisdictions.

Disagreements between the IMO and the USA over emissions regulations emerged in 2007 when the State of California decided to press ahead with ship emission laws that will override IMO regulations (Tradewinds 2007a). Only a few weeks after this came news of problems with applying the new IMO ballast water treatment regulations for which in September 2007, there was no type of approved equipment available (Lloyd's List 2007l). Some new vessels would soon have to comply and yet had no way of doing this—clear example of governments and business failing to cooperate effectively (Tradewinds 2007b). By May 2008, there was what was described a 'pitiful level of ratification of the 2004 Ballast Water Management Convention' with barely 3.5 % of the world's merchant fleet covered, exposing the 'insidious gap between political aspirations and technical and operational feasibilities' (Lloyd's List 2007h). There were also claims that changes by the IMO to MARPOL Annex VI with regard to ship emissions were short-term measures that failed to take the long term into consideration and might lead to decreased pollution from ships but at the expense of increased pollution from refineries.

Meanwhile, the IMO attempted to defend its position with Secretary-General Mitropoulos suggesting that it was pledged to speeding up work on emissions and would not be pressured by the EU to show more urgency. To achieve this, the IMO suggested the introduction of a Panel of Experts in Ship Pollution to examine the issue of airborne pollution holistically rather than pollutant by pollutant (Lloyd's List 2007h). However, as the policy model continued to disintegrate, the EU was examining the possibility of including ship emissions (and especially CO_2) into carbon trading schemes thus instituting a unilateral, regional policy (Tradewinds 2007e).

Issues relating to ship emissions and bunkering became a serious cause for concern from late 2007 onwards with some major disagreements taking place between the IMO, EU and nation-states resulting in almost no progress by late 2010. During this period there were repeated threats of regionalization and unilateralism in the application of industry standards for emissions. The European Commission made clear its frustrations with the IMO and how slow it was moving coupled with what they saw as a flawed global scheme to reduce CO_2 emissions (Lloyd's List 2007k). By early 2008, the USA was commenting that the IMO had reached a crucial point in its consideration of CO_2 reduction (Lloyd's List 2008e). If the proposals were not ready for the April IMO Marine Environment Protection Committee (MEPC) or if they were seen as too weak then the USA would go alone in introducing air pollution standards and as a result would quite likely set the context for others and in particular the EU, to follow. California was already in court defending unilateral imposition of low sulphur rules and it was expected that if they succeeded to resist attempts to block the moves, further unilateral action would come from the Great Lakes Region, the Pacific North–West, Mexico, British Columbia, and the St Lawrence Corridor. The IMO would be heading toward marginalization especially in the North American and European marketplaces.

It was not only CO_2 that featured in this debate about the role of the IMO and its jurisdiction over maritime environmental policy. Complaints were raised the same week at the Bulk Liquids and Gases Sub-committee of the International Chamber of Shipping by Panama and the Marshall Islands about NOx emission reductions and the inadequate information upon which the proposed IMO legislation was to be based (Lloyd's List 2008f). Meanwhile, it was not all bad news as evidence emerged of agreement between both the WSC (representing the major liner shipping companies), and the IMO, and the ICS and the IMO concerning plans to cut ship emissions with the prime aim of reducing the risk of regionalization and unilateralism (Lloyd's List 2008n, 2009qq, rr).

Over the next year the IMO devised an Energy Efficiency Design Index (EEDI) to be combined with the voluntary operational index and Ship Emissions Management Plan (SEMP) as part of its policy to reduce CO_2 emissions. This was to be presented to the Copenhagen Summit on Climate Change in November 2009. The ICS continued to be highly critical describing it as so complex as to be unworkable, while the 'politics of the situation is not making it easy for the industry to move forward' (Lloyd's List 2009m).

Fears of a supranational response to the need to curb ship emissions as a direct response to the inadequacies of the IMO featured in reports in January 2009 which noted the failings of the MEPC (MEPC) in dragging their heels (Lloyd's List 2009nn). The requirement for a global standard was clearly expressed but difficulties remained in identifying the performance of individual vessels, allocating these vessel emissions to specific areas and making allowance for the different emissions that might come from the same ship dependent on operating circumstances. These problems were cited by the IMO as reason for the slow progress but the response from the EU was to suggest that they went ahead alone. The EU had not made up its mind however, and called for shipping to be covered by a global agreement to be completed at Copenhagen late in 2009 (Lloyd's List 2009oo).

Meanwhile in April 2009, the European Commission confirmed that EU-wide legislation for greenhouse gas emissions had been postponed until at least 2010. Originally in June 2008, the Commission had informed the IMO that they had until early 2009 to come up with acceptable proposals for reducing CO_2 blaming the problem of internal politicking within the United Nations between the IMO and the UNFCCC. Both UN bodies were angling for setting the ground rules for standards which should apply to shipping, thus delaying an already very slow process (Lloyd's List 2009n). With Copenhagen on the horizon where there was a possibility that the UN might agree with some sort of deal in principle to be in place by about 2011, the Commission was prepared to delay the introduction of regional policies until the end of that year (Lloyd's List 2009pp). However, concern was also expressed that even if the IMO could agree to a process for controlling CO_2 emissions, these would not be mandatory and consequently, would be unacceptable to the Commission.

The debate between the positions of the IMO and the EU over emissions standards also reflected a fundamental disagreement between the institutions' views on global versus regional measures. The former, preferred by the shipping industry as well as the IMO was not seen as essential by the EU who referred to a similar regional scheme for aviation standards which had been successfully introduced for Europe. The question remained as to who was driving whom;

whether there was need for a global maritime authority or whether some other sort of jurisdictional framework for maritime governance might work better. Criticism of the speed of progress at the IMO continued in October of the same year with pressure from the Carbon War Room, an interest group fronted by Richard Branson and incorporating a number of shipping stakeholders. A spokesman for the group suggested that the 'IMO has a way of bogging the process down' (Lloyd's List 2009o). Meanwhile, issues relating to sovereignty, tax hypothecation and the role of the EU in possibly implementing an EU-wide environmental tax on shipping raised their heads, with a number of member-states making it clear that taxation was not something which they were willing even to discuss (Lloyd's List 2009aa).

By October 2009, three of the EU's biggest players in shipping, Greece, Cyprus, and Malta, had convinced the rest of the EU member states that CO_2 reductions should be the responsibility of the IMO and not allowed to be unilateralized (Lloyd's List 2009p) negotiating another year's grace through the agreement of the EU Council of Environment Ministers Meeting as long as a solution to the issue was agreed by the end of 2010 and approved by 2011. The Council was setting out an EU negotiating position ahead of the Copenhagen Summit and in the words of a Council member, was a result of the 'Presidency (taking) them into one of the private smoke-filled rooms and (coming back) with the agreement for us to vote on'. However, reflecting the chaotic process of negotiation that ensued, by November the Commission was quoted as expecting a 'patchwork of regional measures to enforce CO_2 emissions reduction from shipping' (Lloyd's List 2009q). In the long term the IMO was still considered as best placed to secure a global agreement but in the short term there would be no other alternative but to introduce an EU carbon regime. The EU Environment Directorate commented:

> I find it rather bizarre that we are hearing all this talk about wanting a global action but in the politics and the global discussions at the (UNFCCC) we do not see this pressure coming through... a very likely reality is that (the shipping industry) might be facing a mixture of solutions where you get some measure from the IMO, some measures from the US and something from the EU in the short-term.

The IMO timetable bore no resemblance to that demanded by the EU which wanted everything in place by the end of 2011. The IMO was not expected to agree to a process by 2011, and implementation would take at least 5 years more.

When Copenhagen started the IMO began a fight to retain control of global CO_2 emissions standards and rather than the main opponent being the EU, in fact, the UNFCCC emerged as the greatest obstacle with a political in-fight at the UN over responsibility and power (Lloyd's List 2009r, s, t). In particular, the IMO retained its mandate that all policies should be equally applicable across all member states, while the UNFCCC differed in that its mandate was to apply differentiated policies according to a state's development status. However, despite the legitimate debate between the two organizations concerning application of any rules that could be agreed, it remained the case that the IMO had agreed to develop policy for CO_2 emissions at Kyoto in 1997 and since then had made almost no progress.

Once the conference was underway in Copenhagen, new threats emerged, not least the suggestion from a number of countries that shipping fuel should be taxed globally (along with the aviation industry) and this should be undertaken outside the jurisdiction of the IMO (Lloyd's List 2009u). The industry expressed concern that the IMO might continue to press for their own market-based approach to reducing emissions alongside any global tax.

And the result of all this? Described by Lloyd's List (2009v) as 'opaque'! The industry continued to face the possibility of a twin attack on emissions in addition to regional schemes in the USA and Europe. The IMO's role was unclear although the Secretary General suggested that he was pleased that the Conference had moved things in the direction of a global solution—something similarly unclear to most observers. This was also not agreed by the UNFCCC and the problem of reconciling the two different mandates of the two UN bodies also remained totally unresolved. The result was likely to be continued work by the IMO on emissions reduction, short-term regional standards imposed around parts of the world, and the squabble between the IMO and the UNFCCC would go on. The Secretary-General of the IMO continued in optimistic tone:

> The outcome of (Copenhagen)… gives the IMO more time to make real progress, whilst also creating an increased obligation on IMO to intensify its efforts, and prove to the UNFCCC that it can achieve unified reduction in shipping's emissions.

Some would say that to give IMO more time is always somewhat risky although some commentators, for example ECSA were optimistic, perhaps reflecting the desire of shipowners for a global solution rather than a true belief that this could be achieved by the IMO in good time (Lloyd's List 2009w). Progress was planned to be reported to the MEPC in March 2010.

And so to that meeting where progress was supposed to be made in agreeing to a way forward for the IMO in its policy for reducing greenhouse gas emissions. However, before it began, plans to present a unified EU front were scuppered by a group of member states, centering on Greece, the UK, and Denmark, refusing to agree to a unified Commission position (Lloyd's List 2010i). Along with support from Germany, Cyprus, and Malta this rebellion was a surprise in that it came with these member states failing to support a position paper that was to be presented by the EU to the MEPC and which had already been agreed by all member states during 2009. Meanwhile, questions were being asked by the EU about the IMO's ability to develop and enforce a global emissions-trading system, one of the ways that the IMO was considering to reduce greenhouse gas emissions (Lloyd's List 2010j).

Reflecting the complexity of the horse-trading that was going on behind closed doors, and in the corridors and smoke-filled backrooms and bars of London, the rebel EU member-states suddenly withdrew their opposition to a common EU position at the MEPC with Spain left as the EU representative country to put forward the EU's common position apparently a result of substantial diplomatic pressure on the recalcitrant member-states (Lloyd's List 2010k). To accompany this reversal of positions, the EU common statement then went on to back the IMO as the body to develop a global system for combating greenhouse gases with no mention of the threat of regional legislation (Lloyd's List 2010l). Disputes however, continued between the IMO and its own members, especially those from developing countries which would rather see the differentiated principles of the UNFCCC incorporated in any legislation than the global solution favored by the IMO (Lloyd's List 2010m).

The MEPC produced no conclusions. Lloyd's List (2010n) suggested 'a huge rift has opened up in the IMO over the issue of climate change'. Developing countries in particular expressed 'dismay' at the 'tyranny of the majority' and proceeded to block any attempts at imposing a mandatory Energy Efficiency Design Index (EEDI) for vessels with global pretensions. Despite more than 10 years' debate about the index it was argued that 'measures were insufficiently advanced', suggesting in itself something wrong with international maritime governance. The failure to progress the issue reflected an increasing schism between the developed and developing maritime countries suggested in Chinese attempts to widen the debate beyond climate change into a 'political matter of international development'. Progress was described as 'painfully slow and tortuous' now bogged down in further disputes between member states, blocs of member states, between the EU and the IMO, and between different UN bodies.

The environment has not gone away. January 2010 saw ECSA lobbying the Commission to not go it alone (Lloyd's List 2010hh) and in March, the IMO called for early talks with the Commissioner about climate change (Lloyd's List 2010ii). In May 2010, shipping industry representatives lobbied the European Commission in turn to lobby the IMO to change the proposed amendment to the marine pollution convention and in particular the requirement for operators in Northern Europe to reduce sulphur dioxide emissions through the use of more expensive distillate fuels (Lloyd's List 2010o). However, this in turn generated questions about whether the EU member states really realized what had been agreed at the IMO. The EU member-states readily signed up to this amendment to Annex VI of the convention but were now faced with attempting to convince the EU Commission that actually it was not what was wanted at all. The Commission might have seen this as a piece of good fortune in that it implied that some sort of coordination (perhaps through single EU representation) might prevent such things happening again (Lloyd's List 2010p).

In late May 2010, the European Commission continued to be under pressure to intervene in these new low-sulfur rules that were about to be put into place in Northern Europe, to soften the impact on the grounds that it would drive cargo onto trucks and away from environmentally sensitive shipping (Lloyd's List 2010q). However, the Commission showed little interest citing the fact that all (then) 21 member-states voted unanimously for the measures suggesting that either member-states and the Commission were poorly prepared or that they failed to maintain good communications with their own industries.

The imminent IMO–MEPC meeting in September 2010 generated a flurry of environmental activity as attempts were made to reach a reasonable and politically acceptable solution to the regional/global debate and the location of authority. The European Commission generated surprise and dismay in the shipowning community at announcing that there would be no delaying the 2015 deadline for ultra-low sulfur fuel citing that it would an unprecedented reversal of agreement to an IMO decision (Lloyd's List 2010x). This was despite criticism which had come from the Chairman of the European Parliament Transport Committee generating further confusion as the different EU institutions argued between themselves:

> It's unusual for the IMO to take such a decision. Normally it takes them 50 years. But when they take it quickly they get it wrong. (Lloyd's List 2010aa)

The MEPC met from 27 September 2010, and was addressed by the Secretary-General of the IMO where he emphasized the need for the Committee to put into effect the plan agreed in 2006 in good time and the need to avoid unilateral or regional measures (Lloyd's List 2010y, z, bb). Although commenting that the IMO's review of market-based measures had not reached a single conclusive outcome he was confident that the MEPC could now evaluate the possibilities suggesting it 'was imperative we get it right' and 'rather than rush to choose a half-baked solution we should prudently assess the parameters involved and make a truly balanced decision in the best interests of the environment'. He was clearly aware of the wider issues and the political context.

During the days of the MEPC meeting it emerged that there was considerable disagreement between the developed and developing country representatives over the issue of whether any climate change agreements should be applied multilaterally or should incorporate the principle agreed in Copenhagen of differentiation between the developed and developing world (Lloyd's List 2010cc). In a very unusual move it was proving likely that a vote in the Committee would have to take place—something that was described as 'extremely rare'. Pressure was exerted on the UN Advisory Group on Climate Change by the ICS that any moves should come through the IMO (and thus be multilateral) and not the UNFCCC (and therefore be differentiated) (Lloyd's List 2010dd, ee).

The outcome of the MEPC was nothing short of a disaster for the IMO and for the governance of the maritime sector as a whole. To quote Lloyd's List (2010ff):

> The international Maritime Organization is a house divided. The fault lines have been most visibly drawn during the greenhouse gas emissions debates in recent years, but the schism is not simply a question of rich versus poor.

The clash over climate change within the UN between universal application and differentiation to allow for the differing resources of the rich and poor countries has created a political divide that has seized up any attempts to legislate at the IMO. China, Brazil, and India have made it clear that there is no room for maneuver over the issue of differentiation and lengthy debates repeating intractable positions are now common. The result is little if any policy progress and increasing regionalism epitomized by the EU's attempts to force common regional positions and to gain IMO representation. The whole climate change debate was pushed on by the MEPC to 2011 and in the process risked splitting the IMO (Tradewinds 2010g). The EEDI and Ship Energy Efficiency Management Plan which the developed countries thought had been agreed at least in principle were blocked by the developing countries so that revisions to Annex VI of MARPOL could not go ahead.

By July 2011, and in an atmosphere of some desperation to achieve something tangible in the face of a political stalemate, the MEPC finally agreed to adopt amendments to MARPOL Annex VI to curb CO_2 emissions from ships engaged on international voyages, something publically welcomed with some relief by the IMO Secretary General (Lloyd's List 2011b). This would center upon an EEDI for all new-buildings and a SEMP for all ships (Lloyd's List 2011f). A 4- year waiver period was to be allowed and although the vote by the committee was split (49 in favour; 5 against; 2 abstentions), a sizeable majority was achieved reflecting at least some consensus whatever the motive (Lloyd's List 2011c).

However, commentary on the agreement was less positive from elsewhere. LR's climate change expert Dr Anne-Marie Warris suggested that the vote had been divisive and reflected poorly upon the state of political debate within the IMO. Describing it a 'patched-together' agreement it did at least send a:

> clear signal to the UNFCCC and the UN that the IMO is in charge of this issue and is delivering and is therefore on a very firm foundation to take the work forward into the market-based measures debate next year. (Lloyd's List 2011d)

The EEDI would be mandatory from 2013 following further decisions by intersessional meetings of the MEPC, albeit with the waivers noted above, and had a positive message in that it not only reflected a commitment to reduce emissions but also indicated the way of achieving them.

Debate would continue however, on regional measures that were likely to be imposed on the industry by the EU as the Commission remained committed to prepare proposals for consideration by the Council of Ministers in 2012. It seems unlikely that the agreement upon the EEDI by the MEPC was likely to suffice given that:

> The EEDI is expected to deliver emissions cuts of around 40–50 m tonnes annually against a shipping industry that will (by 2013) be churning out about 1,500 m tonnes. Set that scenario off against the EU's existing commitment to a 20 % reduction by 2020 and the question of whether the EEDI is going to be sufficient by itself looks fairly obvious. (Lloyd's List 2011d)

The issue of waivers by flag states was not expected to be a significant one simply because by waivering; shipowners would be constrained in their future ability to flag-hop and sell ships (Lloyd's List 2011e).

1.1.13 Regional Policy

Regional policy represents a key foundation of the EU (Bache 1998). However, further policy problems were emerging over jurisdictional disputes between the European Commission and the European Conference of Peripheral Maritime

Regions (CRPM) which comprised 154 members stretching from the Baltic to the Black Sea. In October 2006, CRPM President Claudio Martini described it as:

> contradictory that the Commission dedicates so much talent and energy to promoting on a global scale the Europe of the seas and oceans while at the same time forsaking one of the key (regional) policies in the area of maritime transport (Lloyd's List 2006ii)

1.1.14 Logistics

The policy-making problems of the maritime sector are far-reaching and do not just focus on issues centering on flags, registries, and nation-states that have many of their foundations in tradition. The more recent emergence of logistics as a central part of maritime activity has also been characterized by problems of governance and a continued failure of the EU policy-makers to understand and incorporate the sector stakeholder adequately.

For example, initiatives aimed at making freight transport in the EU more efficient and sustainable received a very mixed response from the logistics sector (Lloyd's List 2007j). The debate centered on a package of proposed measures adopted by the Commission in October 2007 including logistics, priorities to rail freight and European ports, a European maritime transport area, and the motorways of the sea. While the Commission saw these proposals as linking logistics with the various transport modes, the comment from the industry was that they could not be taken seriously in particular because the EU had only very limited budgetary control related to the proposals and would rely on the compliance of member-states. This would not be forthcoming. To quote Thomas Cullen from UK based Transport Intelligence, the Commission would be better off focusing on areas where it could facilitate the private sector:

> rather than attempting to direct the industry from a centrally developed and ultimately flawed blueprint. (I.E.Canada 2007)

1.1.15 EU Policy Documents

The generation of maritime policy in the EU has never been easy and it is perhaps facile to suggest that it ever could be. However, repeated attempts in the early years of the twenty first century reflected some serious shortcomings in the approaches taken. In 2006, a new Green Paper on maritime affairs identified the problems that were created by the hierarchical policy-making process that existed (Lloyd's List 2006xx). For once the Paper suggested that the EU was aware that the hierarchical structure of maritime policy, with its variety of jurisdictions, was a significant issue which would be central to policy success and that an extension of

1.1 Failure? What Failure? Are You Sure?

stakeholder involvement in maritime governance was a necessity. However, despite recognizing these issues little fundamentally different was proposed to accommodate them.

By early 2008, the generic EU maritime policy remained in tatters. The attempts to develop a new 'holistic' and coherent policy following the extensive Green Paper consultation throughout 2007 and the subsequent Blue Paper in many ways shows up the failures in approach, in governance, and in policy making that undermine the EU's increasingly frequent and desperate attempts to come to grips with the problems of the maritime sector. Both the Green and Blue Papers were full of good intentions and on the surface provided wide access to all interested parties to have their views put forward and considered, a process emphasized during the Green Paper's launch in late 2006 (Lloyd's List 2006v, nn, qq, rr, ss). The Commission followed all good consultative practice. But what has it achieved? A reiteration of hopes and ideals that have to be put into place by a Commission that has a track record of policy failure. An inadequate inclusion of a selection of those stakeholders that have an interest in the maritime sector. Asking for their opinions and views is not the same as ensuring that all interested parties have adequate access to the policy-making process. It is not adequate 'holistic' governance to publish suggestions, ask for comments, and hold meetings that tend to reinforce the established views of the limited range of interests that are geared up to maritime lobbying in the EU. Where is the networking, the inclusivity, the balancing of pressure groups, the public, labor interests, commercial sector, shipowners and agents, ports, media, politicians, and so on in the generation of ideas and the consideration of proposals? Instead we had the established and traditional hierarchical approach centered around state interests, failing to accommodate public concerns or to include the needs of all those involved in particular if those groups or individuals could not be easily incorporated into the existing governance structures. The Commission (with advice) proposes and suggests; there is a limited response back from the traditional and predictable lobbyists; the Commission decides and legislates. The industry in all its forms and interests rejects, resists, prevaricates. The same old policy failings.

These failings raised their head throughout the public consultation process. Just before the Commission issued the Green Paper, the UK government expressed concern at the leaked proposal that the EU might be looking to extend the power of the European Maritime Safety Agency (EMSA), with the aim of taking power away from national governments (Lloyd's List 2006e). This reflected similar moves by the Commission with regard to the IMO which had generated further resistance to the Green Paper proposals from both a range of national governments and from ECSA (Lloyd's List 2007b). This followed claims by the European Commission Maritime Policy Director, Fotis Karamitsos, that a campaign of misinformation was being waged by those generally opposed to the EU. He suggested that the Commission was looking for no more than 'observer status' at the IMO and was not looking to 'muzzle' member-states at the UN forum (Lloyd's List 2006bb).

Further dismay at the Green Paper and a general feeling by both member states and the shipping community that the Commission was attempting to extend its jurisdiction in maritime affairs and reduce that of the national representatives and industrial stakeholders was expressed in response to proposals for a 'European maritime space' including issues related to maritime cabotage (Lloyd's List 2006jj, 2007a). The Secretary-General of ECSA expressed concern that European maritime cabotage could be interpreted by the Commission as a method of extending European protectionism—restricting intra-European trade to only EU registered vessels rather than at present whereby the cabotage rules were aimed at opening up trades to vessels from all countries (Lloyd's List 2006hh). This stance against this interpretation of the Commission position was supported by BIMCO (Lloyd's List 2006qq) and further attacked by ECSA in early 2007 (Lloyd's List 2007a).

Thus Joe Borg's trumpeted new holistic maritime policy may well have contained the right words, focused heavily on the issue of maritime governance and included the views of many of the traditional maritime interests (Lloyd's List 2007m), but did it change anything? The outcome of the year-long maritime policy consultation was announced in October 2007 and included the withdrawal of proposals for a new EU flag and coastguard in response to stakeholder demands (Lloyd's List 2007d, n) so perhaps this was a good sign? But did it result in a flexible, responsive, meaningful, sensitive, and representative set of policies that will be implemented by those with the responsibility to do so? Or simply lead to another EUROS, another Treaty of Rome anomaly, another neglected or rejected policy for ports, for maritime safety, or even further subsidies for the maritime sector under a supranational governance regime that claims to support policies of competitiveness, efficiency, and a single market for all commercial and industrial sectors? Will shipping companies continue to be able to abuse registries and tax regimes, by trading off member state flags, one against another, using the unstoppable cloak of globalization as a means of blackmailing the EU into retaining a tonnage tax regime that does nothing for global maritime safety, environment, security, social conditions, or even efficiency?

By July 2008, the European Commissioner was continuing to promote the new integrated maritime policy and urged member states to follow suit by accommodating the concept of holistic maritime governance (Lloyd's List 2008g, h).

Early 2009 brought another policy document; a Communication from the Commission to the Council, the European Parliament, the Economic and Social Committee, and the Committee of the Regions which showed little progress in terms of reformulating the structures of governance (Commission of the European Communities 2009b). Aimed at presenting the main strategic goals of maritime policy for the EU up to 2018 and particularly in light of the then, current economic crisis, it retained a narrow focus in the definition of stakeholders confining them to 'experts of the member-states' and 'senior shipping professionals' (Lloyd's List 2009cc, dd). It was breathtaking in its hypocrisy, stressing in sequential paragraphs that tonnage tax should be extended and improved while the Commission continued to support 'fair international maritime trade' and especially the 'liberalization of trade in maritime services'. Support meanwhile was stressed for international maritime organizations (IMO, ILO, WTO, WCO) while in the same breath it was stated that if IMO negotiations on maritime safety and security should fail, then the EU should 'take the lead in implementing measures'. Emphasis was also placed once again on the need to develop a competitive ports policy.

1.1.16 World Trade Organisation (WTO)

It is not only the EU that exhibits maritime policy failure. At an international level, the continued delays and arguments that are taking place throughout the negotiations about world trade at the WTO are symptomatic of further policy failure and the pressing need to address governance in the maritime sector. Continued reports emerge of trade ministers unable to agree on any principles of how to cut subsidies and tariffs for agricultural and industrial goods including the major international trading nations of the UK, the EU as a whole, Japan, India, and Australia. Failure to reach agreement on these fundamental principles would leave little time or chance for agreement on services such as shipping (Fink et al. 2011; Lloyd's List 2006r, t). Despite recognition that gains from agreement on services at the WTO talks would generate benefits twice those of products, little has been achieved. Weak offers were put forward by Brazil, India and South Africa on shipping issues while the USA continued to refuse to discuss shipping, suggesting a flaw in the governance framework within which international shipping policy is supposed to be developed.

1.1.17 Bulker and Tanker Construction

The role of the IMO in global maritime policy-making was again in question over new goal-based standards for shipbuilding which were resisted by almost all stakeholders. Emerging in detail during 2006, the Maritime Safety Committee of the IMO endorsed the view of its working group that over-arching goals should be drafted as amendments to the International Convention for the Safety of Life at Sea (SOLAS). The objective was to agree to a set of standards that could be applied to all shipbuilding and based on goals that need to be achieved. However, although laudable, a major problem emerged as to who should hold verification authority. The work group agreed that this authority needed to be experts under the auspices of the committee and that no further liability issues would arise for the IMO (Lloyd's List 2006d).

Further, to these problems of verification of IMO goal-based standards, the Cyprus Shipping Council and the Union of Greek Shipowners expressed concerns about the new rules to be applied to the construction of tankers and bulk carriers by the International Association of Classification Societies (IACS) (Lloyd's List 2006j). Both the Cyprus and Greece organizations claimed that the new common structural rules were seriously flawed and that the International Chamber of Shipping would represent the concerns of national shipping organizations in further discussions. Further to this development, it emerged that the concerns centered on the inadequacy of standards that had been proposed and the failure to achieve a trail of responsibility leading back to shipowners whose liability covered only the first year after delivery (Lloyd's List 2006p).

These episodes relating to vessel construction for both the IMO and IACS suggest serious governance flaws whereby the industry stakeholders feel that their concerns are inadequately represented resulting in policies that fail to address the problems that exist. In both cases the result can only be inadequate policy-making that consequently cannot achieve its prime aims.

Maritime policy-making should be all about 'progress'. About improving the lives of those served by, dependent upon and working on and living alongside the sea. However, Bailey et al. (1993, p. 49) saw something 'specious in (the)... lop sided focus on progress and growth'. They cited Wilber (1979, pp. 20–21) and quoted Arendt (1948) who considered that 'Progress and Doom are two sides of the same medal'. According to Bailey et al. (1993, p. 49) 'this obsession with progress has provided the necessary technocratic organization for several of (the twentieth century's) worst human and environmental catastrophes':

> As we in the West approach the end of the twentieth century, the 'modern' record – world wars, the rise of Nazism, concentration camps (in both East and West), genocide, worldwide depression, Hiroshima, Vietnam, Cambodia, the Persian Gulf, and a widening gap between rich and poor… makes any belief in the idea of progress or faith in the future seem questionable (Rosenau 1992, p. 5).

Progress cannot be helped by policy failure and failure is undoubtedly endemic in maritime policy-making—so common that its scale, scope, and depth is almost overwhelming. Despite this it is important to point out that there are policy successes that reflect how the existing jurisdictional hierarchy and structures can work together—examples of nation-states and the EU cooperating in attempting to streamline shipping's regulatory framework have been noted from Norway (Lloyd's List 2006gg, 2009y) and the UK (Lloyd's List 2010r, s) and how China and the ILO (representing national and international jurisdictions respectively) have been happy to cooperate in developing a new Labour Convention (Lloyd's List 2006pp). Industrial support for a global approach to shipping regulation has also been frequently expressed (see for example BIMCO (Lloyd's List 2009bb) in addition to continued calls for revision and reduction of shipping regulation (see for example Lloyd's List 2006y).

However, by comparison, the level of disappointment is remarkable. The EU is central to much of this failure and here we have only indicated a sample of the policy problems which have arisen since the turn of the century—others include clear evidence of slow, bureaucratic, and anachronistic policy-making (Lloyd's List 2008p, 2009xx, yy); failure by the Commission to investigate known breaches of rules by member states (Lloyd's List 2009ww); watering down of regulation of passenger rights by the European Parliament despite opposition from the Commission (Lloyd's List 2009vv); claims that UK naval bases must be 'Europeanized' to protect shipping (Lloyd's List 2009uu); further disputes between Commissions responsible for Transport and Competition (Lloyd's List 2010jj), and innumerable others (for example see Lloyd's List 2009ss, tt). However, the EU is not the sole or even main cause of maritime governance and policy collapse. It is simply part of a jurisdictional policy framework that apparently no longer works. Because of its activity and central position in that framework, it features most commonly in stories of policy inadequacy—but other significant institutions—the IMO, industrial representatives (IACS, WSC, ICS, ECSA, ESPO etc.), member state governments—have all played their part in contributing to this dismal scene.

Interestingly, the EU has recognized the policy failures that have been increasingly occurring and turned to the issue of revised governance arrangements and procedures as a solution (Lloyd's List 2006uu, vv, 2007e). The Blue Paper featured significant discussion of the concept of governance and EU Maritime Affairs Commissioner Borg emphasized the need to ensure that the processes of governance are correct so that the 'mechanisms for cooperation, coordination and integration are clearly identified' (Lloyd's List 2007e). However, although recognition of the problem is a stage forward, is it enough and can the generation of new policies which continue to fail in alarming numbers ever produce a satisfactory environment for the maritime sector? It is the contention here that new policies will never resolve the problems faced by the sector until the underlying framework which supports and directs policy generation and implementation is appropriate. This can only happen if all maritime stakeholders are substantively involved in the process of policy making and implementation, if the relationships between the maritime jurisdictions are appropriately organized and changed when necessary, if the appropriate mechanisms to develop and implement maritime policies exist, and if the societal forces that drive the world economy are recognized for what they are and a governance framework designed to accommodate them.

Before we can even begin to reach conclusions on the steps taken by the maritime community to address these failures in policy and to consider the governance of the maritime sector, how its inadequacies have led to these failures and how this situation might be redressed, we must first consider what we mean generically by governance in the wider context of policy making.

References

Arendt, H. (1948). *The origins of totalitarianism.* New york: Harcourt Brace Jovanovich.
Bache, I. (1998). *The politics of European Union regional policy. Multilevel governance or flexible gatekeeping?* Sheffield: Sheffield Academic Press.

Bailey, D., Heon, F., & Steingard, D. (1993). Post-modern international development: Interdevelopment and global interbeing. *Journal of Organizational Change Management, 6*(3), 43–63.

BIMCO. (2003). *Round table press release on concrete measures on places of refuge.* www.bimco.org.

Bredimas, A., & Tzoannos, J. G. (1981). In search of a common shipping policy for the EC. *Journal of Common Market Studies, 20*, 95–114.

Brownrigg, M., Dawe, G., Mann, M., & Weston, P. (2001). Developments in UK shipping: the tonnage tax. *Maritime Policy and Management, 28*(3), 213–223.

Commission of the European Communities. (2009b). *Strategic goals and recommendations for the EU's maritime transport policy until 2018*, Communication from the Commission to the Council, the European Parliament, The European Economic and Social Committee and the Committee of the Regions, COM (2009) 8, Brussels.

De Vivero, J. L. S., & Mateos, J. C. R. (2004). New factors in ocean governance. From economic to security-based boundaries. *Marine Policy, 28*(2), 185–188.

EMSA. (2004). *EMSA takes a look at "places of refuge"*, European Maritime Safety Agency, March 5, 2004 www.emsa.eu.int.

EurActiv. (2009). *Ministers approve maritime safety laws.* www.EurActiv.com.

Farrell, S. (2001). If it ain't bust, don't fix it: The proposed EU directive on market access to port services. *Maritime Policy and Management, 28*, 307–313.

Fink, C., Mattoo, A., & Neagu, I. C. (2011). Trade in international maritime services: How much does policy matter? *World Bank Economic Review, 16*(1), 81–108.

I.E.Canada. (2007). October 24 http://iecanada.com/ietoday/oct_07/07_10_24.htm#10.

IMO. (2005). http://www.imo.org.

Leggate, H., & McConville, J. (2005). Tonnage tax: Is it working? *Maritime Policy and Management, 32*(2), 177–186.

Lloyd's List. (2005a). *Pushing at the legal boundaries to protect lives and human rights.* May 20.

Lloyd's List. (2005b). *Improve public image, says Mitropoulos.* May 24.

Lloyd's List. (2005c). *Cause for concern.* July 8.

Lloyd's List. (2005d). *Brussels wants bigger role for EU in overseeing maritime law.* September 6.

Lloyd's List. (2005e). *Dutch owners back Greeks over Brussels 'interference'.* October 31.

Lloyd's List. (2005f). *US 'ignored' Brussels.* November 7.

Lloyd's List (2005g) *Thought-bubbles get busy in the run-up to IMO assembly.* November 16.

Lloyd's List. (2005h). *Greece calls for full discussions on new laws.* December 7.

Lloyd's List. (2006a). *Brussels denies plans to overthrow national flags.* February 17.

Lloyd's List. (2006aa). *EU states will keep tabs on conference changes.* September 27.

Lloyd's List. (2006b). *IMO audits 'a marketing ploy'.* March 23.

Lloyd's List. (2006bb). *Brussels threatens IMO rebels with court action.* September 29.

Lloyd's List. (2006c). *Brussels poised to intervene in port sector.* March 31.

Lloyd's List. (2006cc). *Tramp trade faces early Brussels probe.* September 29.

Lloyd's List. (2006d). *IMO at forefront of goal-based standards.* May 23.

Lloyd's List. (2006dd). *One month deadline to submit liner proposals.* October 2.

Lloyd's List. (2006e). *Britain will oppose EMSA power grab.* May 26.

Lloyd's List. (2006ee). *Post-conference truce unfolds in EU liner sector.* October 5.

Lloyd's List. (2006f). *EU to push for coastal rights overhaul.* May 26.

Lloyd's List. (2006ff). *European flag plans doomed to failure, says EU official.* October 6.

Lloyd's List. (2006gg). *Norway backs EU's maritime framework.* October 6.

Lloyd's List. (2006hh). *EU's Borg pleads for 'candid' debate on European flag issues.* October 9.

Lloyd's List. (2006i). *Coalition in EU directive legal challenge.* June 7.

Lloyd's List. (2006ii). *Regions say EU in slow lane on sea motorways.* October 11.

Lloyd's List. (2006j). *Cyprus joins Greeks in common rule clash.* June 7.

Lloyd's List. (2006jj). *Shipowners warn against 'European Jones Act'.* October 11.

Lloyd's List. (2006kk). *Conference system: Kroes back in charge for final days.* October 17.

Lloyd's List. (2006m). *QC slams Brussels' anti-pollution plan as a 'muddle'.* June 8.

References

Lloyd's List. (2006n). *Ports fear worst in EU power struggle.* June 9.
Lloyd's List. (2006o). *IMO authority vs the unilateralists.* June 9.
Lloyd's List. (2006g). *EU's Barrot refuses to close door on port laws.* June 5.
Lloyd's List. (2006h). *EU's Barrot refuses to rule out third stab at port law.* June 5.
Lloyd's List. (2006k). *Maritime leaders embark on pollution directive challenge.* June 7.
Lloyd's List. (2006s). *India wants oil blocks back.* July 6.
Lloyd's List. (2006l). *Britain scorns idea of EU flag and coastguard.* June 8.
Lloyd's List. (2006pp). *ILO "bill of rights" will stem tide of seafarer shortages.* November 6.
Lloyd's List. (2006ll). *Bahamas proposes a major IMO shake-up.* October 25.
Lloyd's List. (2006mm). *Brussels to criminalize 'green' offences.* October 26.
Lloyd's List. (2006nn). *Lines win shipper support over post-conference regime.* October 31.
Lloyd's List. (2006oo) *India well placed to fight EU over shipping rules.* November 3.
Lloyd's List. (2006p). *Greek shipowners still in revolt over common structural rules.* June 12.
Lloyd's List. (2006q). *First blood to industry over EU pollution directive.* July 3.
Lloyd's List. (2006qq). *BIMCO voices fears over European protectionism: Extra cabotage regulations could hit shore-based industries.* November 16.
Lloyd's List. (2006r). *WTO chief adopts crisis diplomacy role to salvage troubled Doha talks.* July 5.
Lloyd's List. (2006rr). *Halting erosion of skill base.* November 22.
Lloyd's List. (2006ss). *Tell us what you want from an EU maritime policy, says Joe Borg.* November 29.
Lloyd's List. (2006t). *WTO services talks face gridlock.* July 14.
Lloyd's List. (2006tt). *EC backtracks from creation of a mega maritime ministry. No plans to create 'super directorate for seas and oceans'.* November 30.
Lloyd's List. (2006u). *EU seeks to extend territorial powers.* July 19.
Lloyd's List. (2006uu). *Excellence is key to integrated maritime policy.* November 30.
Lloyd's List. (2006v). *A fair wind at the start for Borg's maritime Green Paper.* September 5.
Lloyd's List. (2006vv). *Europe's maritime policy is making waves.* December 5.
Lloyd's List. (2006w). *Container lines told to work together in supply chain role.* August 14.
Lloyd's List. (2006ww). *Shipping groups express concern about EC Maritime Green Paper, Full EU membership of the IMO would cause 'serious damage'.* December 13.
Lloyd's List. (2006x). *IMO voluntary audit scheme criticised for lack of transparency.* September 8.
Lloyd's List. (2006xx). *Maritime clusters should drive EU policy.* October 25.
Lloyd's List. (2006y). *Weighed down by the burden of the rulebook.* September 8.
Lloyd's List. (2006yy). *EU ministers seal fate of conferences.* September 26.
Lloyd's List. (2006z). *Brussels' 24-hour box rule 'fatally flawed'.* September 21.
Lloyd's List. (2007a). *Why nations need to observe the rules of the global game.* January 3.
Lloyd's List. (2007b). *Europe's 'maritime space' draws fire.* January 30.
Lloyd's List. (2007c). *Tonnage tax must stay says Carnival boss. Dingle says EC must not be protectionist.* February 8.
Lloyd's List. (2007d). *Unions back EU coast guard plans.* February 19.
Lloyd's List. (2007e). *Borg urges military focus in maritime policy.* March 21.
Lloyd's List. (2007f). *Keep out of our ports, ESPO tells EU.* March 26.
Lloyd's List. (2007g). *Brussels expected to opt for 'soft law' on seaports policy.* May 1.
Lloyd's List. (2007h). *Mitropoulos wins breathing space in pollution battle.* May 1.
Lloyd's List. (2007i). *Why there must be some hard thinking over EC's soft options.* November 2.
Lloyd's List. (2007j). *EU plans for greener freight 'struggle to be taken seriously'.* November 5.
Lloyd's List. (2007k). *IMO brings forward its greenhouse gas deadline.* November 19.
Lloyd's List. (2007l). *Mitropoulos seeks to calm fears on ballast water tanks.* November 20.
Lloyd's List. (2007m). *In his own words: Joe Borg on maritime policy.* July 31.
Lloyd's List. (2007n). *Borg scraps EU coastguard and flag.* October 11.
Lloyd's List. (2007o). *Shippers warn EU against liner bias.* September 24.
Lloyd's List. (2008g). *EU maritime lead should be followed, urges Borg.* July 17.
Lloyd's List. (2008b). *IMO heads for its big birthday bash with plenty to talk about.* May 28.

Lloyd's List. (2008h). *Gentle persuasion.* July 18.
Lloyd's List. (2008d). *EC tonnage tax proposal under fire.* February 1.
Lloyd's List. (2008e). *Its time to clear the haze over air emission standards.* February 1.
Lloyd's List. (2008f). *IMO warned of problems over emissions.* February 4.
Lloyd's List. (2008k). *Emec head wants to join yards at IMO table.* September 30.
Lloyd's List. (2008l). *Tajani targets IMO membership for EU.* November 26.
Lloyd's List. (2008i). *Maritime giant's future threatened by back-tax.* May 30.
Lloyd's List. (2008j). *French owners urge IMO to follow lead on Erika3.* October 13.
Lloyd's List. (2008m). *International community must be 'prepared to shoot'.* October 3.
Lloyd's List. (2008n). *Container bosses back IMO bid to cut ship emissions.* December 22.
Lloyd's List. (2008o). *Class finds ally in IUMI as Brussels row hots up.* December 4.
Lloyd's List. (2008p). *Brussels fails to put heat under reluctant ratifiers.* October 20.
Lloyd's List. (2009a). *If Brussels controls European shipping, who controls Brussels?* June 15.
Lloyd's List. (2009aa). *When is a tax not a tax?* May 15.
Lloyd's List. (2009b). *Europe does not need member state approval to join IMO.* January 22.
Lloyd's List. (2009bb). *Shipping's shining light sets highest standards.* May 26.
Lloyd's List. (2009c). *Brussels lashes out at IMO weaknesses.* February 2.
Lloyd's List. (2009d). *Asian owners on course for seat at IMO.* May 20.
Lloyd's List. (2009dd). *Brussels puts 'crisis' measures on agenda.* January 20.
Lloyd's List. (2009e). *Asian owners drop plan for seat at IMO.* May 26.
Lloyd's List. (2009ee). *Cartel practices will not solve liner challenges.* September 9.
Lloyd's List. (2009f). *Owners urge EC to scrap regional cargo rules plan.* May 21.
Lloyd's List. (2009ff). *Brussels 'shying away' from market distortions.* June 4.
Lloyd's List. (2009i). *Grounds for concern.* November 12.
Lloyd's List. (2009j). *Katseli vows to retain Greece tax benefits.* November 24.
Lloyd's List. (2009g). *Nautical Institute lays plans for its IMO voice.* September 8.
Lloyd's List. (2009n). *Brussels drops threat of regional emissions plan.* April 20.
Lloyd's List. (2009o). *Brussels presses IMO on curbing emissions.* October 1.
Lloyd's List. (2009p). *Brussels agrees to hold fire on shipping emissions intervention.* October 21.
Lloyd's List. (2009q). *Brussels says patchwork CO_2 measures likely.* November 16.
Lloyd's List. (2009r). *IMO fights for control over CO_2 cuts.* December 9.
Lloyd's List. (2009h). *Brussels in new push to join IMO.* November 6.
Lloyd's List. (2009t). *IMO in bid to control shipping emissions policy.* December 11.
Lloyd's List. (2009u). *Copenhagen turns its gaze on new taxes for shipping.* December 17.
Lloyd's List. (2009k). *New voices fail to become IMO council members.* December 1.
Lloyd's List. (2009l). *IMO to make controversial audit scheme mandatory.* December 4.
Lloyd's List. (2009m). *IMO emissions move criticised.* March 24.
Lloyd's List. (2009s). *IMO should keep control.* December 10.
Lloyd's List. (2009cc). *Crisis becomes aspect of EU maritime strategy.* January 16.
Lloyd's List. (2009mm). *Asian shipping calls for greater say in international policies.* February 12.
Lloyd's List. (2009pp). *Polemis hits back as EU issues CO_2 ultimatum.* September 10.
Lloyd's List. (2009gg). *European ports demand state aid guidelines.* July 10.
Lloyd's List. (2009hh). *European ports must open up to competition.* September 22.
Lloyd's List. (2009ii). *Is the European Commission really interested in ship safety.* January 14.
Lloyd's List. (2009jj). *IMO under fire over fatigue related accidents.* February 19.
Lloyd's List. (2009kk). *Nautical Institute lays plans for its IMO voice.* September 8.
Lloyd's List. (2009ll). *European Commission considers changing IMO membership status.* January 21.
Lloyd's List. (2009nn). *Global industry searches for universal deal on emissions.* January 14.
Lloyd's List. (2009oo). *Brussels urges inclusion of shipping in climate deal.* January 29.
Lloyd's List. (2009qq). *Backing for IMO as (sic) emissions forum.* September 16.
Lloyd's List. (2009rr). *Industry push to retain IMO leadership on CO_2 reduction.* September 25.
Lloyd's List. (2009ss). *Owners voice doubts over Brussels' scrapping plan.* July 10.
Lloyd's List. (2009tt). *Brussels 'has no plans' for ship scrapping subsidies.* July 22.

References

Lloyd's List. (2009uu). *Report says UK naval bases must be 'Europeanised' to protect shipping.* March 31.
Lloyd's List. (2009v). *Climate talks leave shipping in limbo.* December 21.
Lloyd's List. (2009vv). *European parliament waters down passenger rights law.* March 31.
Lloyd's List. (2009w). *Owners confident that IMO will retain control over emissions.* December 30.
Lloyd's List. (2009ww). *Brussels says it is too busy to investigate alleged breaches of competition rules.* February 12.
Lloyd's List. (2009x). *Asia asks for louder voice in regulatory organizations.* February 13.
Lloyd's List. (2009xx). *Simplified EU customs law held back by infighting.* January 12.
Lloyd's List. (2009y). *Norway supports IMO bid to manage emissions.* December 2.
Lloyd's List. (2009yy). *Only Europe can make laws that are out of date already.* February 16.
Lloyd's List. (2009zz). *IMO review piracy legislation.* March 31.
Lloyd's List. (2010a). *Transport chief to champion EU shipping.* January 14.
Lloyd's List. (2010aa). *Europe emissions control areas a big mistake.* September 27.
Lloyd's List. (2010b). *Treading softly on the road to dock labour reform.* June 9.
Lloyd's List. (2010bb). *MEPC to consider 10 proposals.* September 28.
Lloyd's List. (2010c). *Wanted: Machiavellian diplomat with iron fists.* August 31.
Lloyd's List. (2010d). *Question of life or death.* January 14.
Lloyd's List. (2010dd). *IMO has mandate to find emissions solution.* September 30.
Lloyd's List. (2010e). *Flagging efforts.* July 22.
Lloyd's List. (2010ee). *MEPC could be forced to vote on emissions.* September 28.
Lloyd's List. (2010f). *Major flag state holds back on fees.* March 15.
Lloyd's List. (2010ff). *IMO divided over climate change funding.* October 6.
Lloyd's List. (2010gg). *Mordue quits European maritime safety chief role.* January 25.
Lloyd's List. (2010hh). *IMO will lead emissions plan.* January 4.
Lloyd's List. (2010i). *States scupper EU plan for common stand on emissions.* January 28.
Lloyd's List. (2010j). *Brussels questions IMO's ability to enforce emissions trading scheme.* February 10.
Lloyd's List. (2010n). *No resolution as MEPC ends.* March 26.
Lloyd's List. (2010o). *Industry fury over planned SOx cuts.* May 18.
Lloyd's List. (2010p). *Fair deal on SOx limits.* May 19.
Lloyd's List. (2010q). *Brussels steps away from north Europe ECA protest.* May 31.
Lloyd's List. (2010r). *Posidonia 2010: Unite against Brussels, urges Lord Mayor of London.* June 9.
Lloyd's List. (2010g). *IMO in fees plea to late-paying members.* May 14.
Lloyd's List. (2010h). *IMO should open up.* March 16.
Lloyd's List. (2010k). *Rebel states back Brussels on IMO emissions statement.* March 1.
Lloyd's List. (2010l). *EU backs IMO on emissions reduction.* March 9.
Lloyd's List. (2010m). *IMO facing emissions deal snag.* March 22.
Lloyd's List. (2010s). *Protecting the IMO.* June 10.
Lloyd's List. (2010cc). *ICS warns over UN solutions to emissions.* September 30.
Lloyd's List. (2010ii). *IMO calls for early talks with Brussels climate chief.* March 12.
Lloyd's List. (2010jj). *EU transport czar in bid to control maritime agenda.* January 14.
Lloyd's List. (2010kk). *Leading flag state withholds IMO fees.* March 16.
Lloyd's List. (2010x). *Brussels puts its foot down over low-sulphur fuel.* September 27.
Lloyd's List. (2010y). *Mitropoulos MEPC opening speech.* September 27.
Lloyd's List. (2010z). *Mitropoulos urges workable solution to emissions.* September 27.
Lloyd's List. (2011g). *Kallas unveils port services initiative.* September 8.
Lloyd's List. (2011b). *Lloyd's List cleans up with report on MEPC's CO_2 success.* July 18.
Lloyd's List. (2011c). *Japan's top man at the MEPC is bullish on greenhouse gas rules.* July 19.
Lloyd's List. (2011d). *What happens next?* July 25.
Lloyd's List. (2011f). *Polemis urges European climate chief to work with shipping on CO_2 cuts.* July 29.

Lloyd's List. (2011h). *Brussels still a few elements short of a full toolbox, says new ECSA chief.* September 27.

Maritime Global Net. (2009). *EU scuppers key safety measure at STW.* February 16, http://www.mglobal.com.

Paixao, A., & Marlow, P. B. (2001). A review of the European Union shipping policy. *Maritime Policy and Management, 28*(2), 187–198.

Roe, M.S. (2007a, May 10–11). *European Union maritime policy. the role of new member states and the need for a review of governance.* International Shipping Management Forum 07, Athens, Greece.

Roe, M. S. (2007b). Shipping, policy and multi-level governance. *Maritime Economics and Logistics, 9,* 84–103.

Roe, M.S. (2007c, July 4–6) *Jurisdiction, shipping policy and the "New" states of the European Union.* International Association of Maritime Economists Annual Conference (IAME), Athens.

Roe, M. S. (2007d). Shipping policy governance: A post-fordist interpretation. In I. Visivikis (Ed.), *Trends and developments in shipping management* (pp. 131–142). Athens: T & T Publishing.

Roe, M.S. (2007e). Shipping policy, governance and jurisdictional friction. *Economic outlook, 104,* 8–15 January.

Roe, M. (2008a). Safety, security, the environment and shipping—the problem of making effective policies. *WMU Journal of Maritime Affairs, 7*(1), 263–279.

Roe, M.S. (2008b, April 2–4). *State versus the stakeholder; The governance of the maritime sector.* International Association of Maritime Economists Annual Conference (IAME), Dalian.

Roe, M.S. (2009a, June 24–26). *Maritime capitalist liposuction—a postmodern interpretation of maritime governance failure.* International Association of Maritime Economists Annual Conference (IAME), Copenhagen.

Roe, M.S. (2009b, September 17–18). *Policy Failure in shipping; A global disaster of epic dimensions.* 2nd International Symposium on Maritime Safety, Security and Environmental Protection, National Technical University of Athens, Athens, Greece.

Roe, M. S. (2009c). Multi-level and polycentric governance: effective policymaking for shipping. *Maritime Policy and Management, 36*(1), 39–56.

Roe, M. S. (2009d). Maritime governance and policy-making failure in the European Union. *International Journal of Shipping, Transport and Logistics, 1*(1), 1–19.

Roe, M.S. (2010a, July 7–9). *Globalization and maritime policy-making—the case for governance rehabilitation.* International Association of Maritime Economists Annual Conference (IAME), Lisbon.

Roe, M.S. (2010b). Shipping policy and globalization: Jurisdictions, governance and failure, In C. Grammenos (ed.), *The Handbook of maritime economics and business* (2nd ed., pp. 539–556). Lloyd's List: London.

Roe, M.S. & Selkou, E. (2006). *Multi-level governance, shipping policy and social responsibility.* International Conference "Shipping in the Era of Social Responsibility" in honour of the late Professor Basil Metaxas (1925–1996), Argostoli, Kefalonia, Greece.

Rosenau, J. N. (1992). Governance, order and change in world politics. In J. N. Rosenau & E. O. Czempiel (Eds.), *Governance without government: Order and change in world politics* (pp. 1–29). Cambridge: Cambridge University Press.

Selkou, E., & Roe, M. S. (2004). *Globalization, policy and shipping.* Cheltenham: Policy and Shipping, Edward Elgar.

Selkou, E., & Roe, M. S. (2005). Container shipping policy and models of multi-level governance. *Annals of Maritime Studies, 43,* 45–62.

Smith, B. (1993, pp. 82–101). EUROS: The European community ship register. In P. Hart., G. Ledger., M. Roe & B. Smith (Eds.), *Shipping policy in the European community,* Avebury: Aldershot.

References

Smith, M. (1996). The European Union and a changing Europe: Establishing the boundaries of order. *Journal of Common Market Studies, 34*(1), 5–28.
Tradewinds. (2006a). *IMO authority vs the unilateralists.* June 9.
Tradewinds. (2006b). *Europe snubs liner scheme.* October 6.
Tradewinds. (2006c). *Shippers shun lines.* November 10.
Tradewinds. (2007a). *IMO wake-up call.* August 17.
Tradewinds. (2007b). *Lobbying hard.* September 28.
Tradewinds. (2007c). *Liner trade body to test EU guideline.* September 21.
Tradewinds. (2007d). *Carriers seeking EU ban clarity.* October 26.
Tradewinds (2007e) *Mitropoulos hits back.* December 7.
Tradewinds. (2008d). *Accident process to cast wider net.* February 29.
Tradewinds. (2008e). *A good plan that could go wrong.* February 29.
Tradewinds. (2008f). *Time to reassess the safety regime.* March 14.
Tradewinds. (2010g). *Clash risks IMO split on emissions* October 8.
Van Loon, H. W. (1932). *Van Loon's Geography.* New York NY: Simon and Schuster.
Wilber, K. (1979). *No Boundary: Eastern and Western approaches to personal growth.* Boston, MA: Chambhala.

Chapter 2
Governance

Such a short and simple word but what confusion it causes. Its significance in the context of maritime failure is so substantial that before we go on it is essential that its meaning and interpretation are made clear. Our first task is to define the term governance—but however, easy this might sound, the sheer number of definitions that exist is itself indicative of a concept that is both uncertain and central to the policy-making debate.

Governance is a vague term that has been widely re- and mis-interpreted over many years and examples of the variety of interpretations come from Finkelstein (1995), Gaudin (1998), Heretier (2002), Kooiman (2003), Kersbergen and Waarden (2004) and as far back as Heclo (1972). Jessop (1998, p. 29) indicated that governance had only relatively recently become a 'buzzword' in the social sciences while Krahmann (2003, p. 323) stressed that definitions are 'as varied as the issues and levels of analysis to which the concept is applied'. However, he did suggest that common to them all was the changing focus of political activity. Meanwhile Rhodes (1996) had already observed that there were at least six uses of the term governance in the UK political system alone. The range of meanings and definitions has expanded as the concept has increased in use which between 1980 and 2000, an analysis of the Social Science Citation Index indicated had risen to 20 entirely different areas from virtually none. The significance of arriving at an acceptable definition was stressed by Skelcher et al. (2004, p. 3) who cited Hajer's (2003) conviction that policy-making now proceeds in an 'institutional void' with no clear rules or processes for governance which increasingly addresses multiple problem public domains. At the same time the creeping influence of globalization has meant that policy-making is occurring 'in spaces parallel to and across state institutions and their jurisdictional boundaries'; Jessop (2003a, p. 31) described this as 'looking beyond the state' while Zurn (2003, pp. 348–349) used similar terminology to encompass 'political denationalization' which can take the form of regional, supranational or international government. The need for a meaningful definition could not be more obvious.

Here we can only point to the large body of literature that has considered the issue of definition and suggest a number of interpretations that are acceptable and have particular relevance to the policy-making process in the maritime sector. Krahmann provided support for this approach as he defined four broad areas of use of governance as a concept, one of which he identified with studying a particular policy sector such as education, health or transport. For our purposes this will focus on maritime governance.

Jachtenfuchs (2001, p. 24) took the work of Zurn (1998, p. 12) to provide a generic definition that is useful in establishing the premise upon which governance is based:

> governance can be understood as the intentional regulation of social relationships and the underlying conflicts by reliable and durable means and institutions, instead of the direct use of power and violence.

A similarly very broad definition was given by Kazancigil (1998, p. 70) quoting from the Report of the Commission on Global Governance (1995, p. 2) where governance was seen as:

> the sum of the many ways individuals and institutions, public and private, manage their common affairs. It is a continuing process through which conflicting or diverse interests may be accommodated and co-operative action may be taken. It includes formal institutions and regimes empowered to enforce compliance as well as informal arrangements that people and institutions either have agreed or perceive to be in their interest.

Rhodes (1997, p. 11) suggested an alternative which he applied to the analysis of governmental change:

> Governance refers to self-organizing, interorganizational networks characterized by interdependence, resource exchange, rules of the game, and significant autonomy from the state.

In the light of this he suggested that there were six different ways in which governance can be used:

- In the minimal state.
- As corporate governance.
- As the new public management.
- As good governance.
- As a socio-cybernetic system.
- As self-organizing networks.

It is the contention here that these all interlink and overlap and that in the context of maritime policy-making, 1, 3, 4, 5 and 6 all have their role to play. However, 5 is probably of most interest at this stage as its consideration would inevitably include a variety of issues such as:

- Policy outcomes are not only the products of actions by central government as these actions have to interact with many other authorities.
- Central and national government is no longer supreme.

- The international system is a vital component of governance at all jurisdictions (Rhodes 1997, pp. 25–26).

Rhodes concluded that there were a number of shared characteristics of governance that will always feature regardless of the context:

- Interdependence between organizations. Governance is broader than government, covering non-state actors. Changing the boundaries of the state meant that the boundaries between public, private and voluntary sectors became shifting and opaque.
- Continuing interactions between network members, caused by the need to exchange resources and negotiate shared purposes.
- Game-like interactions, rooted in trust and regulated by rules of the game negotiated and agreed by network participants.
- A significant degree of autonomy from the state. Networks are not accountable to the state they are self-organizing. Although the state does not occupy a sovereign position it can indirectly and imperfectly steer networks.

How many of those are reflected in the current state centric, hierarchical governance framework that characterizes the maritime sector?

Meanwhile, De Alcantara (1998, p. 105) focused more on the process of governance in defining it as involving:

> building consensus, or obtaining the consent or acquiescence necessary to carry out a programme, in an arena where many different interests are in play.

The World Bank (2000), cited in (Brooks and Cullinane 2007a, p. 10) suggested that governance is:

> The traditions and institutions by which authority in a country is exercised for the common good. This includes (i) the process by which those in authority are selected, monitored and replaced, (ii) the capacity of the government to effectively manage its resources and implement sound policies, and (iii) the respect of citizens and the state for the institutions that govern economic and social interactions among them.

Perhaps most interestingly, the World Bank, despite being an international policy-making institution, is clearly nationally focused in its assessment of governance, reflecting the national representation that dominates its proceedings. Meanwhile, Jessop (1998, p. 30) considered governance to be a complex issue but this complexity can be accommodated if it is decentred and pluralistic, the latter reflecting the multi-stakeholder characteristics of those governed (Glagow and Willke 1987). He rejected the traditional dichotomous governance arguments of market vs hierarchy, plan vs market, anarchy vs sovereignty and private vs public which have dominated the literature (see for example Zhang 2005, p. 199 who suggested that governance models which reflect the rise of globalization can be divided neatly into market-economy oriented and civil society oriented) and quoted from Scharpf (1993, p. 57):

Clearly, beyond the limits of the pure market, hierarchical state, and domination-free discourses, there are more—and more effective—co-ordination mechanisms than science has hitherto grasped empirically and conceptualized theoretically. *(translated by Jessop)*

Wang et al. (2004, p. 238) cited Stoker (1998) who saw governance increasingly accommodating market and government failures and needing to recognize the increasing power and influence of non-government bodies. This suggested an ever increasing interdependence of actors leading to the idea that governance was best seen as:

- A set of institutions and actors that are drawn from but also beyond government.
- The blurring of boundaries and responsibilities for tackling social and economic issues.
- Power dependence involved in the relationships between institutions involved in collective action.
- Autonomous self-governing networks of actors.
- The capacity to get things done which does not rest on the power of governments to command or use their authority.

Williams (2005, p. 9) provided a state-orientated definition that although vague, pointed in the direction of governance that many people might first consider:

the ability of the state to impose an order on aspects of social, political and economic life… It is concerned above all with the capacity of the state to govern social, political and economic processes.

Referring back to Krahmann (2003, p. 9), he also provided a national focus where:

the concept of governance has come to represent political systems in which authority is fragmented amongst a multitude of governmental and nongovernmental actors to increase efficiency and effectiveness.

Jessop (1998, p. 29) suggested that governance can refer to 'any mode of coordination of interdependent activities' These included what he termed as the anarchy of exchange (or the market approach to activity); organizational hierarchy (which has particular relevance to maritime governance); and self-organizing heterarchy (which referred to networking by self-interested participants). The choice of which governance mode is applied in any particular context (including maritime) should rest on the characteristics of the sector.

Smouts (1998, p. 83) took a wider view and incorporated explicitly both the private and public sector in his consideration of governance, quoting from the Commission on Global Governance (1995, pp. 2–3):

Governance is the sum of the many ways individuals and institutions, public and private, manage their common affairs. It is a continuing process through which conflicting or diverse interests may be accommodated and co-operative action taken. It includes formal institutions and regimes empowered to enforce compliance, as well as informal arrangements that people and institutions either have agreed or perceive to be in their interest.

Borzel (2010, p. 194) suggested that governance had to consist of both structure and process, the former consisting of the actors and institutions and the latter the social coordination that takes place between them. Gereffi et al. (2005, pp. 83–84) defined the characteristics of governance within value chains while Hess (2008, pp. 456–457) took this further, quoting Allen (1997, p. 64) who suggested that governance was related to power and that:

> power is anchored in institutional space, with different types of disciplinary institutions, the workplace among them, governed by an assembly of practices which map onto one another but also display their own specific, diagrammatic quality.

Other definitions included Malpas and Wickham (1995, p. 40) who took the broadest of views of governance as 'any attempt to control or manage a known object'; and Hirst's (1997b, p. 3) all encompassing attempt:

> the means by which an activity or ensemble of activities is controlled or directed, such that it delivers an acceptable range of outcomes according to some established social standard.

Szczerski (2004, pp. 1–2) took a comprehensive review in outlining Rhodes' (2000, pp. 56–64) seven different types of governance to which we referred earlier. Meanwhile de Senarclens (1998, p. 92) claimed that governance:

> reflects the idea that national governments do not have the monopoly to legitimate power, that there are other institutions and actors that contribute to maintaining order and participate in economic and social regulation. The mechanisms of managing and controlling public affairs involve – at local, national and regional level – a complex set of bureaucratic structures, political powers ranked in a more or less rigid hierarchy, enterprises, private pressure groups and social movements. Governments no longer monopolize the functions of political command and mediation. These functions are now exercised by a broad range of governmental and non-governmental organizations, private undertakings and social movements.

and that of De Alcantara (1998, p. 105) who suggested that:

> governance' involves building consensus, or obtaining the consent or acquiescence necessary to carry out a programme, in an arena where many different interests are at play.

The emphasis on the word 'consensus' is especially significant in the light of the deteriorating relationships between the IMO, the EU and member states over many maritime policy issues and the debates going on within member states about governance and the maritime sector (Lloyd's List 2008).

Jordan (2001, p. 199) attempted to place a definition of governance within the framework of European Union integration—something of direct relevance to the discussion of maritime policies. He quoted Rosenau (1992, p. 4) on governance:

> Governance… is a more encompassing phenomenon than government. It embraces governmental institutions, but it also subsumes informal, non-governmental mechanisms whereby those persons and organizations within its purview move ahead, satisfy their needs and fulfill their wants.

This optimistic view of governance is of direct relevance to the policy failures we have identified here—if successful governance creates a satisfied and fulfilled society, then does this not question the competence and relevance of a governance framework for the maritime sector that seems to generate an over-abundance of failure?

He went on to quote Kooiman (1993, p. 4) whose comments on governance he saw as particularly relevant to the EU in that governance was:

> a pattern or structure that emerges in socio-political systems as a 'common' result or outcome of the interacting intervention efforts of all the involved actors. This pattern cannot be reduced to one actor or group of actors in particular... No single actor, public or private, has all knowledge and information required to solve complex, dynamic and diversified problems; no actor has sufficient overview to make the application of particular instruments effective: no single actor has sufficient action potential to dominate unilaterally in a particular governing model.

This view that governance is far from restricted to state action was backed up by Stoker (1998, p. 17) who saw:

> the essence of governance (as) its focus on governing mechanisms that do not rest on recourse to the authority and sanctions of government.

Stoker also suggested that governance provided a framework for 'understanding changing processes of governing' and quoted Judge et al. (1995, p. 3) who claimed that these frameworks:

> provide a language and frame of reference through which reality can be examined.

Stoker also provided what he termed five 'propositions' that define governance. They comprise a useful summary of the characteristics of governance that have relevance to our discussion as we move on. In some cases the characteristics he pointed out relate directly to issues we shall raise in maritime governance and its potential failure to provide an adequate framework for the policy-making process:

- Governance refers to a set of institutions and actors that are drawn from but also beyond government.
- Governance identifies the blurring of boundaries and responsibilities for tackling social and economic issues.
- Governance identifies the power dependence involved in the relationships between institutions involved in collective action.
- Governance is about autonomous self-governing networks of actors.
- Governance recognizes the capacity to get things done which does not rest on the power of government to command or use its authority. It sees government as able to use new tools and techniques to steer and guide.

Newman and Thornley (1997, p. 968) considered governance in the light of urban policy and planning and believed it involves 'grasping the fragmentation of institutions, the changing role of central government and the emergence of new networks—both public and private—which attempt to co-ordinate policy areas'. Obando-Rojas et al. (2004, p. 299) saw governance as typically dependent upon fulfilling 'multi-layered and multi-agency obligations to achieve intended goals'.

One of the most interesting definitions was provided again by Borzel (2007, p. 3) who, following the work of Mayntz and Scharpf (1995, p. 19; Mayntz 2004; Scharpf 2001) considered that governance could be:

> understood as institutionalized modes of coordination through which collectively binding decisions are adopted and implemented. In this understanding governance consists of both structure and process. Governance structures relate to the institutions and actor constellations while modes of social coordination refer to the processes which influence and alter the behaviour of actors. Governance structures and processes are inherently linked since institutions constitute arenas for social coordination and regulate their access, allocate competencies and resources for actors and influence their action orientations.

Newman (2001, pp. 11–12, 15, 17) wrote extensively defining governance and identifying its main themes. She suggested that it represented shorthand for changes in a number of relationships:

- A set of elusive but potentially deeply significant shifts in the way in which government seeks to govern (Pierre and Peters 2000).
- It denotes the development of ways of coordinating economic activity that transcend the limitations of both hierarchy and markets (Rhodes 1997).
- It highlights the role of the state in 'steering' action within complex social systems (Kooiman 1993, p. 2000).
- It denotes the reshaping of the role of local government away from service delivery towards 'community governance' (Clarke and Stewart 1999; Stewart and Stoker 1988).

These changes are located within the broader process of change exemplified by globalization and later described and analyzed within the Postmodern epoch where the flow of power has been away from the nation-state and both upwards to international and supra-national authorities and downwards to the regions and localities. The new governance that these changes demands must 'rely on a plurality of interdependent institutions and actors drawn from within and beyond government'. Newton (2001, p. 12) cited Pierre and Peters (2000) and Rhodes (1997) in describing governance as a 'promiscuous concept, linked to a wide range of theoretical perspectives and policy approaches'. It was a two-way process following the work of Kooiman (1993a, p. 4), no longer the governed following the governing but one where 'aspects, qualities, problems and opportunities of both the governing system and the system to be governed are taken into consideration'. It operates over all jurisdictions and at each may well operate in different ways. It was complex and multi-facetted accommodating 'complex interactions and interdependencies of government institutions, communities, citizens and civil society' (Newton 2001, p. 17). Table 2.1 summarizes Newton's view of the new governance that needed to emerge as a response to the globalized era which we are witnessing.

Weiss (2000, p. 806) also reflected on the need for a new governance accompanying globalization, noting the fact that international trade, relations and politics are no longer simply composed of states but are also characterized by a growing number of non-state actors. The world has become increasingly decentralized so

Table 2.1 Governance shifts

1.	A move away from hierarchy and competition as alternative models for delivering services towards networks and partnerships traversing the public, private and voluntary sectors
2.	Recognition of the blurring of boundaries and responsibilities for tackling social and economic issues
3.	Recognition and incorporation of policy networks into the process of governing
4.	Replacement of traditional models of command and control by 'governing at a distance'
5.	The development of more reflexive and responsive policy tools
6.	The role of government shifting to a focus on providing leadership, building partnerships, steering and coordinating and providing system-wide integration and regulation
7.	The emergence of 'negotiated self-governance' in communities, cities and regions, based on new practices of coordinating activities through networks and partnerships
8.	The opening-up of decision-making to greater participation by the public
9.	Innovations in democratic practice as a response to the problem of the complexity and fragmentation of authority and the challenge this presents to traditional democratic models
10.	A broadening of focus by government beyond institutional concerns to encompass the involvement of civil society in the process of governance

Source Newton 2001

that the existing 'monolithic and top-down view of governance' with a central sovereign in its midst, is now inappropriate.

Fritz (2010, pp. 4–5) characterized good governance by five main principles derived from the work of the EU:

- *Participation*—which he saw as the most significant characteristic. This is often viewed as the same as pluralism and ensuring that the maximum number of interested parties has the opportunity to take part in the governance process. Fritz divided pluralism into external and internal features with the former only well designed when it provides for widespread opportunities for participation rather than involving pre-selection of familiar experts. Internal pluralism referred to those from the immediate community of the discipline involved. External maritime pluralism would include politicians, environmentalists, media groups etc.; internal pluralism would include traditional maritime interests. Both should encourage participation rather than only inviting it.
- *Accountability*—this is usually equated with traceability and accepted rules of procedure and might be reflected in procedures in meetings, the layout, content and language of documents and discussions etc.
- *Openness*—is more about transparency although it is both recognized and understood that complete openness in any policy-making is not always possible for security and confidentiality reasons. In maritime governance openness raises questions about public accessibility to where decisions about policy are made, the availability and cost of policy documents, the use of language and the clarity of information provided by policy-makers etc.
- *Effectiveness*—centres on the use of resources in the context of the task in hand. It is thus a criterion close to 'efficiency'. The EU has recently come up with

three sub-criteria—saliency (have decision-makers the information they need), credibility (is the information technically correct) and legitimacy (are the interests of the maritime user at its heart)—which might help to make the assessment of effectiveness more precise.
- *Coherence*—is the governance process consistent and does it provide a coordinated and systematic process of policy-making?

The European Commission (Commission of the European Communities 2001, pp. 10–11; Commission of the European Communities 2009, pp. 2–3) provided considerably more detail of these five pointers (dressed up with a little proportionality and subsidiarity) towards good governance and for now we shall take them as indicators of how we might progress the maritime governance process at all jurisdictions. Gore and Wells (2009, p. 160) indicated what they saw as particularly useful approaches to understanding policy-making and governance including adopting post-positivist analytical and methodological approaches as outlined by Mazey (2000, p. 355). These include constructivism, discourse analysis, new institutionalism, policy networks and social movement theory which together can be used to explain 'political and policy phenomena'. Governance has to accept that 'reality is socially constructed and as such, rests upon ideational factors such as norms and embedded belief systems'. Good governance therefore needs to be diffused, dispersed across multiple jurisdictions, essentially superior to any monopoly-based governance determined by a central state (Marks and Hooghe 2004, p. 16). Such systems would reflect better the 'multitude and complexity of individuals' preferences, alluding to the issue of stakeholders in policy-making to which we return much later on, and also to the greater opportunities for jurisdictional competition, experimentation and innovation—something which is wholly absent in maritime governance at present (Weingast 1995; Oates 1999, 2002; Perraton and Wells 2004). Jessop (1997, p. 574) agreed identifying a clear move from government to governance, from a central state apparatus, characterized by 'state-sponsored economic and social projects and political hegemony towards an emphasis on partnerships between governmental, para-governmental and non-governmental organizations in which the state apparatus is often only first amongst equals' (MacLeod and Goodwin 1999, p. 506).

Haas (1992, p. 6) suggested that good governance is only possible if there is a greater understanding of international policy change and he proposed a number of ways in which this understanding could be improved. Table 2.2 summarizes these approaches.

Valaskakis (1999, pp. 163–164) debated the appropriate level of governance that should be applied and hit upon the problem that around 500,000 individual government bodies existed at all jurisdictions world-wide and the extension of globalization had made the selection of suitable jurisdictional locations for policy-making that much more difficult. He saw the principle of subsidiarity very useful but left the reader a little lost as to how to arrive at the optimum jurisdiction for any governance application. Turke (2008, p. 1) meanwhile saw complexity, diversity and dynamics as central features in today's governance referring to the

Table 2.2 Approaches to the study of policy change

Approach	Level of analysis and area of study	Factors that influence policy change	Mechanisms and effects of change	Primary actors
Epistemic communities approach	Transnational; state administrators and international institutions	Knowledge; causal and principled beliefs	Diffusion of information and learning; shifts in the patterns of decision-making	Epistemic communities; individual states
Neo-realist approaches	International; states in political and economic systems	Distribution of capabilities; distribution of costs and benefits from actions	Technological change and war; shifts in the available power resources of states and in the nature of the game	States
Dependency theory-based approaches	International; global system	Comparative advantage of states in the global division of labor; control over economic resources	Changes in production; shifts in the location of states in the global division of labor	States in the core, periphery and the semi-periphery; multinational corporations
Post-structuralist approaches	International; discourse and language		Discourse; the opening of new political spaces and opportunities	Unclear

Source Haas (1992, p. 6)

fact that modern societies were multi-layered and complicated and that multiple actors rather than governments alone, featured prominently. These actors were pluralist and fragmented and governments switched between representation of local/regional to international/supranational constituencies at alarming speed.

The issue of pluralism and its contrast with the traditional hierarchical governance framework that characterizes the maritime sector is one that deserves more explanation especially since any move away from hierarchies and state-centricism will inevitably involve a pluralist shift. McLennan (1995, p. 39) stressed the rise of pluralism and its relevance to today's governance debate:

> with Marxism and neo-corporatism shoved out into the wings, we are now confronted with the rather surprising scenario in which sociopolitical pluralism, somewhat dazed, rusty and hesitant, has been summoned to rise up and take the theoretical centre-stage once again.

McLennan (1995, pp. 28–29) was a fierce proponent of pluralist governance citing Bertrand Russell who in 1907 had experienced a 'revolt into pluralism'. Russell was fearful of the 'monistic theory of truth' when he spoke at the Aristotelian Society in London (Russell 1959). This 'logical atomism' as it became known had negative characteristics, viz:

> the logic that I should wish to combat maintains that in order thoroughly to know any one thing, you must know all its relations and all its qualities, all the propositions in fact in which that thing is mentioned; and you deduce of course from that that the world is an interdependent whole… of course it is clear that since everything has relations to everything else, you cannot know all the facts of which a thing is a constituent without having some knowledge of everything in the universe. (Russell 1972, pp. 59–60)

William James was another proponent who expressed concern at the neglect of pluralism as a philosophical concept:

> It is curious how little countenance radical pluralism has ever had from philosophers. Whether materialistically or spiritualistically minded, philosophers have always aimed at clearing up the litter with which the world is apparently filled… As compared with all their rationalizing pictures, the pluralistic empiricism which I profess offers but a sorry appearance. It is a turbid, muddled, gothic sort of affair, without a sweeping outline and with little pictorial nobility. (James 1909, p. 26)

Quite. Yet no less valid because of it. Pluralist trends have been emerging for decades and the increasing problems faced by maritime policy-makers are indicative of the pluralist conflict with the traditional monolithic, governance hierarchy. Pluralism rests on the suggestion that as McLennan states, 'there are many separate things in the world'. However, its own characteristics mean that it lacks coherence and as a result has been commonly looked upon as an 'oppositional programme' which provides a 'purposeless' framework for society (including governance, policy-making and so on). Ward (1911) also recognized a self-defeating problem with pluralism in that if strictly adhered to, the differentiation it implies can run to infinity. Meanwhile, to be able to make use of its values as a philosophical structure, an overarching framework is needed, a 'unifying move' which is represented by a principle of faith to make sense of the ontological dispersion that pluralism suggests. This is inevitably self-contradictory. The pluralist debate has continued ever since as the concept has risen in significance with the dialect implied by globalization where fragmentation and consolidation work hand in hand to create dilemmas for governance.

Definitions of pluralism abound. Schmitter (1979) suggested that:

> Pluralism can be defined as a system of interest representation in which the constituent units are organized into an unspecified number of multiple, voluntary, competitive, non-hierarchically ordered and self-determined… categories which are not specially licensed, recognized, subsidized, created or otherwise controlled in leadership selection or interest articulation by the state and which do not exercise monopoly of representational activity within their respective categories.

Atkinson and Coleman (1989, pp. 55–57) wrote extensively on a variety of different pluralisms that they identified including 'pressure', 'clientele' and 'parentula' pluralisms each of which are characterized to a different degree by fragmented interests and horizontal power relationships.

Puchala (1972, p. 278) reflected on pluralist interpretations of the nation-state while Kjaer (2004, p. 22) viewed pluralism as opposed to interests vested in powerful organizations with direct and privileged access to state institutions and instead placed in a plurality of individuals. McLennan (1995, p. 34) viewed pluralism as 'premised on a critical approach to both metaphysics and statism', but particularly focusing on the latter. As such it is particularly relevant to our discussion of governance, the role of the nation-state and the problems that seem to have occurred in policy-making in the maritime sector. He supported the significance of pluralism in the newly globalized world through five 'planks':

- Industrial society evolves in an ever-more complex way, interdependent and increasingly well-informed.
- Social groups are the basic units of interaction and socialization—commonly overlapping and relatively small.
- The state remains but should be acting as a 'processor', aiming at striking balances between competing interests and demands.

> The whole political process has been cast by turns as a social physics of pressures and counter-pressures, as an equilibrium curve, as the demand for and supply of political goods, and as an evolutionary sequence of competition and organic adaptation. (Kjaer 2004, p. 22)

- Basic civic democracy is reflected in a pluralist organization of society and this is allied with the individual's failure to agree to the over-arching interests of social harmony.
- Pluralism rejects 'grand theorizing around abstract quasi-philosophical themes', preferring empiricism and qualitative description.

Pluralism as a concept related to governance has generated considerable debate over a considerable period of time. Morgan (1997, pp. 404–405) identified a substantial literature relating to political thought (Bentley 1908; Maitland 1911; Figgis 1913; Follett 1918; Laski 1917, 1919); organizational theory (Follett 1973; Filley 1975; Thomas 1976; Robbins 1978; Burrell and Morgan 1979; Brown 1983; Graham 1995); its contrast to unitary theory (Ross 1958, 1969; Fox 1966, 1974); its relationship to disordered social and cultural cleavages (Merrien 1998, p. 58); and its relationship to conflict (Coser 1956). Atkinson and Coleman (1992, p. 160) stressed the importance of policy pluralism quoting Laumann and Knoke (1987, pp. 378–380) in that policy outcomes 'are the product of decentralized contention among a plurality of organizations'; 'a collection of policy arenas incorporating both governmental and private actors'. Turke (2008, pp. 1–2) summed it up:

> Complexity', 'Dynamics' and 'Diversity' are omnipresent in today's discourse on governance… social conditions in modern societies are perceived as multi-layered and complicated. Social issues are being addressed by multiple actors: governments are not

playing a primary role anymore.... actor and group identities are fragmented and pluralistic.

Jencks' (1992b, p. 11) comments were particularly relevant since he related the rise of pluralism to that of Postmodernity—something to which we return in some detail in a later chapter. The end of a 'single world view' is a key characteristic of the Postmodern and as such pluralism is an essential part whether we are dealing with architecture, governance, literature, philosophy or shipping, the EU, the Far East or developing Africa. Postmodernism embraces pluralism by revisiting single explanations, respecting differences, and celebrating all jurisdictions. The relevance of pluralism to the Postmodern condition that we shall consider more deeply later is evidenced elsewhere. Three quotes will suffice:

> The Postmodern political condition is premised on the acceptance of the plurality of cultures and discourses. Pluralism (of various kinds) is implicit in Postmodernity as a project. (Heller and Feher 1988, p. 5)

> At the heart of a Postmodern culture is the acceptance of the irreducible pluralistic character of social experiences, identities and standards of truth, moral rightness and beauty. (Seidman 1994, p. 324)

> Distinctive features of the new cultural politics of difference are to trash the monological and homogenous in the name of diversity, multiplicity and heterogeneity; to reject the abstract, general and universal in the light of the concrete, specific and particular; and to historicize, contextualize and pluralize by highlighting the contingent, variable, tentative and changing. (West 1993, p. 3)

Kjaer (2004, pp. 84–85) considered pluralism in the light of the governance of international relations. Braudel (1982, pp. 464–466) compared societies from the basis of hierarchies and pluralism and concluded that the hierarchical order is never simple and that plurality is an essential feature of any movement resistant to change. Meanwhile Jessop (1998, p. 30) saw pluralism as an important part of governance directing 'inter-systemic coordination', bringing together the various stakeholders in policy-making.

Jordan and Schubert 1992 considered the role of the pluralist state where authority and autonomy are diluted by the range of interest groups and individuals attempting to access the policy-making process. The state retains a role but reduced from the hierarchical, state-centred governance that characterizes the maritime market place. Pluralism was seen as a response to totalitarianism and a function of the Cold War from 1945 until 1989/1990. Only by dispersing political power over a wide range of interests could individual rights be secured. Despite the ending of the Cold War, this move from a rigid and institutional bound, hierarchical process is one that has continued across many governance scenarios. However, the maritime sector shows little such progression.

King and Kendall (2004, pp. 72–3, 145) commented on the globalization of business and suggested that the business environment of firms is more 'pluralist, diverse and dynamic than is often supposed' (Mathews 2002). Functional differentiation needed to be accommodated by the distribution of political power to accommodate 'multiple identities, cross-cutting group memberships and diverse sources and centres of power'. This meant that there can be no legitimate national 'will'. Such pluralist power was a 'critical democratic resource for preventing despotism' in a global world. Scharpf (2001, p. 20) agreed suggesting that the complexity of policy-making institutional configuration that is ignored by maritime policy-making at present can be accommodated by using a 'plurality of simpler concepts representing different modes of multi-level interaction'. A pluralist approach to governance might overcome the inadequacies of the state-centric hierarchy to provide a Postmodern interpretation of policy-making.

Krasner (1984, pp. 226–230) discussed extensively the issue of pluralism placing it within a political context. Pluralism he saw emphasizing issues of allocation and responsibility rather than 'rule and control'. He quoted Nordlinger (1981, p. 151):

> … (as) portrayed by pluralism, civil society is made up of a plethora of diverse, fluctuating, competing groups of individuals with shared interests. Many effective political resources are available to them.

As such they need to be accommodated in the governance and policy-making framework. They are self-interested, at times autonomous but kept under control by 'cross-cutting cleavages and broad consensus on the rules of the game'.

However, pluralism is not without its problems. Atkinson and Coleman (1992, pp. 157, 162, 167) viewed the complexity that the pluralist approach to policy-making implies as unhelpful with different policy requirements for every discipline (transport, agriculture, monetary policy etc.), actor (trade associations, corporations, universities, private industry, state enterprises etc.), and generating transfers of power from the state to sub-sectors and sub-systems. As such, 'progress toward a multi-level model of the policy process, in which networks and communities play a critical role, will be hampered by a slavish devotion to pluralist images of the state'.

Rhodes (1990, p. 301) quoted Hanf (1978, pp. 1–2) who argued that the characteristic problem of many countries is that 'the problem-solving capacity of governments is disaggregated into a collection of sub-systems with limited tasks, competences and resources'. King and Kendall (2004) implied that governance pluralism has been doomed from birth citing the work of Michels (1911) who had serious reservations about participatory democracy in societies that were increasingly bureaucratized and complex. Inevitably an elite has to emerge (thus reducing the validity of claims of pluralism) because of the need for technical expertise and the result is decision-making and the exercise of power is bound to be restricted to key policy-makers alone.

Pluralism has a particular resonance within the network form of governance which is considered in more detail in a later section of this chapter. Rhodes and Marsh (1992, pp. 200–201) reflected on the neo-pluralist view that some interest groups retain privileged positions and that networks of individuals are a fundamental part of this. This rejection of naïve pluralism is simply a process of being more realistic and in Richardson and Jordan (1979, p. 74) terms, entirely consistent with a pluralist stance. Those in privileged positions do enjoy advantages but despite this network governance allows a wider range of interests to be involved in the policy-making process. As Rhodes and Marsh suggested:

> disaggregation and plurality are seen as synonyms in this species of corporate pluralism. There are many divisions within government, civil society is highly fragmented… and policy-making takes place within a variety of policy networks characterized by close relations between different interests and different sections of government.

Blom-Hansen (1997, p. 670) saw policy networks growing out of earlier traditions of interest group politics and therefore from the literature on pluralism, sub-governments and corporatism (Jordan 1990; Rhodes and Marsh 1992; Smith 1993; Rhodes 1997, p. 30), something supported by Marsh (1995). Atkinson and Coleman (1992, p. 163) discussed the relationship between the emergence of network governance and pluralist theories of the state where authority is seriously fragmented, competition characterizes agencies, interest groups proliferate and 'disjointed incrementalism is the dominant policy style'. This is highly reminiscent of the globalized maritime sector—or at least the pressures to which the globalized maritime sector should be reacting. Research from a wide range of disciplines has reinforced the validity of this view—for example Heinz et al. (1990, pp. 390–391) found little evidence of a policy core in Washington lobbyists where network contacts reflected the pluralist nature of political life, something Rhodes and Marsh (1992) called 'elite interest group pluralism'.

Governance presents difficulties in definition which changes according to context, time, space and a multitude of other factors. This complexity and variability paints the background for the discussion on external cluster governance that follows. However, before we can go on to look at the ways in which governance in the maritime sector is configured and to examine its relationship to policy failure we must first take some time to understand the different models of governance that exist. This in turn will provide us with an insight into the alternative ways that governance frameworks can be applied to the maritime sector.

Governance is a complex and ever changing concept that both reflects and is reflected by societal change—and the failure of policy mechanisms to recognize the need for a sensitive and flexible governance framework can lead to policy inadequacy and failure. Maritime policy problems may well be a reflection of such inadequacies. To quote Eberlein and Kerwer (2004, p. 136):

> Just as there is no one single 'old governance', we should not expect to find a one single 'new governance'.

2.1 Governance Models

While Amin and Hausner (1997) introduced a wide discussion on the issues of governance models and Davies (2005) provided a brief introduction, a more useful and succinct assessment is to be found in Borzel et al. (2005, p. 3) who suggested that they are needed as a way of simplifying the:

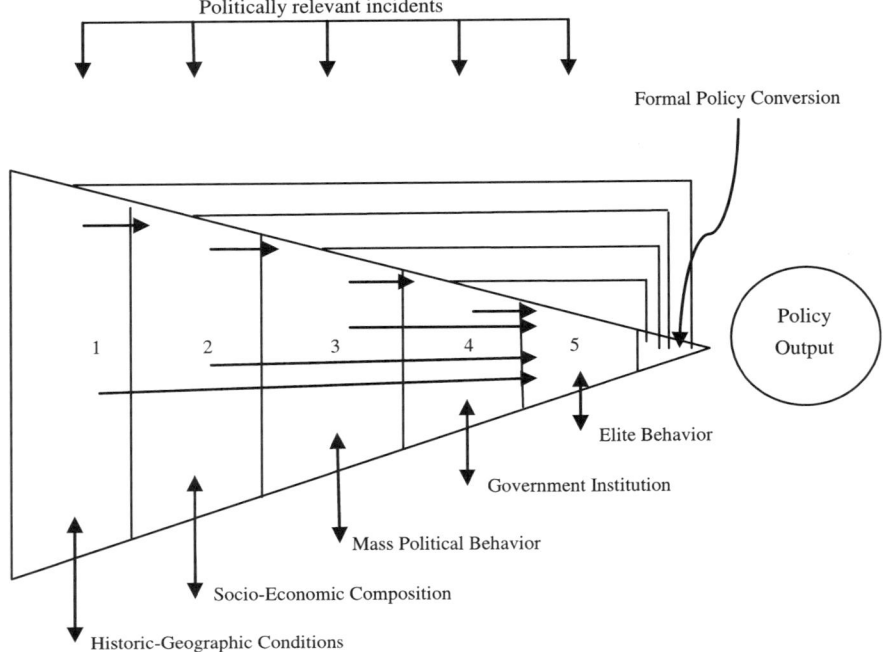

Fig. 2.1 Model of comparative study of policy formation. *Source* Hofferbert 1974

Babylonian variety of definitions and understandings of governance.

Much earlier, Hofferbert (1974) had synthesized a wide range of policy formulation models into a single cohesive model that although not termed as such, clearly has a close relationship to models of governance (Fig. 2.1).

Hofferbert's model of policy formation, although a useful and early contribution to the debate and also giving some insights to the design of good governance, was highly rigid and formulation based in that it relied upon specific and detailed socio-economic and attitudinal measures which may not always be appropriate nor even possible to calculate. The model did however, encourage the inclusion of qualitative measures of government institutional impact, the effect of elite behaviour and the role of citizen participation. Mazmanian and Sabatier (1980, p. 465) suggested that the model was very useful in its consideration of a broad conceptual overview of policy formation that can help to guide governance design. As such it is a useful starting point in reflecting upon the wide range of issues that needs to be considered.

Meanwhile Borzel (1998) claimed that governance draws heavily as a concept on transaction economics, referring to the structures of social order which are conventionally divided into three main forms (Eccles 1981; Mariotti and Cainarca 1986; Jarillo 1988). The earliest identified are markets and hierarchies (Lindblom

2.1 Governance Models

1977; Williamson 1979) which Davies (2005) saw as forming the foundations of the nation-state as a complex compound with economic processes involving 'a mix of market and hierarchy'. Meanwhile, these two have been joined in recent years by networks (alternatively known as communities, clans and associations). The extensive literature on policy networks includes Stoker (1998), Ouchi (1980), Streeck and Schmitter (1985), Atkinson and Coleman (1989, 1992), Jordan (1990), Powell (1990), Schneider (1992), Blom-Hansen (1997), Borzel (1998, 2007), Jessop (1998), Carlsson (2000), Rosenau (2000), and Mikkelsen (2006) amongst very many others. In the words of Borzel et al. (2005, p. 4) hierarchies attempt to 'coordinate social action by using command and control mechanisms'. Meanwhile, markets are 'spontaneous orders that emerge from the self-coordination of autonomous actors' and networks function by 'non-hierarchical coordination based in the exchange of resources and/or trust'. Table 2.3 provides a comparison of the three approaches and indicates how they differ in their generic approach to governance. Kjaer (2004) saw hierarchies as using power as the medium of exchange relying on chains of command and control. They are viewed as inefficient in the allocation of values and generate market failures in the provision of public goods. Markets operate through competition using prices as the medium of exchange. Government is minimized. Individualism is preferred as this maximizes utility, while public bureaucracies are seen as inefficient 'rent-seekers'. Networks meanwhile might be considered as replacing both hierarchies and markets and governments would have to accept the loss of their ability to steer economies and societies and have to learn how to manage networks. We shall look at each of these types of governance model in the following sections, but in particular will focus on hierarchies as the existing governance framework for the maritime sector shows considerable reliance on this type of structure to form the basis for current policy-making.

Nielsen (2001), along with Somerville (2004), Powell (1990) and Rhodes (1997) who applied the concept to the activities of the UK government, also identified three broad modes of governance simplified to hierarchies, characterized by top down control and state regulation; market-based forms of resource allocation; and networks involving a variety of public and private cooperation (Bell and Park 2006). Interestingly he noted an increasing reliance upon network governance models responding to increased complexity in society which, following Jessop (2003b) 'cannot be managed appropriately through more traditional forms of governance such as hierarchy and market'. Hirst (1997a) also suggested that three models of governance existed which he termed Imperative, Exchange and Negotiated Controls. These are Hierarchies, Markets and Networks under a different name.

The network theme is one that has appeared under a number of other guises one of the more emotive of which is the 'iron triangle', emerging from the USA in the 1950s (Maass 1950; 1951), describing how decision-making is commonly segmented between interest groups, the relevant administrative agency and the relevant government committee (Jordan 1981, p. 96; Peterson 1993). Peters (1986, p. 24) quoted in Jordan (1990, p. 324) suggested that:

Table 2.3 Comparing market, hierarchies and network models of governance

	Markets	Hierarchies	Networks
Basis of relationships	Contract and property rights	Employment relationship	Resources exchange
Degree of dependence	Independent	Dependent	Interdependent
Medium of exchange	Prices	Authority	Trust
Means of conflict resolution and coordination	Haggling and the courts	Rules and commands	Diplomacy
Culture	Competition	Subordination	Reciprocity

Source Kjaer (2004, p. 42), taken from Rhodes (1999) in Stoker (ed.) (1999)

> Each actor in the iron triangle needs the other two to succeed, and the style that develops is symbiotic. The pressure group needs the agency to deliver services to its members and to provide a friendly point of access to government, while the agency needs the pressure group to mobilize political support for its programs among the affected clientele... All those involved in the triangle have similar interests. In many ways they all represent the same individuals, variously playing roles of voter, client and organization member.

Iron triangles were actually a response to the popularists who had viewed policy-making as ideally taking place as a response to the interests of the many (Jordan and Schubert 1992, p. 21). In practice 'the flaw in the pluralist heaven was that the heavenly chorus sings with a strong upper class accent' (Schattschneider 1960, p. 35). Yishai (1992, p. 92) cited Wolfe (1977) and Lowi (1979) who stressed that the participants in the iron triangle network became largely autonomous with mutual interests. The result was an insulated policy-making process where the 'irony of democracy, was revealed, with policy decisions restricted to the elite and yet all the formal rules of the democratic game adhered to' (Dye and Ziegler 1970; Rhodes and Marsh 1992, pp. 198–199). Iron triangle policy-making is thus characterized by predictable and closed private relationships between a limited number of interested groups. Further definitions can be found in Adams (1982), Gais et al. (1984), Van Waarden (1992, p. 45), Marsh (1995) and Rhodes (1997, p. 34) and including that of Cerny (2001, p. 400) who felt that the traditional iron triangle had now mutated into 'issue networks', and 'policy networks' less well defined by Euclidean geometry (King 1978). Meanwhile Dowding (1995, pp. 136–142) and Jordan and Schubert (1992, pp. 12–14) along with Van Waarden (1992, pp. 29–31), citing much of the work of Heclo (1972, 1978) indicated how iron triangles are just part of a range of policy models including policy community, policy network, issue network, sub-governmentalism, corporatism, cabinet government and the like. Jordan (1981, p. 98) provided a direct comparison in Table 2.4.

Iron triangles have not been without their critics. Sabatier (1988, p. 131) suggested that there is a need to broaden traditional notions of iron triangle policy-making with their limited actors, and to increase stakeholder involvement to journalists, researchers, policy analysts and any others who play significant parts in the 'generation, dissemination and evaluation of policy ideas (Heclo 1978; Dunleavy 1981; Milward and Wamsley 1984; Sharpe 1984).

2.1 Governance Models

Table 2.4 Characteristics of images of the policy process

	Iron triangles	Issue network	Cabinet government	Corporatism
Political alignments	Stable	Not stable	Stable	Stable
Decision-making arenas	Segmented	Fragmented	Segmented (under cabinet)	Segmented
Number of participants	Limited	Unlimited	Limited	Limited
Central authority power	None disaggregated	None. Very disaggregated	Present. aggregated by political channels	Present
Final point of decision	Yes (in each sector)	No	Cabinet	
Groups	Voluntary	Voluntary	Voluntary (unimportant). closed	Compulsory
Access to decision process	Closed	Open	Closed	Closed
Issue resolution	Issues resolved	Issues often not resolved	Issues resolved	Issues resolved

Source Jordan (1981, p. 98)

Gais et al. (1984, p. 162) were particularly critical questioning the utility of such a policy model as a guide to the process of policy-making. Referring once again to Heclo (1978, pp. 94–105), and echoing Sabatier, they identified the presence of fluid issue networks which go far beyond the traditional three of the iron triangle and include technical specialists, administrators and political entrepreneurs. King (1978, p. 391) viewed these new and looser networks as 'like trying to build coalitions out of sand' while Heclo saw these as seedbeds for coalitions advocating new legislation:

> Based largely on early studies of agriculture, water and public works policies, the iron triangle concept is not so much wrong as it is disastrously incomplete... the conventional view is especially inappropriate for understanding changes in politics and administration during recent years... Looking for closed triangles of control, we tend to miss the fairly open networks of people that increasingly impinge upon governments. (Heclo 1978, p. 88)

Gais et al. (1984, p. 163) concluded that many iron triangles were still identifiable in 1984 but their influence was 'less pervasive than in the 1940s and 1950s'.

Rhodes (1990, p. 297) was equally as dismissive, and although recognizing the significance of sub-governmental governance (of which iron triangles are one example) he felt that this image of the 'policy-making process is altogether too rigid and stereotypes the literature on sub-government (Freeman and Stevens 1987, pp. 12–13)'. This view was confirmed by Scholzman and Tierney 1986 cited in Jordan (1990, p. 330) who favoured the idea of 'a less clearly focused issue network rather than ossified sub-government'. Jordan went on to describe the iron triangle as 'a fixed and forceful metaphor of rigidity'.

Finally Cerny (2001, p. 402) pointed the way forward, suggesting that iron triangles, once valuable as a policy model, were now superseded by 'golden pentangles'. Iron triangles were characteristic of giant industries, with national markets and 'close corporatist relationships with the state'. Meanwhile the new pentangles reflect global financial markets, cross-border activity, international capital flows and an increased market orientation (Cerny 1994). Three of the pentangle sides are the same as the iron triangle—politicians, bureaucrats and interest groups. The other two are international/transnational actors (typified by the IMO, OECD, WTO etc.) and cross-cutting structural factors composed of the 'interlocking webs of governance'—mixed private/public quasi-institutions and interest groups.

Zhang (2005) provided a rather different view on governance models but still one closely related to the tripartite framework noted above. He saw governance transformed by globalization into two dimensions—relating to the market economy and the civil society—but then expanded to include both the private and public sector in each dimension. Effective governance thus requires the reconciliation of market, state and civil dimensions. While retaining their own features these three characteristics reflect the market, hierarchy and network governance models noted already.

Ronit and Schneider (1999, p. 243) provided a more extensive discussion of the significance of markets, hierarchies and networks to governance. They suggested that recent emphasis on the role of markets above all other governance mechanisms is misplaced as:

> Not all of the resources necessary for the production and maintenance of social systems… can be supplied by the interactions of individuals though the marketplace. Some goods and services can only be provided through such non-market devices such as hierarchies and networks.

The significance of public goods and non-market externalities in the maritime sector is worth noting here and the failure of any governance framework to recognize this would in turn make it ineffective. Ronit and Schneider went on to discuss this in relationship to governance models generally and the implications of mixed private and public organizations that exist.

Podolny and Page (1998, p. 59) also recognized the hierarchy, market and network configuration of governance but suggested that too much emphasis had been placed upon the former two derived from transaction cost economics and principal-agent theory which envisaged a dichotomous view of sociological and economic organization. Network governance emerged as an alternative to this view, not as a hybrid of hierarchical or market governance but as a framework in its own right (Jones et al. 1997; Powell 1990; Dicken et al. 2001). Podolny and Page also went on to stress that from a structural viewpoint, all governance models are networks and that the hierarchical and market forms are 'simply two manifestations of the broader type'. What differentiates the three is that network governance assumes 'enduring exchange relations'; in hierarchies:

relations endure for long episodes and there is a clearly recognized, legitimate authority to resolve disputes.

… while in markets:

relations are not enduring, but episodic, formed only for the purpose of a well-specified transfer of goods and resources and ending after their transfer.

They continued by emphasizing that the moves away from nation-state dominated governance will have a significant effect upon the importance of networks, moving governance away from other models.

Stewart (1995, p. 36) was emphatic that linear governance (epitomized by the hierarchy) was anachronistic, stressing Rhodes' (1997, p. 78) work illustrating the importance of networks. The move from hierarchies to an appreciation of the value of market and network based governance models was outlined by Kjaer (2004) who suggested that the original Weberian principles that generated the hierarchical approach no longer held a monopoly. The appropriateness of what Kjaer termed the 'hands-on control' has been replaced by a move towards 'steering through policy networks'. The significance of these comments was not so much that a variety of governance models existed, but that there were circumstances where hierarchical models were still applied despite their origins in social and economic history under circumstances which may have well changed significantly. This of course hints at where we are going in this discussion in looking at the dominant governance model at work in maritime policy-making and to see how appropriate it now is. Perhaps the noted policy failures of recent years have some connection with the approach to the governance of the maritime sector that has been taken.

Blatter (2003) has also suggested that there has been a transformation of 'political order' from hierarchies and what he termed market anarchy towards networks. Society is no longer controlled by the state central unit but the controlling devices that do exist are further dispersed and 'material resources and information are shared by a multiplicity of divergent actors'. He stressed that this trend away from hierarchical and market governance towards network governance was reflected in the moves towards federalism, changes in local politics, in international trade and relations and much more.

2.2 Networks

Today, the role of the government in the process of governance is much more contingent. Local, regional and national political elites alike seek to forge coalitions with private businesses, voluntary associations and other societal actors to mobilize resources across the public–private border in order to enhance their chances of guiding society towards politically defined goals. (Pierre and Stoker 2002, p. 29, quoted in Davies 2005, p. 313)

What we have, in reality, is a variety of developmental trajectories and a spectrum of different forms of governance, that are best captured by the notion of *networked interrelationships structured by different degrees and forms of power and influence.* (Dicken 1994, p. 105) (emphasis original)

Cooke and Morgan (1993, p. 543) noted the emergence of networks in organizations across a range of activity including 'organizational theory, business administration, economic theory, sociology, political science and regional studies to name but a few (Antonelli 1988; Freeman 1990; Imai 1989; Imai and Baba 1989; Powell 1990; Sabel 1989). Miles and Snow (1992, p. 33) had earlier commented on the 'organizational revolution' from centrally controlled hierarchies towards flexible networks which were an ocean away from the traditional pyramid.

Stalder was wholly convinced:

> in all sectors of society we are witnessing a transformation in how their constitutive processes are organized, a shift from hierarchies to networks... Through deep flows of information and people along networks that span the globe, innovation (or tradition) travels from its place of origin to where it appeals to people and their agendas. In the process it is transformed, adapted, and becomes an essential part in the constitution of the very networks along which it flows. This applies as much to production methods as to social movements, to efforts to save the planet as to attempts to destroy it. (Stalder 2006, pp. 1, 2)

While Rhodes (1990, 1997, p. 11) suggested that policy networks are 'meso-level' concepts which describe interest group and government intermediation (Schmitter and Lehmbruch 1979; Marsh 1983; Benson 1982), Jordan (1990) along with Marsh and Smith (2000), Carlsson (1996) and Heinz et al. (1990) outlined the development and application of policy networks. However, it is Dowding (1995, pp. 138–139) who provided more detail of the history and development of network models of governance noting the earliest examples from Griffiths (1939) (whirlpool); Cater (1964), Freeman (1965), Truman (1971) (sub-governments, triangles); and Jones (1979) (sloppy hexagons). Heclo (1978) suggested that 'issue network' was a more appropriate term than 'policy network' building on earlier work defining 'community networks' (Heclo and Wildavsky 1974). However, all commentators were agreed on the issues of policy-making in principle; the distinction between private and public was flexible, patterns of linkages affected outcomes, and that underlying sub-linkages were fundamental to policy success and outcomes.

Deleuze and Guattari (1988, p. 17) outlined the principles of the network as a governance framework with:

> all individuals interchangeable, defined only by their state at a given moment – such that the local operations are coordinated and the final, global result synchronized without a central agency. Transduction of intensive states replaces topology and the 'graph regulating the circulation of information is in a way the opposite of the hierarchical graph'.

Kjaer (2004, p. 34) outlined the importance of policy networks in the context of the activities of the UK government suggesting that this involved the greater use of non-government actors and agencies to undertake service delivery. Rhodes and Marsh (1992, pp. 182–183) discussed the types of policy network model that existed and developed a new approach to understanding how they worked. Alba (1982), Thorelli (1986), Schneider (1992), Van Waarden (1992), Dowding (1995, pp. 150–158), Marsh (1995), Rhodes (1997, pp. 4, 9–13) and Hirst (1997a, pp. 7–15) provided general and wide-ranging discussion about the role of networks in governance of all types and contexts while de Senarclens (1998, p. 97) was more forthright:

> The political system is no longer structured by government decisions alone but by a dense and complex system of networks and systems of functional cooperation.

Altman and Petkus (1994, p. 39) took a stakeholder approach to policy-making that is in effect network structured while Cox (1998, pp. 2–3) encouraged greater attention to be placed on networks if the 'politics of space' are to be understood better and emphasized their unevenness and porous boundaries. Picciotto (1997, p. 1036) even considered that the modern international state system comprised networks of state activity rather than hierarchical arrangements of which the state was a part. Borzel (2010, p. 192) saw network governance as where 'the authoritative allocation of values is negotiated between state and societal actors' (Kohler-Koch and Eising 1999; Ansell 2000; Schout and Jordan 2005). Skelcher et al. (2004, p. 3) described how network forms of governance have been encouraged by the 'fragmentation of organizational structure, political control and accountability that resulted from the hollowing out of the state' which in turn has occurred as a result of globalization and the changing fortunes of alternative jurisdictions.

MacLeod (1997, p. 300) was confident that the growth of socio-economic networks and territorial embedding which he grouped together under the term 'institutional thickness' were fundamental to creation of institutional and socio-cultural factors that underpin sustained economic success (Amin and Thrift 1994a, b; Granovetter and Swedberg 1992; Cooke and Morgan 1993; Putnam 1988, 1993). This institutional thickness:

> establishes legitimacy and nourishes relations of trust… (and)… continues to stimulate entrepreneurship and consolidate the local embeddedness of industry… what is of most significance here is not the presence of a network of institutions *per se*, but rather the process of institutionalization; that is the institutionalizing processes that both underpin and stimulate a diffused entrepreneurship—a recognized set of codes of conduct, supports and practices which certain individuals can dip into with relative ease. (Amin and Thrift 1995)

The network approach to governance has not been without its critics (see for example Dowding 1995; MacLeod and Goodwin 1999, p. 512; Gore and Wells 2009, p. 161) where the main disappointment lies in the failure to provide a means of understanding how networks work and how bargaining within them is structured. However, support for the analysis of governance using a network approach has grown to the point where it is now widely recognized to have considerable validity over and above that of the traditional hierarchy. Obando-Rojas et al. (2004) for example, use the maritime sector to emphasize the long-standing globalized characteristics of the sector and the inadequacies of a state-centric hierarchical approach to understanding the way it works.

Table 2.5 Rhodes' model of policy networks

Network type	Network characteristics
Policy community/territorial community	Stability, highly restricted membership, vertical independence, limited horizontal articulation
Professional network	Stability, highly restricted membership, vertical independence, limited horizontal articulation, serves interests of the profession
Intergovernmental network	Limited membership, vertical independence, extensive horizontal articulation
Producer network	Fluctuating membership, limited vertical interdependence, serves interests of producer
Issue network	Unstable, large number of members, limited vertical interdependence

Source Rhodes and Marsh (1992, p. 183)

Rhodes (1986) distinguished five types of policy network lying on a continuum from integrated policy communities, through professional networks, intergovernmental networks, and producer networks to loose issue networks (Table 2.5).

Etzioni and Lawrence (1991), Granovetter and Swedberg (1992), Perrow 1986 and Zukin and Di Maggio (1990) considered their role in socially embedding economic decision-making, and Hess (2004) looked at the relationship between spatial activity, embeddedness and networks. Meanwhile Forsgren and Johanson (1992) Johanson and Mattson (1987), Hakansson (1987) and Hagg and Johanson 1992 have argued that production systems as a whole as well as the activities of all actors within them, should be viewed from a network perspective. Brandes et al. (1999, pp. 83–84) argued that political networks could be interpreted from a visualization perspective concentrating on the nodes (organizations or individuals) and links (communications, participation, resource exchange etc.) and that it was possible to understand the policy-making process by analyzing these structures and relationships (Knoke 1990). Bair (2008, p. 340) cited Castells (1996) in considering the social connectivity issues of networks, Boltanski and Chiapello (2007) and the relationship between new capitalism and networks, and Harvey (1990) in terms of networks, space-time compression and flexible accumulation.

Peters (1998) provided a comprehensive introduction to networks as models for governance, Peterson (1994, p. 156) emphasized their importance within EU governance activity; Zurn (2003, p. 354) suggested that any transnational activity requires a network governance approach; while Grabher and Stark (1997, pp. 534, 537) were unashamed in their belief in networks as all-embracing. Taking East Europe as their example they stressed that the only way of understanding economic activity was to view it as a network linking firms and connecting individuals. Governance of this activity has to be similarly structured (Powell 1990; Grabher 1993; Stark and Bruszt 1995). Meanwhile Scharpf (2001, pp. 3–4) stressed the suitability of a network governance model for the EU, describing the 'European polity as a condominio, consortio or a fusion of governing functions' (Marks et al. 1996; Schmitter 1996; Wessels 1997; Kohler-Koch and Eising 1999).

2.2 Networks

Skelcher et al. (2004, p. 3) felt that globalization's 'hollowing out of the state' had generated more 'collaborative arrangements' to cope with the fragmentation of organizational structure, political control and accountability (Agranoff and McGuire 2003; Milward and Provan 2000; Lowndes and Skelcher 1998). King and Kendall (2004, pp. 129–130) described 'epistemic communities' consisting of all manner of governance stakeholders that form a 'maze of public and private interconnections'. These networks form the key to effective governance rather than the 'power of legal violence or surveillance' (Giddens 1985). Altman and Petkus 1994 discussed the relationship between stakeholders, networks, governance and policy-making. Kenis and Schneider (1989, pp. 6–9), cited in Jordan and Schubert 1992 noted networks had proliferated because of a number of factors:

- The emergence of an organized society.
- Sectorization in policy-making, generating cross-jurisdictional and contextual interaction.
- Overcrowding in the policy-making arena.
- Increased scope of state policy-making.
- Decentralization and fragmentation of the state.

The implication is that hierarchies do not have much of a role to play. Network relationships are temporary, sporadic and all-embracing but nevertheless highly significant especially as Klijn and Koppenjan (2000, p. 136) suggested 'government is not actually the cockpit from which society is governed and that policy-making processes rather are generally an interplay among various actors'. Klijn and Koppenjan (2000, p. 142) also went on to provide a description of the main characteristics of policy networks (Table 2.6) which provided a good indication of how networks differed as an approach to governance from hierarchies and markets. Thompson (2005, p. 6) commented on the four types of networks that he identified particularly emphasizing their socio-economic context and policy-making implications (Table 2.7). Significance was emphasized by the number of publications that emerged on this topic so quickly between 1985 and 1998 including, Milward and Wamsley (1985) Wilks and Wright (1987), Rhodes (1988), Marin and Mayentz (1991), Marsh and Rhodes (1992), Glasbergen (1995), Provan and Milward (1995), O'Toole (1997), Kickert et al. (1997) and Yeung (1998).

Definitions of networks and their relationship to governance abound. Evans (2001, p. 542) described a policy network as 'a metaphorical term characterizing group-government relations, while Dowding (1995) suggested they are common within the policy sciences. O'Toole (1997, p. 45) saw them as 'structures of interdependence involving multiple organizations or parts thereof, where one unit is not merely the formal subordinate of the others in some larger hierarchical arrangement'. Kassim (1994, p. 19) felt that the network governance model grew out of international relations and was characterized by a 'multiplicity of linkages and interactions connecting a large number and a wide variety of actors from all levels of government and society'. Power is believed to be 'widely dispersed between the wide number of actors that influence the determination of policy'.

Table 2.6 Theoretical assumptions of the policy network approach

	Theoretical assumptions
Networks	Actors are mutually dependent for reaching objectives
	Dependencies create sustainable relations between actors
	Dependencies create some veto power for various actors. The sustainability of interactions creates and solidifies a distribution of resources between actors
	In the course of interactions, rules are formed and solidified which regulate actor behavior
	Resource distribution and rule formulation lead to a certain closeness of networks for outside actors
Policy processes	Within networks, interactions between actors over policy and issues take place focused on solving the tension between dependencies on the one hand and diverging and conflicting interests on the other
	In doing so actors depart from perceptions they hold about the policy area, the actors and the decisions at stake
	Actors select specific strategies on the basis of perceptions
	Policy processes are complex and not entirely predictable because of the variety of actors, perceptions and strategies
Outcomes	Policy is the result of complex interactions between actors who participate in concrete games in a network
Network management	Given the variety of goals and interests and—as a result—the actual and potential conflict over the distribution of costs and benefits, co-operation is not automatic and does not develop without problems
	Concerted action can be improved through incentives for co-operation through process and conflict management, and through the reduction of risks linked to co-operation

Source Klijn and Koppenjan (2000, p. 142)

Meanwhile Podolny and Page (1998, p. 59) in their extensive review of network governance in organizations defined a network as a:

> form of organization as any collection of actors (n2) that pursue repeated, enduring exchange relations with one another and, at the same time, lack a legitimate organizational authority to arbitrate and resolve disputes that may arise during the exchange.

Keating and Hooghe (1996, p. 218) quoted in King and Kendall (2004, p. 176) noted the growth in significance of networks with globalization and the decline of the nation-state:

> Policy-making retreats into complex networks that do not correspond to formal institutions; and new and rediscovered forms of identity emerge at the sub-national level and even the supranational.

Dicken et al. (2001, p. 97) provided a clear interpretation of what is needed if governance in an age of globalization is to be relevant. They began by suggesting that 'social actors' and 'business networks' have to be identified before a workable governance framework can be created. The global economy is made up of 'spaces of network relations'. The social actors are represented by a wide variety of stakeholders—for example individuals, households, firms, industrial sectors, trades

Table 2.7 Macro political structures/models involving network formations of various kinds that affect policy-making

Policy networks	Policy community
	Issue networks
	Fragmentation of the policy-making framework
	Sectoralization of policy-making
	Organizational explosion
	Increased scope for policy-making
	Democratization/fragmentation of the state
	Blurring of boundaries between public and private
	Over-crowded policy-making
	Increase in policy turnover rate
Corporatism	Private interest governance
	Large-scale interest groups/social partners
	Macro-level network of policy governance operating in parallel to state apparatus
	Return of social compacts
Associationalism	Pluralistic version of corporatism
	Political associations with dispersed sovereignties (commissions, churches, trusts, trade unions, friendly societies, professional societies, places of work, QUANGOs)
	Negotiated state
	Normative mutualism
NGOs of dissent	Similar to associationalism/private interest governance
	Campaigning and propaganda for a particular purpose
	Persuasion, mobilization, pressure politics
	Bottom-up rather than top-down networks
Elites	Small groups with a particular access to power
	Based on common economic, political and cultural outlook/formation
	Exclusive top down network
	Undemocratic and non-egalitarian
Multi-level governance and commitology	Vertical networks
	OMC
	Distribution of jurisdictions
	Experts

Source Thompson (2005, p. 19)

unions, interest groups etc.—and to generate meaningful governance structures then the 'intentions and motives' and the consequent 'emergent power' of these groups needs to be considered. These relationships are to be found in specific, varying and overlapping spaces. The spaces vary through time, dependent on jurisdiction and context and also the actors who are taking part. There is no one single representative framework. Governance can then be designed, described, analyzed and understood in many different ways—for example spatially, sectorally, organizationally and so on. The key is to:

> recognize the fundamental interrelatedness of all of these phenomena, not in some abstract sense but in seriously grounded form.

Dicken et al. went on (2001, p. 106) suggesting that this network approach will undermine 'the image of the faceless juggernaut of globalization' and instead will ground governance in a web of social, political and economic relationships. Unfortunately maritime governance is currently far removed from such a conceptual model, failing to reflect either the participants or their relationships in how it is designed or operated.

Sabatier (1991, p. 148) was insistent that the traditional focus of decision-makers on single institutions and a single level of governance may be helpful in understanding decisions made but are 'inadequate for understanding the policy process over any length of time' (Jones 1975; Heclo 1978; Kingdon 1984; Sabatier 1988). Kassim (1994, p. 16) reiterated the value of networks particularly in understanding policy-making in the EU and the 'important role played by interest groups'. Merrien (1998, p. 58) stressed how good governance can only occur when the state operates in a network configuration utilizing the private sector interests and groups as partners. Meyer et al. (1997, p. 148) suggested that the governance situation is even more important to get right:

> For realist perspectives the world is either anarchic …or networked.

Jones et al. (1997, p. 911) provided a variety of definitions of network governance in a wide-ranging discussion of the issue summarized by: 'coordination characterized by informal social systems rather than by bureaucratic structures within… formal contractual relationships between them—to coordinate complex products or services in uncertain and competitive environments' (Piore and Sabel 1984; Powell 1990; Ring and Van de Ven 1992; Snow et al. 1992). Jones et al. (1997, p. 915) continued by providing an extensive categorization of network governance definitions which clarifies the boundaries and helps to be more specific in dealing with this area (Table 2.8).

Carlsson (2000, p. 502) suggested that although network governance remains lacking in theoretical construct it does provide what Kenis and Schneider (1991) saw as a viable alternative to the 'textbook' version of policymaking (Nakamura 1987). He went on to quote, Benson (1977) in Rhodes (1990, p. 304) in that the term policy network could be understood as a broad generic categorization of a large number of subcategories. Policy networks are seen as 'cluster(s) or complexes of organizations connected to each other by resource dependencies and distinguished from other clusters or complexes by breaks in the structure of resource dependencies'. Kenis and Schneider (1991, pp. 41–42) were more explicit:

> A policy network is described by its actors, their linkages and its boundary. It includes a relatively stable set of mainly public and private corporate actors. The linkages between the actors serve as channels for communication and for the exchange of information, expertise, trust and other policy resources. The boundary of a given policy network is not in the first place determined by formal institutions but results from a process of mutual recognition dependent on functional relevance and structural embeddedness. Policy networks should be seen as integrated hybrid structures of political governance.

Table 2.8 Differing terms and definitions for network governance

References	Term	Definition of network governance
Alter and Hage (1993)	Inter-organizational networks	Unbounded or bounded clusters of organizations that, by definition, are non-hierarchical collectives of legally separate units
Dubini and Aldrich (1991)	Networks	Patterned relationships among individuals, groups and organizations
Gerlach and Lincoln (1992)	Alliance capitalism	Strategic, long-term relationships across a broad spectrum of markets
Granovetter (1994, 1995)	Business groups (some)	Collections of firms bound together in some formal and/or informal ways by an intermediate level of binding
Kreiner and Schultz (1993)	Networks	Informal inter-organizational collaborations
Larson (1992)	Network organizational forms	Long-term recurrent exchanges that create interdependencies resting on the entangling of obligations, expectations, reputations and mutual interests
Liebeskind et al. (1996)	Social networks	Collectivity of individuals among whom exchanges take place that are supported only by shared norms of trustworthy behavior
Miles and Snow (1986, 1992)	Network organizations	Clusters of firms or specialized units coordinated by market mechanisms
Powell (1990)	Network forms of organization	Lateral or horizontal patterns of exchange, independent flows of resources, reciprocal lines of communication

Source Jones et al. (1997, p. 915)

The significance of network governance over and above that of markets or hierarchies is drawn out by a number of commentators. Grabher and Stark (1997, p. 534) considered that alternatives to the hierarchy and market are needed to understand social organization (Powell 1990). O'Toole (1997, p. 45) viewed networks as independent structures made up of multiple organizations or parts thereof which are not necessarily placed in an hierarchical arrangement (Grabher 1993; Stark and Bruszt 1995). Nielsen (2007, pp. 4–5) suggested that network governance differs substantially from markets and hierarchies. Network governance was characterized by:

- Relationships between actors, pluri-centric and interdependent (Kersbergen and Waarden 2004, p. 148). Hierarchies are mono-centric and based on dependency and subordination. Market governance is multi-centric, consisting of a virtually infinite number of independent actors.
- Decision-making. Networks are characterized by reflexive rationality, continuing negotiation and the pursuit of collective solutions. Hierarchies are centralized and top-down; markets directed by Adam Smith's 'invisible hand'.
- Compliance. Networks rely on trust. Hierarchies on rules and laws; markets on the fear of economic loss.

Jachtenfuchs (2001, p. 254) was equally as supportive of networks in his consideration of governance models calling it a 'fruitful heuristic device for empirical analysis', significant in understanding EU governance in particular despite criticisms of fuzziness (Borzel 1998). The move towards supranationalism that has characterized maritime policy-making since 1945, with its weakness of power centre and fragmented and fluid institutional structure, has changed the context for governance from that state-orientated where an hierarchical approach was more relevant. Networks lie in the governance spectrum between hierarchy and anarchy/markets (Benz 1998, 2000) with a loose coupling of elements contrasting with the rigidity of hierarchies and the complete freedom from coupling of anarchy/markets. Networks have 'flexible mandates' from their 'constituents' which can accommodate the decline in formalism and rigidity that characterizes international and domestic relationships, not least in the maritime sector. 'Hierarchical governance in such a setting is not a very promising endeavour' (Jachtenfuchs 2001, p. 255). Yet maritime governance remains hierarchical.

Hill and Lynn (2004, p. 174) emphasized how the decline in hierarchical domination has affected all jurisdictions quoting Frederickson and Smith (2003, p. 208)—'The administrative state is now less bureaucratic, less hierarchical and less reliant on central authority to mandate action', with the 'emphasis now on horizontal, hybridized and networked aspects of governance' (Hill and Lynn 2004, p. 174; Kettl 2002; Salamon 2002).

Peters (1998, p. 302) was equally as sceptical of hierarchical governance stressing that strong vertical linkages within governance makes horizontal coordination that much more difficult. 'Less pluriform networks are less likely to coordinate effectively'. Castells (2000, p. 12) concurred commenting:

> The work process is interconnected between firms, regions, and countries, in a stepped up spatial division of labour, in which networks of locations are more important than hierarchies of places.

He went on (Castells 2000, p. 19):

> Networks dissolve centres, they disorganize hierarchy, and make materially impossible the exercise of hierarchical power without processing instructions in the network, according to the network's morphological rules.

Castells actually provided a long and robust argument in favour of the network approach considering that new technological developments had retained the advantages of networks of flexibility and adaptability, while overcoming traditional difficulties of co-ordination, management and focusing resources. They are therefore vital to modern decision-making and governance particularly in a globalizing environment where social organization has changed dramatically.

O'Toole (1997, pp. 46–47) suggested that any policies for ambitious or complex issues require networked structures, a trend that was encouraged by the continued growth of limited and liberal government and the recognition of the secondary (even tertiary) impacts of policy initiatives. Brugha and Varvasovszky (2000, p. 240) provided indirect support for the network approach in relating it to

2.2 Networks

the significance of stakeholders and how they form a major part of this form of governance. There is almost universal agreement that the role of stakeholders needs to be raised across all governance, not least maritime and so an approach which overtly recognizes this has to have an advantage. The need to move governance in the maritime sector from elitism (Lasswell 1958; Bacharach and Baratz 1962) where power is constrained to the very few; professionalism, where power rests with professional elites who may use it to further their own interests; and technocracy where decision-making is dominated by scientific rationalism; to pluralism (Lindblom 1959; Dahl and Lindblom 1976) where power is distributed across relevant groups in society, is not only a necessary trend but also one that is helped by adopting a network approach.

Dicken et al. (2001, pp. 91–92) also saw networks as significant. They saw networks as 'relational' and as a mechanism suitable for application to the global economy (Gulati et al. 2000; Hassard et al. 1999; Latour 1993; Law 1994; Sack 1997; Thrift 1996). They represented an organizational form which is far more appropriate for modern governance than the existing structural systems that dominate the maritime sector, described by Dicken et al. as 'atomistic'. The global economy is made up of 'spaces of network relations' and modern governance must reflect the changes that have taken place largely as a consequence of globalization and its core characteristics relating to communication, speed and flexibility. Powell and Smith-Doerr (1994) described them as analytical tools that subscribe to a form of governance, while many other commentators rave about their contribution to governance in the business sector (for example; Doz and Mahel 1998; Dunning 1997; Nohira and Goshal 1997).

However, Nohira (1992, p. 3) was less convinced:

> [a]nyone reading through what purports to be network literature will readily perceive the analogy between it and a terminological jungle in which any newcomer may plant a tree.

So there is no universal support for network governance and it is not without both critics and criticisms. Rhodes and Marsh (1992, p. 183) were critical of the Rhodes model of network governance in particular and its attempt to place networks on a logical continuum of interpretation of policy-making. Kassim (1994, pp. 16, 20–25) was particularly sceptical citing three major drawbacks. Firstly he claimed that policy-making at the supranational level (referring particularly to the EU but applicable to other policy-makers) was characterized by variety at the micro-level with considerable fragmentation taking place so that no one process is appropriate. Network governance is a meso-level approach requiring consistency across this level to be meaningful and this in turn cannot readily accommodate micro-level variation (Marsh and Rhodes 1992). Secondly, the continued significance of institutions in policy-making (including those of the nation-state) contrasted with the network governance approach that aims to move away from this. Finally, network governance demanded the definition of a boundary to the network under consideration—but increasingly this was impossible to achieve meaningfully. Without clear boundaries it becomes very difficult to manage, interpret or manipulate the process of governance.

Stoker (2004, p. 24) quoted in Davies (2005, p. 314) claimed that:

> There is nothing to suggest that networked community governance should be any less susceptible to conflict regarding goal definitions and defining priorities, than the traditional view of planning.

Meanwhile Hess (2008, p. 452) noted that a number of commentators had expressed unease at 'network essentialism' including Thompson (2003) and Sunley (2008). He summarized their concerns that the term network had become a 'ubiquitous metaphor, lacking precision, being applied to everything and therefore becoming void of explanatory power'. Curiously they also contradicted Kassim's claim that networks were too meso-level orientated, suggesting instead that they could deal only with micro-level governance and are inadequate when considering higher level dynamics. Others have supported this criticism including Granovetter (1985), Peck (2005), Yeung (2005), and Sunley (2008).

Podolny and Page (1998, p. 60) suggested that the network approach to governance remained unclear as it was always difficult to map formal organizational arrangements, which remain dominant within the maritime sector, onto a network structure. However the clearest criticisms of the network approach came from Klijn and Koppenjan (2000, p. 137) who identified five main drawbacks that had been commonly cited (Table 2.9). While they argued against many of these criticisms, there remained some notable questions to be answered. Skelcher (2005, p. 90) considered that the polycentric nature of networks made democratic decision-making almost impossible as conventional democratic 'constitutional engineering' required a strict hierarchical jurisdictional arrangement for it to work while networks lack this jurisdictional integrity:

> Authority is diffuse and ill-defined because of the complexity of spatial patterning, functional overlays between jurisdictions, variable density of political spaces, and differential coupling between organizations.

2.3 Other Governance Models

Rosenau (2000, pp. 11–13) provided a detailed analysis of a governance model which brought together many of the issues discussed. Table 2.10 provides a summary of his typology.

Rosenau considered that this model might accommodate all the diversity, horizontality and number of steering mechanisms that characterize the world of globalized governance. The formal, top-down process is typically that of nation-state governments and is characterized by the existing maritime governance process. Bottom-up governance reflects the growth of power in institutions and sectors which are now pressing to be active ingredients in the policy-making process. They are sadly neglected by traditional maritime policy-making and governance frameworks. Market governance attempts to summarize the rising role of

2.3 Other Governance Models

Table 2.9 Criticisms of the network approach

1. Lack of theoretical foundations and clear concepts. The network approach is not based on a solid theoretical body of knowledge and as a result there is no coherent theoretical framework (Borzel 1998).
2. Lack of explanatory power. Networks are primarily 'metaphoric', highly descriptive and provide no explanation of the policy process (Dowding 1995; Salancik 1995; Blom-Hansen 1997; Borzel 1998)
3. Neglect of the role of power. Too much emphasis on the role of co-operation and consensus and ignores conflict, power and power differences (Brans 1997)
4. Lack of clear evaluation criteria. They are vague and lack a substantive norm. The approach insufficiently acknowledges the goals of government (Propper 1996; Brans 1997)
5. Normative objections against networks and the role of public actors within them. Network approaches consider governments as only just another actor and ignore their social role. As a result network approaches to governance seriously jeopardize public sector policy innovation, the pursuit of the common good and the primacy of politics (Ripley and Franklin 1987; Marin and Mayentz 1991; Rhodes 1996; De Bruijn and Ringeling 1997)

Source Derived from Klijn and Kopperjan (2000, p. 137)

Table 2.10 Six Types of Governance

Structure	Process	Process
Formal	Top-down governance (states, TNCs, IGOs)	Network governance (states, business, alliances, IGOs)
Informal	Bottom-up governance (mass public, NGOs, INGOs)	Side-by-side governance (NGO and INGO elites, state officials)
Mixed formal and informal	Market governance (states, IGOs, elites, mass public, TNCs)	Mobius-web governance (states, elites, mass public, TNCs, IGOs, NGOs, INGOs)

Source Rosenau (2000, p. 12–13)
TNC Trans-national corporation, *IGO* International Governmental Organization, *NGO* National and Sub-national non-profit Non-governmental Organization, *INGO* International non-profit Non-governmental Organization

horizontal linkages and the decline of the hierarchy as a formal governance structure. Once more the maritime sector seems sadly lacking.

Rosenau went on to explain that while some model categories are obvious, that of the Mobius web refers to the highly intricate networking that can occur across structures (jurisdictions) and between private and public, interest group and individual, company and elites, mass public and governments and which is increasingly the case. Maritime governance structures show little appreciation of these trends that are 'pervaded by nuance, by interactive and multiple flows of influence' which make them difficult to understand but no less important to accommodate.

Turke (2008, pp. 3–8) identified three models of governance that provide a structure to the discussion of governance problems and policy-making inadequacies that feature so prominently in the maritime sector today:

- The *structure-oriented* view—where the social system is viewed as entirely independent and where it governs itself in a way through 'self-reverential' processes (autopoiesis). Thus they are operationally closed systems, with only the system's internal configuration and skills determining the governance process that exists (Kickert 1993).
- The *actor-oriented* view—whereby social systems are considered 'Gebilde' and the system is viewed as constituting both the communications that take place but also the actors involved. Human actions and actor constellations are the focus (Mayntz 1976, 2004).
- The *synthetic* view—attempts to combine the best of both of these models by striking a middle ground between organization and actor. Kooiman (2003, p. 4) provides a central focus in which governing (and by implication governance) is 'the totality of interactions in which… actors participate'. Turke even attempts to summarize his view on these interactions (Fig. 2.2)—and although this may be only one view, and indeed simplified at that, it remains a pertinent reminder of the process of governance in a globalized world that is replacing the existing traditional, simplistic, hierarchical model.

Gibbon et al. (2008, pp. 317, 319) provided an alternative model placing the discussion on governance in the context of what they called Global Value Chains (GVC) (Rabach and Kim 1994; Bair 2005). They identified two traditional governance models—that of mainstream international political economy and radical political economy. Mainstream international political economy governance looks at how powerful and effective the major international institutions are (for example the IMO, OECD, WTO etc.) in contrast to local, regional and national institutions. A radical political economy interprets governance as a relationship between international finance, the same international institutions, and major private corporations. GVC meanwhile sees the commodity chain as a 'network of labor and production processes whose end result is a finished commodity' (Hopkins and Wallerstein 1994, p. 9). It thus focuses specifically on the firm and its organization of production networks rather than on the capitalist process and the institutions that characterize it, and that underlie it. This organization is not spontaneous, automatic or systematic but emerges from the more significant actor firms that manage the access to markets globally.

Termeer (2008) also suggested a three tier structure for governance which she terms first, second and third generation governance. First generation governance focuses upon the tools and instruments of government which can be used to change people's behavior and therefore resolve societal problems. She identified a number of typologies of instruments typified by 'carrots, sticks and sermons' or 'legislative, communicative and economic'.

She went on to criticize this sort of approach to governance as insensitive to the fact that firms and individuals will avoid disagreeable effects of policies, that substantial information and interference is required of government to achieve impacts and that it is necessary to over-simplify real-life by bureaucrats to have a chance of influencing the market-place.

2.3 Other Governance Models

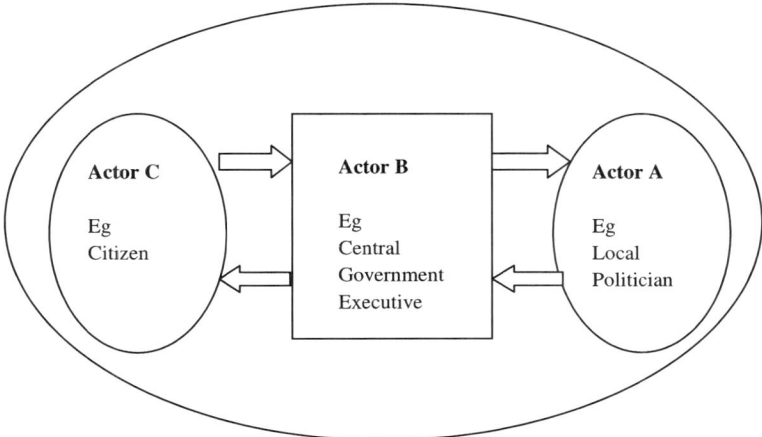

Fig. 2.2 Governance as a process of interaction. *Source* Turke (2008, p. 7)

Second generation governance is a reaction to first generation governance and its lack of information, problematic compliance, self-referentiality and its lack of scalability and reflexivity. Here the focus shifts from single instruments to instrumental mixtures or policy arrangements. More indirect measures are employed and attempts are made to work through networks to help those involved solve problems themselves.

Third generation governance tries to 'leave behind the idea of a government who knows what kind of behaviour is required from people'. It relies more on standing-back, small steps, observing, planning, self-reliance, adaptive management and self-organization and as such is a far cry from the hierarchical, top-down, nation-state approach we have identified in the maritime sector. It is likely to be firmly resisted by policy-makers who are more used to problem identification and solution rather than guidance, stimulating interest, uncovering opportunities and then standing back and waiting to see. Termeer's view was that all three generations have value but that the third is most appropriate where the environment is complex and dynamic and it is these characteristics of the maritime environment that make the discussion most relevant.

Treib et al. (2007, p. 6) suggested an alternative model of governance derived from their work with the EU NEWGOV project (Table 2.11) looking at the model structure as formulated around actors and steering modes. They also provided an alternative viewpoint using policy, polity and politics as central features guiding the approaches available to governance (Table 2.12).

Meanwhile Ramachandran et al. (2009, p. 343) suggested a far simpler model of global governance that may help to clarify the significant issues involved which have been outlined by Jachtenfuchs (1995, p. 124), Rosenau (2000, p. 4) and Martin (2004, p. 147) amongst many others. Good governance consisted of achieving a balance between three components—people, economics and nation-

Table 2.11 NEWGOV modes of governance

		ACTOR 1	ACTOR 2	ACTOR 3
		Public actors only	Public and private actors	Private actors only
STEERING MODES	Hierarchical top-down/ legal sanctions	Traditional nation-state, supranational institutions		
	Non-hierarchical bargaining/positive incentives	Intergovern-mental bargaining	Delegation of public functions to private actors; neo-corporatism	Private interest Government.
	Non-hierarchical, non-manipulative; persuasion; learning and arguing; diffusion	Institutional problem-solving across levels; European agencies	Public–private networks; benchmarking	Private –public partnership (NGOs)

Source Derived from Treib et al. (2007, p. 8)

states (Fig. 2.3) an approach adopted in principle for example, by the UN Security Council (Dervis and Ozer 2005) and which although it skims over essential governance ingredients such as non-state actors, military capacity, interest group representation and more, these are actually subsumable within the three elements in this model.

Finally models of good governance need in some ways to be treated with caution for as Smouts (1998, p. 87) pointed out:

> international regulation exists among a small number of states, with private and elitist companies sharing the same communication code (that of free trade and the Western conception of human rights). It pertains more to what Hirst and Thompson (1999) (after Yarborough and Yarborough 1992) called 'minilateralism' than to global construction.

The concept of working towards good global governance is thus one confined to those with the financial, economic and political resources to even conceive of it; a luxury reserved for the few even though it might impinge on the many.

2.4 Governance Failure

The problems of policy-making and their potential relationship to governance are documented outside the maritime sector, and there has been considerable interest for some time in the more general impact of the relationship between governance and policy effectiveness. Nielsen (2007, p. 2) cited Rosenau (2003, p. 5) suggesting that global governance networks commonly face problems and like all governance approaches, can also be inadequate as they are 'prey to dilemmas,

2.4 Governance Failure

Table 2.12 Existing conceptions of modes of governance

State intervention	Societal autonomy
POLICY	Soft law
Legal bindingness	Flexible approach to implementation
Rigid approach to implementation	Absence of sanctions
Presence of sanctions	Procedural regulation
Material regulation	Malleable norms
Fixed norms	
POLITICS	Only private actors involved
Only public actors involved	
POLITY	
Hierarchy	Market
Central locus of authority	Dispersed loci of authority
Institutionalized interactions	Non-institutionalized interactions

Source Treib et al. (2007, p. 6)

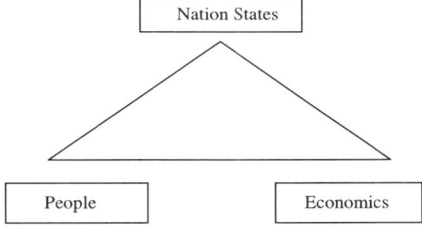

Fig. 2.3 Model of good global governance. *Source* Ramachandran et al. (2009, p. 343)

contradictions, paradoxes and failures'. This indicates the need for coordination and management of governance using systems of metagovernance (Sørensen 2006, p. 100). Rosenau (2000, p. 8) also suggested that governance was a social process that transcends 'state and social boundaries so thoroughly as to necessitate a reinvention of the wheel'. This would require a complete overhaul of 'rule systems and processes' through which authority is exercised, and would have to be effective across national boundaries. Rosenau went even further noting that new 'terminal entities' were needed to act as replacements for national loyalties and provide an affiliation of the new groups emerging under globalization. Rather than identifying and differentiating between these groups in terms of territory, boundary and location, the need was to refocus on processes, flows and the structures that sustain them. Rosenau saw people (and organizations) now 'converging electronically as equals, or at least not as superiors and subordinates'. The diffusion of authority that has occurred makes redundant any thought of hierarchical governance centred on the nation-state. And yet what do we have in the maritime sector?

Ruggie (1993, p. 143) expressed concern at the 'fundamental institutional discontinuity in the system of states' which was causing repeated governance failures in the ecological field while Sutherland and Nichols (2006, pp. 7–8) identified hierarchical governance models, enacted through policies, laws and regulations (Hoogsteden et al. 1999; Paquet 1999; Savoie 1999) which exhibit considerable inadequacies. A hierarchical approach which currently characterizes maritime governance, assumes that 'organizations operate in 'a world of deterministic, well-behaved mechanical processes' (Paquet 1999). Life however, is actually 'full of paradoxes, contradictions and surprises' making the hierarchical approach inadequate faced with 'ill-defined, uncertain, unstable or unreliable' situations (Sutherland and Nichols 2006, p. 7).

Jessop (1998, p. 43, 2004, p. 16) considered that failure was almost endemic in governance particularly if it is based wholly on market forces as it forms an inherent part of the market process. However, the definition of governance failure when it occurs is more difficult. There is no pre-given reference point, nor a substantive criterion against which to judge failure. Jessop suggested that the point of governance is that goals are modified over time and as circumstances change and hence failure is essentially elusive. Malpas and Wickham (1995, p. 40) supported his view suggesting that governance remains permanently incomplete and therefore must fail. To quote Jessop (1998, p. 43):

> Given the growing structural complexity and opacity of the social world, indeed, failure becomes the most likely outcome of most attempts to govern it with reference to multiple objectives over extended spatial and temporal horizons – whether through markets, states, partnerships or some other mechanism.

He went on to stress the failure of governance as he saw it manifesting itself as an evaporation of authority with the existing hierarchical arrangements no longer effective as new, network-based horizontal flows (exemplified by the growth of NGO power, the increased role of interest groups and their facilitation by the process of globalization) have come to predominate. Meanwhile the existing structural flows of policy-making remained largely unchanged. Rosenau's argument built on his earlier observations (Rosenau 1997, p. 363) citing Zurn (1993, p. 40) that globalization has created extended boundary-crossing activities leading to 'uneven denationalization' resulting in the alienation of people from the policy-making process.

Arrighi (1994, p. 2) noted the twentieth century failure of governance to be associated as an inevitable consequence of the structural crisis that has affected the Fordist-Keynesian regime of accumulation where investment in fixed capital was expected to create the potential for regular increases in productivity and consumption (Boyer 1990; Tickell and Peck 1992). For this to occur, adequate government policies and actions, social institutions and norms of behavior had to be sustained—in other words an appropriate governance framework. As the structural crisis has dug deep and the process of globalization changed the rules, so a new governance framework is needed or failure will occur (Aglietta 1979; De Vroey 1984; Lipietz 1987, 1988).

Lash and Urry (1987, pp. 6, 300–301) concurred, seeing the rise of a Postmodern era requiring a new form of governance because of the major transformations that this has generated. These transformations manifest themselves as internationalization from above—global corporations, international monetary reorganization and global governments; decentralization undermining national-based

2.4 Governance Failure

societies from below; and growth of the service sector, transforming society from within. In the face of such change in governance scale and jurisdictional relationships, governance must change too. Held (1991, p. 148) agreed:

> by taking the nation-state for granted, and by essentially reflecting on democratic processes within the boundaries of a nation-state, nineteenth and twentieth century democratic theory has contributed very little to understanding some of the most fundamental issues confronting modern democracies and the fate of democracy in the modern world.

This is taken up as well by Taylor (1994, p. 159) who considered that the old 'wealth containers are no longer operative but politicians have not given up on territoriality. New economic blocks represent another attempt at creating wealth containers but at a scale that may stem at least some of the leaks'.

Stewart (1996) identified policy failure over a wide range of activities particularly associated with problems not of generation but implementation. This may sound familiar. He identified earlier work by Pressman and Wildavsky (1973) which showed how policies created did not necessarily flow into action—something they termed an implementation gap. Barrett and Fudge (1981, p. 9) suggested that this occurs because of the separation of those making policies and those actually implementing them. This may also sound familiar. The need was to view policy-making and implementation as a single process that encompasses a complex assembly job fitting together the different partners that have a stake. The process has to be interactive and recursive and cannot rely upon hierarchies. Crosby (1996, p. 1404) further emphasized the problems of policy implementation caused by the structural failure of governance frameworks that were characterized by a top-down process.

Stoker (1998, p. 24) felt that tensions within civil society in addition to 'inadequacies in the organizations that bridge the gaps between public, private and voluntary sectors may lead to governance failure'. He cited Orr and Stoker (1994) in suggesting that failure of leadership, the lack of appreciation of various timescales and horizons of key partners and the existence of social conflict can all derail what might appear to be on the surface, an admirable policy. Consequently there was a need to 'think beyond the retooling of government to a broader concern with the institutions and social and economic fabric beyond government'. The situation had not been helped by the blurring of boundaries between the state and private sector, between the official and the voluntary, which has occurred increasingly from the 1970s.

Brenner was clear about what he thinks has gone wrong, citing Smith (1997, pp. 50–51) in the process. He criticized what he termed the 'habitual spatial assumptions' of the existing state-centric epistemology of contemporary governance. Globalization continues to upset, deconstruct and rework the foundations of society and as a consequence there is a need for new, analytical frameworks that do not 'imprison the social sciences within timeless, territorialist and unhistorical representations of social space'. However, the response so far has been either to 'transpose state-centric mappings of space onto a global scale or to assume that globalization is borderless and non-territorial and the state is meaningless. While the role of the nation-state and its borders may well be changing and governance approaches need reformulating, simply abandoning the notion of space is not an option that makes sense.

Cerny (1995, pp. 620, 625) considered that the failure of governance was a result of the state's own reaction to globalization, commonly cheating or free-riding on opportunities created by 'autonomous transnational market structures' emanating from similarly autonomous transnational unaccountable structures—something that shipping will recognize. The interaction between these new, unrepresentative organizations is complex and non-linear, tending towards chaos or at best 'plurilateral stability'.

Hajer (2003) associated the failure of current governance to the inadequacies in policy-making of what he termed 'classical-modernist' institutions, defined as 'codified arrangements that provide the official setting of policy-making and politics in the post-war era in Western societies'. In later chapters this will emerge as a fundamental feature that has been created by the seismic move in society as capital is chased and the structure of economic, social and political life has changed with globalization. New political institutions have emerged, referred to by Hajer as an 'ensemble of mostly unstable practices' aimed at resolving the problems that the established institutions are failing to address successfully.

> the constitutional rules of the well-established classical-modernist polities do not tell us about the new rules of the game. In our world the polity has become discursive; it cannot be captured in the comfortable terms of generally accepted rules, but is created through deliberation. The polity, long considered stable in policy analysis, thus becomes a topic for empirical analysis again. (Hajer 2003, p. 176)

Modernist (nation-state dominated, institutionalized) governance has been eroded by an institutional void generated by the move to a Postmodern, globalized society of which shipping is a central part.

Ramachandran et al. (2009) suggested that no progress at all is being made towards achieving 'an effective, coherent system of global governance' with the existing arrangements involving institutions such as the United Nations and World Bank, failing to keep pace with change in the world. Efforts to reform have been partial and most strikingly, national membership of international organizations is an inadequate guarantee of effective governance. Zurn (2000, p. 184) was equally as dismissive of attempts to incorporate citizens in the process on governance in an effective way provided by international organizations such as the WTO, dominated by hundreds of state representatives and tens of thousands of pages of agreements. While the aims of maritime governance (security, safety, efficiency and environmental improvements) can be 'better achieved with international organizations than without them, the mere existence is no guarantee of good governance'.

The legitimacy of international organizations like the UN, WTO, IMO and others is considered by Coicaud (2001, pp. 523–524) who firmly believed that this is achieved only through the nation-state members, from whom they receive their mandate and agenda. In so-doing they also open up the possibilities for governance over and beyond the member states themselves. However, this seeming panacea

2.4 Governance Failure

for globalized governance cannot work where the international organizations themselves lack 'convergence and consistency' creating 'quasi-institutionally embedded, disorganized courses of ideas and actions... generating disagreements over values and policies'. Coicard identified these problems existing in international organizations at present and suggested that the relationships with member states, private stakeholders, non-governmental organizations and individuals need to change.

> rather than being global institutions with worldwide effective operational reach, they tend to be headquarters organizations. In this context the head is likely to be remote from the rest of the organization and its activities on the ground. At times, the two hardly recognize each other as parts of the same entity. As internal deficiencies usually result in poor power projection, this situation largely accounts for the inadequate cohesion of decision-making processes and erratic implementation of operations in the field. As a consequence what international organizations have built so far is less a thick multi-directional web or matrix than a thin network with a relatively meager normative, operational and political grip on or 'pull power' over developed and developing countries. (Coicaud 2001, p. 527)

De Senarclens (2001, p. 509, 510) noted the increasing influence of international organizations since 1945 generated by the ever-increasing globalization of markets and encouraged by the hegemonic traits that they permit or encourage. International organizations act as an intermediary to enable a dominant superpower to retain influence over other nation-states (Petit and Soete 1999, p. 179). However, De Senarclens (2001, p. 514) went on to suggest that nation-state representation at these international organizations was inadequate. Decision-making tended to be in the hands of a very limited number of member nation-states and the mass of stakeholders in any particular process are largely unrepresented. Despite this, most nation-states do not want to abandon their international representation, thus retaining their political audience despite their ineffectual policies, lack of democratic representation and debatable legitimacy. Nation-states remain at the heart of all the international organizations despite the best efforts of globalization and consequently sustain the hierarchical notion of governance that currently persists. Petit and Soete (1999, pp. 178, 180) suggested that only nation-states can co-ordinate and maintain the international organizations and the global policies that have to be agreed. 'Nation-states will be the principal agents forging the new institutions'.

Meanwhile Woods and Narlikar (2001, p. 569) noted how significant the international organizations had become, intruding 'deeply into the national politics of member states':

> each of the international economic institutions is now involved in decision-making which directly affects local communities, interest groups, national domestic and political arrangements, and also specific groups of countries. Many of these groups are today claiming that they should be treated as stakeholders, and that the international economic institutions should be more directly accountable to them. (Woods and Narlikar 2001, p. 569)

Borzel (2007, p. 16) was confident that governance within the European Union generally fails because of the lack of representation and involvement of the private sector—to which much policy-making is directed. This is highly inappropriate. The problem is then intensified by the hierarchical nature of the EU policy-making process, something designed and sustained by the EU itself and which fails adequately to accommodate the multi-level governance characteristics which are needed.

2.5 Maritime Governance

The literature considering European Union governance is immense and reflects the substantive issues raised by the growth of new supra-national authorities and their relationship to nation-states. The EU has shown considerable interest in governance in the early twentyfirst century in response to some of the comments made but with only a marginal impact upon the processes adopted. Evidence of the new interest comes from a number of policy documents from the Commission (for example Commission of the European Communities 2001a, b, c, 2008a, 2009), however, clearly, much is left to be done.

We cannot review all that has been discussed with respect to the EU and governance but some of the main issues can be noted. As a result, before we go on to examine maritime governance in more detail, it is important to have a closer look at EU governance and the issues that characterize it, since many of the maritime governance problems that exist worldwide, manifest themselves most clearly in this region.

Ruggie (1993, p. 140) quoted The Economist (1991, p. 16) when considering the governance of the EU:

> Eurocrats speak of overlapping layers of European economic and political 'spaces' tied together in the words of EC Commission President Jacques Delors, by the community's 'spiderlike strategy to organize the architecture of a Greater Europe'.

Peterson (1994, pp. 152–153) outlined EU policy-making and governance. EU policies were 'products of bargaining between a diverse array of national and supranational, public and private and political and administrative actors'. Governance was characterized by networking and actors wishing to influence the process needed to be involved early on when control by EU politicians was most tenuous (Mazey and Richardson 1993). Policy-making, and hence governance, is complex. Peterson gave an example of the levels of analysis in decision-making that characterize this process (Table 2.13).

Scharpf (1994, pp. 220, 221) had identified some years before what he called a 'democratic deficit' in EU governance (later to be reiterated by Hofmann and Toller 1998), and particularly the failure of the nation-states which make up the EU to relinquish any power. The plan seemed to be that they would continue to enlarge the competencies of the EU while also controlling decision-making at a nation-state level through the Council of Ministers. Policy-making had to be improved and this Scharpf saw as only possible by improving the governance of

2.5 Maritime Governance

Table 2.13 Levels of analysis in EC decision-making

Level	Type of decision	Dominant actors	Rationality
Super-systemic	'history-making'	European council: ECJ	Political; legalistic
Systemic	'policy-setting'	Council of ministers; COREPER	Political; administrative
Meso-level	Policy-formulating, shaping and implementing	Commission; council; secretariat; committees; private actors	Technocratic; consensual; administrative

Source Peterson (1994, p. 153)

negotiations between politically autonomous governments. Since the European institutions are completely dependent upon the nation-state members for finance and political power, then this has to be recognized in the governance model adopted. This contrasts somewhat with our earlier suggestions that in fact the way ahead is to do exactly the opposite (especially in the globalized world of shipping) and detach the two jurisdictions, or even derive new relationships across jurisdictions or avoiding jurisdictional interference. Scharpf (1994, p. 238) was confident that the existing EU hierarchical arrangement must fail. Consequently a new model was needed requiring 'complementary adjustments of the forms of governance at both the European level and that of member states'.

Zurn (2000, p. 183) was similarly as depressed by the EU as a democratic institution, citing its inadequate governance as a major factor:

> If the EU were to apply for membership in the EU, it would not qualify because of the inadequate democratic content of its constitution.

The maritime sector is also far from immune to governance failures—and it is the contention that this in turn has undermined any attempts at coherent and meaningful maritime policy-making. Sletmo (2002b, p. 5) noted how national shipping policy is an 'oxymoron' as globalization of the industry eliminates the need for national policies. National shipping governance would be similarly ineffective, encouraging distortion of policy-making more generally although alternative views were beginning to emerge in the context of severe worldwide economic depression in 2011 (Tradewinds 2011). Sletmo cited Canada, China and West Africa as examples of where different shipping policies were emerging as a response to the different circumstances that existed making an appreciation of the variation internationally in maritime governance significant. He also noted the importance of maritime policy institutions and that they need to be effective if the policies themselves were to work. Sletmo's contribution also built on earlier work by Zacher and Sutton (1996, p. 38) and Hosseus and Pal (1997) who considered the boundaries of policy that exist for shipping and therefore the types of issues that comprise shipping policy. Some 473 topics were derived, consolidated to 135 categories along with accompanying policy instruments. Although they did not consider issues of policy governance, their work was valuable in providing boundaries around a highly diffuse and difficult to define area.

Gibson and Donovan (2000) provided an extensive review of USA maritime policy up to the end of the twentieth century and in the process could not avoid the issues of governance. Similarly Cafruny (1987) and Sturmey (1975) each considered world shipping policies and the governance frameworks that lay behind them even though much of this was unrecognized at the time. Bennett (2000, p. 879) however, was more open in his appreciation of the importance of governance to maritime policy-making and the issues of international regulation in particular quoting Young (1996, p. 2) in support:

> Governance… is a social function whose performance is crucial to the viability of all human societies; it centres on the management of complex interdependencies among actors (whether individuals, corporations, interest groups, or public agencies) who are engaged in interactive decision-making and therefore, taking actions that affect each other's welfare.

Bloor et al. (2006) meanwhile placed the maritime industry into the governance of supply chains characterizing the shipping industry as part of Gereffi' et al.'s (2005) global value chains, 'divided among multiple enterprises spread across wide swathes of geographical space'. They went on to reflect upon the poor record of labour standards in the maritime sector and suggested that there were governance issues at the root and not just regulatory failings.

Chircop (2009, p. 361) was firmly convinced of the importance of governance and jurisdictional integrity in particular in his consideration of shipping in the Arctic identifying multiple levels of hierarchical policy-making. De Vivero and Mateos (2010) assessed ocean governance issues and the management of ocean space in some detail while Gekara (2010) provided a more detailed examination of the role of tonnage taxation in the UK and its jurisdictional and governance integrity.

Bennett (2000, p. 893) emphasized the significance of maritime governance by indicating that the problems associated with substandard shipping were not caused by too few or inadequate regulations but by a 'result of a lack of responsibility and enforcement'. In this he had political support:

> With the body of laws that now exist… I believe that less emphasis needs to be put on developing requirements relating to new regulatory technical standards for ships and crews and more resources, time and energy needs to be given to fair and effective enforcement. (Speech by former EU Transport Commissioner, Neil Kinnock, June 4th, 1998).

Bloor et al. (2006, p. 535) considered the problems of shipping governance created by the growth in globalization with 'markets in vice' developing as 'competitors seek price advantage without regard to the health and welfare of the workforce, the pollution of the environment, or the protection of the consumer'.

A range of other maritime governance issues have been considered in some detail and these include piracy (Galgano 2009), seafarer certification (Obando-Rojas et al. 2004), port state control (Bloor et al. 2006; Perepelkin et al. 2010), the role of the IMO (Bloor and Sampson 2009), flag states (Kovats 2006), ports

(Brooks 2004, 2007; Slack and Fremont 2005), and also issues relating more specifically to EU maritime policy (Wakefield 2010).

Placing the whole maritime governance debate in its wider context Chlomoudis et al. (2000) considered the movement from the recognition of a Modernist genre for ports to one that we shall see is commonly termed Postmodern. This theme was driven from the work of Rodrigue et al. (1997) and was continued by Van de Loo and Van de Velde (2003, pp. 6–8) who noted Notteboom and Winkelman's (2001b) cultural shift from a Fordist to a Post-Fordist scenario exemplified by the rise in outsourcing which together with globalization provides for a fragmentation and specialization in the ports industry which in turn generates the need for new governance rules and methods. Both Koivurova (2009, pp. 171–173) and his 'creeping jurisdiction', and Sletmo (2002a) and his recognition of the need to revise maritime policy-making to accommodate the national/international movements that were taking place, consider these issues further.

In shipping the emergence of alliances, logistics service providers, dedicated terminals and the like require governance adaptation to new Postmodern ideals. Van de Loo and Van de Velde (2003, p. 14) in particular saw this occurring in the reinvention of port authorities worldwide and Hatzaras (2005, pp. 4–5) noted the applicability of Radaelli's (2003) concepts of inertia, absorption, transformation and retrenchment to the maritime sector.

Thompson (2003, p. 140) placed this discussion into the wider governance debate where under globalization, a new framework of 'disembedded economic relationships' exists, autonomous from national economies and agents largely determining what can be done at the national level, dominated by transnational corporations seeking competitive advantage by 'roaming the globe for cheap but efficient production locations'—something clearly close to the heart of every shipowner. Globalization was viewed by most post-war governments and policy-makers as manageable both domestically and internationally and the result in Thompson's (2003, p. 143) words was 'complacency'. The result was actually the growth of a series of unstable and unmanageable institutions and interests (that) now inhabited national economies'. Shipping is a formidable example. They refuse to be 'managed or controlled and indeed (end) up by dictating the terms of their economic activity to the policy-makers'. Ruggie (1993, p. 164) added the problems of governing international waters to those of shipping as the globalization event has unfolded.

Furger (1997, p. 445) took us beyond the borders of the EU in suggesting that maritime policy, particularly maritime environmental policy, has been 'informed by a command-and-control approach to regulation'. This has relied on all those actively involved being largely self-interested, and economically self-optimizing actors (Sugden 1991). It is assumed that groups of rational actors can rarely maintain commonly defined rules. This failure to self-regulate justifies giving strong regulatory powers to agencies responsible for protecting the public interest. The classic state-centric, hierarchical and institution bound maritime governance model fits the bill sweetly. The industrial and social stakeholder is a nuisance and the role of intermediaries, acting as facilitators between the self-interested industry

Table 2.14 Defining characteristics of policy-making

Characteristics	Policy-making
Goals	Deal with immediate problems
Valued action	Practical action, decision-making, problem solving, legislation, regulations, decisions and constituency satisfaction
Time frame	Immediate short term (days, weeks, months dependent on the nature of the problem or crisis)
Basis for decisions	Information commonly extracted from science but generally reinterpreted within the political context of values, public opinion and economics
Expectations	Clear advice expected and specific answers to questions. More science and information do not necessarily lead to better decisions
Values	Practical experience valued more highly than qualifications. Public and political opinion valued
Intellectual direction	Balance of resource use and development. Economic efficiency. Sustainable development. Multiple use, jobs and pragmatism
Focus	Focus on problems at hand. Multi-disciplinary approach usually required
World view	Primacy of political, social, interpersonal and economic mechanisms. A mixture of objective and subjective views

Source Derived from Plasman (2008, p. 812)

and the protectionist state is an anathema. This is a Modernist vision which is confronting a Postmodern phenomenon.

Plasman (2008, p. 811) in the context of marine spatial planning, considered policy-making and governance to be one and the same thing and to be typically hierarchical. Its defining characteristics are shown in Table 2.14 and from this she derived a series of governance ingredients needed if policy is to be effective. These included:

- The use of all available and relevant authorities of all jurisdictions.
- Recognition of the importance of leadership and the significance of the right individual, with appropriate powers and influence, at the top.
- Extend all activities to all relevant stakeholders and instruments.
- Use science effectively to guide policy options.
- Be interactive and holistic. Build short-term accomplishments into longer-term strategy.
- Communicate transparently and building trust and public support.

Although hardly ground-breaking it is interesting to see how few of these are adequately served by existing maritime governance.

Braithwaite and Drahos (2000) considered extensively the governance issues that emerge from the activities of the IMO. Setting the scene they insist that IMO standards are not a problem and nor is their international acceptance by governments of nation-states:

> Their problem is non-enforcement by states and classification societies which supposedly subscribe to them. It is not a leading down of global standards. Rather, it is a problem of a market niche that allows a minority of the industry – albeit a substantial minority – to profit by evading enforcement of global standards and falling back on limited liability and

bankruptcy whenever loss-bearers attempt litigation to internalize the externalities they inflict on the world. (Braithwaite and Drahos 2000, p. 424)

However, the IMO is not wholly innocent when it comes to governance failure. Van Leeuwen (2008, pp. 9–10) assessed the inadequacies which emerged over the issue of double-hulled tankers between the IMO and some of its member-states and the difficult relationships with the EU which have followed. Meanwhile, with contributions dependent upon the tonnage of merchant fleets, Liberia and Panama for example, are major providers, yet by the late 1970s, neither had ever been represented on the IMO Secretariat nor been elected to Council (Silverstein 1978). Despite some representation since then the imbalance remains apparent. Meanwhile, although Braithwaite et al. see the IMO as representing a 'decisive triumph of global harmonization of standards over national sovereignty', at the same time Hindell's (1996, p. 371) opinion was that:

> When the member states of the IMO put on their rule-making hats they are at their most benign and most constructive. It is when they go home to exercise their 'national sovereignty' by applying these international regulations that the facts of day appear.

Raaymakers (2003, p. 22) identified a number of governance failures surrounding the work of the IMO exemplified by the unilateral action of the French and Spanish governments after the 2003 *Prestige* accident, when they deployed warships to escort transit tankers outside their respective EEZs. This violated the right of innocent passage and freedom of navigation and resulted in the generation of policy from below in direct contradiction to the hierarchical, top-down structure of the IMO and other jurisdictions. The IMO was also seen to be lacking with respect to the ineffectiveness of ballast water regulation and the application of MARPOL to the problem of maritime debris. Raaymakers identified examples of duplication of regulation by national and regional regimes to be applied to the international shipping industry. However, despite these problems, his support was for international policy-making and the retention of the existing hierarchical framework, but clearly its strengthening was a necessity. The IMO was not at fault but the nation-states that make it up were:

> UNCLOS... 'provides the most complete treaty that exists on the maritime environment.' The principle problem hindering efforts to prevent maritime accidents is the failure by states to implement UNCLOS laws (and IMO instruments) relating to navigational safety and the marine environment. Too often (states are lax and) happy just to react to accidents related to the use of the seas.

Raaymakers' conclusions were that integrated and meaningful maritime policies need to be multi-sectoral and holistic as well as ensuring that related jurisdictions work effectively together. Meanwhile Hinds (2001, p. 415) suggested that the UN overall exhibits problems because of the structure it has adopted and in particular the specialized agencies such as the IMO. The agency structure has resulted in global fragmentation and there has been evidence of resistance to coordination. As Williams (1987, p. 48) commented:

the 'Founding Fathers' of the system decided that the system should be polycentric, with connected but essentially separate organizations responsible for political matters (the UN), health matters (WHO), labour matters (ILO), and so on (e.g. *the IMO for maritime matters*). The dangers and disadvantages of having one monolithic organization were considered to outweigh the dangers and disadvantages of having a dispersed network. So the various specialized Agencies were set up as separate institutions, each with its own membership, intergovernmental institution, budget, staff, Executive Head and policies. (*section in italics added*)

Hinds suggested that a fundamental problem that this structure created is that these agencies have failed to accommodate regional needs and ambitions and that to be more effective they need to transfer more of their programme leadership to the regions and nation-states that make up the agencies. At present policy-making at this level is realized through a 'dialogue process within the institutional architecture of individual specialized agencies' in combination with many special conferences which together has led to a 'multitude of action programmes', 'excessive demands on the UN system and their member states', and diffusion of responsibility and competition for scarce resources (Hinds 2001, p. 419). There are no external monitors of what goes on—the process of control is limited to family agencies of the UN system.

The result was that member states fail to assume ownership rights to UN sponsored activities and as such:

> it is not surprising then, that most recipient governments display little urgency to economize in the use of grant-financed technical co-operation... the prevailing incentives contribute to proliferation of projects, minimal local ownership, haphazard linking of technical co-operation with national priorities, and wasteful retention of unsatisfactory or little-used technical assistance personnel. (Berg 1993, p. 174).

In Hinds' (2001, p. 420) words, 'in the case of the UN system in general, the necessary resources and the political will to effectively implement its mandate has been elusive over the past five decades'. Additionally, the problems of the IMO as a Specialized Agency were further emphasized by Childers and Urquhart (1994, p. 189).

> It is now seldom recalled that the founders of the UN family intended, and in fact the first General Assembly provided, that the headquarters of Specialized Agencies should be located at the seat of the UN itself. There were supposed to be only 'very strong reasons' for any exception to this general principle. Fifty years of experience demonstrate what was lost by the total ignoring of this key feature of the original design.

Evidence of concern over maritime governance failure comes from all quarters, not just the IMO, and builds upon the discussion on maritime policy failure in the previous chapter. Of course, the two are intrinsically linked as the possibility of meaningful and enforceable policies without a robust and appropriate governance structure is slim.

The focus of many of the maritime policy failures which we have identified has been the European Union, and although maritime failure is clearly apparent elsewhere, the EU provides the most serious and concentrated examples of the problems that exist. Recent discussion is exemplified by Wegge's (2011)

2.5 Maritime Governance

consideration of Norway's role in the EU maritime policy process. Meanwhile we have noted the five main good governance principles that the EU identified in 2001 and which Fritz (2010, pp. 4, 5) summarized as openness, participation, accountability, effectiveness and coherence, reinforced by proportionality and subsidiarity. However, although fine in principle, there remains much to be done to actually improve EU governance and quite possibly this has to involve substantial reform. An indication of the changes needed comes from the 2001 White Paper (Commission of the European Communities 2001a, p. 25) where the emphasis is clearly placed upon the nation-state to ensure effectiveness of governance, overlooking the shift that has occurred towards globalization—maritime or otherwise. The hierarchical nature of traditional governance is emphasized through the Commission's insistence on EU law and the significance this takes in ensuring good governance and the hierarchy is re-emphasized where the Commission continues by stressing its need to extend supranational representation across all the jurisdictional levels—especially international and regional. In fact it is only in the Commission's acceptance of the increasing extended role of stakeholders from wider backgrounds that any form of move away from the traditional hierarchy model can be seen. This was followed by more specific maritime governance proposals in 2006 and 2009 (Commission of the European Communities 2006, 2009).

Meanwhile, the European Union Integrated Maritime Policy emerged in 2007 (Commission of the European Communities 2008a), and made specific reference on a number of occasions for the need for proper governance measures if it was to succeed in its aims of integrating increased economic activity and environmental protection. In the EU the maritime governance problem is extremely complex as the region includes a number of highly sensitive enclosed seas (Mediterranean, Baltic and Black Seas) as well as considerable areas of High Sea (Commission of the European Communities 2009, p. 1). The Commission identified two significant problems that achieving effective maritime governance faced:

- Each sectoral policy is pursued by its own national administration making effective international policy very difficult to develop. This problem we have already seen relating to the work of the IMO in particular, where nation-states tend to work to their own agendas rather than those globally optimal.
- The large area of High Seas which lie near to the EU or within which EU vessels operate makes the planning, organization and regulation of activities very difficult within a national-bound governance regime.

In addition, EU maritime governance would need to comply with the ideal principles of good governance identified in 2001 (Commission of the European Communities 2001a) viz. stakeholder participation, transparency of decision-making, and effective implementation of agreed rules. Almost none are currently upheld.

To move towards a position of good maritime governance the Commission has proposed a number of changes in the maritime policy-making process specifically for the Mediterranean region and which might be applied elsewhere more generally:

- Encourage stakeholders and administrations in the Mediterranean to address maritime affairs in a more integrated manner and to engage in priority-setting for maritime governance.
- Examine ways to foster further cooperation among stakeholders and administrations from across all maritime related sectors and the Mediterranean.
- Assist Mediterranean EU Member states to exchange best practice in integrated maritime policy.
- Make available assistance under the European Neighbourhood and Partnership Instrument for non-EU countries.
- Promote ratification and implementation of UNCLOS.
- Set up a basin-wide Working Group on integrated maritime policy to foster dialogue.
- Enhance multi-lateral, cross-sectoral cooperation through specific studies and regional agreements.

The focus was clearly on technical competence within the confines of the existing governance jurisdictional structure—reliant upon the nation-state and the convergence of national and supranational/global priorities. Essentially the right words were being used (stakeholders, integration, transparency, dialogue etc.) but little is being done to move away from the traditional hierarchical, state-centred focus.

These policy trends were re-emphasized in 2008 with the publication of a Communication from the Commission to the Parliament and the Economic and Social Committee (Commission of the European Communities 2008b). This document stressed the need for holism in policy-making in the maritime sector and many of the proposals put forward were clear and well-defined with examples of good practice in the generation of maritime policy by member states that attempted (and succeeded) in looking at the martiitme sector as a whole. However, there remained an assumption that the esisting hierarchical governance structure would be unchanged with no discussion of the need at least to review the institutional and decision-making framework, and in addition although the important role of stakeholders in policy-making was clearly flagged up, there was nothing to suggest that those involved or who needed to be involved would be any more than the traditional enthusiasts who characterized maritime policy-making to date.

Discussion of the problems of maritime policy-making dates back some years [see for example Yannopoulos (1989) and Cafruny (1987, 1991)] but without any explicit linkage to maritime governance. However, more recently the issues have clarified. For example, Vallega (2001, p. 399) began a wide-ranging discussion of ocean governance by outlining the maritime origins of the word. Derived from the Greek 'kybernan', meaning 'to hold the helm', its roots are clearly in navigation and implies conducting a structure/vessel towards a destination, considering the organization of a system as a whole in a way consistent with that organization's aims. While governance's maritime roots may be co-incidental to our discussion, it remains a neat reminder of the relevance of governance to the sector as a whole.

2.5 Maritime Governance

Vallega goes on to stress how the rise of the Postmodern ocean has required a Postmodern approach to governance and questions whether this has been achieved.

Van Leeuwen and Bruyninckx (2006) considered the broader issues of maritime governance in relation to the dumping of waste by shipping, while Van Leeuwen (2008, p. 2) discussed the application of new governance principles in the maritime sector and stressed early on how multiple actors are now seen as a vital ingredient in the general governance mix, including non-state and private stakeholders. Governance has also recognized the importance of incorporating multi-level practices through the interlinking of international, supranational, national and regional representatives across traditional jurisdictional boundaries, along with the development of more informal policy regulations and norms. However, how much of this has actually happened and more importantly, how much in the maritime sector?

Van Leeuwen (2008, pp. 7–8) continued to stress the significance of the nation-state in maritime policy-making to date. She notes what she terms the importance of the 'actor dimension' of states—reflected in the terms port state (when a ship visits a port); flag state (reflecting registration) and coastal state (when the ship passes through a state's coastal waters). This state domination of maritime governance is further apparent from the work of the IMO which is designed by and operationalized at a state level. Power struggles over maritime policy are almost always between 'flag states and ship owners' in some direct or indirect way.

Alderton and Winchester (2002, p. 39) saw a direct link between sovereignty and maritime governance failure, citing flags of convenience as an excellent example. Here they claimed, the problem was that whatever international policy-making organization you have (for example the IMO), policy is always enacted on a state-by-state basis and transforming these universally accepted goals and rules into a binding legal obligation was each state's sovereign privilege (Cairola and Chiarabini 1999). This presented the freedom both to act and not to act and this in turn meant that the 'path between the flag state and the ship owner is at best, obscure and minimal' (Alderton and Winchester 2002, p. 40). This governance failure was reinforced by the increased standards implied by port state control policies which hunted down the lower quality vessels to the point that a new market was created for those shipowners who could afford to maintain standards.

Van Tatenhove (2008) provided a general discussion of maritime governance and its links to maritime policy-making in the context of fisheries, commercial shipping and offshore activities relating these to the changing role of nation-states and the rise of the private sector in comparison to the public. He stressed the significance of achieving an integrated and effective maritime governance framework and the inertia identifiable in the existing governance institutions along with their institutionalized maritime policy domains. In an earlier paper, along with Van Tatenhove et al. (2006, p. 10) he stressed the importance of informal networks and the emergence of the 'network state'. This followed the work of Castells (1997, p. 332) who described:

... the new form of state epitomized by European institutions; *the network state. It is a state characterized by the sharing of authority... along a network.* A network by definition has nodes not a centre. Nodes may be of different sizes, and may be linked by asymmetrical relationships in the network, so that the network state does not preclude the existence of political inequalities among its members. (*emphasis in original*)

The relationship between ports and governance has generated much interest in recent years, stimulated in part by the proliferation of privatization that has taken place and the changes in ownership structure and the process of organizing and administering the port sector that has resulted.

De Langen (2002) defined port cluster governance as the 'mix of and relations between different modes of governance' and identified four main variables—trust, intermediaries, leader firms and the quality of solutions to collective action problems which dictated the success of otherwise of the governance model adopted. Port cluster success was dependent upon an appropriate governance framework. Wang et al. (2004, p. 237) suggested that port governance modeling had gained increased credibility in the early years of the twentyfirst century with particular reference to port reforms in developing economies (Baltazar and Brooks 2001; Wang and Slack 2002; Wang and Olivier 2003) citing particular examples from China. After some discussion of the different ways that port governance had been viewed, they went on to propose that a greater emphasis was needed on stakeholder involvement (something we return to in a later chapter) and also an increase in the weight placed on social and cultural variables, citing Stoker's five propositions of governance (Stoker 1998). Their adapted framework for port governance is shown in Fig. 2.4 where the three axes consider scale, function and stakeholder. While far from ideal (or even very clear!), Wang et al.'s emphasis on the significance of port governance and its relationship to scale, stakeholder, function and jurisdiction is obvious.

Brooks and Cullinane (2007a) placed considerable emphasis on the ownership issues relating to governance and the devolution process that has occurred widely in ports. While ownership and its impact remain significant for port governance there are many other aspects that need consideration as well. The models of port governance described in Brooks and Cullinane (2007b) relate almost entirely to port responsibilities in relation to types of devolution and the balance achieved between private and public and there is no discussion of the cultural and societal issues nor of the flows of information, power and influence nor of the alternative structures that might develop in jurisdiction and government. While Cullinane and Brook's discussion is wholly relevant to port governance, it remains only partial and over-emphasizes the impact of the process of privatization that has taken place.

Brooks and Pallis (2008, pp. 411–413) took a similar line with continued emphasis on ownership as the fundamental port governance issue in their interpretation of the relationship between governance models and port performance. The emphasis is clear from their discussion of the 34 different models of ownership derived by Brooks and Cullinane (2007b) which were considered in themselves models of governance and not of ownership. Meanwhile Notteboom and

2.5 Maritime Governance

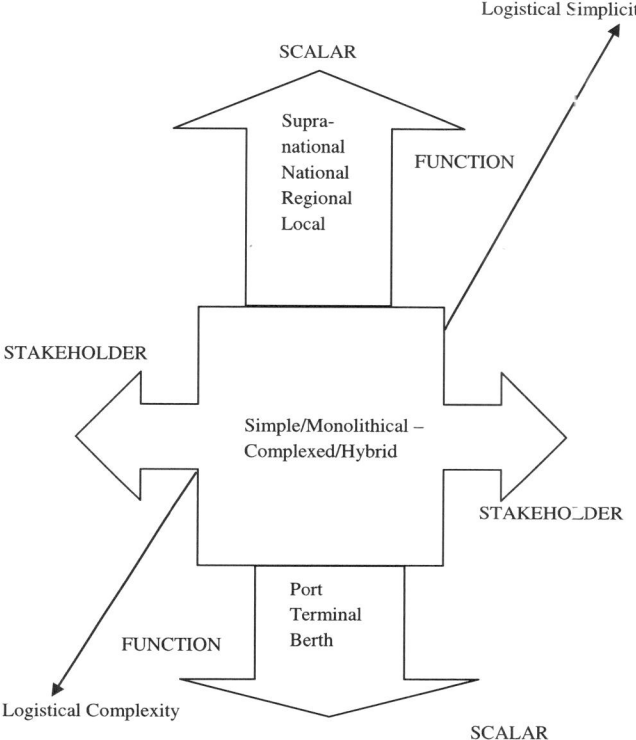

Fig. 2.4 Three dimensional model of port governance. *Source* Derived from Wang et al. (2004)

Winkelmans (2001a) provided a broad introduction into the idea that the public sector must provide good governance for seaports especially as they face continuous reform in the light of further globalization pressures. As part of this they widened the discussion to issues such as stakeholder relations and agreed with Wang and Slack (2002) in that social and cultural variables were a vital part. As emphasized in Table 2.14, there is much more to port governance than ownership and failure in ports is evidenced by the relationship that exists between performance and governance in its widest sense

However, in contrast Verhoeven (2009, p. 79) saw the changing ownership patterns in ports challenging existing governance models especially in the context of repeated (and failed) attempts by the EU to introduce a ports policy centred around competition. Verhoeven went on to define port governance as confined between markets and regulation with a large number of positions that could be taken up between these two. No mention was made of network models as an alternative nor of the significance of hierarchies, nation-state changes or jurisdictional chaos.

Three challenges that new port governance models need to face were identified as arising from sustainability, the rise of integrated logistics and the increased role of markets but these were then interpreted only in terms of ownership issues and there was limited space given to the need to address representation, stakeholder involvement or jurisdictional integrity. However, Verhoeven did finish with some more diverse ideas about what needs to be considered to encourage 'good' port governance. These reflected on the need for enhanced autonomy and management culture of port authorities; improvements in the interaction between government, port authorities and market actors; increasing the effectiveness of reform programmes including political influence and finally increasing diversity in port governance systems.

2.6 Conclusions

Governance is at the heart of maritime failure and it is clear from this discussion just how complex the area is and how significant an issue it has become in the debate about policy-making in general both within and outside the maritime sector. Before we can go on to assess the ways that something might be done to overcome the plague of policy failure that has affected shipping and ports, it is appropriate first to understand the fundamental principles that lie behind current and future potential governance structures. To do this we first turn to the notion of the hierarchy, followed in turn by the nation-state and globalization. Alternative configurations to the existing Modernist maritime governance model can then be considered.

References

Adams, G. (1982). *The politics of defense contracting: The iron triangle*. New Brunswick, NJ: Transaction Books.
Aglietta, L. (1979). *A theory of capitalist regulation: The US experience*. London: New Left Books.
Agranoff, R., & McGuire, M. (2003). *Collaborative public management: New strategies for local governments*. Washington, DC: Georgetown University Press.
Alba, R. D. (1982). Taking stock of network analysis: A decade's results. *Research in the Sociology of Organizations, 1*, 39–74.
Alderton, T., & Winchester, N. (2002). Globalization and de-regulation in the maritime industry. *Marine Policy, 26*, 35–43.
Allen, J. (1997). Economics of power and space. In R. Lee & J. Wills (Eds.), *Geographies of economies* (pp. 59–70). London: Hodder Arnold.
Alter, C., & Hage, J. (1993). *Organizations working together*. Newbury Park, CA: Sage.
Altman, J. A., & Petkus, E., Jr. (1994). Towards a stakeholder-based policy process: An application of the social marketing perspective to environmental policy development. *Policy Sciences, 27*, 37–51.
Amin, A., & Hausner, J. (1997). *Beyond market and hierarchy*. Cheltenham: Edward Elgar.

References

Amin, A., & Thrift, N. (1994a). Living in the global. In A. Amin & N. Thrift (Eds.), *Globalization, institutions and regional development in Europe* (pp. 1–22). Oxford: Oxford University Press.

Amin, A., & Thrift, N. (1994b). Holding down the global. In A. Amin & N. Thrift (Eds.), *Globalization, institutions and regional development in Europe* (pp. 257–260). Oxford: Oxford University Press.

Amin, A., & Thrift, N. (1995). Globalization, institutional thickness and the local economy. In P. Healey, S. Cameron, S. Davoudi, S. Graham, & A. Madani-Pour (Eds.), *Managing cities: The new urban context* (pp. 91–108). Chichester: Wiley.

Ansell, C. (2000). The networked polity; regional development in Western Europe. *Governance, 13*(3), 303–333.

Antonelli, C. (1988). *New informational technology and industrial change*. Dordrecht: Kluwer.

Arrighi, G. (1994). *The long twentieth century*. London: Allen and Unwin.

Atkinson, M. M., & Coleman, W. D. (1989). Strong states and weak states; sectoral policy networks in advanced capitalist countries. *British Journal of Political Science, 19*(1), 47–67.

Atkinson, M. M., & Coleman, W. D. (1992). Policy networks, policy communities and the problems of governance. *Governance, 5*(2), 154–180.

Bacharach, P. S., & Baratz, M. S. (1962). Two faces of power. *American Political Science Review, 56*, 1947–1952.

Bair, J. (2005). Global capitalism and commodity chains: Looking back, going forward. *Competition and Change, 9*(2), 153–180.

Bair, J. (2008). Analysing global economic organization: Embedded networks and global chains. *Economy and Society, 37*(3), 339–364.

Baltazar, R., & Brooks, M. R. (2001). *The governance of port devolution: A tale of two countries*. Seoul, Korea: World Conference on Transport Research.

Barrett, S., & Fudge, C. (Eds.). (1981). *Policy and action*. London: Methuen.

Bell, S., & Park, A. (2006). The problematic metagovernance of networks; water reform in New South Wales. *Journal of Public Policy, 26*(1), 63–83.

Bennett, P. (2000). Environmental governance and private actors: Enrolling insurers in international maritime regulation. *Political Geography, 19*, 875–899.

Benson, J. K. (1977). Organizations: A dialectical view. *Administrative Science Quarterly, 22*, 1–21.

Benson, J. K. (1982). A framework for policy analysis. In D. L. Rogers & D. A. Whetten (Eds.), *Interorganizational coordination: Theory research and implementation* (pp. 137–170). Ames, IA: Iowa State University Press.

Bentley, A. F. (1908). *The process of government*. Cambridge, MA: Harvard University Press.

Benz, A. (1998). Politikverflechtung ohne Politikverflechtungsfalle. Koordination und Strukturdynamik im europaischen Mehrebenensystem. *Politische Vierteljahresschrift, 39*, 558–589.

Benz, A. (2000). Entflechtung als Folge von Verflechtung. Theoretische Uberlegungen zur Entwicklung des europaischen Mehrebenesystems. In E. Grande & M. Jachtenfuchs (Eds.), *Wie problemlosingsfahig ist die EU? Regieren im europaischen Mehrebenesystem*. Baden-Baden: Nomos.

Berg, L. (1993). Between modernism and postmodernism. *Progress in Human Geography, 17*(4), 490–507.

Blatter, J. (2003). Beyond hierarchies and networks: institutional logics and change in transboundary spaces. *Governance, 16*(4), 503–526.

Blom-Hansen, J. (1997). A 'new institutional' perspective on policy networks. *Public Administration, 75*, 669–693.

Bloor, M., & Sampson, H. (2009). Regulatory enforcement of labour standards in an outsourcing globalized industry. *Work, Employment and Society, 23*(4), 711–726.

Bloor, M., Datta, R., Gilinskiy, Y., & Horlick-Jones, T. (2006). Unicorn amongst the cedars: On the possibility of effective 'smart regulation' of the globalized shipping industry. *Social and Legal Studies, 15*(4), 534–551.

Boltanski, L., & Chiapello, E. (2007). *The new spirit of capitalism*. London: Verso.

Borzel, T. A. (1998). Organizing Babylon—on the different conceptions of policy networks. *Public Administration, 76*, 253–273.

Borzel, T. A. (2007). *European governance—negotiation and competition in the shadow of hierarchy*. Montreal, Canada: European Studies Association Meeting.

Borzel, T. A. (2010). European governance: negotiation and competition in the shadow of hierarchy. *Journal of Common Market Studies, 48*(2), 191–219.

Borzel, T.A., Guttenbrunner, S., & Seper, S. (2005). *Conceptualizing new modes of governance in EU enlargement*. Report on the European Commission New Modes of Governance Project NEWGOV, Berlin: Free University of Berlin.

Boyer, R. (1990). *The regulation school: A critical introduction*. New York: Columba University Press.

Braithwaite, J., & Drahos, P. (2000). *Global business regulation*. Cambridge: Cambridge University Press.

Brandes, U., Kenis, P., Raab, J., Schneider, V., & Wagner, D. (1999). Explorations into the visualization of policy networks. *Journal of Theoretical Politics, 11*(1), 75–106.

Brans, M. (1997). Challenges to the practice and theory of public administration in Europe. *Journal of Theoretical Politics, 9*(3), 389–415.

Braudel, F. (1982). *The wheels of commerce*. New York: Harper and Row.

Brooks, M. R. (2004). The governance structure of ports. *Review of Network Economics, 3*(2), 168–183.

Brooks, M. R. (2007). Port devolution and governance in Canada. In M. R. Brooks & K. P. Cullinane (Eds.), *Devolution, port governance and port performance* (pp. 237–258). London: Elsevier.

Brooks, M. R., & Cullinane, K. P. (2007a). Introduction. In M. R. Brooks & K. P. Cullinane (Eds.), *Devolution, port governance and port performance*. London: Elsevier.

Brooks, M. R., & Cullinane, K. P. (2007b). Governance models defined. In M. R. Brooks & K. P. Cullinane (Eds.), *Devolution, port governance and port performance* (pp. 1–28). London: Elsevier.

Brooks, M. R., & Pallis, A. A. (2008). Assessing port governance models: Process and performance components. *Maritime Policy and Management, 35*(4), 411–432.

Brown, L. D. (1983). Managing conflict among groups. In D. A. Kolb, I. M. Rubin, & J. McIntyre (Eds.), *Organizational psychology*. Englewood, NJ: Prentice Hall.

Brugha, R., & Varvasovszky, Z. (2000). Stakeholder analysis: A review. *Health Policy and Planning, 15*(3), 239–246.

Burrell, G., & Morgan, G. (1979). *Sociological paradigms and organizational analysis*. London: Heinemann.

Cafruny, A. (1987). *Ruling the waves*. Berkeley, CA: University of California Press.

Cafruny, A. (1991). Toward a maritime policy. In L. Hurwitz & C. Lequesne (Eds.), *The state of the European Community* (pp. 285–299).

Cairola, E., & Chiarabini, A. (1999). *International labour standards: A trade union training guide, international training centre: Torino*. Geneva: International Labour Office.

Carlsson, L. (1996). Non-hierarchical implementation analysis. An alternative to the methodological mismatch in policy analysis. *Journal of Theoretical Politics, 8*(4), 527–546.

Carlsson, L. (2000). Policy networks as collective action. *Policy Studies Journal, 28*(3), 502–520.

Castells, M. (1996). *The rise of the network society*. Cambridge, MA: Blackwell.

Castells, M. (1997). *The information age*. Oxford: Blackwell.

Castells, M. (2000). Materials for an exploratory theory of the network society. *British Journal of Sociology, 51*(1), 5–24.

Cater, D. (1964). *Power in Washington*. New York: Random House.

Cerny, P. G. (1994). The dynamics of financial globalization: technology, market structure and policy response. *Policy Sciences, 27*(4), 319–342.

Cerny, P. G. (1995). Globalization and the changing logic of collective action. *International Organization, 49*(4), 595–625.

References

Cerny, P. G. (2001). From 'iron triangles' to 'golden pentangles'? Globalizing the policy process. *Global Governance, 7*, 397–410.

Childers, E., & Urquhart, B. (1994). *Renewing the United Nations system*. Uppsala: Dag Hammarskajold.

Chircop, A. (2009). The growth of international shipping in the Arctic. Is a regulatory review timely? *The International Journal of Marine and Coastal Law, 24*, 355–380.

Chlomoudis, C. I., Karalis, A. V., & Pallis, A. A. (2000). *Transition to a new reality: Theorising the organizational restructuring of ports*. Infomare, International Workshop, Special Interest Group on Maritime Transport and Ports, Genoa, June 8–10.

Clarke, M., & Stewart, J. (1999). *Community governance, community leadership and the new local government*. York: Joseph Rowntree Foundation.

Coicaud, J.-M. (2001). Reflections on international organizations and international legitimacy: Constraints, pathologies, and possibilities. *International Social Science Journal, 170*, 523–536.

Commission of the European Communities. (2001a). *European governance. A white paper*, COM (2001) 428, Brussels.

Commission of the European Communities. (2001b). *Strengthening Europe's contribution to world governance*. Report of Governance Working Group 5, Brussels.

Commission of the European Communities. (2001c). *Multi-level governance: Linking and networking the various regional and local levels*. Report of Governance Working Group 4c, Brussels.

Commission of the European Communities. (2006). *A future maritime policy for the Union: A European vision of the oceans and seas*. COM (2006) 275, Brussels.

Commission of the European Communities. (2008a). *An integrated maritime policy for the European Union*. Brussels.

Commission of the European Communities. (2008b). *Guidelines for an integrated approach to maritime policy: Towards best practice in integrated maritime governance and stakeholder consultation*. COM (2008) 395 final, Brussels.

Commission of the European Communities. (2009). *Towards an integrated maritime policy for better governance in the Mediterranean*. Communication from the Commission to the Council and the European Parliament, COM (2009) 466 Final, Brussels.

Commission on Global Governance. (1995). *Our global neighbourhood*. Oxford: Oxford University Press.

Cooke, P., & Morgan, K. (1993). The network paradigm: new departures in corporate and regional development. *Environment and Planning D, 11*, 543–564.

Coser, L. A. (1956). *The functions of social conflict*. New York: Routledge and Kegan Paul.

Cox, K. R. (1998). Spaces of dependence, spaces of engagement and the politics of scale, or: Looking for local politics. *Political Geography, 17*, 1–23.

Crosby, B. L. (1996). Policy implementation: The organizational challenge. *World Development, 24*(9), 1403–1415.

Dahl, R., & Lindblom, C. E. (1976). *Politics, economics and welfare*. New York: Harpers.

Davies, J. S. (2005). Local governance and the dialectics of hierarchy, market and network. *Policy Studies, 26*(3/4), 311–335.

De Alcantara, C. H. (1998). Uses and abuses of the concept of governance. *International Social Science Journal, 50*(155), 105–113.

De Bruijn, J. A., & Ringeling, A. B. (1997). Normative notes. Perspectives on networks. In W. J. M. Kickert, E. H. Klijn, & J. F. M. Koppenjan (Eds.), *Managing complex networks* (pp. 152–165). London: Sage.

De Langen, P. W. (2002). Clustering and performance: The case of the maritime clustering in the Netherlands. *Maritime Policy and Management, 29*(3), 209–221.

De Senarclens, P. (1998). Governance and the crisis in the international mechanisms of regulation. *International Social Science Journal, 50*(55), 91–104.

De Senarclens, P. (2001). International organizations and the challenges of globalization. *International Social Science Journal, 170*, 509–522.

De Vivero, J. L. S., & Mateos, J. C. R. (2010). Ocean governance in a competitive world. The BRIC countries as emerging maritime powers—building new geopolitical scenarios. *Marine Policy, 34*, 967–978.

De Vroey, M. (1984). A regulation approach interpretation of the contemporary crisis. *Capital and Class, 2*, 45–66.

Deleuze, G., & Guattari, F. (1988). *A thousand plateaus: Capitalism and schizophrenia*. London: The Athlone Press.

Dervis, K., & Ozer, C. (2005). *A better globalization: Legitimacy governance and reform*. Washington, DC: Center for Global Development.

Dicken, P. (1994). Global-local tensions: Firms and states in the global space-economy. *Economic Geography, 70*(2), 101–128.

Dicken, P., Kelly, P. F., Olds, K., & Yeung, W. (2001). Chains and networks, territories and scales: Towards a relational framework for analysing the global economy. *Global Networks, 1*(2), 89–112.

Dowding, K. (1995). Model or metaphor? A critical review of the policy network approach. *Political Studies, 43*, 136–158.

Doz, Y. L., & Mahel, G. (1998). *Alliance advantage: The art of creating value through partnering*. Boston, MA: Harvard Business School Press.

Dubini, P., & Aldrich, H. (1991). Personal and extended networks are central to the entrepreneurial process. *Journal of Business Venturing, 6*, 305–313.

Dunleavy, M. (1981). *The politics of mass housing in Britain 1945–1975*. Oxford: Clarendon Press.

Dunning, J. H. (1997). *Alliance capitalism and global business*. London: Routledge.

Dye, T. R., & Ziegler, H. (1970). *The irony of democracy. An uncommon introduction to American politics*. Belmont, CA: Wedsworth.

Eberlein, B., & Kerwer, D. (2004). New governance in the European Union: A theoretical perspective. *Journal of Common Market Studies, 42*(1), 121–142.

Eccles, R. G. (1981). The quasifirm in the construction industry. *Journal of Economic Behaviour and Organization, 2*(4), 335–357.

Economist. (1991). *Inner space*. May 18.

Etzioni, A., & Lawrence, P. R. (Eds.). (1991). *Socio-economics: Towards a new synthesis*. Armonk, NY: M.E. Sharpe.

Evans, M. (2001). Understanding dialectics in policy network analysis. *Political Studies, 49*, 542–550.

Figgis, J. N. (1913). *Churches in the modern state*. London: Longmans.

Filley, A. C. (1975). *Interpersonal conflict resolution*. Glenview, IL: Scott Foresman.

Finkelstein, L. S. (1995). What is global governance? *Global Governance, 1*, 367–372.

Follett, M. P. (1918). *The new state*. London: Longmans.

Follett, M. P. (1973). *Dynamic administration: The collected papers of Mary Parker Follett*. In E. M. Fox & L. Urwick (Eds.), London: Pitman.

Forsgren, M., & Johanson, J. (1992). Managing in international multi-centre firms. In M. Forsgren & J. Johanson (Eds.), *Managing networks in international business*. Philadelphia, PA: Gordon and Breach.

Fox, A. (1966). *Industrial Sociology and Industrial Relations*. Royal Commission on Trades Unions and Employers' Associations, HMSO: London.

Fox, A. (1974). *Beyond Contract: Work, Power and Trust Relations*. Faber and Faber: London.

Frederickson, H. G., & Smith, K. B. (2003). *The public administration theory primer*. Boulder, CO: Westview Press.

Freeman, J. L. (1965). *The political process*. New York: Random House.

Freeman, C. (1990). *Networks of innovators*. Montreal: International Workshop on Networks of Innovators.

Freeman, J. L., & Stevens, J. P. (1987). A theoretical and conceptual re-examination of subsystem politics. *Public Policy and Administration, 2*, 9–24.

References

Fritz, J. (2010). Towards a 'new form of governance' in science-policy relations in the European maritime policy. *Marine Policy, 34*, 1–6.

Furger, F. (1997). Accountability and systems of self-governance: the case of the maritime industry. *Law and Policy, 19*(4), 445–476.

Gais, T. L., Peterson, M. A., & Walker, J. L. (1984). Interest groups, iron triangles and representative institutions in American national government. *British Journal of Political Science, 14*(2), 161–185.

Galgano, F. A. (2009). The borderless dilemma of contemporary maritime piracy: Its geography and trends. *Pennsylvania Geographer, 47*(1), 3–33.

Gaudin, J. (1998). Modern governance, yesterday and today: Some clarifications to be gained from French government policies. *International Social Science Journal, 50*(155), 47–56.

Gekara, V. O. (2010). The stamp of neo-liberalism on the UK tonnage tax and the implications for British seafaring. *Marine Policy, 34*(3), 487–494.

Gereffi, G., Humphrey, J., & Sturgeon, T. (2005). The governance of global value chains. *Review of International Political Economy, 12*(1), 78–104.

Gerlach, M. L., & Lincoln, J. R. (1992). The organization of business networks in the United States and Japan. In N. Nohria & R. G. Eccles (Eds.), *Networks and organizations: Structures, form and action*. Boston, MA: Harvard Business School Press.

Gibbon, P., Bair, J., & Ponte, S. (2008). Governing global value chains: An introduction. *Economy and Society, 37*(3), 315–338.

Gibson, A., & Donovan, A. (2000). *The abandoned ocean*. Columbia, SC: University of South Carolina Press.

Giddens, A. (1985). *A contemporary critique of historical materialism. The nation-state and violence* (Vol. 2). Cambridge: Polity Press.

Glagow, M., & Willke, H. (Eds.). (1987). *Dezentrale Gesellschaftssteurung: Probleme der Integration polyzentristischer Gesellschaft*. Pfaffenweiller: Centraurus-Verlagsgesellschaft.

Glasbergen, P. (1995). Managing environmental disputes. In P. Glasbergen (Ed.), *Network management as an alternative*. Dordrecht: Kluwer.

Gore, T., & Wells, P. (2009). Governance and evaluation: The case of EU regional policy horizontal priorities. *Evaluation and Program Planning, 32*, 158–167.

Grabher, G. (1993). Rediscovering the social in the economics of interfirm relations. In G. Grabher (Ed.), *The embedded firm: On the socio-economics of industrial networks* (pp. 1–31). London: Routledge.

Grabher, G., & Stark, D. (1997). Organizing diversity: Evolutionary theory, network analysis and postsocialism. *Regional Studies, 31*, 533–544.

Graham, P. (Ed.). (1995). *Mary Parker Follett: Prophet of management*. Boston, MA: Harvard Business School Press.

Granovetter, M. (1985). Economic action and social structure: The problem of embeddedness. *American Journal of Sociology, 91*(3), 481–510.

Granovetter, M. (1994). Business groups and social organization. In N. J. Smelser & R. Swedberg (Eds.), *The handbook of economic sociology* (pp. 429–450). Princeton, NJ: Princeton University Press.

Granovetter, M. (1995). Coase revisited: Business groups in the modern economy. *Industrial and Corporate Change, 1*, 93–130.

Granovetter, M., & Swedberg, R. (1992). *The sociology of economic life*. Boulder, CO: Westview Press.

Griffiths, E. S. (1939). *The impasse of democracy*. New York: Harrison-Wilton.

Gulati, R., Nohira, N., & Zaheer, A. (Eds.), (2000). Special issue: Strategic networks. *Strategic Management Journal, 21*, 191–425.

Haas, P. M. (1992). Introduction: Epistemic communities and international policy coordination. *International Organization, 46*(1), 1–35.

Hagg, L., & Johanson, J. (Eds.). (1992). *Firms in networks*. Stockholm: SNS Forlag.

Hajer, M. (2003). Policy without polity? Policy analysis and the institutional void. *Policy Sciences, 36*, 175–195.

Hakansson, H. (1987). *Industrial technological development. A network approach.* London: Croom Helm.
Hanf, K. (1978). Introduction. In K. Hanf & F. W. Scharpf (Eds.), *Interorganizational policymaking* (pp. 1–15). London: Sage.
Harvey, D. (1990). *The condition of postmodernity.* Cambridge, MA: Blackwell.
Hassard, J., Law, J., & Lee, N. (1999). Special themed section on actor-network theory and managerialism *Organization, 6,* 387–497.
Hatzaras, K. (2005). *Multi-level governance, Europeanization and the poorest EU region.* Epirus and the 2nd Hellenic Community Support Framework, 2nd LSE Symposium on Modern Greece.
Heclo, H. H. (1972). Policy analysis. *British Journal of Political Science, 2*(1), 83–108.
Heclo, H. H. (1978). Issue networks and the executive establishment. In A. King (Ed.), *The new American political system* (pp. 87–124). Washington, DC: American Enterprize Institute.
Heclo, H. H., & Wildavsky, A. (1974). *The private government of public money.* London: Macmillan.
Heinz, J. P., Laumann, E. O., Salisbury, R. H., & Nelson, R. L. (1990). Inner circles or hollow cores? Elite networks in national policy systems. *Journal of Politics, 52*(2), 356–390.
Held, D. (1991). Democracy, the nation-state and the global system. *Economy and Society, 20*(2), 138–172.
Heller, A., & Feher, F. (1988). *The postmodern political condition.* Cambridge: Polity Press.
Heretier, A. (2002). New modes of governance in Europe: Policymaking without legislating? In A. Heretier (Ed.), *Common goods: Reinventing European and international governance* (pp. 185–206). Lanham, MD: Rowman and Littlefield.
Hess, M. (2004). Spatial' relationships? Towards a reconceptualization of embededness. *Progress in Human Geography, 28*(2), 165–186.
Hess, M. (2008). Governance, value chains and networks: An afterword. *Economy and Society, 37*(3), 452–459.
Hill, C. J., & Lynn, L. E., Jr. (2004). Is hierarchical governance in decline? Evidence from empirical research. *Journal of Public Administration Research and Theory, 15*(2), 173–195.
Hindell, K. (1996). Strengthening the ship regulating regime. *Maritime Policy and Management, 23*(4), 371–380.
Hinds, L. (2001). Policy implications and reconstruction of marine institutional architectures: The regional indigenous organization (RIO) option. *Marine Policy, 25,* 415–426.
Hirst, P. (1997a). *From statism to pluralism.* London: UCL Press.
Hirst, P. (1997b). *Democracy, civil society and global politics.* London: UCL Press.
Hirst, P. Q., & Thompson, G. (1999). *Globalization in question.* Cambridge: Polity Press.
Hofferbert, R. (1974). *The study of public policy.* Indianapolis, IN: Bobbs-Merrill.
Hofmann, H. C. H., & Toller, A. E. (1998). Zur Reform der Komitologie—Regeln und Grundsatze fur die Verwaltungskooperation im Ausschussßystem der Europaischen Gemeinschaft. *Staatswissenschaften und Staatspraxis, 1*(2), 209–239.
Hoogsteden, C. B., Robertson, G., & Benwell, G. (1999). Enabling sound marine governance. Regulating resource rights and responsibilities in offshore New Zealand. In Proceedings of the New Zealand Institute of Surveyors and FIG Commission 7th Conference and Annual General Assembly, October 9–15.
Hopkins, T., & Wallerstein, I. (1994). Commodity chains: Construct and research. In G. Gereffi & M. Korzeniewicz (Eds.), *Commodity chains and global capitalism* (pp. 17–19). Westport, CT: Praeger.
Hosseus, D., & Pal, L. A. (1997). Anatomy of a policy area: The case of shipping. *Canadian Public Policy, 23*(4), 399–415.
Imai, K. (1989). *Potential of information technology and economic growth in Japan.* Paris: OECD.
Imai, K., & Baba, Y. (1989). *Systemic information and cross-border networks.* Paris: OECD.
Jachtenfuchs, M. (1995). Theoretical perspectives on European governance. *European Law Journal, 1*(2), 115–133.

Jachtenfuchs, M. (2001). The governance approach to European integration. *Journal of Common Market Studies, 39*(2), 245–264.
James, W. (1909). *A pluralistic universe*. Cambridge, MA: Harvard University Press.
Jarillo, J. C. (1988). On strategic networks. *Strategic Management Journal, 9*(1), 31–41.
Jencks, C. (1992b). The post-modern agenda. In C. Jencks (Ed.), *The post-modern reader* (pp. 10–39). London: Academy Editions, (1992a).
Jessop, B. (1997). Capitalism and its future: remarks on regulation, government and governance, *Review of International Political Economy, 4*(3), 561–581.
Jessop, B. (1998). The rise of governance and the risk of failure: the case of economic development. *International Social Science Journal, 50*(155), 29–45.
Jessop, B. (2003a). The future of the state in an era of globalization. *International Politics and Society, 3*, 30–46.
Jessop, B. (2003b). *Governance and metagovernance: On reflexivity, requisite variety and requisite irony*. Lancaster: Department of Sociology, University of Lancaster.
Jessop, B. (2004). Multi-level governance and multi-level metagovernance. Changes in the EU as integral moments in the transformation and reorientation of contemporary statehood. In I. Bache & M. Flinders (Eds.), *Multi-level governance* (pp. 49–74). Oxford: Oxford University Press.
Johanson, J., & Mattson, L. G. (1987). Interorganizational relations in industrial systems—a network approach compared with the transaction cost approach. *International Studies of Management and Organization, 17*, 34–48.
Jones, C. (1975). *Clean air*. Pittsburgh, PA: University of Pittsburgh Press.
Jones, C. O. (1979). American politics and the organization of energy decision-making. *Annual Review of Energy, 4*, 99–121.
Jones, C., Hesterly, W. S., & Borgatti, S. P. (1997). A general theory of network governance: Exchange conditions and social mechanisms. *Academy of Management Review, 4*, 911–945.
Jordan, G. (1981). Iron triangles, woolly corporatism and elastic nets: Images of the policy process. *Journal of Public Policy, 1*, 95–123.
Jordan, A. G. (1990a). Sub-governments, policy communities and networks. Refilling the old bottles? *Journal of Theoretical Politics, 2*, 319–338.
Jordan, G. (1990b). Sub-governments, policy communities and networks. *Journal of Theoretical Politics, 2*(3), 319–338.
Jordan, A. G. (2001). The European Union: An evolving system of multi-level governance…or government? *Policy and Politics, 29*(2), 193–208.
Jordan, G., & Schubert, K. (1992). A preliminary ordering of policy network issues. *European Journal of Political Research, 21*, 7–27.
Judge, D., Stoker, G., & Wolman, H. (1995). Urban politics and theory: An introduction. In D. Judge, G. Stoker & H. Wolman (Eds.), *Theories of urban politics* (pp. 1–12). London: Sage.
Kassim, H. (1994). Policy networks, networks and European Union policy-making: A sceptical view. *West European Politics, 17*(4), 15–27.
Kazancigil, A. (1998). Governance and science: Market-like modes of managing society and producing knowledge. *International Social Science Journal, 50*(155), 69–79.
Keating, M., & Hooghe, L. (1996). By-passing the nation-state? In J. J. Richardson (Ed.), *European Union: Power and policy-making* (pp. 216–229).
Kenis, P., & Schneider, V. (1989). *Policy networks as an analytical tool for policy analysis*. Max Planck Institut Conference, December 4–5, Koln.
Kenis, P., & Schneider, V. (1991). Policy network and policy analysis; scrutinizing a new analytical toolbox. In B. Marin & R. Mayntz (Eds.), *Policy networks: Empirical evidence and theoretical considerations*. Frankfurt am Main: Campus Verlag.
Kersbergen, K. V., & Waarden, F. V. (2004). 'Governance' as a bridge between disciplines; cross-disciplinary inspiration regarding shifts in governance and problems of governability, accountability and legitimacy. *European Journal of Political Research, 43*, 143–171.
Kettl, D. F. (2002). *The transformation of governance: Globalization, devolution and the role of government*. Washington, DC: Brookings Institution.

Kickert, W. J. M. (1993). Complexity, governance and dynamics. In J. Kooiman (Ed.), *Modern governance* (pp. 191–204). London: Sage.

Kickert, W. J. M., Klijn, E. H., & Koppenjan, J. F. M. (Eds.). (1997). *Managing complex networks*. London: Sage.

King, A. (Ed.). (1978). *The new American political system*. Washington, DC: American Enterprize Institute.

King, R., & Kendall, G. (2004). *The state, democracy and globalization*. Basingstoke: Palgrave Macmillan.

Kingdon, J. (1984). *Agendas alternatives and public policies*. Boston, MA: Little Brown and Co.

Kjaer, A. M. (2004). *Governance*. Cambridge: Polity Press.

Klijn, E. H., & Koppenjan, J. F. M. (2000). Public management and policy networks. *Public Management, 2*(2), 135–158.

Knoke, D. (Ed.). (1990). *Political networks: The structural perspective*. Cambridge: Cambridge University Press.

Kohler-Koch, B., & Eising, R. (Eds.). (1999). *The transformation of governance in the European Union*. London: Routledge.

Koivurova, T. (2009). A note on the European Union's integrated maritime policy. *Ocean Development and International Law, 40*, 171–183.

Kooiman, T. (1993a). Social and political governance. In T. Kooiman (Ed.), *Modern governance*. London: Sage Publications.

Kooiman, T. (Ed.). (1993b). *Modern governance*. London: Sage Publications.

Kooiman, J. (2003). *Governing as governance*. London: Sage.

Kovats, L. (2006). How flag states lost the plot over shipping's governance. Does a ship need a sovereign? *Maritime Policy and Management, 33*(1), 75–81.

Krahmann, E. (2003). National, regional and global governance: One phenomenon or many? *Global Governance, 9*, 323–346.

Krasner, S. D. (1984). Approaches to the state: Alternative conceptions and historical dynamics. *Comparative Politics, 16*(2), 223–246.

Kreiner, K., & Schulz, M. (1993). Informal collaboration in R & D: The formation of networks across organizations. *Organization Studies, 14*, 189–209.

Larson, A. (1992). Network dyads in entrepreneurial settings: A study of the governance of exchange relationships. *Administrative Science Quarterly, 37*, 76–104.

Lash, S., & Urry, J. (1987). *The end of organized capitalism*. Cambridge: Polity Press.

Laski, H. J. (1917). *Studies in the problem of sovereignty*. New Haven, CT: Yale University Press.

Laski, H. J. (1919). *Authority in the modern state*. New Haven, CT: Yale University Press.

Lasswell, H. (1958). *Politics. Who gets what, when and how*. New York: Meridian Books.

Latour, B. (1993). *We have never been modern*. Hemel Hempstead: Harvester Wheatsheaf.

Laumann, E. O., & Knoke, D. (1987). *The organizational state: Social choice in national policy domains*. Madison, WI: University of Wisconsin Press.

Law, J. (1994). *Organising modernity*. Oxford: Blackwell.

Liebeskind, J. P., Oliver, A. L., Zucker, L., & Brewer, M. (1996). Social networks, learning and flexibility: Sourcing scientific knowledge in new biotechnology firms. *Organization Science, 7*, 428–443.

Lindblom, C. E. (1959). The science of muddling through. *Public Administration Review, 19*, 78–88.

Lindblom, C. E. (1977). *Politics and markets*. New York: Basic Books.

Lipietz, A. (1987). *Mirages and miracles: The crisis of global Fordism*. London: Verso.

Lipietz, A. (1988). Reflection on a tale; the Marxist foundations of the concepts of regulation and accumulation. *Studies in Political Economy, 26*, 7–36.

Lloyd's List. (2008). *Italy's maritime coalition in governance plea*, March 19.

Lowi, T. R. (1979). *The end of liberalism*. New York: Norton.

Lowndes, V., & Skelcher, C. (1998). The dynamics of multi-organizational partnerships: an analysis of changing modes of governance. *Public Administration, 76*(2), 313–334.

Maass, A. (1950). Congress and water resources. *American Political Science Review, 44*(3), 756–793.
Maass, A. (1951). *Muddy waters: The army engineers and the nation's rivers.* Boston, MA: Harvard University Press.
MacLeod, G. (1997). Institutional thickness and industrial governance in Lowland Scotland. *Area, 29,* 299–311.
MacLeod, G., & Goodwin, M. (1999). Space, scale and state strategy: Rethinking urban and regional governance. *Progress in Human Geography, 23*(4), 503–527.
Maitland, F. W. (1911). *Collected papers.* Cambridge: Cambridge University Press.
Malpas, J., & Wickham, G. (1995). Governance and failure: On the limits of sociology. *Australian and New Zealand Journal of Sociology, 31*(3), 37–50.
Marin, B., & Mayntz, R. (Eds.), (1991). *Policy networks: Empirical evidence and theoretical considerations.* London: Free Press.
Mariotti, S., & Cainarca, G. C. (1986). The evolution of transaction governance in the textile-clothing industry. *Journal of Economic Behavior & Organization, 7,* 354–374.
Marks, G., & Hooghe, L. (2004). Contrasting visions of multi-level governance. In I. Bache & M. Flinders (Eds.), *Multilevel governance* (pp. 15–30). Oxford: Oxford University Press.
Marks, G., Hooghe, L., & Blank, K. (1996). European integration from the 1980s: State-centric v. multi-level governance. *Journal of Common Market Studies, 34*(3), 342–378.
Marsh, D. (1983). *Pressure politics.* London: Junction Books.
Marsh, D. (1995). State theory and the policy network model, Paper 102, University of Strathclyde Papers on Government and Politics, Glasgow.
Marsh, D., & Rhodes, R. A. W. (1992). *Policy networks in British government.* London: Clarendon Press.
Marsh, D., & Smith, M. (2000). Understanding policy networks: Towards a dialectical approach. *Political Studies, 48,* 4–21.
Martin, R. (2004). Editorial: Geography: Making a difference in a globalizing world. *Transactions of the Institute of British Geographers NS, 29,* 147–150.
Mathews, J. (2002). *Dragon multinational.* Oxford: Oxford University Press.
Mayntz, R. (1976). Conceptual models of organizational decision-making and their application to the policy process. In G. Hofstede & S. Kassem (Eds.), *European contribution to organization theory* (pp. 114–125). Amsterdam: Van Gorcum.
Mayntz, R. (2004). Governance im modernen Staat. In A. Benz (Ed.), *Governance—regieren in komplexen regelsystemen* (pp. 65–76). Wiesbaden: VS Verlag fur Sozialwissenschaften.
Mayntz, R., & Scharpf, F. W. (1995). Steuerung und Selbstorganization in staatsnahen Sektoren. In R. Mayntz & F. W. Scharpf (Eds.), *Gesellschaftliche Selbstregelung und politische Steuerung* (pp. 9–38). Frankfurt am Main: Campus.
Mazey, S. (2000). Introduction: Integrating gender—intellectual and 'real world' mainstreaming. *Journal of European Public Policy, 7,* 333–345.
Mazey, S., & Richardson, J. (Eds.). (1993). *Lobbying in the European Community.* Oxford: Oxford University Press.
Mazmanian, D. A., & Sabatier, P. A. (1980). A multivariate model of public policy-making. *American Journal of Political Science, 24*(3), 439–468.
McLennan, G. (1995). *Pluralism.* Buckingham: Open University Press.
Merrien, F.-X. (1998). Governance and modern welfare states. *International Social Science Journal, 50*(155), 57–67.
Meyer, J. W., Boli, J., Thomas, G. M., & Ramirez, F. O. (1997). World society and the nation-state. *American Journal of Sociology, 1,* 144–181.
Michels, R. (1911). *Political parties.* New York: Dover Publications.
Mikkelsen, M. (2006). Policy network analysis as a strategic tool for the voluntary sector. *Policy Studies, 27*(1), 17–26.
Miles, R. E., & Snow, C. C. (1986). Organizations: new concepts for new forms. *California Management Review, 28*(3), 62–73.

Miles, R. E., & Snow, C. C. (1992). Causes of failures in network organizations. *California Management Review, 34*(4), 53–72.

Milward, H. B., & Provan, K. G. (2000). Governing the hollow state. *Journal of Public Administration Research and Theory, 10*(2), 359–379.

Milward, H. B., & Wamsley, G. (1984). Policy subsystems, networks and the tools of public management. In R. Eyestone (Ed.), *Public policy formation and implementation* (pp. 3–25). New York: JAI Press.

Milward, H. B., & Wamsley, G. (1985). Policy subsystems, networks and the tools of public management. In K. Hanf & Th. A. J. Toonen (Eds.), *Policy implementation in federal and unitary systems*. Dordrecht: Martinus Nijhoff.

Morgan, G. (1997). *Images of organization*. Thousand Oaks, CA: Sage.

Nakamura, R. T. (1987). The textbook policy process and implementation research. *Policy Studies Review, 7*(1), 142–154.

Newman, J. (2001). *Modernising governance*. London: Sage.

Newman, P., & Thornley, A. (1997). Fragmentation and centralization in the governance of London. *Urban Studies, 24*, 867–888.

Newton, K. (2001). *The politics of the new Europe*. Harlow: Addison Wesley Longman.

Nielsen, L. D. (2001). Introduction. In L. D. Nielsen & H. H. Oldrup (Eds.), *Mobility and transport* (pp. 9–11). Aalborg: Transportradet, Aalborg University, Department of Development and Planning.

Nielsen, T. W. (2007). *Metagovernance in the global compact—regulation of a global governance network*. Center for Democratic Network Governance, Roskilde, Working Paper, p. 2.

Nohira, N. (1992). Is a network perspective a useful way of studying organizations? In N. Nohira & R. G. Eccles (Eds.), *Networks and organizations: Structure form and action* (pp. 1–22). Boston, MA: Harvard Business School Press.

Nohira, N., & Goshal, S. (1997). *The differentiated network: Organizing multinational corporations for value creation*. San Fransisco, CA: Jossey-Bass.

Nordlinger, E. (1981). *On the autonomy of the democratic state*. Cambridge, MA: Harvard University Press.

Notteboom, T. E., & Winkelmans, W. (2001a). Reassessing public sector involvement in European seaports. *International Journal of Maritime Economics, 3*(2), 242–259.

Notteboom, T. E., & Winkelmans, W. (2001b). Structural changes in logistics: How will port authorities face the challenge? *Maritime Policy and Management, 28*(1), 71–89.

O'Toole, L. J., Jr. (1997). Treating networks seriously: Practical and research based agendas in public administration. *Public Administration Review, 57*(1), 45–52.

Oates, W. E. (1999). An essay on fiscal federalism. *Journal of Economic Literature, 37*, 1120–1149.

Oates, W. E. (2002). Fiscal and regulatory competition: theory and evidence. *Perspektiven der Wirtschaftspolitik, 3*, 377–390.

Obando-Rojas, B., Welsh, I., Bloor, M., Lane, T., Bodigannavar, V., & Maguire, M. (2004). The political economy of fraud in a globalised industry. *Sociological Review, 52*(3), 295–313.

Orr, M., & Stoker, G. (1994). Urban regimes and leadership in Detroit. *Urban Affairs Quarterly, 30*(1), 48–73.

Ouchi, W. G. (1980). Markets, bureaucracies and clans. *Administrative Science Quarterly, 25*, 129–141.

Paquet, G. (1999). *Governance through social learning*. Ottawa: University of Ottawa Press.

Peck, J. (2005). Economic sociologies in space. *Economic Geography, 81*(2), 129–175.

Perepelkin, M., Knapp, S., Perepelkin, G., & de Pooter, M. (2010). An improved methodology to measure flag performance for the shipping industry. *Marine Policy, 34*, 395–405.

Perraton, J., & Wells, P. (2004). Multi-level governance and economic policy. In I. Bache & M. Flinders (Eds.), *Multi-level governance* (pp. 179–194). Oxford: Oxford University Press.

Perrow, C. (1986). *Complex organizations: A critical essay*. New York: Random House.

Peters, B. G. (1986). *American public policy*. Basingstoke: Macmillan.

Peters, B. G. (1998). Managing horizontal government: The politics of coordination. *Public Administration, 76*, 295–311.

Peterson, M. A. (1993). Political influence in the 1990s: From iron triangles to policy networks. *Journal of Health Politics, Policy and Law, 18*(2), 395–438.

Peterson, J. (1994). Policy networks and governance in the European Union. In P. Dunleavy & J. Stanyer (Eds.), *Contemporary political studies* (Vol. 1). Proceedings of the Annual Conference of the Political Studies Association, Swansea.

Petit, P., & Soete, L. (1999). Globalization in search of a future. *International Social Science Journal, 160*, 165–181.

Picciotto, S. (1997). Networks in international economic integration. *Northwestern Journal of International Law and Business, 17*(2/3), 1014–1056.

Pierre, J., & Peters, R. G. (2000). *Governance, politics and the state.* London: Macmillan.

Pierre, J., & Stoker, G. (2002). Towards multi-level governance. In P. Dunleavy, A. Gamble, R. Heffernan, & I. Holliday (Eds.), *British politics 6.* Basingstoke: Palgrave.

Piore, M. J., & Sabel, C. F. (1984). *The second industrial divide: Possibilities for prosperity.* New York: Basic Books.

Plasman, C. (2008). Implementing marine spatial planning: A policy perspective. *Marine Policy, 32*, 811–815.

Podolny, J. M., & Page, K. L. (1998). Network forms of organization. *Annual Review of Sociology, 24*, 57–76.

Powell, W. P. (1990). Neither market nor hierarchy: network forms of organization. *Research in Organizational Behavior, 12*, 295–336.

Powell, W. W., & Smith-Doerr, L. (1994). Networks and economic life. In N. J. Smelser & R. Swedberg (Eds.), *The handbook of economic sociology* (pp. 368–402). Princeton, NJ: Princeton University Press.

Pressman, J., & Wildavsky, A. (1973). *Policy analysis implementation.* Berkeley, CA: University of California Press.

Propper, I. M. A. M. (1996). Success and failure in the management of policy networks. *Beleidswetenschap, 10*(4), 345–365.

Provan, K. G., & Milward, H. B. (1995). A preliminary theory of interorganizational network effectiveness; a comparative study of four community mental health systems. *Administration Science Quarterly, 40*, 1–33.

Puchala, D. J. (1972). Of blind men, elephants and international integration. *Journal of Common Market Studies, 10*(3), 267–284.

Putnam, R. D. (1988). Diplomacy and domestic politics: The logic of two-level games. *International Organization, 42*(3), 427–460.

Putnam, R. D. (1993). *What makes democracy work? Civic infrastructure* (pp. 101–7). Spring.

Raaymakers, S. (2003). *Maritime transport and high seas governance—regulation, risks and the IMO regime.* International Workshop on Governance of High Seas Biodiversity Conservation, June 17–20, Cairns.

Rabach, E., & Kim, E. M. (1994). Where is the chain in commodity chains? The service sector nexus. In G. Gereffi & M. Korzeniewicz (Eds.), *Commodity chains and global capitalism* (pp. 123–161). Westport, CT: Praeger.

Radaelli, C. M. (2003). The Europeanization of public policy. In K. Featherstone & C. M. Radaelli (Eds.), *The politics of Europeanization* (pp. 27–56). Oxford: Oxford University Press.

Ramachandran, V., Rueda-Sabater, E. J., & Kraft, R. (2009). Rethinking fundamental principles of global governance: How to represent states and populations in multilateral institutions. *Governance, 22*(3), 341–351.

Rhodes, R. A. W. (1986). *The national world of local government.* London: Allen and Unwin.

Rhodes, R. A. W. (1988). *Beyond Westminster and Whitehall: The subsectoral governments of Britain.* London: Unwin Hyman.

Rhodes, R. A. W. (1990). Policy networks, a British perspective. *Journal of Theoretical Politics, 2*(3), 293–317.

Rhodes, R. A. W. (1996). The new governance: Governing without government. *Political Studies, 44*(4), 652–667.

Rhodes, R. A. W. (1997). *Understanding governance. Policy networks, governance, reflexivity and accountability*. Buckingham: Open University Press.

Rhodes, R. A. W. (1999). Foreword. In G. Stoker (Ed.), *The new management of British local level governance*. Basingstoke: Palgrave Macmillan.

Rhodes, R. A. W. (2000). Governance and public administration. In J. Pierre (Ed.), *Debating governance authority: Steering and democracy* (pp. 54–90). Oxford: Oxford University Press.

Rhodes, R. A. W., & Marsh, D. (1992). New directions in the study of policy networks. *European Journal of Political Research, 21*, 181–205.

Richardson, J. J., & Jordan, G. (1979). *Governing under pressure*. Oxford: Martin Robertson.

Ring, P. S., & Van de Ven, A. H. (1992). Structuring cooperative relationships between organizations. *Strategic Management Journal, 13*, 482–498.

Ripley, R. B., & Franklin, G. (1987). *Congress, the bureaucracy and public policy*. Homewood, IL: Dorsey.

Robbins, S. P. (1978). Conflict management and conflict resolution are not synonymous terms. *California Management Review, 21*, 67–75.

Rodrigue, J.-P., Comtois, C., & Slack, B. (1997). Transportation and spatial cycles: Evidence from maritime systems. *Journal of Transport Geography, 5*(2), 87–98.

Ronit, K., & Schneider, V. (1999). Global governance through private organizations. *Governance, 12*(3), 243–266.

Rosenau, J. N. (1992). *Post-modernism and the social sciences: Insights, inroads and intrusions*. Princeton, NJ: Princeton University Press.

Rosenau, J. N. (1997). The complexities and contradictions of globalization. *Current History, 96*, 360–364.

Rosenau, J. N. (2000). *The governance of fragmegration: Neither a world republic nor a global interstate system*. Quebec: Congress of the International Political Sciences Association.

Rosenau, J. N. (2003). Globalization and governance: Bleak prospects for sustainability. *Internationale Politik und Gesellschaft, 3*, 11–29.

Ross, N. S. (1958). Organized labour and management: The UK. In E. M. Hugh-Jones (Ed.), *Human relations and modern management*. North Holland: Elsevier.

Ross, N. S. (1969). *Constructive conflict*. Edinburgh: Oliver and Boyd.

Ruggie, J. G. (1993). Territoriality and beyond: Problematizing modernity in international relations. *International Organization, 47*(1), 139–174.

Russell, B. (1959). *My philosophical development*. London: George Allen and Unwin.

Russell, B. (1972). The philosophy of logical atomism. In D. Pears (Ed.), *Russell's logical atomism* (pp. 54–61). London: Fontana.

Sabatier, P. A. (1988). An advocacy coalition framework of policy change and the role of policy-oriented learning therein. *Policy Sciences, 21*, 129–168.

Sabatier, P. A. (1991). Toward better theories of the policy process. *Political Science and Politics, 24*(2), 147–156.

Sabel, C. (1989). The re-emergence of regional economies. In P. Hirst & J. Zeitlin (Eds.), *Reversing industrial design?* (pp. 17–70). Leamington Spa: Berg.

Sack, R. D. (1997). *Homo geographicus: A framework for action, awareness and moral concern*. Baltimore, MD: John Hopkins University Press.

Salamon, L. M. (2002). *The tools of government: A guide to the new governance*. Oxford: Oxford University Press.

Salancik, G. R. (1995). Wanted: A good network theory of organization. *Administration Science Quarterly, 40*, 345–349.

Savoie, D. J. (1999). *Governing from the centre: The concentration of power in Canada*. Toronto: University of Toronto Press.

Scharpf, F. W. (Ed.). (1993). *Games in hierarchies and networks. Analytical and empirical approaches to the study of governance institutions*. Frankfurt-am-Main: Campus Publications.

Scharpf, F. W. (1994). Community and autonomy: Multi-level policy-making in the European union. *Journal of European Public Policy, 1*(2), 219–242.

Scharpf, F. W. (2001). Notes toward a theory of multilevel governing in Europe. *Scandinavian Political Studies, 24*(1), 1–26.

Schattschneider, E. E. (1960). *The semi-sovereign people*. New York: Holt, Rinehart and Winston.

Schmitter, P. (1979). Still the century of corporatism. In P. Schmitter & G. Lehmbruch (Eds.), *Trends towards corporatist intermediation* (pp. 7–52). London: Sage.

Schmitter, P. (1996). Imagining the future of the Euro-Polity with the help of new concepts. In G. Marks, F. W. Scharpf, P. C. Schmitter, & W. Streeck (Eds.), *Governance in the European union* (pp. 121–150). London: Sage.

Schmitter, P., & Lehmbruch, G. (Eds.). (1979). *Trends towards corporatist intermediation*. London: Sage.

Schneider, V. (1992). The structure of policy networks. *European Journal of Political Research, 21*, 109–129.

Scholzman, K., & Tierney, J. T. (1986). *Governance organized interests and American democracy*. New York: Harper and Row.

Schout, A., & Jordan, A. (2005). Coordinated European governance; self-organizing or centrally steered. *Public Administration, 83*(1), 201–220.

Seidman, S. (1994). *Contested knowledge: Social theory in the postmodern era*. Oxford: Blackwell.

Sharpe, L. J. (1984). National and subnational government and coordination. In F. X. Kaufmann, V. Ostrom, & G. Majone (Eds.), *Guidance control and evaluation in the public sector*. de Gruyter: Berlin.

Silverstein, H. B. (1978). *Superships and nation-states: The transnational politics of the intergovernmental maritime consultative organization*. Colorado, CO: Westview Press.

Skelcher, C. (2005). Jurisdictional integrity, polycentrism and the design of democratic government. *Governance, 18*(1), 89–110.

Skelcher, C., Mathur, N., & Smith, M. (2004). *Negotiating the institutional void*. Political Studies Association Annual Conference, Lincoln.

Slack, B., & Fremont, A. (2005). Transformation of port terminal operations: From the local to the global. *Transport Reviews, 25*(1), 117–130.

Sletmo, G. K. (2002a). The rise and fall of national shipping policies. In C. Grammenos (Ed.), *The handbook of maritime economics and business* (pp. 471–494). London: Lloyd's of London Press.

Sletmo, G. K. (2002b). *National shipping policy and global markets: A retrospective for the future*. International Association of Maritime Economists Annual Conference (IAME), Panama City, November 13–15.

Smith, M. J. (1993). *Pressure power and policy, state autonomy and policy networks in britain and the United states*. New York: Harvester Wheatsheaf.

Smith, N. (1997). Antinomies of space and nature in Henri Lefebvre's the production of space. *Philosophy and Geography, 2*, 50–51.

Smouts, M. (1998). The proper use of governance in international relations. *International Social Science Journal, 50*(155), 81–89.

Snow, C. C., Miles, R. E., & Coleman, H. J., Jr. (1992). Managing 21st century network organizations. *Organizational Dynamics, 20*(3), 5–20.

Somerville, P. (2004). *Governance and democratic transformation*. Political Studies Association Annual Conference, Lincoln.

Sørensen, E. (2006). Metagovernance. The changing role of politicians in processes of democratic governance. *American Review of Public Administration, 36*(1), 98–114.

Stalder, F. (2006). *Manual Castells*. Cambridge: Polity Press.

Stark, D., & Bruszt, L. (1995). *Network properties of assets and liabilities: Inter-enterprize ownership networks in Hungary and the Czech republic, Working Paper on Transitions from State Socialism, Einaudi Center for International Studies*. Ithaca, NY: Cornell University.

Stewart, J. (1996). A dogma of our times—the separation of policy-making and implementation. *Public Money and Management, 16*(3), 33–40.

Stewart, J., & Stoker, G. (1988). *From local administration to community, government.* Fabian Research Series 351, London: The Fabian Society.

Stoker, G. (1998). Governance as theory: Five propositions. *International Social Science Journal, 50*(155), 17–28.

Stoker, G. (Ed.). (1999). *The new management of British local level governance.* Basingstoke: Palgrave Macmillan.

Stoker, G. (2004). *Transforming local governance: From Thatcherism to new labour.* Basingstoke: Palgrave.

Streeck, W., & Schmitter, P. C. (1985). Community, market, state and associations? The prospective contribution of interest governance to social order. In W. Streeck & P. C. Schmitter (Eds.), *Private interest government. Beyond market and state* (pp. 1–29). London: Sage.

Sturmey, S. G. (1975). *Shipping economics: Collected papers.* London: Macmillan Press.

Sugden, R. (1991). Rational choice: A survey of contributions from economics and philosophy. *Economic Journal, 101*, 751–785.

Sunley, P. (2008). Relational economic geography: A partial understanding or a new paradigm? *Economic Geography, 84*(1), 1–26.

Sutherland, M., & Nichols, S. (2006). Issues in the governance of marine spaces. In M. Sutherland (Ed.), *Administering marine spaces: International issues* (pp. 6–20). Copenhagen: International Federation of Surveyors.

Szczerski, K. (2004). *The EU multi-level governance in post-communist countries—challenge in governability for the new member states.* 12th NISPAcee Annual Conference, Vilnius, Lithuania, May 13–15.

Taylor, P. J. (1994). The state as container: Territoriality in the modern world system. *Progress in Human Geography, 18*(2), 151–162.

Termeer, K. (2008). *Third generation governance.* Dies Natalis Symposium, Netherlands: Wageningen University.

Thomas, K. W. (1976). Conflict and conflict management. In M. D. Dunnette (Ed.), *Handbook of organizational and industrial psychology* (pp. 889–935). Chicago, IL: Rand McNally.

Thompson, G. F. (2003). *Between markets and hierarchy.* Oxford: Oxford University Press.

Thompson, G. (2005). *Networks and public management.* International Workshop on New Developments in Institutional Theory and the Analysis of Institutional Changes in Capitalism, Roskilde: Roskilde University.

Thorelli, H. B. (1986). Networks: Between markets and hierarchies. *Strategic Management Journal, 7*, 37–51.

Thrift, N. J. (1996). *Spatial formations.* London: Sage.

Tickell, A., & Peck, J. A. (1992). Accumulation, regulation and the geographies of post-fordism; missing links in regulationist research. *Progress in Human Geography, 16*(2), 190–218.

Tradewinds, (2011). *It's time for new ways of thinking.* October 14.

Treib, O., Bahr, H., & Falkner, G. (2007). Modes of governance; towards a conceptual clarification. *Journal of European Public Policy, 14*(1), 1–20.

Truman, D. (1971). *The governmental process.* New York: Knopf.

Turke, R. (2008). *Governance. Systemic foundation and framework.* Heidelberg: Physica-Verlag.

Valaskakis, K. (1999). Globalization as theatre. *International Social Science Journal, 160*, 153–164.

Vallega, A. (2001). Ocean governance in a post-modern society—a geographical perspective. *Marine Policy, 25*, 399–414.

Van de Loo, B., & Van de Velde, S. (2003). *Key success factors for positioning small-island sea ports for competitive advantage.* Rotterdam: Faculteit der Bedrifskunde/Rotterdam School of Management, Erasmus University.

Van Leeuwen, J. (2008). *Policy change and spheres of authority in the global environmental governance of shipping in the North Sea*. International Studies Association's Annual Conference, San Francisco, CA.

Van Leeuwen, J., & Bruyninckx, H. (2006). The *use of governance typologies in the analysis of the global environmental governance of the dumping of waste during shipping. maritime risks and vulnerability*. 3rd International Conference on the Centre for Maritime Research 'People and the Sea III: New Directions in Coastal and Maritime Studies', Amsterdam, July 7–9 2005. Amsterdam, Netherlands: Centre for Maritime Research.

Van Tatenhove, J. (2008). *Innovative forms of marine governance: A reflection.* www.dies.wur.nl

Van Tatenhove, J., Mak, J., & Liefferink, D. (2006). The inter-play between formal and informal practices. *Perspectives on European Politics and Society, 7*(1), 8–24.

Van Waarden, F. (1992). Dimensions and types of policy networks. *European Journal of Political Research, 21*, 29–52.

Verhoeven, P. (2009). European ports policy: Meeting contemporary governance challenges. *Maritime Policy and Management, 36*(1), 79–101.

Wakefield, J. (2010). Undermining the integrated maritime policy. *Marine Policy, 60*(3), 323–333.

Wang, J. J., & Olivier, D. (2003). *Port governance and port-city relationships in China*. Research Seminar: Maritime Transport Globalization, Regional Integration and Territorial Development, Le Havre, France, June 3–5.

Wang, J. J., & Slack, B. (2002). *Port governance in china: A case study of Shanghai*. Occasional Paper Series, Paper No. 9, The Centre for China Urban and Regional Studies, Hong Kong: Hong Kong Baptist University.

Wang, J. J., Ng, A. K., & Olivier, D. (2004). Port governance in China: A review of policies in an era of internationalizing port management practices. *Transport Policy, 11*, 237–250.

Ward, J. (1911). *The realm of ends, or pluralism and theism*. Cambridge: Cambridge University Press.

Wegge, N. (2011). Small state, maritime great power? Norway's strategies for influencing the maritime policy of the European union. *Marine Policy, 35*, 335–342.

Weingast, B. (1995). The economic role of political institutions: Market preserving federalism and economic development. *Journal of Law Economics and Organization, 11*, 1–31.

Weiss, T. G. (2000). Governance, good governance and global governance: conceptual and actual challenges. *Third World Quarterly, 21*(5), 795–814.

Wessels, W. (1997). An ever closer fusion? A dynamic macropolitical view on integration processes. *Journal of Common Market Studies, 35*, 267–299.

West, C. (1993). The new cultural politics of difference. In C. West (Ed.), *Keeping faith: Philosophy and race in America*. New York: Routledge.

Wilks, S., & Wright, M. (1987). *Comparative government industry relations*. Oxford: Oxford University Press.

Williams, D. (1987). *The specialized agencies, the United Nations: The system in crisis*. London: C. Hurst and Company.

Williams, D. (2005). *Good governance and global governance*. Political Studies Association Annual Conference, Leeds.

Williamson, O. E. (1979). Transaction cost economics: The governance of contractual relations. *Journal of Law and Economics, 22*, 233–261.

Wolfe, A. (1977). *The limits of legitimacy*. New York: Free Press.

Woods, N., & Narlikar, A. (2001). Governance and the limits of accountability: The WTO, the IMF, and the world bank. *International Social Science Journal, 170*, 569–583.

World Bank. (2000). http://www.worldbank.org/wbi/wbigf/governance.

Yannopoulos, G. N. (Ed.). (1989). *Shipping policies for an open world economy*. London: Routledge.

Yarborough, B. V., & Yarborough, R. M. (1992). *Cooperation and governance in international trade*. Princeton, NJ: Princeton University Press.

Yeung, H. W. (1998). The socio-spatial constitution of business organizations: A geographical perspective. *Organization, 5*, 101–128.

Yeung, H. W. (2005). Rethinking relational economic geography. *Transactions of the Institute of British Geographers NS, 30*, 37–51.

Yishai, Y. (1992). From an iron triangle to an iron duet? *European Journal of Political Research, 21*, 91–108.

Young, O. R. (1996). The effectiveness of international governance systems. In O. R. Young, G. J. Demko, & K. Ramakrishna (Eds.), *Global change and international governance* (p. 1027). Hannover, NH: University Press of New England.

Zacher, M. W., & Sutton, B. A. (1996). *Governing global networks*. Cambridge: Cambridge University Press.

Zhang, X. (2005). Coping with globalization through a collaborative federate mode of governance. *Policy Studies, 26*(2), 199–209.

Zukin, S., & Di Maggio, F. (Eds.). (1990). *Structures of capital: The social organization of the economy*. Cambridge: Cambridge University Press.

Zurn, M. (1993). *What has changed in Europe? The challenge of globalization and individualization*. What Has Changed? Competing Perspectives on World Order, Copenhagen, May 14–16th.

Zurn, M. (1998). *Regieren jenseits des Nationalstaates. Globalisierung und Denationalisierung als Chance*. Frankfurt: Suhrkamp.

Zurn, M. (2000). Democratic governance beyond the nation-state; the EU and other international institutions. *European Journal of International Relations, 6*(2), 183–221.

Zurn, M. (2003). Globalization and global governance; from societal to political denationalization. *European Review, 11*(3), 341–364.

Chapter 3
Hierarchies

> Arborescent systems are hierarchical systems with centers of significance and subjectification, central automata like organized memories. In the corresponding models, an element only receives information from a higher unit, and only receives a subjective affectation along pre-established paths… In a hierarchical system, an individual has only one active neighbor, his or her hierarchical superior… The channels of transmission are pre-established; the arborescent system pre-exists the individual, who is integrated into it at an allotted place (Rosenstiehl and Petitot 1974).

In this section, we will concentrate on just one of the three governance models that have been discussed; not because, it is intrinsically more robust but simply because the characteristics of maritime sector governance at present would appear to be firmly hierarchical. We need to begin by understanding the framework for governance in the maritime sector before we can go on to assess its significance for the processes of policy making and operationalization.

We have seen already that the hierarchical model of governance is one characterized by formal structures, with sets of rules to govern the processes by which decisions are taken and imposed. They are hierarchical in that the model always incorporates a process of top-down authority, commonly starting with broad policies emerging at the higher levels which then cascade down the hierarchy to lower jurisdictions. These lower levels have the responsibility of then operationalizing the broad policies derived at the upper levels. The whole system relies upon coordination between jurisdictions so that there is minimal conflict between them as the policies descend through the framework. Each jurisdiction must inherit the decisions from above with at least a modicum of good will so that the broad policies can be interpreted and operationalized effectively.

Governance hierarchies of this type commonly have at the apex an international jurisdiction which formulates broad and generic policy for the sector. Representatives from all (or more commonly a large number of) nation-states would make up this jurisdiction. Decisions would then be taken by representatives from a lower jurisdiction giving them ownership and helping in the process of coordination among the levels.

Since 1945, there has been an increase in the involvement of a second jurisdictional level—that of the supranational authority characterized by the European Union but also including The Association of Southeast Asian Nations (ASEAN) and The North American Free Trade Agreement (NAFTA) as further examples—which would have the responsibility of taking up the policy baton from the international level. Their task is to take on the policies derived at international level and then to interpret them at their own spatial/organizational level in a more specific context. This task is made more complex by the fact that they may (or may not) have representation at the international level of jurisdiction and it is possible that failure to be represented means that they perceive that they have less ownership of the international decisions taken. Supranational bodies commonly have observer status at the international jurisdiction, but there are many examples of where this is viewed as increasingly inadequate, especially as supranational bodies grow in significance. We have seen an example of this with the deterioration in the IMO and EU relationship noted earlier and the policy failings that this has implied.

The supranational jurisdiction now passes on the policy decisions inherited from above to the nation-states at the lower level for further operationalization. Once again this requires coordination and broad acceptance of what is being received by the nation-states. We have already seen how this may not always be the way that the system works, in practice, but theoretically at least, a hierarchical governance model requires such agreement for policies to be operationalized effectively. Further interaction with regional and subsequently local jurisdictions, requires similar processes of coordination and agreement before policies are fully operationalized at the industry/activity/service/individual level.

Hierarchical governance models are formal, structured, organized, and require considerable discipline and respect from participants to work something which Blaney and Inayatullah (2000, p. 30) felt can only certainly be the case when we have an 'age of empire.' Hierarchies are also inherently state-centric in that membership of the international and supranational levels is by nation-state (apart from a limited number of bureaucrats there are no representatives of international or supranational bodies, only member nations), and regional and local operationalization is largely controlled through national budgetary and political pressure. Failure to achieve adequate coordination and cooperation between jurisdictions and to retain the focused state-centric characteristics of the process might well be a serious problem for governance effectiveness.

Hierarchies have strong traditional support from Burch (2000, p. 190) who identified them as a major 'type' of rule, citing Weber who saw hierarchies as a central feature of sovereignties. Directive-rules, informing agents how to behave, can only work well in a nation-state context, however, as inter-state relations remain informal by comparison. Despite this much globally continues to follow a hierarchical pattern of command.

3.1 The Hierarchical Model

The significance of hierarchies was stressed by Dickens and Ormrod (2007, pp. 22–23) in their discussion of the relationship between cosmic and social hierarchy. Originating with the Ancient Greeks, the 'Great Chain of Being' emerged as a ranking of beings in the universe from God at the top, to worms at the bottom including a variety of angels, kings, peasants, and rocks in between (Lovejoy 1960). The essence of this hierarchy continued to be sustained through the Italian Renaissance when the hierarchical order was determined by the amount of 'spirit' and 'matter' contained at each level. The order was seen to be preordained by the Creator thus making the hierarchy a *scala naturae*, something that resounds with many current hierarchical governance assumptions relating to the nation-state and also something clearly interrupted by the growth of globalization.

Figure 3.1 shows a sixteenth century version of the Great Chain. It is only one example of many but is indicative of the significance of hierarchical thought over a considerable period of time.

The significance of hierarchies to governance has been well evidenced for some time, a clear example being the work of Epple and Zelenitz (1981) which emphasized their significance in the context of jurisdictional competition. Their long history is also noted by Braudel (1982, pp. 376–379) who identified trade hierarchies as central to commerce and its development from the eleventh century onwards. He went on to analyze the existence of social hierarchies (461–463) which mirrored those in trade.

Meanwhile, Brenner (1998) introduced the concept of hierarchies from a Marxist geographical perspective emphasizing that scales were largely perceived as relatively stable and nested arenas within which the 'production of space' took place—and by implication how most activities were ordered. Brenner, however, questioned this order suggesting that scale (and by implication jurisdictional hierarchies) was 'socially constructed rather than ontologically pre-given' and thus implicated in the constitution of social, economic, and political processes. The importance of this debate is that it implies that hierarchies are not a necessity for governance in any sector (including the maritime) but have been chosen along with the institutions and processes that make them up. Consequently, they have the potential to be questioned, altered, and possibly discarded if alternative arrangements can be shown to be more effective.

Brenner went on to analyze the relationship between capital's necessary dependence upon territory and place, and by implication its relationship to hierarchical governance which in turn has a close relationship to state centrism and the territorial implications of the state. He also questioned the implications of capital's moves toward deterritorialization and 'time-space' compression derived from the work of Harvey (1990) all of which has significant implications for the design of governance mechanisms. We shall return to all of this much later but for the moment it is important simply to note that hierarchies have been questioned in fundamental terms and that the potential for conflict between the existing

Fig. 3.1 The great chain of being. A sixteenth century example. *Source* Dickens and Ormrod (2007, p. 23)

God
Angels
Kings/Queens
Archbishops
Dukes/Duchesses
Bishops
Marquises/Marchionesses
Earls/Countesses
Viscounts/Viscountesses
Barons/Baronesses
Abbots/Deacons
Knights/Local Officials
Ladies-in-Waiting
Priests/Monks
Squires
Pages
Messengers
Merchants/Shopkeepers
Tradesmen
Yeomen Farmers
Soldiers/Town Watch
Household Servants
Tenant Farmers
Shepherds/Herders
Beggars
Actors
Thieves/Pirates
Gypsies
Animals
Birds
Worms
Plants
Rocks

state-centric hierarchical governance framework that dominates the maritime sector and the process of effective policy making and implementation is significant.

Hirst (1997, p. 4) also provided an extensive discussion of the nature of hierarchical governance—which he terms 'imperative control'. In this model, he saw governance operating through dominant agencies which exercise exclusive control over the activity in question and coordinates and directs through hierarchical transmission of orders to subordinate agencies. Citing the classic model of bureaucracy:

> subordinates receive orders and transmit evidence of compliance up the chain of command, thus ensuring whether subordinates know whether the outcomes have been attained and providing the informational conditions for continued governance.

Hierarchy is thus:

> a bounded information system, in which the conditions of control are internalised in a structure of command.

3.1 The Hierarchical Model

De Senarclens (1998, p. 92) provided a neat analysis of the hierarchy as an institutional framework while also hinting at the changes taking place that could well bring its downfall. He suggested that national governments no longer had a monopoly on legitimate power and that many other actors contribute to maintaining order and participate in social and economic regulation. He saw the:

> mechanisms of managing and controlling public affairs involve – at local, national and regional level – a complex set of bureaucratic structures, political powers ranked in a more or less rigid hierarchy, enterprises, private pressure groups and social movements.

De Alcantara (1998) provided a cogent argument for why despite the inexorable rise of globalization and its implications for the nation-state, there remains a need for a hierarchical arrangement for governance. This rests on the assumption that hierarchical governance provides an effective mechanism for creating structures of authority at the various levels of societal jurisdiction. As a result, hierarchies are indispensable in the process of managing trans-jurisdictional processes. Consequently, moving away from hierarchical governance is not easy. Rosenau (2000, p. 1) agreed in that:

> our analytical capacities are rooted in methodological territorialism, in a long-standing, virtually unconscious habit of probing problems in a broad geographical or spatial context.

Meanwhile, Borzel (2007, p. 6) provided further discussion of the role and nature of hierarchical governance characterizing hierarchical modes of coordination as incorporating 'authoritative decisions' which as Scharpf (1997) suggested, participants must obey even against their own self-interests reflecting the jurisdictional coordination that ought to apply to existing maritime governance. Borzel indicated that nonhierarchical governance can also exist requiring only voluntary compliance and it is this approach that characterizes network and competitive systems of governance. In effect:

> actors engage in processes of non-manipulative persuasion... through which they develop common interests and change their preferences accordingly... (or)... they pursue a common goal or some scarce resources of which they want to obtain as much as possible by performing better than their competitors.

These ideas are supported in earlier work by Borzel et al. (2005) who proposed a framework to illustrate how hierarchical governance neglects significant non-public actors although they did emphasize that all governance frameworks need some hierarchical injection to be effective and that the development of new governance frameworks needs to combine the characteristics of hierarchy with a wider interpretation of stakeholders and their relationships. Meanwhile, Borzel (2007) continued to see problems with the hierarchical governance structure within which the EU places itself in that it remains dominated by (national) governmental actors, therefore, neglecting many stakeholders with a justifiable interest in the policy making process, something she termed the 'shadow of hierarchy.'

Soja (2000, pp. 199–200) questioned what he saw as the reification of the nation-state as part of the hierarchy of spatial scales that have been manufactured by society. 'This nested hierarchy of scales, as integral parts of human spatiality, is not naively given but is socially constructed, as a vital part of what Henri Lefebvre described as the production of perceived, conceived and lived spaces' a view which had earlier been emphasized by Soja (1989, p. 151) where he analyzed the significance of the hierarchy of differentiated nodal locales. Consequently, anything that is socially constructed and not a 'natural given' can not only be created but also destroyed and must be as society changes and the given hierarchy become less relevant. Human action can change the privileged status of the nation-state and it is this inconsistency between society and hierarchy that creates the potential for maritime policy failure. Soja went on to emphasize that globalization has caused a 'reconstitution of spatial scales' and can be identified as such as the cause of governance problems in the maritime sector which cannot be overcome unless the new spatial constructs that have emerged are understood and recognized.

Finally, Harvey (2000, pp. 75–77) discussed in some depth the relevance of spatial scales and their hierarchical nature before going on to emphasize the significance of globalization in changing these scales and the relationships between them. Harvey saw human beings as commonly organizing life through nested hierarchies of spatial scales and it is widely accepted that different scalar views will give different results and opinions. However, significantly, Harvey saw this as a dangerous move as such scalar hierarchies appear immutable and natural which brings substantial problems when technologies, human organization and political struggle cause change elsewhere.

Hierarchies are thus a product of human affairs and are a reflection of the unevenness of geographical developments (Smith 1990, 1992). 'Plainly, the hierarchical scales at which human activities are now being organized are different from say, 30 years ago.' And globalization is a major cause of this.

The policy failures identified in the maritime sector provide good reason to look closely at hierarchies and how effective they are as a governance framework. Immanuel Kant suggested that authority is not simply a matter of power but is subject to alternation and dispersal, and that there is a greater chance of 'perpetual peace' through the dispersal of authority within and among republics than in the development of a single world republic (Kant 1795).

3.2 The Effectiveness of Hierarchies

> In effect, the geopolitics of nations that yesterday still pre-supposed the hierarchical privilege of the center over its peripheries, of the summit over the base, the 'radioconcentrism' of exchanges and horizontal communications, loses its value in the same way as does the extreme vertical densification to the benefit of an inapparent morphological configuration. The NODAL succeeds the CENTRAL in a preponderantly electronic environment, 'tele-localizing' favoring the deployment of a generalized eccentricity, endless periphery, forerunner of the overtaking of the industrial urban form, but especially of the decline of the sedentary character of the metropolis to the advantage of an obligatory *interactive confinement*, a sort of inertia of human populations for which the name of *teleconcentricism* may be proposed, while waiting for that of 'homeland' to replace that of

3.2 The Effectiveness of Hierarchies

the large suburb. The secular opposition city/country is being lost while the geomorphological uniqueness of the state is dissipating. (Paul Virilio, quoted in Der Derian 1999, p. 219).

Scales evolve relationally within tangled hierarchies and dispersed interscalar networks… and (as such) scalar hierarchies constitute mosaics not pyramids. (Brenner 2001, pp. 605–606, quoted in Stubbs 2005, p. 76)

Brenner's doubts about the validity of organized and structured hierarchies are clear and mirrored by many others. Puchala (1972) questioned the relevance of hierarchies in the context of the debate about pluralism while Ostrom (1985, p. 18, 1986) was one of the earlier commentators to suggest that hierarchies were not inevitable to produce effective policies for what she termed 'modern economic growth and stability.' She described the convention to be one of integration and centralization—two prime characteristics of hierarchies—and fragmentation and decentralization (their opposites) to be 'pathological.' While accepting that institutional hierarchies are suited for many tasks she also suggested that they had their limits. Ostrom saw that hierarchical institutional designs:

tend to eliminate redundancy from institutional arrangements. Given the fallibility of humans, however, failures are to be expected… Malfunction will occur somewhere in a system.

As Campbell (1982, p. 73) stressed:

Redundancy is a means of keeping the system running in the presence of malfunction.

To clarify at an early stage, Ostrom's claim is that hierarchies have their place but have the unfortunate characteristic of glossing over failures within the system by eliminating redundancy as this threatens the very existence of the institutions that dominate within the hierarchy. Redundancy (and by implication change and the recognition of what is needed to be changed) is needed to maintain the system (in our case that of maritime policy-making) in the presence of inevitable failure. Effective governance systems need to recognize the existence of failure and the need to accommodate redundancy. Hierarchical institutional frameworks work in entirely the opposite way. And the result is a crisis of governance and a plethora of policy failure.

Aglietta (1982, p. 21) agreed that hierarchies have value and suggested that international relations depended upon a hierarchical structure. However, these hierarchies needed to be dynamic rather than static and flexible rather than inflexible—the latter particularly referring to the need to accommodate both the nation-state and the international community.

Mayntz (1976, p. 122) analyzed the ineffectiveness of a hierarchical policy arrangement in the Federal Republic of Germany and concluded that it was unworkable. Jordan (1990, p. 324) implied considerable criticism of the traditional hierarchical, policy-making framework when considering the period from the 1930s through the war years in the USA suggesting that it:

neither accounted for enough of what happened on the governmental scene nor showed in systematic terms how the system actually operated.

Scharpf (1994a, p. 231) agreed that the member states of the EU formed a hierarchically integrated system but one in which solutions to problems could be put into effect without the agreement of all those involved. In addition, the nation-state was seen as inadequate as the main functionary within this hierarchy particularly after the completion of the internal market. Consequently, the EU was constructed hierarchically, yet could not govern hierarchically and at the same time could not rely on nation-state leadership and control.

Frey and Eichenberger (1996) presented an extensive discussion of how jurisdictions should be organized along functional lines rather than territorial ones—therefore, upsetting the basic principle of a state-centric hierarchical governance arrangement that characterizes the maritime sector. Meanwhile, McGinnis (2005) suggested that governance frameworks could function around standards set across jurisdictions rather than by them.

Crosby (1996, p. 1404) saw hierarchical structures in the same way as Kahler (1989) representing a 'top-down, non-participative process, confined to a narrow set of decision-makers.' The main problem is that those with implementation responsibility for policies do not normally participate and although the EU's move toward subsidiarity might be construed as a deliberate attempt to move implementation and participation together, it can also be seen as a further process whereby generic policy is divorced from those responsible for operationalization. Meanwhile, Morgan (1997, pp. 117–118) was sympathetic toward the hierarchical approach to governance as moves away to more flexible forms of structure would inevitably lead to a redistribution of power within an organization or society, sometimes with negative effects. Imposed hierarchies could not work, however, and need to 'emerge and change as different elements of the system take a lead in making their various contributions.'

Stewart (1996) considered that the essential relationship between policy formulation and its implementation was impossible to optimize using a 'top-down' approach—or in our terms a hierarchy—as they could not be viewed as separate processes whereby a superior level could develop policies that inferior levels would accept and implement. Meanwhile, Meyer et al. (1997, p. 145) were particularly convinced of the inadequacies of the nation-state and its relationship with hierarchical governance structures that the same nations make up. Referring to Meyer (1980):

> The almost feudal character of parcelized legal-rational sovereignty in the world has the seemingly paradoxical result of diminishing the causal importance of the organized hierarchies of power and interests celebrated in most 'realist' social science theories.

Or to put it in another way, the excessive state-centrism of society makes any hierarchical governance societal structure unworkable.

Brenner (1997) also questioned the role of the hierarchy in his discussion of Lefebvre's classic writings (1974, 1976a, b, 1977) suggesting that capitalism has consistently 'territorialized, deterritorialized, and reterritorialized' in its desperate search to accumulate more capital to the point where we have now created a 'world spatial system' around which differentiated spaces are arranged. These may

3.2 The Effectiveness of Hierarchies

(or may not) be hierarchically defined. Lefebvre (1976a), quoted in Brenner (1997, p. 145) saw capitalism as:

> a multi-layered scaffolding of intertwined, co-evolving spatial scales upon which historically specific inter-linkages between processes of capital accumulation, forms of state territorial organization and patterns of urbanization have been crystallized.

And nothing characterizes capital accumulation as a process better than the shipping industry.

Brenner (1997, p. 137) also stressed that the state has been seen as a 'neutral container within which temporal development unfolds,' but that increasing global interdependence has changed this to a 'constitutive, historically produced dimension of social practices' and as such something that can be questioned and changed. This he termed, quoting Soja (1989, 1996) a 'reassertion of space in critical social theory.'

Criticism of hierarchies did not stop them. Stoker (1998, p. 19) suggested that in terms of governance:

> there are many centers and diverse links between many agencies of government at local, regional, national and supranational levels... a complex architecture to systems.

going on to say:

> Government in the context of governance has to learn an appropriate operating code which challenges past hierarchical modes of thinking.

This implied that the strictly hierarchical approach to governance may not be always the best. Rhodes (1997, p. 1) was highly critical of the hierarchical approach to governance in an environment characterized by networks of organizations which resist central direction. He quoted Luhmann (1995, pp. 253–255) who suggested we live in a 'centreless society' and hence one that cannot be hierarchically directed. There is no longer a:

> mono-centric or unitary government; there is not one but many centers linking many levels of government – local, regional, national and supranational.

Perhaps we should add international as well. Meanwhile, de Senarclens (1998, p. 98) flagged up the role of globalization in the degeneration of the hierarchy as an institutional unit suggesting that in the context of social issues there has grown a dominant international dimension, whilst the 'political institutions to deal with (them) are not up to the task.' This may well be equally applicable to the maritime context.

Merrien (1998) supported this view seeing the nation-state as not so much at the peak of a hierarchy of stakeholders but more occupying an essential place in the development of a 'relevant action network.' This process he allied with the decline of the nation-state as a significant force as a consequence of increasing globalization—something to which we shall turn later. Meanwhile, Borzel (1998, p. 260) stressed how governments have become 'increasingly dependent upon the co-operation and joint resource mobilization of policy actors outside their hierarchical control.' She also identified that the traditional hierarchical system was 'dysfunctional' in that it always produces 'losers' who have to bear the cost of political decisions taken by those superior in the hierarchy (Scharpf 1992, 1994b). Hierarchical coordination thus becomes increasingly difficult as interactions across sectoral, organizational, and national borders become more common.

Peters (1998, p. 302) placed the discussion of hierarchies firmly in that of networks and their value to governance. He suggested that the existence of strong vertical linkages in and between any organizations makes the coordination of horizontal linkages much more difficult. In other words, a hierarchical (otherwise known as vertical) model has an inevitable detrimental effect on the horizontal networks that characterize much of commerce and society today. Horizontal agreements among players are difficult to amend or adjust by a governance system that is designed in a distinctly different way. As Peters said: 'failures of horizontal coordination can be understood through the success of vertical coordination.' The significance of this clearly rests upon how much it is agreed that modern, globalized society is network-bound, horizontally characterized, and cross functional. Peters suggested that modern society (of which the maritime sector is a significant part) is pluriform and the less pluriform the governance, the less effective it can be. The traditional maritime hierarchy is about as nonpluriform as it is possible to be.

Despite this criticism, Neyer (2003) suggested that hierarchies remain a valid part of European integration theory which are a necessity, but which can prove problematic in getting to work. Neyer went on to look at an alternative approach which he would substitute for the two governance options he saw currently available—anarchy and hierarchy, the latter of which has been preferred in the maritime sector. He chose a heterarchical approach which lay somewhere between centralized political authority (hierarchy) and decentralized (anarchy) authority. Governance interaction is undertaken both vertically (hierarchy) and horizontally (anarchy) and also cross-border and cross-level/sector.

Meanwhile, Rhodes (1997) provided an analysis of the merits of hierarchies (or 'bureaucracies' as he calls them). He interpreted such structures in the way that Weber (2004) did with unified line bureaucracy as a service delivery system suggesting that there are some conditions under which they are appropriate. These include where fragmentation produces suboptimal outcomes; where it encourages poor communications through the proliferation of agencies; and where it unifies accountability. They represent reliability, integrity, predictability, probity, cohesion, and continuity in the face of politicization and personalization. Bureaucracies provide 'hands-on control of services through hierarchical, rule-based disciplinary structures' (Rhodes 1997, p. 30).

Jachtenfuchs (1995, p. 124) commented on the failures of the hierarchy in the context of governance on a number of occasions. He suggested that the idea of governance beyond the state (for example through globalization):

> does not necessarily mean governance above the state, thus simply reconstituting the state with all its constituent elements simply on a higher institutional level. On the contrary, the idea of governance beyond the state has to stop relying on the state as the institutional form and the hierarchical center of an integrated society.

3.2 The Effectiveness of Hierarchies

Jachtenfuchs (2001, pp. 254–255) continued by questioning the effectiveness of hierarchical models of governance in the context of the existence of what he termed 'consociation.' This is the co-existence of political relations, on one hand, and the pursuit of individual interests (as opposed to the common good) on the other. He cited modern systems theory [and Luhmann (1995) in particular] and the claim that society consists of a number of subsystems that:

> largely function according to their own autonomous logic. For efficiency as well as for normative reasons, the autonomy of these sub-systems should be respected.

Hierarchical governance in this context is not 'a very promising endeavour' (Jachtenfuchs 2001), something exacerbated by the territorial subsystems of the EU (i.e. the member nation-states) that exist.

Hirst (1997) went on to use the practices of Fordist workplaces as a good example of governance through hierarchy. Fordism was a central feature of the Modernist organization. However, the conditions suited for hierarchical planning and societal control have become less common as markets increasingly refuse to accept undifferentiated output.

> Rapidly changing product mixes and an emphasis on customization and the quality (*rather than the quantity*) of products weaken the capacity of imperative control from top to bottom. (Hirst 1997, p. 5) (*author's emphasis and contribution in brackets*)

Hirst saw three major changes following from this:

- The decentralization of decision making to levels where appropriate information is available and not necessarily the jurisdiction that the formal hierarchy dictates.
- The granting of greater autonomy to the producers of goods and services.
- The development of more complex and multi-centered methods of monitoring product quality and productive performance.

Each of these features has characterized the maritime industry in recent years and questions the hierarchical governance model that remains a central feature of the sector.

Hierarchical governance was viable throughout much of the twentieth century because products and services that were demanded were relatively simple in character and requirements for delivery were fairly uniform. With the emergence of globalization, the demand for greater variety and flexibility in both products and services has made hierarchical governance an anachronism. Individuals delivering products and services need to have more knowledge of clients and discretion to make choices and thus are more difficult to control. Governance models need to adapt to this increased flexibility, individualism, and variety, and to reconcile the conflicts that emerge from the forces of localization (variety and flexibility) and globalization (quantity, predictability, and uniformity) that this implies. In fact, the pressures of hierarchical obsolescence may lead to losing the benefits of localization, flexibility, and variety which should emerge, but without the effective bureaucratic control which once kept the whole governance system stable—a consequence of what Rosenau (2000) termed 'fragmegration.' The result is chaos.

The extent of the debate about the validity of the hierarchical model in the light of globalization is notable. Cooke and Morgan (1993, p. 545) felt that hierarchical economic governance models were polarized and neglected what we now term the network model incorporating joint ventures, strategic alliances, and corporate consortia—all of which are prevalent in the shipping sector. Hierarchies were described as classically Fordist and as such were inappropriate for the Post-Fordist (and Postmodern) world. Rhodes (1997, p. 3) suggested that there was no longer a single, identifiable center to government or governance and that there were now multiple centers crossing jurisdictions. The hierarchy was thus untenable. MacLeod (1997, p. 299) was convinced that globalization had:

> mutually articulated, with a selective re-emergence of regional economies premised upon industrial clusters, districts of economic agglomeration and innovation-mediated production complexes (Porter 1990; Amin and Thrift 1992; Sabel 1994; Florida 1995). Such processes are inextricably linked to an organic reorganization of the firm beyond the hierarchical Fordist model. (Best 1990)

Mathews (1997, p. 52) identified a clear shift in power and responsibility from downwards (from the state through foreign relations, military strength, etc.) to upwards (from individuals and groups) caused by the development of information technologies, together disrupting the traditional hierarchy. Raco (1998) considered that hierarchies could be simply dismissed as an appropriate framework for governance; Martin (2004, p. 150) emphasized the local credentials of globalization, not just the top-down, international features and as such discarded the hierarchical model; Gereffi et al. (2005, p. 80) stressed that modern logistics, supply chain management and international trade all required 'intense interactions across enterprise boundaries' and flexibility in the design of associated systems. The implication is that a one size fits all hierarchy is bound to be inadequate (Fine 1998; Langlois and Robertson 1995). Hatzaras (2005, p. 5) cited Radaelli (2003, p. 30) in that governance in the contemporary era is characterized by construction, diffusion, institutionalization of informal rules, shared beliefs and norms, discourse, dialog, values, and identities each of which makes the traditional hierarchy redundant.

Brenner (1999, p. 51), along with Cox (1998, p. 1) saw a need to accommodate the process of globalization into our wider conceptions of space and suggested that we needed to rescale territoriality. Quoting Smith (1997, pp. 50–51) he provided a strong theoretical argument for the need to reexamine how we consider space in society—and by implication whether institutional hierarchies have a role left to play:

> The solidity of the geography of twentieth century capitalism at various scales has melted; habitual spatial assumptions about the world have evaporated... It is as if the world map as jig-saw puzzle has been tossed in the air these last two decades, leaving us to reconstruct a viable map of everything from bodily and local change to global identity. Under these circumstances, the taken-for-grantedness of space is impossible to sustain. Space is increasingly revealed as a richly political and social product, and putting the jig-saw puzzle back together – in practice as well as in theory – is a highly contested affair.

3.2 The Effectiveness of Hierarchies

The conventional institutional hierarchy that characterizes the governance of the maritime sector is founded on the conventions of space as they were defined before the acceleration of globalization in the past 50 years. Rosenau (2000) also suggested that the current global situation has placed questions besides the maintenance of hierarchical governance mechanisms. Parties affected by changes in (for example) international trade, commonly find themselves wandering through differing jurisdictions which require negotiation and discourse with a range of diverse authorities with differing interests and goals threatening hierarchical stability. Instead of the established institutions and responsibilities:

> diverse local, provincial, national and supranational governments, along with even more diverse… bureaucratic or nongovernmental agencies operate as steering mechanisms for regulatory systems.

Rather curiously, he confined corporations, industries, stock exchanges, international organizations, and other financial instruments to market systems and idea systems such as democracy, human rights, and environmental issues to other collectivities and publics rather than seeing them as part of a range of stakeholders which need to be integrated within a single (and quite possibly nonhierarchical) governance framework.

Peters and Pierre (2001, p. 2) questioned the significance of hierarchical ordering in inter-governmental relations and in particular, the need to think of institutional levels as vertically ordered with a requirement to work through intermediary layers. Instead, and citing a number of others to back their argument including Kohler-Koch (1996), Marks et al. (1996), Scharpf (1997), and Puchala (1999), they felt that institutional relationships could work just as well (if not better) 'directly between, say the transnational and regional levels, thus by-passing the state level.' The hierarchy would become redundant. As they went on to say:

> decentralization and European integration have jointly reshuffled institutional relationships and created a system where institutions at one level can enter into exchanges with institutions at any other level and where the nature of the exchange is characterized more by dialogue and negotiation than command and control.

We have already seen Brenner's (1998, p. 463) criticism of the hierarchical approach to the organization of space using Harvey's (1982) suggestion that one of the implications of a society that is dominated by capital is that it presents contradictions for territorial-organizational structures and in particular for the nation-state. This process of globalization which sees a redefinition of the role of the state and the territorial boundaries that are associated with it has enormous implications for a maritime governance structure that is essentially state-defined and hierarchical suggesting that new governance mechanisms are necessary to reflect the process of capital accumulation that has generated a new spatial framework. Following Harvey again, Brenner suggested that capital requires a 'spatial fix' which allows it to maximize capital accumulation that implies new scales and boundaries between individual and institutional stakeholders. This is:

secured through the coordination of social processes articulated upon multiple overlapping scales and through the 'meshing' of various 'hierarchical arrangements' such as transnational corporations, monetary regimes, legal codes, interurban networks and state regulatory institutions. (Harvey 1982, p. 423).

…which is a far cry from the neat, state-centric ordering that is implied by the existing maritime governance framework.

However, hierarchies have developed and continue to exist because as Harvey (1982, p. 422) suggested:

> The tensions between fixity and motion in the circulation of capital, between concentration and dispersal, between local commitment and global concerns, put immense strains upon the organizational capacities of capitalism. The history of capitalism has, as a consequence, been marked by continuous exploration and modification of organizational arrangements that can assuage and contain such tensions. The result has been the creation of nested hierarchical structures of organization which can link the local and particular with the achievement of abstract labor on the world stage. Crises are articulated and class and factional struggles unfold within such organizational forms while the forms themselves often require dramatic transformation in the face of crises of accumulation.

These hierarchical arrangements and their interpretation we shall discuss further in a later chapter when we look at spatial fix and its relationship to the maritime sector. For the time being, the important thing to note is that the society characterized by capital accumulation within which the maritime sector sits, is also one that has developed an hierarchical governance system to ameliorate organizational tension which at the same time has to increasingly face up to a process of 'time-space compression' (read globalization) which makes the same hierarchical arrangement irrelevant. The policy implications are not good. As Brenner (1998, p. 464) stated:

> Each geographical scale (*or hierarchy*) under capitalism must be viewed as a complex, socially contested territorial scaffolding upon which multiple overlapping forms of territorial organization converge, coalesce and interpenetrate. (*section in italics added*)

This inherent complexity suggests that as time passes and as globalization progresses, a simplified, hierarchical arrangement for governance and policy-making cannot be adequate.

Skelcher (2005) provided an interesting discussion concerning the value of jurisdictional integrity—the political and legal competence of a unit of government to operate within a spatial and functional realm—and the decline of conventional hierarchical governance alongside that of the influence of the state. This in turn has implications for jurisdictional integrity as the rise of what he saw as polycentric systems of governance removes the formalized structures inherent in the established hierarchical frameworks. With the dispersal of political authority across jurisdictions each of which has confusing and overlapping authority and interests, governance of any sector becomes more complex.

Polycentrism presents challenges as its jurisdictional integrity is lower even though its reflection of stakeholder relationships can be far more effective. The decline in hierarchical competence as globalization takes place thus, presents not only a problem for the existing governance frameworks but also for what might replace them, something we return to later. However, just because the alternative system is frighteningly complex does not mean that the existing system can be considered adequate.

The debate about the relevance of institutional hierarchies for governance continues to this day and although the hierarchical jurisdictional format centring on the nation-state retained support in some quarters for some time (see for example Slaughter [1997, pp. 183–184]) more typical are the comments of Zhang (2005, p. 207) suggesting that:

> the unitary pattern of government centred governance and traditional hierarchical public administration... needs to be replaced with a new pattern of governance alternatives to meet the challenge of economic adaptability, resilience and diversity.

Although referring specifically to the situation in China, Zhang's comments remain intensely relevant to the international context.

3.3 The European Union, Governance and Hierarchies

The emphasis of the discussion on governance in the maritime sector has focused upon the EU as a supranational jurisdiction of significance in maritime sector policy making. Consequently, we now look at discussion that has taken place looking at the role of hierarchies and governance in the EU.

Scharpf (1994a) had stressed how the EU once considered hierarchies as a way of both improving policy making and also as a means of integrating the 'problem-solving capabilities' of the member states into this process. This required increasing coupling of policy processes at all jurisdictions. However, member states are central to the process of effective policy-making of the EU (thus stressing its state-centric nature), working within the jurisdictional structure that exists—in this case a hierarchy, describing it as a multi-level system. In practice, this creates difficulties, as integration within the Single Market has gone far beyond the capacity for hierarchical coordination among nation-states. As a consequence, the EU is no longer in a position to exercise effective powers through a hierarchical structure.

Benz and Eberlein (1999) anticipated European governance moving away from a traditional hierarchical form to one characterized by 'separated, but loosely coupled arenas.' In addition, in contrast to hierarchies these would be linked by communications rather than by resource dependency and control—something that is more characteristic of governance of the maritime sector at present. Balme and Jouve (1996) term these linkages 'cognitive rather than political' This ties in well with views of the development of governance through forming 'advocacy coalitions' which significantly, are informal networks cutting across the boundaries of institutionalized arenas (Heinz 1988; Heinz and Jenkins-Smith 1988; Jenkins-Smith 1988; Sabatier 1988). Perhaps maritime governance has something to learn here?

Zurn (2000) questioned the democratic credentials of the EU and its jurisdictional relationships with member states within a hierarchical framework. His main thrust was that just because fundamental aims of good governance—for example, social welfare and security—should normally be better achieved through a hierarchical framework, the existence of the appropriate institutions is no guarantee that this will be the case. He saw no certain relationship between system effectiveness and citizen participation in the EU and therefore, concluded that an alternative to the existing governance approach was needed.

However, Marks (1993, p. 392) had already seen moves away from hierarchical governance models (characterized by strict jurisdictional discipline) occurring within the EU and evidenced by:

> A system of continuous negotiation among nested governments at several territorial tiers – supranational, national, regional, and local – as the result of a broad process of institutional creation and decisional reallocation.

This view was supported by Anderson (1996) who sensed that the traditional hierarchical arrangement in which the EU lay was no longer nested, as stakeholders were often directly members of the international bodies not through national organizations; local groups would deal directly with supranational institutions; and regional groups might deal directly with their counterparts in other nations without working through their own national representatives. Anderson's view was that hierarchies still exist but not with the same 'chain of command' which was assumed when they were established. All this implied greater complexity. The dominance of the nation-state within a hierarchical framework was no longer as significant, but they remain important players although interacting with a wide range of stakeholders from a varied assembly of associations and networks.

Hix (1998, p. 54), along with Nugent (2006, p. 556) similarly noted changes taking place:

> The EU is transforming politics and government at the European and national levels into a system of multi-level, nonhierarchical, deliberative and apolitical governance via a complex set of public/private network and quasi-autonomous agencies, which is primarily concerned with the reregulation and deregulation of the market.

Blatter (2003) also questioned the continued existence of hierarchies as a central focus for governance in the context of cross-border relationships particularly noting a move from hierarchical to network governance patterns of interaction, although rather less so in modes of interaction. He went on to stress that this implied a tightening of jurisdictional relationships rather than a loosening, but clearly also, one that suggested considerable changes in governance mechanisms that would be appropriate.

Eberlein and Kerwer (2004, pp. 123–125) discussed new governance approaches within the EU and contrasted their effectiveness with established (hierarchical) processes suggesting that they possessed neither democratic legitimacy nor effectiveness. The 'Open Method of Coordination (OMC)' emerged from the European Employment Strategy laid down in the Amsterdam Treaty of 1997 (Hodson and Maher 2001). It has four main principles:

3.3 The European Union, Governance and Hierarchies

- Fixed guidelines set for the EU with short, medium, and long-term goals.
- Quantitative and qualitative benchmarks and indicators.
- European guidelines translated into national, regional and local policies, and targets.
- Periodic monitoring, evaluation, and peer review.

They indicated that this process:

> seeks to initiate an iterative process of mutual learning on the basis of diverse national experiences with reform experiments. It avoids strict regulatory requirements and allows experiments that are adapted to local circumstances, while fostering policy improvements, and possibly policy convergence through institutionalised mutual learning processes.

Further details can be found in Hodson and Maher (2001) and Scott and Trubeck (2002). It clearly contrasted with traditional top-down and command and control-type regulation backed by 'hard-law' sanctions. However, it retained clear hierarchical foundations based in jurisdictional tradition and as such may not offer a way forward to meet current governance inadequacies. Further examples can be found in Tommel (2000), Hodson and Maher (2001), Kerwer and Teutsch (2001), Radaelli (2001), Lenschow (2002), Zeitlin and Caviedes (2002), Eberlein (2003) and Goetschy (2003) covering areas such as regional policy, the environment, taxation, transport, utility regulation, European monetary policy and social policy, but little seems to have rubbed off onto the maritime area.

Grande (2000), Hooghe and Marks (2001a, p. 11) and Eberlein and Kerwer (2004) together saw the EU as a multilevel governance system in which a large number of decision-making arenas are differentiated along both functional and territorial lines, and in which these arenas are interlinked in a nonhierarchical way, resulting in a 'dispersion of authoritative decision-making across multiple territorial levels.' In this context, what role does a hierarchical governance framework have to play?

Aalberts (2004, p. 29), with reference to Ruggie (1998) and Wallace (1999), also questioned the legitimacy of the hierarchical governance model citing it as misleading since it suggests:

> some sort of linear, pyramid-like structure of sub/supra-relations... (In contrast) The authority structures seem far more complex, flexible, cross-cutting... far more Postmodern.

He saw hierarchies as just representing one form of governance that contrasts with the anarchical vision of Waltz (1979) and which fitted into Hooghe and Marks' (2001b, 2003) multi-level definition of governance as of two types (I and II). The former included hierarchical structures associated with federalism and a passion for vertical authority; the latter was characterized by the anarchy of polycentrism.

The debate about multi-level governance as an approach to accommodate the changes generated in policy making by globalization and the wider Postmodern movement has been long and intense and wide recognition of its broader characteristics has been achieved. Jachtenfuchs (1995, pp. 122–123) provided an early commentary followed by Marks, Hooghe and Blank (1996, pp. 342–347), Rhodes (1997, p. 33) who introduced some of the drawbacks to it as an approach, and Smouts (1998, p. 83). Benz and Eberlein (1999, pp. 329–333) suggested that multi-level governance was replacing the hierarchy as an effective and meaningful governance model. Rosenau (2000, pp. 4–5) further contributed, Benz (2000, pp. 21–22) criticized the approach as ill-defined, whilst Jordan (2001, pp. 194–196, 200–201) discussed the application of the approach to the governance of the EU. Peters and Pierre (2001, pp. 131–132, 2004) assessed the moves that they suggested have occurred from nation-state to multi-level governance. Hooghe and Marks (2001a, 2003, pp. 233–235) remain the greatest proponents. More recent contributions come from Neyer (2003, p. 689) who considered multi-level governance in the EU as heterarchical, Aalberts (2004, pp. 23–30) who looked at it as acting 'beyond the state,' McGinnis (2005, p. 6) who considered it in the light of moves toward subsidiarity in the EU, Eberlein and Kerwer (2004, p. 128), Szczerski (2004, pp. 1–3) who placed it in the context of globalization, Stubbs (2005, pp. 66–71) who emphasized the close relationship among multi-level governance and stakeholders and how opposed to the state-centric model it was, and Pallis (2006, pp. 138–139) who applied the principles to the maritime sector.

Meanwhile, the debate on hierarchies and the EU has not gone away. Borzel (2010, pp. 192, 199) stressed how the EU remains committed to hierarchical policy-making permitting supranational institutions to make policy without recourse to either member states or private stakeholders. EU rule structures offer ample opportunity for hierarchical policy-making but little for the development and application of other models.

3.4 Conclusions

> In troubled homes one nearly always finds a 'dead horse' in the middle of the dining room table. The family may refuse to recognize or speak about the dead horse. They may try to pretend it is not there. But the lifeless animal's presence nonetheless weighs heavily on the family and disrupts its daily interactions. Recognizing the dead horse and acknowledging its existence is the first step toward dealing with the family's problems. (Lake 2003, p. 303)

Lake viewed hierarchies as 'dead horses,' plaguing governance with their continued presence in spite of widespread condemnation. Despite this, there remains support for the hierarchical approach to governance. Ronit and Schneider (1999, p. 246), for example, did not view 'the capacities of private actors at the national level… decimated' and felt that 'national and regional perspectives remain important,' There is also clear acceptance that the hierarchical image of society remains dominant. Ferguson and Gupta (2002, p. 983) described a 'topography of stacked, vertical levels' with the 'state reaching down into communities, intervening in a top-down manner to manipulate or plan society.'

3.4 Conclusions

However, the widespread view is that hierarchies have passed their sell-by date. Campbell (1990, p. 279) cited in Luke (1991, p. 318) summarized many of the doubts about hierarchical governance in a globalized world seeing nation-states defining themselves spatially against a range of landscapes in a 'conjunctive, centralizing hierarchy.' Globalization has taken these 'fictive constructs of linear space' (Luke 1991, p. 319) that the nation-state attempts to impose both within and outside its territory and has altered the power dynamics by 'generating new organizational logics nested in flexible accumulation's rapid and intense flow of ideas, goods, symbols, people, images and money.' These are disjunctive, fragmentary, anarchical, and disordered. He saw power flowing placelessly beneath, behind, between, and beyond spatial boundaries. These boundaries are characteristic of, in fact vital to, hierarchical governance as without them the jurisdictions that epitomize hierarchical governance cannot function. We shall return later to this move to the Postmodern vision, from 'place to flow, spaces to streams' but for now its impact upon traditional hierarchical governance is becoming clear.

Dicken et al. (2001, p. 95) took a geographical slant on the whole process of structure, hierarchy and networks in understanding society. They considered that no analysis of any global, social activity is sufficient at one scale because what actually exists is a complex intermingling of jurisdictions, something emphasized by many others (Jessop 1999; Amin 1997, 1998; Brenner 1998, 1999; Kelly 1999; MacLeod 2001; MacLeod and Goodwin 1999; Swyngedouw 1997, 2000). Capitalism needs each of the traditional scales (local, regional, national, supranational, global) to maximize its potential and none is more important than any other. All the scales are 'mutually constitutive parts of a globalizing economy.' A hierarchical approach to governance makes the mistake of privileging one scale above another, and instead we should be thinking of 'networks of agents (such as individuals, institutions or objects) acting across various distances and through diverse intermediaries.'

They went on to suggest that no institution should be privileged above any other as well—regardless of its jurisdictional location. Thus in practice, major international companies can influence global maritime policy just as much as national governments—take Maersk and Denmark as an example.

Dicken and Thrift (1992, pp. 285–286), with specific reference to production chains but equally applicable elsewhere, further emphasized how coordination and organization had not been helped by a 'dichotimization' of organizational relationships into hierarchies and markets (or networks as we might otherwise know them). They saw a need to move toward a single interpretation of organization—and this would also refer to the broader aspects of organization such as governance—which is more likely to be akin to markets than hierarchies (although keeping some aspects of each). Referring to corporations rather than governments, they suggested that the:

> intra-firm structure of large corporations is, increasingly better represented as a network than as a hierarchy... The organizational and the spatial forms that result are infinitely more complex than the simplistic hierarchies envisaged in earlier studies.

Castells (2000, p. 12) concurred seeing work activities of all types as 'interconnected between firms, regions, and countries... in which networks of locations are more important than hierarchies of places.' He went on: 'networks dissolve centres, they disorganize hierarchy, and make materially impossible the exercise of hierarchical power without processing instructions in the network, according to the network's morphological rules.'

Caporaso (1996, p. 32) cited Rosenau (1992) in suggesting that international governance (and surely nothing is more international than shipping) had traditionally created a hierarchy at the international level in the image of domestic politics. Caporaso believed that this is unnecessary and to a large extent unworkable. Duchacek (1988, pp. 12–14) emphasized the opportunities that exist (and in some cases have been taken) to operate outside the traditional hierarchy in cross-jurisdictional fashion, citing examples from Canada, Switzerland, and the Seychelles, where subnational authorities have worked closely with international bodies. The traditional hierarchical framework is thus in some doubt as an appropriate and viable structure upon which policy making can be hung. This is the case both in the context of EU governance and across other jurisdictions and as a consequence may be a fundamental reason why maritime policy-making exhibits the inadequacies that it does. To quote Hardt and Negri (2004, p. 323):

> ...we have passed from national government to imperial governance, from the hierarchy of fixed national powers to the mobile and multilevel relations of global organizations and networks.

At this stage, it is essential to understand the reason why maritime governance is characterized by hierarchical structures and then to go on to examine what might have changed to make these structures inadequate. However, before we can do that we also need to spend a little more time in focusing upon a central issue within hierarchies—the nation-state and the significant role that it continues to play despite the pressures of globalization that exist.

References

Aalberts, T. E. (2004). The future of sovereignty in multilevel governance Europe—A constructivist reading. *Journal of Common Market Studies, 42*(1), 23–46.

Aglietta, M. (1982). World capitalism in the eighties. *New Left Review, 136*(1), 3–41.

Amin, A. (1997). Placing globalization. *Theory, Culture and Society, 14*, 123–137.

Amin, A. (1998). Globalization and regional development: A relational perspective. *Competition and Change, 3*, 145–166.

Amin, A., & Thrift, N. (1992). Non-marshallian nodes in global networks. *International Journal of Urban and Regional Research, 16*, 571–587.

Anderson, J. (1996). The shifting stage of politics: New medieval and postmodern territorialities? *Environment and Planning D, 14*, 133–153.

Balme, R., & Jouve, B. (1996). Building the regional state: Europe and territorial organization in France. In L. Hooghe (Ed.), *Cohesion policy and European integration: Building multi-level governance*. Oxford: Oxford University Press.

Benz, A. (2000). Two types of multi-level governance: Intergovernmental relations in German and EU regional policy. *Regional and Federal Studies, 10*(3), 21–44.

References

Benz, A., & Eberlein, B. (1999). The Europeanization of regional policies: Patterns of multi-level governance. *Journal of European Public Policy, 6*(2), 329–348.
Best, M. (1990). *The new competition*. Cambridge: Institutions of Industrial Restructuring, Polity Press.
Blaney, D. L., & Inayatullah, N. (2000). The Westphalian deferral. *International Studies Review, 2*(2), 29–64.
Blatter, J. (2003). Beyond hierarchies and networks: Institutional logics and change in transboundary spaces. *Governance, 16*(4), 503–526.
Borzel, T. A. (1998). Organizing Babylon—On the different conceptions of policy networks. *Public Administration, 76*, 253–273.
Borzel, T. A. (2007). *European governance—Negotiation and competition in the shadow of hierarchy*. Montreal, Canada: European Studies Association Meeting.
Borzel, T. A. (2010). European governance: Negotiation and competition in the shadow of hierarchy. *Journal of Common Market Studies, 48*(2), 191–219.
Borzel, T.A., Guttenbrunner, S. & Seper, S. (2005) Conceptualizing new modes of governance in EU Enlargement, Report on the European commission new modes of governance project NEWGOV, Free University of Berlin, Berlin.
Braudel, F. (1982). *The wheels of commerce*. New York: Harper and Row.
Brenner, N. (1997). Global, fragmented, hierarchical: Henri lefebvre's geographies of globalization. *Public Culture, 10*(1), 135–167.
Brenner, N. (1998). Between fixity and motion: Accumulation, territorial organization and the historical geography of spatial scales. *Environment and Planning D, 16*(4), 459–481.
Brenner, N. (1999). Beyond state-centrism? Territoriality and geographical scale in globalization studies. *Theory and Society, 28*(1), 39–78.
Brenner, N. (2001). The limits to scale? Methodological reflections on scalar structuration. *Progress in Human Geography, 25*(4), 591–614.
Burch, K. (2000). Changing the rules: Reconceiving change in the Westphalian system. *International Studies Review, 2*(2), 181–210.
Campbell, J. (1982). *Grammatical man. Information, entropy, language and life*. New York: Simon and Schuster.
Campbell, D. (1990). Global inscription: How foreign policy constitutes the United States. *Alternatives, 15*, 280.
Caporaso, J. A. (1996). The European Union and forms of state: Westphalian, regulatory or post-modern? *Journal of Common Market Studies, 34*(1), 29–52.
Castells, M. (2000). Materials for an exploratory theory of the network society. *British Journal of Sociology, 51*(1), 5–24.
Cooke, P., & Morgan, K. (1993). The network paradigm: New departures in corporate and regional development. *Environment and Planning D, 11*, 543–564.
Cox, K. R. (1998). Spaces of dependence, spaces of engagement and the politics of scale, or: Looking for local politics. *Political Geography, 17*, 1–23.
Crosby, B. L. (1996). Policy implementation: The organizational challenge. *World Development, 24*(9), 1403–1415.
De Alcantara, C. H. (1998). Uses and abuses of the concept of governance. *International Social Science Journal, 50*(155), 105–113.
De Senarclens, P. (1998). Governance and the crisis in the international mechanisms of regulation. *International Social Science Journal, 50*(55), 91–104.
Der Derian, J. (1999). The conceptual cosmology of Paul Virilio. *Theory, Culture and Society, 16*(5–6), 215–227.
Dicken, P., & Thrift, N. (1992). The organization of production and the production of organization. *Transactions of the Institute of British Geographers New Series, 17*(3), 279–291.
Dicken, P., Kelly, P. F., Olds, K., & Yeung, W. (2001). Chains and networks, territories and scales: Towards a relational framework for analysing the global economy. *Global Networks, 1*(2), 89–112.

Dickens, P., & Ormrod, J. S. (2007). *Cosmic society. Towards a sociology of the universe.* London: Routledge.
Duchacek, I. D. (1988). Multi-communal and bi-communal polities. In I. D. Duchachek, D. Latouche, & G. Stevenson (Eds.), *Perforated sovereignties and international relations* (pp. 3–28). New York: Greenwood Press.
Eberlein, B. (2003). Formal and informal governance in single market regulations. In T. Christiansen & S. Piattoni (Eds.), *Informal governance in the EU* (pp. 150–172). Cheltenham: Edward Elgar.
Eberlein, B., & Kerwer, D. (2004). New governance in the European Union: A theoretical perspective. *Journal of Common Market Studies, 42*(1), 121–142.
Epple, D., & Zelenitz, A. (1981). The implications of competition among jurisdictions: Does need Tiebout politics? *Journal of Political Economy, 89*(6), 1197–1217.
Ferguson, J., & Gupta, A. (2002). Spatializing states: Towards an ethnography of neo-liberal governmentality. *American Ethnologist, 29*(4), 981–1002.
Fine, C. H. (1998). *Clockspeed: Winning industry control in the age of temporary advantage.* Reading MA: Perseus.
Florida, R. (1995). Toward the learning region. *Futures, 27*, 527–536.
Frey, B. S., & Eichenberger, R. (1996). FOCJ: Competitive governments for Europe. *International Review of Law and Economics, 16*, 315–327.
Goetschy, J. (2003) The European employment strategy: multi-level governance and policy coordination; past, present, future, In J. Zeitlin & D.M. Trubeck (Eds.) *Governing Work and Welfare in a New Economy; European and American Experiments.* Oxford University Press: Oxford, pp. 59–87.
Gereffi, G., Humphrey, J., & Sturgeon, T. (2005). The governance of global value chains. *Review of International Political Economy, 12*(1), 78–104.
Grande, E. (2000). Multi-level governance: Institutionelle besonderheiten und funktionsbedingungen europaischen mehrebenensystems. In E. Grande & M. Jachtenfuchs (Eds.), *Wie problemlosungsfahig ist die EU?* Baden Baden: Regieren im europäischen Mehrebenensystem, Nomos.
Hardt, M., & Negri, A. (2004). *Multitude.* London: Hamish Hamilton.
Harvey, D. (1982). *The limits to capital.* Chicago, IL: University of Chicago Press.
Harvey, D. (1990). *The condition of postmodernity.* Cambridge MA: Blackwell.
Harvey, D. (2000). *Spaces of hope.* Edinburgh: Edinburgh University Press.
Hatzaras, K. (2005) *Multi-level governance, Europeanization and the poorest EU region*: Epirus and the 2nd Hellenic Community Support Framework, 2nd LSE Symposium on Modern Greece.
Heinz, H. T., Jr. (1988). Advocacy coalitions and the OCS leasing debate. *Policy Sciences, 21*, 213–238.
Heinz, H. T., Jr, & Jenkins-Smith, H. C. (1988). Advocacy coalitions and the practice of policy analysis. *Policy Sciences, 21*, 263–277.
Hirst, P. (1997). *From statism to pluralism.* London: UCL Press.
Hix, S. (1998). The study of the European union II: the 'new governance' agenda and its rival. *Journal of European Public Policy, 5*(1), 38–65.
Hodson, D., & Maher, I. (2001). The open method as a new mode of governance. The case of soft economic policy coordination. *Journal of Common Market Studies, 39*(4), 719–746.
Hooghe, L. & Marks, G. (2001a) *Multi-level governance and European integration,* Rowman and Littlefield: Lanham MD.
Hooghe, L. & Marks, G. (2001b) *Types of multi-level governance,* European integration on-line paper 5 (11), http://eiop.or.at/eiop/texte/2001-011a.htm.
Hooghe, L., & Marks, G. (2003). Unravelling the central state. But how? Types of multi-level governance. *American Political Science Review, 97*(2), 233–243.

Jachtenfuchs, M. (1995). Theoretical perspectives on European governance. *European Law Journal, 1*(2), 115–133.
Jachtenfuchs, M. (2001). The governance approach to European integration. *Journal of Common Market Studies, 39*(2), 245–264.
Jenkins-Smith, H. C. (1988). Analytical debates and policy learning: analysis and change in the federal bureaucracy. *Policy Sciences, 21*, 169–211.
Jessop, B. (1999) Some critical reflections on globalization and its illogic(s). In K. Olds, P. Dicken, P. K. Kelly, L. Kong & H.W-C. Yeung (Eds.), *Globalization and the Asia Pacific: Contested Territories* (pp. 19–38). London: Routledge.
Jordan, A. G. (1990). Sub-governments, policy communities and networks. Refilling the old bottles? *Journal of Theoretical Politics, 2*, 319–338.
Jordan, A. G. (2001). The European union: An evolving system of multi-level governance…or government? *Policy and Politics, 29*(2), 193–208.
Kahler, M. (1989). International financial institutions and the politics of adjustment. In J. Nelson (Ed.), *Fragile coalitions: The politics of economic adjustment* (pp. 139–159). Washington, DC: Overseas Development Council.
Kant, I. (1795) Perpetual peace. In H. Reiss (Ed.), (1991) Kant: Political writings. Cambridge: Cambridge University Press.
Kelly, P. F. (1999). The geographies and politics of globalization. *Progress in Human Geography, 23*(3), 379–400.
Kerwer, D. & Teutsch, M. (2001) Transport policy in the European union. In A. Hertier, D. Kerwer, C. Knill, D. Lehmkuhl, T. Teutsch & A-C. Douillet (Eds.), *Differential Europe: The European Union impact on national policymaking* (pp. 23–56). Lanham MD: Rowman and Littlefield.
Kohler-Koch, B. (1996). Catching-up with change: The transformation of governance in the EU. *Journal of European Public Policy, 3*, 359–381.
Lake, D. A. (2003). The new sovereignty in international relations. *International Studies Review, 5*, 303–323.
Langlois, R., & Robertson, P. (1995). *Firms, markets and economic change.* London: Routledge.
Lefebvre, H. (1974). *La production de l'éspace.* Paris: Éditions Anthropos.
Lefebvre, H. (1976a). *De l'état: L'etat dans le monde moderne* (Vol. 1). Paris: Union Générale d'Editions.
Lefebvre, H. (1976b). *De l'état: De hegel a marx par staline* (Vol. 2). Paris: Union Générale d'Editions.
Lefebvre, H. (1977). *De l'état: Le mode de production étatique* (Vol. 3). Paris: Union Générale d'Editions.
Lenschow, A. (2002). New regulatory approaches in 'greening' EU policies. *European Law Journal, 8*(1), 19–37.
Lovejoy, A. (1960). *The great chain of being.* Cambridge MA: Harvard University Press.
Luhmann, N. (1995). *Social systems.* Stanford CA: Stanford University Press.
Luke, T. W. (1991). The discipline of security studies and the codes of containment: Learning from Kuwait. *Alternatives, 16*, 315–344.
MacLeod, G. (1997). Institutional thickness and industrial governance in Lowland Scotland. *Area, 29*, 299–311.
MacLeod, G. (2001). Beyond soft institutionalism: Accumulation, regulation and their geographical fixes. *Environment and Planning A, 33*(7), 1145–1167.
MacLeod, G., & Goodwin, M. (1999). Space, scale and state strategy: Rethinking urban and regional governance. *Progress in Human Geography, 23*(4), 503–527.
Marks, G. (1993). Structural policy and multi-level governance in the EC. In A. Cafruny & G. Rosenthal (Eds.), *The state of the European Community: Volume II. The Maastricht debates and beyond* (pp. 391–410). London: Longman.
Marks, G., Hooghe, L. & Blank. K. (1996) European integration from the 1980s: State-centric v. multi-level governance, *Journal of Common Market Studies*, 34(3), 342–378.

Martin, R. (2004). Editorial: Geography: Making a difference in a globalizing world. *Transactions of the Institute of British Geographers New Series, 29*, 147–150.
Mathews, J. T. (1997). Power shift. *Foreign Affairs, 76*(1), 50–66.
Mayntz, R. (1976). Conceptual models of organizational decision-making and their application to the policy process. In G. Hofstede & S. Kassem (Eds.), *European contribution to organization theory* (pp. 114–125). Amsterdam: Van Gorcum.
McGinnis, M. D (2005). *Costs and challenges of polycentric governance*. Workshop on analyzing problems of polycentric governance in the growing EU, Humboldt University, Berlin.
Merrien, F.-X. (1998). Governance and modern welfare states. *International Social Science Journal, 50*(155), 57–67.
Meyer, J. W. (1980). The world polity and the authority of the nation-state. In A. J. Bergesen (Ed.), *Studies of the modern world system* (pp. 109–137). New York: Academic.
Meyer, J. W., Boli, J., Thomas, G. M., & Ramirez, F. O. (1997). World society and the nation-state. *American Journal of Sociology, 1*, 144–181.
Morgan, G. (1997). *Images of organization*. Thousand Oaks CA: Sage.
Neyer, J. (2003). Discourse and order in the EU: A deliberative approach to multi-level governance. *Journal of Common Market Studies, 41*(4), 687–706.
Nugent, N. (2006). *The government and politics of the European Union*. Durham NC: Duke University Press.
Ostrom, E. (1985) Formulating the elements of institutional analysis. Institutional analysis conference, Washington, DC.
Ostrom, E. (1986). An agenda for the study of institutions. *Public Choice, 48*, 3–25.
Pallis, A. A. (2006). Institutional dynamism in EU policy-making: Evolution of the EU maritime safety policy. *Journal of European Integration, 28*(2), 137–157.
Peters, B. G. (1998). Managing horizontal government: The politics of coordination. *Public Administration, 76*, 295–311.
Peters, B. G., & Pierre, J. (2001). Developments in intergovernmental relations: Towards a multi-level governance. *Policy and Politics, 29*(2), 131–135.
Porter, M. (1990). *The competitive advantage of nations*. London: Macmillan.
Puchala, D. J. (1972). Of blind men, elephants and international integration. *Journal of Common Market Studies, 10*(3), 267–284.
Puchala, D. J. (1999). Institutionalism, intergovernmentalism and European integration. *Journal of Common Market Studies, 37*, 317–332.
Raco, M. (1998). Assessing inheritance thickness in the local context: a comparison of Cardiff and Sheffield. *Environment and Planning A, 30*, 975–996.
Radaelli, C. M. (2001). *The code of conduct against harmful tax competition: Open coordination method in disguise*. Paris: NOTRE Europe.
Radaelli, C. M. (2003). The Europeanization of public policy. In K. Featherstone & C. M. Radaelli (Eds.), *The politics of Europeanization* (pp. 27–56). Oxford: Oxford University Press.
Rhodes, R. A. W. (1997). *Understanding governance. Policy networks, governance, reflexivity and accountability*. Buckingham: Open University Press.
Ronit, K., & Schneider, V. (1999). Global governance through private organizations. *Governance, 12*(3), 243–266.
Rosenau, J. N. (1992). Governance, order and change in world politics. In J. N. Rosenau & E. O. Czempiel (Eds.), *Governance without government: Order and change in world politics* (pp. 1–29). Cambridge: Cambridge University Press.
Rosenau, J. N. (2000). *The governance of fragmegration: Neither a world republic nor a global interstate system*. Quebec: Congress of the International Political Sciences Association.
Rosenstiehl, P., & Petitot, J. (1974). Automate asocial et systèmes acentrés. *Communications, 22*, 45–62.
Ruggie, J. G. (1998). What makes the world hang together? *Neo-utilitarianism and the social constructivist challenge, International Organization, 52*(4), 855–885.

References

Sabatier, P. A. (1988). An advocacy coalition framework of policy change and the role of policy-oriented learning therein. *Policy Sciences, 21*, 129–168.

Sabel, C. (1994). Flexible specialization and the re-emergence of regional economies. In A. Amin (Ed.), *Post-fordism: A reader* (pp. 101–156). Oxford: Blackwell.

Scharpf, F. W. (1992). Europaisches demokratiedefizit und deutscher föderalismus. *Staatswissenschaften und Staatspraxis, 3*, 293–306.

Scharpf, F. W. (1994a). Community and autonomy: Multi-level policy-making in the European union. *Journal of European Public Policy, 1*(2), 219–242.

Scharpf, F. W. (1994b). Games real actors could play. *Journal of Theoretical Politics, 6*(1), 27–53.

Scharpf, F. W. (1997). *Games real actors play. Actor centred institutionalism in policy research*. Boulder CO: Westview Press.

Scott, J., & Trubeck, D. (2002). Mind the gap; law and new approaches to governance in the EU. *European Law Journal, 8*(1), 1–18.

Skelcher, C. (2005). Jurisdictional integrity, polycentrism and the design of democratic government. *Governance, 18*(1), 89–110.

Slaughter, A.-M. (1997). The real new world order. *Foreign Affairs, 76*(5), 183–197.

Smith, N. (1990) *Uneven Development; Nature, Capital and the Production of Space*, Blackwell: Oxford, 2nd Ed.

Smith, N. (1992). Geography, difference and the politics of scale. In J. Doherty, E. Graham, & M. Malek (Eds.), *Postmodernism and the social sciences* (pp. 57–79). London: Macmillan.

Smith, N. (1997). Antinomies of space and nature in Henri Lefebvre's the production of space. *Philosophy and Geography, 2*, 50–51.

Smouts, M. (1998). The proper use of governance in international relations. *International Social Science Journal, 50*(155), 81–89.

Soja, E. W. (1989). *Postmodern geographies; the reassertion of space in critical social theory*. New York: Verso.

Soja, E. W. (1996). Margin/alia: Social justice and the new cultural politics. In A. Merrifield & E. Swyngedouw (Eds.), *The urbanization of injustice*. London: Lawrence and Wishart.

Soja, E. W. (2000). *Postmetropolis: Critical studies of cities and regions*. Oxford: Blackwell.

Stewart, J. (1996) A dogma of our times—the separation of policy-making and implementation, Public Money and Management, July–September, 33–40.

Stoker, G. (1998). Governance as theory: Five propositions. *International Social Science Journal, 50*(155), 17–28.

Stubbs, P. (2005). Stretching concepts too far? Multi-level governance, policy transfer and the politics of scale in South-East Europe, *Southeast European politics, VI*, 2, 66–87.

Swyngedouw, E. (1997). Neither global nor local: 'Glocalization' and the politics of scale. In K. R. Cox (Ed.), *Spaces of globalization: Reasserting the power of the local* (pp. 137–188). New York: Guilford.

Swyngedouw, E. (2000). Authoritarian governance, power and the politics of rescaling. *Environment and Planning D, 18*, 63–76.

Szczerski, K. (2004) *The EU multi-level governance in post-communist countries—Challenge in governability for the new member states*. 12th NISPAcee Annual Conference, Vilnius, Lithuania, May 13–15th.

Tommel, I. (2000). Jenseits von regulative und distributive: Policy-Making der EU und die transformation von staatlichkeit. In E. Grande & M. Jachtenfuchs (Eds.), *Wie problemlosungsfahig ist die EU? Regieren im europaischen mehrebenensystem*. Baden Baden: Nomos.

Wallace, W. (1999). The sharing of sovereignty: The European paradox. *Political Studies, 47*(3), 503–521.

Waltz, K. N. (1979). *Theory of international politics*. Reading: Addison-Wesley.

Weber, M. (2004). Politics as a vocation. In M. Weber (Ed.), *The vocation lectures*. Indianapolis IN: Hackett Publishing Company.

Zeitlin, J., & Caviedes, A. (2002). *The open method of coordination in immigration policy. A tool for prying open fortress Europe?* Madison WI: University of Wisconsin.

Zhang, X. (2005). Coping with globalization through a collaborative federate mode of governance. *Policy Studies, 26*(2), 199–209.

Zurn, M. (2000). Democratic governance beyond the nation-state; the EU and other international institutions. *European Journal of International Relations, 6*(2), 183–221.

Chapter 4
The Nation-State

> Hazel's obsession with Hoosiers around the world was a text book example of a false *karass*, of a seeming team that was meaningless in terms of the ways God gets things done, a text-book example of what Bokonon calls a *grandfalloon*. Other examples of *grandfalloons* are the Communist party, the Daughters of the American Revolution, the General Electric Company, the International Order of Odd Fellows – and any nation, anytime, anywhere. Kurt Vonnegut, *Cat's Cradle* (1963, p. 61)

> Europe has a lot to answer for. The creation of the nation-state, with its ideology of domination, its centralism, arrogant bureaucracy and latent capacity for repression, must figure high on the list. So must the nurturing and propagation of capitalism, which found in the nation-state an ideal ally, ready to identify a country's fortunes with those of its capitalists.

> Now capitalism has shifted its ground. Organized world-wide, it escapes those checks and balances built up over the years, in the nation-state framework, by workers' movements and parties of the Left. The chances of exerting controls at the world level, which would require a political framework and enforceable decisions, are totally remote… Politics alone has continued to thrive cosily within national boundaries. Lambert (1991, pp. 9, 13)

Lambert in many ways summed up much of what we are about to say. But we should not pre-empt things. In this section we shall explore the role of the nation-state within hierarchical governance systems. The nation-state appears to play a central role within hierarchies not least those associated with maritime policy-making (see for example Sturmey 1975) and there is also clear evidence of changes taking place in the relationship between the nation-state and other jurisdictions and also in its significance as the process of globalization develops. Perhaps these changes can help us to understand better the problems that maritime policy-making faces and the need for changes in the governance system that accompanies this process. Firstly, however, some words on the significance of the state more generally and about how this significance and power developed.

Jordan (2001) and Decker (2002) extensively discussed the role of the nation-state within the governance of the European Union without reaching any definite conclusions except to agree that it remained a significant issue. This suggests little progress in resolving many arguments that have raged over the nation-state's continued existence which were clear over 30 years ago in Dear and Clark (1978) who provided an extensive discussion of these issues and stressed that the literature was both confusing and diverse. Croucher (2003, p. 2) was equally as convinced of the difficulties of even defining a nation-state. She suggested that Connor (1978) struggled to do just that whilst as early as 1939, The Royal Institute of International Affairs commented that 'among other difficulties which impede the study of nationalism that of language holds a leading place' (Royal Institute of International Affairs 1939, p. xvi). Seton-Watson (1977, p. 5) claimed that he was 'driven to the conclusion that no scientific definition of the nation can be devised: yet the phenomenon has existed and exists'. Tishkov (2000, p. 627) suggested that 'all attempts to develop terminological consensus around *nation* (have) resulted in grand failure'. Milliband (1969) is quoted as suggesting that the state stands for 'a number of institutions, including government, administration, judiciary and police'. Althusser (1971) agreed that the state is not a single body and suggested that it can be divided between its repressive (police) and ideological apparatus (political parties). However, Harvey (1976) along with Poulantzas (1969) placed emphasis on process with power exercised through institutional arrangements. Overall, Dear and Clark (1978, p. 174) settled on Mandel's (1975, p. 474) interpretation where the state's role in capitalist economies is:

> the protection and reproduction of the social structure (the fundamental relations of production) in so far as this is not achieved by the automatic process of the economy.

While not seeing this as the only definition of a state it does have some distinct relevance. Wide disagreement exists over the degree of intervention necessary in an economy by the state with five main potential activities clear:

- Supplier of public goods.
- Regulator and facilitator of the market place.
- Social engineer in the economy.
- Arbiter between social groups and classes.
- Agent in society and the economy on behalf of a ruling elite.

Opello and Rosow (2004, p. 3) provided an all embracing definition:

> The nation-state is a type of politico-military rule that, first, has a distinct geographically defined territory over which it exercises jurisdiction; second, has sovereignty over its territory. Which means that its jurisdiction is theoretically exclusive of outside interference by other nation-states or entities; third it has a government made up of public offices and roles that control and administer the territory and population subject to the state's jurisdiction; fourth, it has fixed boundaries marked on the ground by entry and exit points and, in some cases, by fences patrolled by border guards and armies; fifth, its government claims a monopoly on the legitimate use of physical coercion over its population; sixth, its population manifest, to a greater or lesser degree, a sense of national identity; and seventh, it can rely, to a greater or lesser degree, on the obedience and loyalty of its inhabitants.

While possibly disputing some of the claims under six and seven, this definition is helpful in our conception of the classic interpretation of the nation-state.

To understand the characteristics and role of the modern state, it is valuable to examine its origins something that Giddens (1985, pp. 83–121) and Braudel (1982, pp. 514–516) consider in depth. Walker (1991, p. 451) provided a neat summary. It begins with:

> tribes: progresses to the Greek city states; becomes complicated with the age of empires, especially in the case of Rome; becomes muddied with the strange geography of European feudalism; flairs into life with the emergence of the Renaissance and the early-modern struggle for autonomy from empire; then becomes increasingly refined as the principle of sovereignty is codified, and as the state meshes with the organization of capitalist economic life on the one hand, and with the fusion of cultural and social differences into national solidarity on the other. As a story it is not clear whether it will just go on and on, in which case boredom is tempered with a sense of tragedy, or whether it will come to a sudden glorious or catastrophic end when patterns of fragmentation give way to those of integration.

The dissolution of feudal hierarchies in late medieval Europe led to the creation of state organized, self-enclosed territorial domains. These developments found themselves formalized through the 1648 Treaty of Westphalia which defined the end of the Thirty-Years War. Morgenthau (1967) claimed that by then 'sovereignty as supreme power over a certain territory was a political fact'. Brenner (1999, p. 47) commented that this:

> recognized the existence of an interstate system composed of contiguous, bounded territories ruled by sovereign states committed to the principle of non-interference in each other's internal affairs.

The Treaty of Westphalia is widely recognized as one of the fundamental events of the second millennium [for example see Polanyi (1944, p. 7) citing Morgenthau (1967, p. 299); Herz (1957, pp. 455–457, 1959, pp. 43–44); Gross (1948, 1968, p. 47); Bull (1977, pp. 27–38); Zacher (1992, p. 59); Spruyt (1994b, pp. 191–192); Brown and Ainley (1997, p. 116); Ruggie (1998, p. 188); Linklater (1998, pp. 23–24); Van Creveld (1999, p. 86); Opello and Rosow (2004, pp. 245–246); Burch (2000); Caparaso (1996, pp. 34–35; 2000); Blaney and Inayatullah (2000, p. 30); Blatter (2001, p. 176); Stalder (2006, pp. 104–105)]. It was concerned in the main with recognizing the right of a ruler within a territory to choose the religion of their choice, and once this was formalized, conflict which had traditionally focused on dispute within territories now began to focus on disputes between territories (Cohen 2001). The state could begin to assert more power than ever before over its territorial area and consequently could attempt to exert that power in negotiation between states, acquiring a formal position of authority over and within its own defined territory. As Hirst (1997b, p. 220) suggested:

> As enemies became increasingly external, states were able to call forth new forms of loyalty on the part of their members, and the ruling elites could begin the project of identifying the subjects with a territory and with the state.

The effect was the enclosure of power (economic, political, military) within a 'global patchwork of mutually exclusive yet contiguous state territories' which remains the situation we have today where state territorial authority is supposedly inviolable and a state's internal affairs remain very largely unquestionable, something Brenner saw as 'bundling territoriality to state sovereignty' (Gottmann 1973; Ruggie 1993; Jackson 1999; Skelcher 2005). Onuf (1991, p. 430) quoted Morgenthau (1967, p. 299) to emphasize the significance of Westphalia—'by the end of the Thirty Years War (1648) sovereignty as supreme power over a certain territory was a political fact'. Taylor (1994) suggested the principle of 'exhaustive multiplicity' referring to:

- The territorialization of state power, through which each state strives to exercise exclusive sovereignty over a delineated self-enclosed geographical space.
- The globalization of the state form, through which the entire globe is subdivided into a single geopolitical grid composed of multiple, contiguous state territories.

Caparaso (2000, p. 3) indicated that not everyone agreed that Westphalia was the originator of the nation-state (for example Hinsley 1986; Spruyt 1994b) and that it had emerged much earlier with the Concordat of Worms (1122 AD) which defined sovereignty as exclusive property rights over territorial space, and continued its development for many centuries after. However, Hirst (1995) emphasized the growth in the power of the state as a consequence of Westphalia, a situation that has continued until today [see for example Dicken (1994, p. 102) and his discussion of the importance of the location of power to the state] but which may no longer be sustainable in the light of globalized tendencies. States constructed 'de-politicization' of their subjects and military externalization of conflict which enabled them to create relatively uniform administrations within their territory and to subject their subjects to internal hierarchical control and organizational hegemony. The means of violence could then be 'controlled, appropriated and monopolized' (Sorel 1908a, b; Giddens 1985; Hirst 1997b).

Croucher (2003, pp. 8–9), along with Deutsch (1966) and Kohn (1967) went on to suggest that new ideas on nations emerged with the development of Modernist thought through the 1950s seeing the modern (at that time) nation emerging from the impact of the French Revolution in the late eighteenth century along with the economic, industrial and social trends of the nineteenth century (Gellner 1983; Giddens 1985; Hobsbawn 1990; Anderson 1991).

March and Olsen (1998) noted that new states were still being formed at the end of the twentieth century and others disintegrating but even so, the majority of commentators would agree that the world was by then 'partitioned into mutually exclusive and exhaustive territorial units called states'. The state imposes a form of unity and coherence based on a supposed national identity utilizing rules and institutions and creating shared meanings and political legitimacy. However, globalization creates tensions within this system of world order that would seem to need to be reflected in the approach to governance (and hence policies) that is taken (Berger and Dore 1996; Zhang 2005). Meanwhile, the world remains 'a spatially demarcated array of political identities fated to clash in perpetual contingency or to converge somewhere over the distant horizon at a time that is always deferred' (Walker 1991, p. 452). Perhaps globalization is bringing that time nearer?

Scholte (2004) also noted that at times Westphalian principles were violated with military invasions and indirect financial and social interventions in foreign jurisdictions by superior territories (Helleiner 1999; Krasner 1999), weak administration of domestic territory by inferior administrations, and a general rejection of Westphalian principles on moral grounds by Marxists and the like. However, in the majority of cases Westphalian principles were held up as the guidelines for governance throughout the world even if in some cases this was more hypothesis than reality (see recent examples from Kuwait, Iraq, Afghanistan and Libya).

March and Olsen (1998, p. 944) thus saw a two-stage conception of organization that focused on the state and international relations. The first is the domestic stage whereby the territorial nation-state creates 'coherent state actors out of the conflicts and inconsistencies of multiple individuals and groups living within the boundaries of a single state'. In the second, these coherent systems compete and cooperate, pursuing state interests in international spheres that 'recognize few elements of collective coherence beyond those that arise from the immediate self-interests of the actors'. Globalization goes a stage further than this, making it possible to avoid, ignore and intervene within other states at will and thus putting severe strain on a system of global governance that has emerged from a state dominated framework.

Szczerski (2004, p. 3) saw the role of the nation-state under globalization as substantial but with 'different expectations to fulfill' and the significance of the state in a globalizing world is a central issue in maritime governance, something reiterated by Weber (1992). Martin (2004, pp. 147–148) stressed this close relationship whilst Zurn (2003) saw the rise of what he termed 'governance with many governments' as having little obvious impact upon the nation-state's retention of the role it inherited from Westphalia. States remain intergovernmental in that they are represented to other states through their governments thus 'laying the ground for an international society' (Bull 1977). Hence the Westphalian concept of the state and the principles it upholds remain in effect. However, March and Olsen (1998) were less convinced questioning whether we are seeing a major transformation of 'constitutive principles and practices of international political life and the beginnings of a new form of political order and governance' (Krasner 1983).

Tilly (1994) provided an extensive history of the development of the nation-state whilst Held (1991) considered that there is nothing new about the relationship between the state and global interconnectedness although he did accept that modern globalization bears little relationship to world trade in the past. Holloway (1994), along with Held (1991), discussed the nation-state as a basic and largely unquestioned category, the existence of which is assumed. Meanwhile Zurn (2000, p. 187) defined a nation as:

> a political community sustained by intensified interactions, which stands in a mutually constitutive relationship.

King and Kendall (2004, p. 10) saw the state as something stemming from the concept of sovereignty. It comprised those:

permanent institutions within a country through which supreme authority is exercised. Its scope – its jurisdiction – is defined and limited by territorial borders.

They also noted that although there are international organizations typified by the United Nations or the WTO, which can influence and constrain what the state can do, they serve primarily as fora for the 'interplay of interests and values between national governments'. They went on to consider the 'stretching' of notions of democracy which would then include wider international bodies and termed this a 'cosmopolitan democracy' responding to the challenge of globalization. However, interestingly they felt that such a democracy would still need strong nation-states to maintain itself partly at least because it seemed unlikely that such states would compromise their own sovereignties too much. Quoting Jessop (2002, p. 9), the state was being:

> re-imagined, redesigned and reoriented in response to these challenges rather than withering away.

Recent events and the speed of progress of globalization would perhaps put this conclusion into doubt. However, Axtmann (2004) also confirmed that states can remain successful in performing a key role in governance. To do this they must meet six requirements as defined by Parekh (2002, p. 182):

> First, it should be territorially distinct, possess a single source of sovereignty, and enjoy legally unlimited authority within its boundary. Secondly, it should rest on a single set of constitutional principles and exhibit a singular and unambiguous identity... Third... (it) represents a homogenous legal space within which its members move about freely, carrying with them a more or less identical basket of rights and obligations. Fourth... all citizens are directly and identically related to the state, not differentially or through their membership of intermediate communities. Fifth, members of the state are deemed to constitute a single and united people... Sixth and finally, if the state is federally constituted, its component units should all enjoy the same rights and powers.

Agnew (1994, p. 56) also emphasized the 'orthodoxy' of the state as equivalent to a rational individual exercising free choice constrained by the 'presence of anarchy beyond state borders'. Each nation-state can be seen to be 'unitary actors' pursuing a 'calculus of status maximization relative to others' (Agnew 1994, p. 57).

> State territories have been reified as set or fixed units of sovereign space. This has served to dehistoricize and decontextualize processes of state formation and disintegration. (Agnew 1994, p. 58)

The result is that the existing configuration of nation-states has been frozen apart from the very occasional (and dramatic) reconfiguration as exemplified by the disintegration of East Europe and the USSR between 1989 and 1992.

In addition a common thread in defining the state is the emphasis on its monopolization of the legitimization of force as suggested by Ruggie (1993). The state normally is the only institution with the right to imprison, physically harm or even kill individuals, something derived from the progressive imposition of the 'King's Peace' manifesting itself in the sole right of the King's authority to impose the law (which subsequently became the right of that of the state) (Norbert 1983).

This relationship between legitimate force and the state was something that Giddens (1985) and Hardt and Negri (2004, pp. 25–32) considered in some detail and Deleuze and Guattari (1988, pp. 447–448) felt was a necessity. Harvey (1990, p. 108) meanwhile viewed the state as a 'coercive authority' with a monopoly over 'institutionalized violence' and was effectively a mechanism through which the 'ruling class can seek 'to impose its will upon the anarchical flux, change and uncertainty to which capitalist modernity is… prone'. The state therefore has been a fundamental part of the process by which, (for example) shipping governance has been imposed and regulated (Lloyd's List 2007). Perhaps the clear changes in the state's role suggest a reason for the maritime governance problems we have identified.

The methods by which the state has traditionally exercised its influence and provided the mechanism for capital accumulation which itself is the guiding principle of modern business, include financial regulation, contract rules, fiscal intervention, credit provision, tax redistribution, social and physical infrastructure provision and direct control of wages and prices. States thus exist for the exercise of power and the ease of administration (Johnston 2001) whilst Sack (1986) considered the issue of territory as a spatial strategy to 'affect, influence and control'. However, to retain control over these tools of influence the state has to legitimize itself using what Harvey terms an 'alternative sense of community to that based on money' alone. In achieving this, the state has to face up to the:

> tension between the fixity (and hence stability) that state regulation imposes, and the fluid motion of capital flow. (Harvey 1990, p. 102)

These tensions begin to hint at the problems faced by maritime governance. The state has formed a significant part of the structure that has held the governance of all commercial activity together at least for the past 200 years. Scholte (2004, p. 50) termed this 'statist', where 'governance is more or less equivalent to the regulatory operations of territorial national governments'. However, the role of the state has changed dramatically in the past few decades with the development of accelerated time-space compression—exemplified by the rise in significance of globalization—characterized by competition in the world economy, exchange rates, capital movements, migration, and direct and indirect political intervention by superior powers. Perhaps the existing governance framework is now inappropriate for the industry to which it is applied. Johnston (2001, p. 684) reached a similar conclusion:

> Territoriality (is) thus a major component of state sovereignty, of its exercise of power over people through its control of bounded space; indeed a territorial definition is a characteristic of all states and territoriality has been viewed as a necessary if not sufficient, strategy in the exercise of state power.

This implies that the decline in effectiveness of territoriality as a consequence of globalization also implies a decline in power. Since modern maritime governance retains the features of state territoriality and originates almost entirely from state-centric organizations such as the EU, IMO, OECD and national governments, a decline in the relevance of state territoriality is highly significant for the efficiency and efficacy of the maritime governance process and consequential policies. Opello and Rosow (2004, p. 235) emphasized the territorial nature of the modern state despite the military, economic and social forces which are questioning its central role. These forces are seen as emanating from the inexorable progress of globalization of the capitalist economy whereby capitalism relies on the nation-state at the same time as undermining its territorial credentials. Obando-Rojas et al. (2004, p. 296) saw shipping as providing a classic example, whilst Opello and Rosow (2004, pp. 253, 255–263) viewed states as relatively ineffectual managers of economic activity within their own boundaries, and at the same time important players in the governance of economies outside their territories and across their borders, something they refer to as the neo-liberalist impact upon nation-state activity.

Territoriality and its relationship to space has a considerable history of debate and is a central theme amongst geographers, sociologists, political scientists and the like (see for example Parker and Bell 2009a, b). Soja (1989, p. 150) suggested that it is a diffuse term, difficult to define (unlike territory) but encompasses notions such as 'sovereignty, property, discipline, surveillance and jurisdiction'. He referred to the 'production and reproduction of spatial enclosures that not only concentrate interaction but also intensify and enforce its boundedness. Boundaries can change over time in permeability and shape and along with territory are expressions of power. Oates (1999, pp. 1130–1131) suggested that state boundaries are historically and culturally defined and may be irrelevant for the current situation. He felt that if all state boundaries were to be dissolved and then reinvented almost none would resemble those that exist. Soja identified almost limitless layers of territorial jurisdictions—from the ego upwards—which are constrained in conventional governance wisdom to the local, regional, national, supranational, international and which represent a spatialization of the ego and a representation of spatial cognition. They are a manifestation of territory and its sociological boundaries 'hingeing around the portable bubbles of personal space zonation and proxemic behavior, a non-verbal and unwritten ordinary language of spatial inter-subjectivity'.

Territoriality and the state is considered by many commentators. Deleuze and Guattari (1988, p. 385) for example saw one of the fundamental tasks of the state as striating space over which it 'reigns', establishing a 'zone of rights over an entire exterior, over all of the flows traversing the ecumenon'. This concentration on territoriality, space and the state has recently focused particularly on its relationship to globalization. Agnew (1994, p. 55) outlined how spatialities relating to the state in particular tended to assume the appropriateness of the spatiality that existed, and an idealized set of 'representations of territorial or structural space… irrespective of historical context'. Space, territory and the state were seen as one, an assumption that is increasingly questionable with globalization. Agnew (1994, p. 73) went on to quote Luke (1991, p. 326) at some length to show the simplistic assumption that the state is represented by its territory and that the two always are synonymous is far from the case:

> The essentially fictive nature of many contemporary nation-states... is exposed by the Kuwaiti and Iraqi experience in the Gulf War [of 1990-1991]. As a classically styled authoritarian state, using modernist myths of military conquest, supreme leadership, national mission and chiliastic global change, Iraq – like fascist Spain, Portugal, Argentina, Japan, Italy or Germany before it – demonstrated the bankruptcy of spatial expansion, place domination and territorial imperialism in the informational flows of contemporary world systems. Kuwait, on the other hand, as a bizarrely Postmodern fusion of pre-modern feudalism with informational capitalism, is more of a place-oriented stream within the global flow of money, ideas, goods, symbols, and power. As a point of production and consumption in the flow, however, Kuwait far outclassed Iraq in global significance, even though it had fewer people, less territory, and a smaller military force. Iraq took Kuwait's real estate but failed to capture its hyperreal estate.

Agnew suggested that contrary to the arguments of Virilio (1977, 1995), space is certainly not necessarily represented by state territory and Kuwait as an example of this, has a serious spatial identity as a 'node in the network of informational capitalism'—similar to that of Hong Kong, Singapore or even cities such as London, New York and Shanghai whose sheer economic and social presence far outshines their spatially confined territory. The danger here is that the emphasis on the eradication of time and space fails to recognize that specific territoriality does still exist but that its boundaries may well be variable, confused and temporally erratic.

Agnew (1999, p. 513) also suggested that the relationship of states to territory is clearest when issues of power monopoly, property ownership and citizenship are discussed. Globalization has placed pressure upon these issues challenging the circulation of assets and people beyond state boundaries and undermining the close link that has long existed between state and territory. The vastly increased scale and scope of these mobilities has partly at least derived from the transfer of power from the state to firms which themselves are increasingly less influenced by state rules and priorities. States actively compete to attract 'mobile property to their territories and the shipping industry is one part of the burgeoning communications and trading markets that globalization has liberalized. Johnston (1995, p. 219) agreed seeing territory as a necessary requirement for the state to fulfill its basic functions as identified by Clark and Dear (1984)—securing social consensus (order, stability); securing the environment (internal disorder is minimized and external threats few); and securing conditions of production (an appropriate infrastructure). Without meeting these three conditions capitalism cannot flourish and to achieve them, territorial reliability is a necessity.

Dicken et al. (2001, pp. 95–96) emphasized that scales and territory are heavily intermixed in network governance in that in reality the traditional jurisdictional framework (global, supranational, national, regional, local) in effect is composed of a complex intermingling through the 'relativatization of scale' (Jessop 1999; Amin 1997, 1998; Brenner 1998, 1999; Kelly 1999; MacLeod 2001; MacLeod and Goodwin 1999; Swyngedouw 1997a, 2000). No one scale is privileged over another in the way that capitalism operates. They saw all scales as 'mutually constitutive parts of a globalizing economy'. The issue of scale and territory is thus both linked and complex as networks of 'agents (such as individuals, institutions and objects) act across various distances and through diverse intermediaries'. Latour (1993, p. 122), cited in both Thrift (1996, p. 5) and Dicken et al. (2001, p. 95), suggested this 'offers points of view on networks that are by nature neither global nor local but are more or less long and more or less connected'.

The significance of territory clearly emerges from this particularly in the context of globalization. Amin (2002, p. 385) for example saw globalization as about the 'spatiality of contemporary social organization, about meanings of place and space associated with intensified world forces'. Amin along with Gottdiener (1987) reinforced this message by noting Marston's (2000) work on spatial scales and globalization which suggested that all jurisdictional scales have no 'pregiven or fixed ontological status, but are socially produced and continually transformed by the imperatives of capitalism and the resulting struggles and conflicts' (Amin 2002, p. 386). Hudson (2000, p. 270) questioned the whole concept of states and sovereignty, with the latter increasingly seen as a 'socially constructed and historically specific socio-spatial institution which changes as social processes and their spatial organization change' (Agnew 1994; Barkin and Cronin 1994; Bartelson 1995; Biersteker and Weber 1996; Burch 1994; Camilleri 1990; Camilleri and Falk 1992; Ferguson and Mansbach 1996; Lapidoth 1992; Philpott 1995; Ruggie 1983; Sassen 1997, 2003; Spruyt 1994a, b; Taylor 1994, 1995; Thomson 1994; Walker 1991, 1993). He continued by suggesting that globalization is all about shifts in scale with a need for the economic changes in scale that have occurred to be matched by those in the political and institutional structures (Hudson 2000, p. 272). 'As the scale of goods and assets produced, exchanged and used diverges from the scale of the state, the authority, legitimacy and effectiveness of the state as a regulatory authority is brought into question, much as is happening today with maritime governance. New structural differentiations need new 'political economies of scale'. Duncan and Savage (1989, p. 182) concurred in that they saw 'all social objects exist(ing) in a spatial arena and hence will be affected by spatially contingent relations'. Newman and Paasi (1998, p. 188) agreed whilst Relyea (1998, p. 30) quoted Albert Einstein (Einstein 1971, p. 31) to make the point:

> Physical concepts are free creations of the human mind and are not, however it may seem, uniquely determined by the external world.

This is clearly relevant for our consideration of maritime governance and policy-making where the jurisdictional structure of this highly capitalized sector (IMO; EU; nation-state, etc.) is socially constructed, historic rather than contemporary and characterized by endemic failure.

MacLeod and Goodwin (1999, pp. 505–506) agreed that spatial scales are never fixed and suggest that the concept of 'process' is far more useful than the immovable jurisdictional structure that is found at the centre of maritime policy-making.

Globalization has changed jurisdictional definition and arrangements, internationalizing the state's role.

Held and McGrew (2002, pp. 1–2) emphasized the significance of global governance, its relationship to globalization and its consequences for the nation-state. They outlined a number of features that characterize its 'institutional architecture':

- *Multilayered*—with multiple jurisdictions.
- *Pluralistic*—with no single focus of authority.
- *Variable geometry*—across the globe these jurisdictional institutions vary in political and regulatory influence.
- *Structurally complex*—consisting of 'diverse agencies with overlapping (functional and/or spatial) jurisdictions, resources and competencies'.
- *Nation-states remain crucial*—legitimizing regulation beyond the state.

Maritime governance as formulated since the 1940s has taken on-board only some of these features of globalized governance—in particular the central role of the nation-state. Even there, global governance requires a new role for the nation-state rather than one preserved from former times. Held and McGrew identified a 'new medievalism' in this configuration of global governance that has emerged with globalization, since it 'resembles the complexity of competing jurisdictions, porous administrative boundaries and multiple levels of political authority which characterized medieval Europe'. This creates ambiguity in the location of authority (and policy-making) which the maritime sector has yet to recognize let alone accommodate.

Kratochwil (1986, p. 27) had much earlier emphasized the contrast that was emerging between what he termed the 'universal recognition of territorial sovereignty' and the 'tendency toward erosion of the exclusivity associated with traditional notions of territoriality'. This was a disjunction of globalization and the state that in turn created 'tensions between bounded political systems and unbounded exchanges such as economic, ideological and informational transactions' (Kratochwil 1986, p. 43), something clearly relevant to maritime governance and policy-making.

Meanwhile, despite the assumption that the networks which have characterized the growth of globalization are able to cross jurisdictional boundaries almost at will, the influence of territory (space, boundaries) on governance and the operation of a network remains evident if complex. Networks, despite crossing jurisdictions do not eliminate boundaries, territories and particularly nation-states and there has been an increasing tendency to assume the insignificance of territory as the role of networks has grown in governance—moving from Thrift and Olds' (1996) 'topological presupposition of the bounded region' and towards a network syndrome suggests the denigration of the nation-state and its territorial significance. Crossing jurisdictional borders still presents 'qualitative disjunctures' between different 'regulatory and socio-cultural environments' (Dicken 1998; Sassen 2000; Smith et al. 1999; Taylor 1994). To quote Dicken et al. (2001, p. 96):

National regimes of regulation continue to create a pattern of 'bounded regions' and networks of economic activity are not simply superimposed upon this mosaic, nor is the state just another actor in economic networks.

Relyea (1998, p. 31) also saw nation-states as retaining much (if changing) significance despite globalization and even suggests that boundaries in some circumstances have become more important stressing that:

> The result of technological advances in travel and information collection and dissemination, the compression of both time and space intrinsic to the contemporary manifestation of globalization, paradoxically heightens both the importance and the irrelevance of borders surrounding states and a variety of other politico-territorial entities, as well as simultaneously serving as a catalyst for the reorientation of existing physical boundaries and the psychological concepts attendant to them.

Swyngedouw (2000, p. 69) placed the nation-state at the centre of changing spatial scales and their 'nested articulation' (Smith 1984, 1993; Swyngedouw 1997a, b). One feature he pointed out was that of 'glocalization' whereby governance structure and processes jump jurisdictions and in particular leave out the nation-state (Fig. 4.1). This process of glocalization tends to be 'autocratic, undemocratic and authoritarian' (Morgan and Roberts 1993) and demands a new, practical engagement with the formation of new scalar relations between governance and civil society (Cerny 1996; Hirst 1997a). Swyngedouw (2000, p. 70) emphasized this 'continuous reshuffling and reorganization of spatial scales (that) are an integral part of social strategies and struggles for control and empowerment'. Power shifts create changes in jurisdictional nesting, interrelationships and spatial extent. The connectivity between spatial scale, territory and the nation-state is thus complex, flexible, ever-changing and dynamic and governance processes need to reflect this. In maritime governance this is rarely the case. As Smith (1987, p. 71) said, 'scales were jumped, (and) a new gestalt of scale is being wrought in ways that has profound and often disturbing consequences for the geometrics of socio-spatial power'.

Agnew (1994, p. 75) agreed that globalization leads to fragmentation of space as well as the re-formulation of the space that the nation-state reflects. He emphasized the rise of the regional and local in economies, placed within an overall global framework. Quoting Said (1979, p. 18) he saw increasing numbers of people living in 'a generalized condition of homelessness', whereby a world is created where 'identities are less clearly bonded to national territories' (Agnew 1994, p. 75).

As Gupta and Ferguson (1992, p. 13) indicated:

> In a world of diaspora, transnational flows and mass movements of populations, old-fashioned attempts to map the globe as a set of culture regions or homelands are bewildered by a dazzling array of post-colonial simulacra, doublings and redoubling, as India and Pakistan apparently re-appear in post-colonial simulation in London, pre-revolution Iran rises from the ashes in Los Angeles, and a thousand similar dramas are played out in urban and rural settings all across the globe.

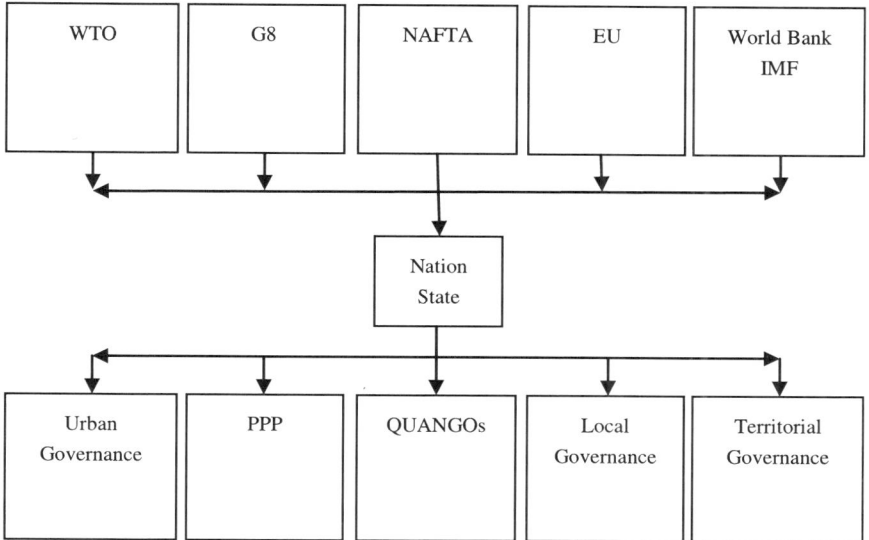

Fig. 4.1 The glocalization of governance. *Source* Swyngedouw (2000, p. 70)

Newman and Paasi (1998, p. 188) brought together a lot of these ideas concerning the relationship of globalization to scale, space, territory, boundaries and the nation-state. They suggested that the recognition of these relationships is a reflection of the Postmodern—something we turn to in more detail later. They also suggested that there is a view that boundaries are becoming not only different in how they work but also less important (Dodds 1994; Sibley 1995; O'Tuathail 1996). Perhaps more significant to our discussion, they implied that boundaries and their relationship to nation-states are more related to transport and communications than physical characteristics. Quoting Morley and Robins (1995, p. 1):

> Patterns of movement and flows of people, culture, goods and information mean that it is now not so much physical boundaries – the geographical distances, the seas or mountain ranges – that define a community or nation's 'natural limits'. Increasingly we must think in terms of communications and transport networks and of the symbolic boundaries of language and culture – the 'spaces of transmission' defined by satellite footprints or radio signals – as providing the crucial and permeable boundaries of our age.

This interpretation of process in boundaries and territories hints of Castells' concepts of space of flows which we shall consider later, and also the debate about the end of the nation-state (as boundaries disappear) which we consider in this chapter (Milnar 1992; Kuehls 1996; Shapiro and Alker 1996). However, Newman and Paasi (1998, p. 192) also suggested that both boundaries and the nation-state survive. Quoting Hirst and Thompson (1995), most people still live in what they term 'closed worlds' 'trapped by the lottery of their birth'. The meaning of sovereignty changes but the state retains a close relationship to territory and boundary. On the other hand they are not unaware of the new Postmodern society where a new 'hyperspace' is emerging where globalization threatens the existing 'particularity of places, borders and territoriality' and hence the nation-state itself.

Over 40 years ago Hoffmann (1966, pp. 862–863) was already questioning the continued role of the nation-state but found substantial emotional evidence of why it retained a central place in an increasingly globalized world:

> A claim to sovereignty based on historical tradition and dynastic legitimacy alone has never had the fervor, the self-righteous assertiveness which a similar claim based on the idea and feelings of nationhood presents.

Something which the highly traditional shipping industry could well understand. He went on:

> The nation-state is still here, and the new Jerusalem has been postponed because the nations in Western Europe have not been able to stop time and to fragment space.

And:

> Nation-states – often inchoate, economically absurd, administratively ramshackle, and impotent yet dangerous in international politics – remain the basic units in spite of all the remonstrations and exhortations.

Hoffmann returned to these issues in 1982 reflecting on the 'bankruptcy of the nation-state' particularly in terms of economic efficiency. However, by the twenty first century the state was still with us although its role perhaps has changed. Meanwhile Mann (1993) emphasized how states were surviving and although weakening of nation-states in Western Europe could be seen, he described this as 'slight, ad hoc, uneven and unique'.

Atkinson and Coleman (1992, pp. 154, 156) recognized the strength of the nation-state in policy-making defining legitimate interests, shaping political organizations and incorporating societal actors. Johnston (1991, pp. 196–197) outlined the nation-state's significant functions. Jachtenfuchs (1995, p. 120) emphasized the significance of the state in governance even beyond the state. Looking at the EU situation he saw the state's role at all levels of jurisdiction as being the most significant. Taking the neo-realist view, he saw international regimes such as the IMO, OECD, WTO or the EU as existing only as the 'rational' decision of sovereign states to co-operate with other states and having the ability to revise this decision at any time they wish. International institutions therefore 'facilitate the pursuit of individual interests by states as compared to autonomous action.' Fundamentally:

> states do not co-operate in order to pursue jointly defined goals which might even change domestic definitions of state interests. Instead, states co-operate in order to pursue their own interests. These interests exist before co-operation and operate independently of it. (Jachtenfuchs 1995)

If this is the case it helps to explain the continued significance of the nation-state in policy-making in sectors such as shipping where its role appears at first sight to be an anachronism, but also hints at where inadequacies in governance may be causing some of the policy problems that clearly exist as states continue to interfere even when the system of capital accumulation within which shipping operates may have changed fundamentally. The question is whether an increasingly globalized society needs nation-states to sustain the process of globalization, or whether the nation-state will be eroded—with consequent implications for governance.

Brenner (1997, p. 138) suggested that most attempts at analyzing globalization assume that the 'power and significance of the territorial state are declining' citing Gupta and Ferguson (1992), Robertson (1992), Shapiro (1994), Appadurai (1996), Castells (1996) and Scott (1996). This he saw as a methodological 'backlash' against Agnew's (1994) 'territorial trap'—which he defined as an:

> ahistorical state centrism in which the only possibility for mapping global processes is in terms of the fixed territorial boundaries of states.

Globalization experts commonly react to the suggestion that the world had to be mapped as states by denying the state any status at all with the consequence that its decline was assumed inevitable. Players in this global-national dichotomous game would have to take sides resulting in a failure to understand that the nation-state might have a place within a globalized society—albeit a different one.

Brenner (1998, p. 469) also considered that the nation-state is traditionally viewed as the main organizational-territorial focus for capital circulation and political ideology. Following the work of Lefebvre (1978) and Harvey (1990, 2000) the state was seen as providing 'preconditions, arenas and outcomes of capital's contradictory dynamic of de- and reterritorialization' playing a crucial role as a 'relatively fixed and immobile territorial configuration upon which each round of capital circulation is grounded'. Lefebvre's 'fixity' of state territorial organization provided a:

> stabilized scaffolding for the increasing 'mobility' and 'transience' of labour power, commodities and capital.

Brenner thus saw the state as a central part of the process of territorialization of capital which in turn is a major precondition for capitalist activity. Quoting Lefebvre (1991, p. 389):

> Political power as such harbours an immanent contradiction. It controls flows and it controls agglomerations. The mobility of the component parts and formants of social space is constantly on the increase, especially in the 'economic' realm proper; flows of energy, of raw materials, of labour, and so on. But such control, to be effective, calls for permanent establishments, for permanent centers of decision and action.

The state therefore has a crucial role to play even in times of intense globalization and the mobility of capital, labour, goods and information that this brings as it provides a stable location from which capital accumulation can be organized. Hence one might conclude that a hierarchical form of governance would be best suited to an increasingly globalized society since it too is inherently state-centric. However, this may be too simple an assumption to make as perhaps it is the continuing significant role of the state that has changed dramatically (and not that it has lost its role) and that new governance models are needed to both accommodate the state as a territorial given, but also to reflect its revised role within the jurisdictional framework and changing pattern of policy stakeholders.

The state has also ensured its own continued significant role in society by 'establishing scalar fixes for capital through its continual reconfiguration and regulation of social space' (Brenner 1998, p. 470). Thus the state by retaining control of economic and social activity within national boundaries as much as it can, has helped to make itself indispensable for longer than globalization might imply. However, perhaps the failings in maritime policy which we have identified are an indication that the hierarchical models used to sustain this level of artificial domination by the state are beyond their date stamp as the processes of capital accumulation begin to become territorially more and more elastic. The state may remain a significant player, but perhaps the rise of globalization has made its role somewhat different and one which the current governance models find unsustainable.

Cohen (2001, p. 76) suggested that the state has been the dominant form of political organization for over 400 years but is now threatened by the 'emerging global economy' characterized by 'interdependent systems of production and consumption, dramatic flows of currency across national borders, and increasingly sophisticated technologies of information gathering and processing' features that are central to the international maritime business community.

Cohen also saw states as becoming less able to control economic activity and losing their relevance as 'facilitators or obstacles to the function of the global economy'. He describes them as heading for 'oblivion'.

After such a depressing message it comes as some surprise that he goes on to emphasize how the state is actually surviving with 'vigor and relevance'. He cites Bairoch and Kozul-Wright (1996) who attacked the notion of an inevitable decline in the nation-state, Helleiner (1994) who claimed that globalization is only possible if it is facilitated by nation-states and Hirst and Thompson (1999) and Weiss (1998) who saw states as playing significant roles in regulating economic activity as it crosses national borders. As Cohen (2001, p. 77) commented:

> the state remains as central as ever in constituting and regulating the terms of global economic exchanges, and shows no signs of disengagement from this position.

A view reaffirmed by Somerville (2004) who felt that despite the increasing significance of both sub- and supranational governance, the nation-state retained much of its importance. The state may well be seeing increasing dispersal of power but this does not necessarily infer a reduction of influence.

The nation-state is therefore in a difficult situation. Stalder (2006, pp. 109–111) was in no doubt terming it a 'crisis' since the nation-state was unquestionably no longer a sovereign political actor, although he continued to cite Castells who rejected the notion of the end of the state. To quote Walker (1991, p. 459), the state retains 'an enormously powerful grip on the contemporary political imagination' but we are:

no longer so easily fooled by the objectivity of the ruler, by the Euclidean theorems and Cartesian coordinates that have allowed us to situate and naturalize a comfortable home for power and authority.

Cooper (2000, p. 16) summed it up. If too strong in its resistance to globalization and change, then it will stop society from functioning. If too weak, it will lose its purpose altogether. Quoting the fencing master to his pupil in the film *Scaramouche* (1959)—'the rapier is like a bird. Grasp it too loosely and it will fly away, too tight and you will crush it'.

So is the state in inevitable decline or is its predicted demise unwarranted?

4.1 The Decline of the Nation-State

> The partitioning of the world was coming in our time to be established on the recognition of the co-existence of many individualized sovereign states, each within its territory. In most international agencies every member state enjoyed a vote, irrespective of its size. Just as this order was reaching global propositions, after centuries of evolution, winds of change began blowing into the political arena, disturbing once more an apparently forthcoming stabilization. New currents in technology, economic demands and political ideas have injected into the arena disquieting forces that expand accessibility, modify operating structures and call for the revision of established concepts and principles. The resulting fluidity has turned the present situation into a rather chaotic state of affairs insofar as the meaning of territory and sovereignty is concerned. (Gottmann 1973, p. 154)

The debate about the decline of the nation-state has been on-going for some years and can also be linked to the identifiable failure of the Westphalian system of nation-state governance. Westphalia relied upon the nation-state being able to contain the significance of global governance and thus to be able to fall back on nation-state sovereignty. Globalization has changed much of this (Zacher 1992, p. 61). Osiander (2001, pp. 260–281) questioned the fundamental assumptions of the impact of Westphalia. Mathews (1997, p. 50) saw the Westphalian framework underlying international governance 'dissolving'. Agnew (1999), cited in Blatter (2001, p. 178) saw the emergence of the hierarchical system of international governance threatening the assumptions of Westphalia. Brown and Ainley (1997, p. 172) described what they saw as 'Westfailure' typified by the 'inability of our present global political system to cope with the problems caused by globalization' (Strange 1999). Miyoshi (1993, pp. 726, 731–732) saw nation-state decline as an inevitable consequence of the decline of colonialism as the state was a creation of West European colonialists:

> the belief in the shared community ruled by a representative government... the myth of racial superiority over the heathen barbarians... citizens bound by kinship and communality.

However, in the modern era, the nation-state finds itself 'in a vacant space that is ideologically uncontested and militarily constabularized'. Langford (1999, pp. 61, 64) was confident of the failure of the state characterizing it with civil strife, political corruption, economic collapse, societal degradation, domestic chaos, human rights abuse, crumbling state infrastructure and governmental failure. 'The central authority through which laws are made and enforced is inoperative; laws are not made, order is not preserved and societal cohesion is not enhanced'. Failure is complex and political. Although shipping may not be associated with all these, the potential for failing states and the impact upon maritime governance is clear. In Hudson's (2000, p. 274) view, the nation-state was fundamentally a Modernist beast dominating regulated landscapes, located in Ruggie's (1993, p. 144) 'historically unique configuration of space'; Valaskakis (1999, p. 155) viewed the nation-state as a 'falling star' where transnational mobility combined with technological progress can force taxation and regulation downwards to the lowest common denominator' something which had been re-emphasized by Schmidt (1995). Think tonnage tax! The state retained roles simply as umpire and re-distributor of income (Valaskakis, 1999, p. 163). Meanwhile, 'the unbundling of sovereignty facilitates the articulation of a global capitalism based on increasingly mobile property with an inter-state system based on immobile property rights or sovereignty'. Caparaso (1996, p. 45) contrasted the Postmodern state with that of the Westphalian and emphasized the anachronism that is characterized by the latter. Cooper (2000, p. 7) was happy to declare the Westphalian nation-state dead and a new system emerging. Modern capitalism was dead; Postmodern capitalism was alive; and the governance of the maritime sector has yet to wake up to this new world.

For example Brown (1973) suggested that we now had a 'world without borders', Deutsch (1981) felt that there was a crisis of the state, Aronowirz (1988, p. 47) considered nationalism was in retreat existing only as a reactionary force to trans-nationalism, whilst Rosenau (1988) described the state as a 'withering colossus'. Walker (1991, p. 448) stressed how the state is not a 'permanent principle of international order' as the impression of permanence is simply a mechanism to 'shift disruptions and dangers to the margin'. Del Rosso (1995) analyzed in some detail the relationship of state decline to security issues. Strange (1996, pp. 4, 5) viewed the nation-state as no longer as important as world markets typified by private enterprise in industry, finance and trade and that this was alarming as it was those same states that were supposed to exercise political authority over those very organizations. This was even more of a concern because the fundamental reason for a nation-state was to exercise political authority 'legitimated either by coercive force or by popular consent' and that nation-states were losing the ability to do this. This increasing weakness was caused by the failure of linkages between national enterprises and nation-states in turn a result of globalization. Petit and Soete (1999, pp. 165–166) felt that policy-makers were increasingly aware of the importance of the international implications of domestic policy actions which can make them 'unsustainable', 'blocking their policy means' and making the 'traditional policy toolbox' inadequate and inappropriate. The importance of the nation-state remains in their view but is 'constrained by its place in the concert of nations', commonly governed by an hegemon (Petit and Soete

4.1 The Decline of the Nation-State

1999, p. 179). Gritsch (2005) summarized many of the arguments suggesting that the nation-state was dead including those from Ruggie (1993), Holloway (1995), Drache (1996), O'Tuathail (1997), Rodnik (1997), Agnew (1988) and Habermas (1999). Reich (1992, p. 3) meanwhile, was convinced that:

> We are living through a transformation that will rearrange the politics and economics of the coming century. There will be no national products or technologies, no national corporations, no national industries. There will no longer be national economies… Each nation's primary political task will be to cope with the centrifugal forces of the global economy… As borders become ever more meaningless in economic terms, those citizens best positioned to thrive in the world market are tempted to slip the bonds of national allegiance.

Evans (1997, p. 63) agreed suggesting that transnational economic gain had created an 'anachronistic state 'marginalised as an economic actor' where transnational corporations were the economically empowered 'citizens' of today (Evans 1997, p. 65). Attempts to block the flow of knowledge and money across borders are increasingly futile since the threads of the emerging global webs are barely visible and consequently elusive (Reich 1992, p. 111). Held (1991) was one of the earlier commentators to question the relationship between the nation-state and globalization, suggesting that no longer could national communities assume that they alone would determine what is right and appropriate for their own citizens. He put forward five ways in which nation-states have had to change as a response to globalization:

- Reduction in the number and effectiveness of political instruments available to national governments—for example customs regulations, border controls and selective taxation.
- Expansion of transnational forces—including the free movement of capital.
- Increased necessity for international collaboration—including defence, economic management, legal systems, etc.
- Increased political integration—including the EU, GATS, WTO etc. This was further emphasized by Benz and Eberlein (1999, p. 330) who observed that European integration 'challenged domestic patterns of territorial interaction'.
- Growth in international and supranational organizations laying a basis for global governance.

Beck (1992) outlined the dangers to the nation-state presented by what he calls the 'risk society' which inherently destroys boundaries. Mann (1997, pp. 473–474) commented that 'for the old nation-state, we find largely epitaphs', and goes on to suggest four indicators that the nation-state is past its sell-by date:

- Capitalism, now global, transnational, post-industrial, 'informational', consumerist, neoliberal and 'restructured' is undermining the nation-state—its macroeconomic planning, its collectivist welfare state, its citizens' sense of collective identity, its general caging of social life.

- New 'global limits' especially population and environmental threats, producing perhaps a new 'risk society' have become too broad and too menacing to be handled by the nation-state alone (Franklin 1998).
- 'Identity politics' and new 'social movements', using new technology increase the salience of diverse local and transnational identities at the expense of both national identities and those broad class identities which were traditionally handled by the nation-state.
- Post-nuclearism undermines state sovereignty and 'hard geo-politics' since mass-mobilization welfare underpinned much of modern state expansion yet is now irrational.

To quote Held (1991, p. 147) the nation-state finds itself in:

> the context of a complex, multinational, multilogic international society, and a huge range of actual and nascent regional and global institutions which transcend and mediate national boundaries... by taking the nation-state for granted, and by essentially reflecting on democratic processes within the boundaries of a nation-state, nineteenth and twentieth century democratic theory has contributed very little to understanding some of the most fundamental issues confronting modern democracies...

Held's model of the relationship between states, borders and globalization is shown in Fig. 4.2. Meanwhile Newman and Paasi (1998, p. 186) assessed the relationship between states, boundaries and globalization from a political geography viewpoint suggesting that we had seen 'territorial transformation' at a global scale. They cited a large number of other authors who have also considered these issues in recent years as evidence of the significance of globalization to boundaries, scale, territory and the state. These include Blake and Schofield (1987), Sahlins (1989), Grundy-Warr (1990), Rumley and Minghi (1991), Dodds (1994), Donnan and Wilson (1994), Galluser (1994), Girot (1994), Johnson (1994), Schofield (1994), Schofield and Schofield (1994), Biger (1995), Forsberg (1995), Newman (1995), Anderson (1996b), Gradus and Lithwick (1996), Krishna (1996), Paasi (1996a, b), Shapiro and Alker (1996), Welchman (1996), and Yiftachel and Meir (1997). It is safe to assume that there have been considerably more since then as well focusing on how boundaries define states and how globalization has effectively changed state boundaries, if not physically, certainly in practice and in the way they work. Without them (or with ones less well-defined) governance of any activity and the generation of effective policies is impossible.

Yeung (1998a, pp. 293, 299) placed considerable emphasis on the relationship of capitalism to space, place, territory, boundaries and the nation-state. He considered that states would not disappear as they adapt by internationalizing themselves, losing traditional capital accumulation functions but gaining through globalization networking. He cited a number of researchers who have looked at the close relationship between capital and space (and incidentally the nation-state). These included Gottdiener (1987), Harvey (1989), Lefebvre (1991), Swyngedouw (1992) and Yeung (1998a). Globalization intensifies the relationship as 'space is seemingly commanded and consumed by capital for further accumulation'. Final completion of this consumption should create the 'borderless world' but Yeung is

4.1 The Decline of the Nation-State

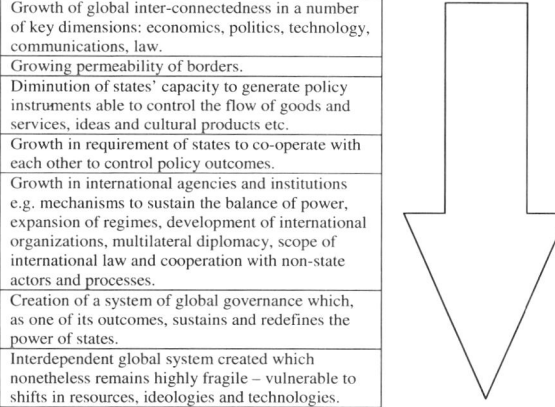

Fig. 4.2 States, borders and international co-operation. *Source* Derived from Held (1991, p. 147)

Growth of global inter-connectedness in a number of key dimensions: economics, politics, technology, communications, law.
Growing permeability of borders.
Diminution of states' capacity to generate policy instruments able to control the flow of goods and services, ideas and cultural products etc.
Growth in requirement of states to co-operate with each other to control policy outcomes.
Growth in international agencies and institutions e.g. mechanisms to sustain the balance of power, expansion of regimes, development of international organizations, multilateral diplomacy, scope of international law and cooperation with non-state actors and processes.
Creation of a system of global governance which, as one of its outcomes, sustains and redefines the power of states.
Interdependent global system created which nonetheless remains highly fragile – vulnerable to shifts in resources, ideologies and technologies.

not convinced as capital actually enjoys the presence of states which enable them to exploit national differences (*cf* shipping) resulting in a 'diverse mosaic of geographical patterns'. The existing national borders might become more porous and changed in the way that they communicate with economic, cultural and social life but they do not disappear. Global corporations emerge which are both placed and placeless—the classic shipping company with assets constantly moving between different nation-states as financial, regulatory and legal incentives change.

Jessop (1997, pp. 573–576) and Meyer et al. (1997, p. 145) were all quite certain that the role of states had changed dramatically with world models shaping what they are able to do especially in the post-war era as 'the cultural and organizational development of world society has intensified at an unprecedented rate'. Their role continues but is redefined. They also saw a linkage with hierarchical frameworks of governance:

> The almost feudal character of the parcelized legal-rational sovereignty in the world (Meyer 1980) has the seemingly paradoxical result of diminishing the causal importance of the organized hierarchies of power and interests celebrated in most 'realist' social science theories.

This suggests that globalization has altered the role of the nation-state and in particular reduced the influence of the hierarchical governance frameworks that characterize policy-making at present. Evans (1997, p. 64) reflected that Nettl's (1968, p. 591) conclusion that '(t)here remains only the one constant—the invariant development of stateness for each national actor in the international field' had been inverted. The invariant has become the international arena's negative effect on states whereby the international system continues to be driven by the nation-state but trans-national corporations now provided the economic power. Evans continued to admit that these same corporations (and this includes shipping interests as much as any other) need nation-states to protect their interests but in turn, this does not protect those same nation-states from decline and change (Evans 1997, p. 72). Meanwhile Falk (1997, pp. 125, 129) felt that:

> territorial sovereignty is being diminished on a spectrum of issues in such a serious manner as to subvert the capacity of states to control and protect the internal life of society, and non-state actors hold an increasing proportion of power and influence in the shaping of world order.

He went on:

> the state itself has been 'globalized'… the policy orientation of the state has been pulled away from its territorial constituencies and shifted outwards, with state actors characteristically operating as an instrumental agent on behalf of non-territorial regional and global market forces.

As a consequence, business elites and transnational corporations are now the most significant policy players—and shipping is no exception. The state's main influence is as a repository for 'ultra-modern military technology' (Sakamoto 1994; Falk 1996).

Rhodes (1997, p. 13), citing Held (1991) also stressed what he termed the 'external hollowing out' of the nation-state which limits their autonomy. Four processes were identified lying behind this trend:

- The internationalization of production and financial transactions.
- The development of international organizations.
- The growth of international law.
- The emergence of hegemonic powers and power blocs.

The result has been a weakening of the nation-state's capacities for governance. Mathews (1997, p. 65) saw the 'clash between the fixed geography of states and the non-territorial nature of today's problems and solutions, which is only likely to escalate, and strongly suggested that the relative power of states will continue to decline'. Gottmann (1973, p. 81) saw evidence of this in the continued geographical regionalization of political systems exemplified by the emergence of the EU, created to reinforce security and 'broadening the framework of economic opportunity'. As such it was a 'logical consequence of the consciousness… states achieve of their relative weakness and frailty'. Despite this Rhodes viewed the state as still pivotal (Hirst and Thompson 1995, p. 408) as it plays an essential role 'suturing' power upwards to the international (and supranational) level and downwards to sub-national agencies.

Meanwhile Jessop (2003a, p. 30) identified four main problems for states:

- The rise of an uncontrolled and possibly uncontrollable global capitalism;
- The emergence of a global risk society;
- The challenge to national politics from the politics of identity and new social movements based on local and/or transnational issues;
- The threat of new forms of terrorism and decentralized network warfare.

However, he gave little indication if the state would survive these issues.

4.1 The Decline of the Nation-State

Hirst and Thompson (1995) continued to see the state as a source of constitutional ordering in:

> A world composed of diverse political forces, governing agencies and organizations at both international and national levels [which] will need an interlocking network of public powers that regulate and guide action in a relatively consistent way, providing minimum standards of conduct and relief from harm.

Podolny and Page (1998) suggested that the state and its hierarchical characteristics will inevitably decline as the need for more complex forms of governance (characterized by networks) becomes more obvious as the process of globalization continues. Simply the continued existence of nations suggests cultural and societal differences which are being eroded by globalization. The result is that the nations themselves are under threat. Meanwhile Arrighi (1994, p. 329) thought that the nation-state would be retained but at the same time its status abused, quoting Lomnitz (1988, pp. 43, 54) who commented on national economies in general in that:

> (t)he more a social system is bureaucratically formalized, regulated, planned, and yet unable to fully satisfy social requirements, the more it tends to create informal mechanisms that escape the control of the system. (They) grow in the interstices of the formal system, thrive on its inefficiencies, and tend to perpetuate them by compensating for shortcomings and by generating factions and interest groups within the system.

The relevance of these interstices in the context of shipping and its policy failures is clear. Taylor (2000) suggested that if globalization can be interpreted as posing a threat to the survival of the state then how can its continued existence be assumed as well? The state becomes explicit rather than the implicit beast which it has assumed. Weiss (2000, p. 796) continued the theme in his discussion of global governance, suggesting that the organization of the world is based upon 'unquestioned national sovereignty', but quoting the former UN General-Secretary Boutros-Ghali, that 'the time of absolute and exclusive sovereignty has passed' (Boutros-Ghali 1992, para. 17). As Weiss put it (Weiss 2000, p. 803), 'sovereignty is not dead, but it is hardly as sacrosanct as it once was'. To quote Gottmann (1973, pp. 126–127):

> The sovereign state, based on exclusive territorial jurisdiction, may have been the evolution's purpose from the sixteenth to the mid-twentieth century. By 1970 sovereignty has been by-passed, and a new fluidity has infiltrated the recently shaped map of multiple national states.

Peters and Pierre (2001, p. 132) saw the state as no longer the 'unrivalled king of the hill' challenged by supranational institutions such as the EU. A common trend was seen as state-centred societies becoming less so, opening up new patterns of interactions between different tiers of government and key actors.

Johnston (2001, p. 685) described the decline of the nation-state which by the end of the twentieth century had become 'leaking containers'. He cited Taylor (1994) as suggesting that the state would remain with us for some time because of vested interests, inertia and the fact that the 'territorial container was so full'.

King and Kendall (2004, p. 168) provided much support for the continued role of the state but at the same time questioned how long it can survive as it finds itself like:

> ricocheting billiard balls on the international pool table.

Scholte (2004) emphasized in some detail how following 50 years of globalization, statism no longer characterized governance. Although this did not mean the end of the nation-state as such, it did suggest that both supra and sub-state authorities can now play a major part in governance that was previously confined to the nation. Scholte also discussed the reasons for the decline of the state in governance and in particular the impact of globalization.

Meanwhile Gottmann (1973, p. 157), in comments that have direct relevance to the shipping industry, emphasized how national sovereignties have been weakened by the networks created by large, international corporations, commonly based in one country under the flag of one nation, but with scattered and diverse operations whereby it is impossible to determine the nationality of any of these activities. De Vivero et al. (2009, pp. 624, 628) also stressed that a necessary transition to new concepts of maritime sovereignty was occurring from when the main actor was the traditional state to one where there was a new form of state related political activity. The state remained the dominant actor but with a more international vision exemplified by the EU's Blue Policy Paper and reflected in the transfer of power upwards to supranational and downwards to regional organizations. Partly this was a response to a recognition that maritime policies are unlike other, conventional state policies and are special because of the domain to which they relate.

The changes affecting the nation-state are thus wide-ranging. Van Hoof and Van Tatenhove (2009, p. 727) provided considerable evidence that maritime policies have to take these changes into account and in particular, the declining significance of the nation-state where they see that 'politics and policy are no longer framed within the nation-state alone, but within a diversity of society-centered forms of governance'. Speaking in broader terms, De Senarclens (1998, p. 101) provided a clear summary of what he saw states have lost and retained within a political framework and although he referred only to OECD countries, his comments provided a useful commentary on what has been discussed here. He saw states as having:

> lost some of their fiscal, monetary and financial independence because budgetary deficits and policies likely to encourage inflation are heavily penalized by the markets.... security and international affairs are frequently influenced by the ambitions and decisions of a few major powers. As a result governments' political independence can be restricted. Essentially however, they retain the monopoly of legitimate violence. They offer an institutional framework for political and social cohesion, settling disputes, sharing resources and political guidance. Political disputes continue essentially to take place within their political space. Thus the expansion of markets and communications does not prevent governments from continuing to be the main centers of political power, purveyors of standards and regimes and prime sources of political allegiances. They take on an essential role in implementing a stable legal framework, maintaining internal security, bringing about conditions favourable to economic activity and employment, controlling poverty,

> protecting the environment, and managing questions of education, training and health. Governments also guide national policies in the fields of trade, investment and industry. The fact that they are engaged in a substantial network of foreign cooperation does not affect their citizen's perception of their sovereignty or their legitimacy. They continue to claim to exercise authority over people entitled to live within their borders, whether temporarily or permanently. The sovereignty of the state continues to embody the myth of an unbreakable link between a nation and its territory.

Phew! But although a long quote it does summarize nicely many of the arguments we have seen discussed in this section and has particular relevance in the light of events of 2011 in the Eurozone. Like De Senarclens, Soja (2000, pp. 194–195) did not deny the continued existence of the nation-state despite the emergence of what he saw as a new Postmodern era where growth is associated with global and world markets commonly outsell domestic. By the third quarter of the twentieth century, the nation-state had become entrenched as a 'container of political power' and community and alternative forms of governance had 'disappeared from the global political imaginary' (Soja 2000, p. 204). Hobsbawm (1998, p. 4) observed that as far as the nation-state goes, what we now have is a 'dual system' made up of an official national economy and an unofficial transnational one. Apart from territory and limited domestic power, all other facets of the nation-state can be (and may well have already been) overridden by globalization.

The nation-state has clearly changed and tried to adapt to the new conditions—but it has certainly not gone away. Despite the pressures of globalization it retains a governance role—but a different one from that envisaged when the hierarchical model which characterizes the maritime sector was devised and also one that possibly it cannot fulfill in a society where the nation-state contrasts markedly with global networks, locational fluidity and virtual economies. Could this be a reason for continued maritime policy failure?

4.2 State Survival

> It is also said that the power of the state has been undermined when in fact the state has increasingly been restructured politically and economically as 'an executive committee for the ruling class' as Marx long ago suggested… the neoliberal phase of globalization has been characterized by a reconfiguration of state powers and the geographical concentration and centralization of political-economic powers within regional alliances of immense strength. Harvey (2001, p. 29)

Harvey clearly appreciates that the nation-state has changed—and this in itself suggests that maritime policy-making might need to change with it—but he also does not see it necessarily in terminal decline. Although we may not agree entirely with his Marxist interpretation, the argument that the state remains with us albeit in a new form, is appealing and something that Swyngedouw (2000, p. 68) emphasized:

This century, it has undoubtedly been the national state that became iconized as the pre-eminent expression of political forms of territorial organization.

Walker (1991, p. 445) was equally as confident in the significance of the modern state:

> The presumed anarchy among states has been an anarchy of a select few, and considerable resources have been procured to delineate and defend the borders. But it is this proliferation, affirmed by accounts of the modern state as institution, container of all cultural meaning and site of sovereign jurisdiction over territory, property and abstract space, and consequently over history, possibility and abstract time that still shapes our capacity to affirm both collective and particular identities. It does so despite all the dislocations, accelerations and contingencies of a world less and less able to recognize itself in the fractured mirror of Cartesian coordinates.

Wright-Mills (1959, pp. 135–136) had stressed the position of the nation-state back during Modernist times and the suggestion that a Postmodern era might emerge had been suspected. He saw the nation-state as the 'most inclusive unit of social structure' and as such was a 'dominating force in world history'. The organization of politics, the military, culture, economics and power were all concentrated in state hands along which all institutions and 'specific milieux'. Wright-Mills saw the state as the frame within with all studies took place—both those relating to larger and smaller spatial structures—the global and the regional for example. In 1959, most, if not all economists and political scientists took the nation-state as their unit of analysis.

Commenting on the nation-state some years later, Nettl (1968, p. 559) stated that: 'The thing exists and no amount of conceptual restructuring can dissolve it'. Picciotto (1997, pp. 1015, 1019, 1021) felt that even though new forces of globalization were at work, the nation-state retained much of its political structure and power although increasingly threatened by the breakdown of international arrangements for allocating competencies between states. Opello and Rosow (2004, p. 2) agreed:

> Although they are being challenged by the forces of globalization… nation-states, having eclipsed all other types of politico-military rule that have existed on the planet, are and will continue to be for the foreseeable future, the basic building blocks of the global order… so ubiquitous that its existence is taken for granted, rarely noticed even by scholars of international relations.

Duchacek (1988, p. 43) deemed sovereign states as the only actors that really counted on the international scene whilst Cohen (2001, pp. 79–80) suggested that we tend to view the sovereign state as a permanent and unchanging feature, an unquestionable, with fixed boundaries and as such power is devolved to them because it is logical, safe and predictable. The nation-state is thus 'a boundary-setting and maintaining institution'. Unbundling of sovereign states and their boundaries is unthinkable. MacLeod and Goodwin (1999, p. 508) quote Harding (1997, p. 308): 'little can happen subnationally without the [national state's] cooperation, acquiescence or benign ignorance'. Alamgir (1999, p. 110) in his review of Weiss's *The Myth of the Powerless State* (1998) was clear in his

4.2 State Survival

assessment that the state is not powerless but a 'conglomeration of varied crystallizations' which has enormous abilities to transform and adapt. The idea that its role and format may well have changed, is a view re-emphasized by Le Gales and Harding (1998). Krasner (1984, pp. 234–235) considered that all institutional structures struggle to respond fluidly or quickly to environmental change and that 'incongruence between the needs and expressed demands of the state and various societal groups is the norm'. In the words of Bagwell and Staiger (2001, p. 519) nations may well need to 'cede control over their economic and social policies if global efficiency is to be achieved' but this does not mean they have no future. Crises and dramatic change are seen as central—something that suggests the importance of a new Postmodern epoch and the primary role of disorganized capitalism. Once crises are passed the new institutional structure tends to become rigidified to the detriment of the consumers of state policy—something that may be reflected in shipping policy. The state is particularly poorly prepared for such change generated by crisis—institutions are self-perpetuating—both domestic and international, and this is reinforced by inertia which focuses not only on sunk costs but also in terms of networks and trust (Skowronek 1982, pp. 10, 20).

Puchala (1972) was an early commentator on the ability of the state to survive. Commentating on international integration and the role of functionalism, the predicted death of the nation-state had just not happened, remaining 'conspicuous and pivotal' particularly in international integration. Functionalists expected states by now to be wholly peripheral. Meanwhile, Hoffmann (1982, p. 21) stressed the fact that the state has survived despite what he terms its 'frequent and well-noted impotence'. Emphasizing a massive decline in political confidence, the challenge to order of consumer and capitalism, the 'desacralization' of state institutions in the 1960s typified by the events in Paris in 1968, repeated assaults in the form of inflation and recession, the decline of traditional industries, despite all this and more, the state is still there. Particularly relevant to the shipping sector, he suggested that the state has survived:

> insidious and perverse effects of economic interdependence, which entail a loss of control by the state over various essential instruments of policy... and because of the capricious role of uncontrolled forces – public or private external actors – on which no state has a grip. Hoffmann (1982, p. 22)

He went on to imply that the nation-state was on the one hand bankrupt but on the other hand still central to policy-making. Thus the state survives (as we have seen in maritime policy-making) but should it, or is its continued significance an anachronism?

Mann (1993) was confident of the survival of the nation-state whilst Scharpf (1994, pp. 220–221) was sure that the institution of the state was thriving despite the increasing strength of the EU as the latter's decisions were almost always the consequence of multilateral discussions between member state representatives and as such, nation-states would always retain their legitimacy and political viability. Scharpf actually viewed the process of harmonization as strengthening rather than weakening the nation's position as its progress was dependent upon national interests being provided for at each stage. Arrighi (1994, p. 331) considered that the state had lost much power and in addition had now inherited unprecedented responsibilities and restrictions on sovereignty rights. Despite this it retained a central, societal position.

Scharpf (1994, p. 220) continued convinced of the role of the nation-state especially in relation to the development of the European Union. Member states he suggested continued to resist any reduction in their powers whilst they controlled the EU's constitution. The 'democratic deficit' of the EU—exemplified by the inadequacies of a European media, political party or genuine public awareness ensured the continued survival of the nation-state despite the rise of the supra-national (Grimm 1992; Kielmansegg 1992; Scharpf 1992).

Gordenker and Weiss (1995, p. 373) were clear in their emphasis of the dominance of the nation state in transnational cooperation suggesting it was apparent in governments, through international officials and academic research. Anderson (1996b, p. 353), quoting Hoffmann in Hanreider (1967, p. 8) noted that the nation-states of the EU were:

> widely seen as being eroded 'from below' by regionalism and 'from above' by EU institutions and globalization – a pincer movement transforming traditional conceptions of territorial sovereignty and national identity.

This he suggested made the modern nation-state an anachronism in many people's minds, but this was implausible as states retained strength through nationalism. This he felt was a consequence of the emergence of new forms of integration which differed substantially from the conventional models of inter-governmentalism (retaining existing nations but in a new collaborative form) and supra-nationalism (where some form of super state evolves). These new forms he termed 'medievalism' so-called because the new form reflected earlier states of authority in Europe with overlapping and vertically segmented structures. This view is in part supported and in part denied by Ruggie (1993) who agreed that aspects of medievalism could be identified but that a process was going on akin to Postmodernism and Harvey's (1990) conception of a 'modern-Postmodern transition' involving accelerating compression of space-time characterized by globalization (of which much more later) leading to the inevitable unbundling of the nation-state.

Anderson went on to suggest that although Harvey's arguments were convincing and that the nation-state was in decline under the forces of space-time compression, suggestions that the nation-state was in terminal decline and were to be replaced by borderless worlds and spaces of flows remained some way off. The 'unbundling of territoriality' was highly selective and instead we need to begin to think of the state and its sovereignty in the plural rather than the singular and then look at how developments might be varied between different contexts.

Duncan and Savage (1989, p. 181) similarly stressed the significance of the nation-state in developing national boundaries, the spatial pattern of which produced a 'physical fixity' which tended to persist regardless of mounting pressures. Their opinion was backed by Sayer (1985a, b) who claimed that spatial relations

4.2 State Survival   165

were always caused by social objects but cannot be reduced to them once set in place.

Cerny (1995) stressed that the state 'constitutes the key unit of collective action' although he did consider that the traditional state had changed considerably and now incorporated civil associations, enterprises and government, whilst Hirst (1997b, p. 238) insisted that nation-states will retain three key roles whatever 'functions they may gain or lose'. To summarize:

- To legitimize and promote international economic regulation;
- To organize social cohesion and economic cooperation between social interests;
- To guarantee the rule of law and enable pluralism to flourish.

Meanwhile, Johnston (1995, p. 217), following the work of Clark and Dear (1984, p. 43) similarly suggested that states were necessary for three reasons:

- To secure social consensus so that the majority of groups in society agree to accept the way in which it operates.
- To secure the conditions of production, by investment that increases production and regulating patterns of consumption.
- To secure social integration to ensure the welfare of all groups.

With or without globalization, these functions are likely to remain in some form and as such the nation-state will retain a role. This is especially the case because a capitalist society needs them to be performed. The relationship between globalization and the nation-state is thus dichotomous with the trends towards a global society both demanding a national structure whilst at the same time tending towards its destruction. The characteristics of the residual national role and its implementation in the context of a growing global society may well be different. However, in all scenarios, territoriality is a necessary strategy for the nation-state (Johnston 1995, p. 219). The result is a changed nation-state but one still with a territorial basis, with identifiable if fluid boundaries and a new relationship to economic, political and social activity. Maritime governance take note. To quote Bell (1988, B3) in Kettl (2002, pp. 12–13):

> The common problem, I believe is this: the nation-state is becoming too small for the big problems of life, and too big for the small problems of life. It is too small for the big problems because there are no effectual international mechanisms to deal with such things as capital flows, commodity imbalances, the loss of jobs and the several demographic tidal waves that will be developing in the next twenty years. It is too big for the small problems because the flow of power to a national political center means that the center becomes increasingly unresponsive to the variety and diversity of local needs. In short there is a mismatch of scale.

Following the work of Reich (1992), Hirst saw a positive role for national governments:

> to ensure that their societies are internationally competitive and that they can offer to international capital, attractive locations based on efficient infrastructure at low cost and a highly trained labour force. National governments are the municipalities of the global system, they offer locally the public goods that business needs.

As such they can be considered vital to the process of globalization rather than the victim (Alesina and Spolaore 1997).

Bulmer (1983, p. 353) quoted Hoffmann (Hanreider 1967, p. 8) who had earlier affirmed this view:

> An international system is a pattern of relations between the basic units of world politics... This pattern is largely determined by the structure of the world, the nature of the forces which operate across or within the major units, and the capabilities, pattern of power, and political culture of those units.

The basic units were what Bulmer saw as the nation-states still possessing the power to 'defend their interests' in considering the impact of global classics such as oil supply and pricing and monetary instability and to choose at what level to do this—national, supranational, international or a combination of any of them. The domestic tier thus remains fundamental within the global economy. This significance he stressed through emphasizing the role of the nation-state within the EU identifying four main factors:

- The nation-state remains the basic political unit and is the level that governments, parliamentary bodies, interest groups and political parties gain legitimacy.
- Each state has its own ideological cleavages, relationships with the outside world and distinct social and economic conditions.
- EU and national policies may look the same but policies relating to other issues often do not. Therefore nations remain distinctive.
- National governments remain the key players in EU political activity and policy-making.

Bulmer concluded that in particular, it is 'the national level where the decisions are taken as to which forum is the most appropriate for the state's interests' and that national governments remain resilient in the face of continued global erosion of their role.

Picciotto (1998, p. 4) cited Slaughter (1997, p. 195) in that the national state was still the 'primary arena for legitimation and enforcement of social norms, in contrast to the more conventional view of globalization as involving a shift of power away from the state' (Mathews 1997). Taylor (1994, p. 151), citing Gottmann (1973), emphasized the social role of the state considering that:

> Across the whole of our modern world, territory is directly linked to sovereignty to mould politics into a fundamentally state-centric process.

Taylor (1996, p. 1919) took a specifically spatial view of the state seeing sovereign territory as the key unit of spatiality.

> As the spaces defining the limits of recognized governmental authority, these territorial entities are of obvious political significance but they have been treated as much more than mere polities. Nearly all social science has assumed that these political boundaries fix the limits of other key interactions.

4.2 State Survival

He considered this 'social formation' was widely assumed to coincide in space with political boundaries and that this goes for economic transactions as well which even with globalization, are confined to the cultural territory defined by the USA (or UK or Chinese, etc.) economic system. Citing Wallerstein (1984):

> This is all premised on a spatiality that divides the world's land surface (plus adjacent sea and air) into spatial segments under the control of states. In this sense, social science is very much a creature, if not a creation of the states... rather than being either inert or a contextual residual, there is a spatial framework at the very heart of social science knowledge. Social science has been a fundamentally state-centric activity and therefore the knowledge so produced has been spatially guided in its creation. (Taylor 1996, p. 1919)

Taylor (2000, p. 1106) was also convinced that the assumption of the state as a basic unit in analysis meant that its consideration was often negligible and:

> whether studying the economy, society or government, the theory of the state was only ever implicit and therefore very rudimentary – in one critique C.B. MacPherson (1977) famously asked 'Do we need a theory of the state?'.

Agnew (1999, p. 503) was a firm believer in the continued role of the state suggesting it was the 'geographical container of modern society' and that statehood was the 'unique source and arena of political power in the modern world'. He went on to analyze a number of models of 'spatialities of power' that stress the central role of the state and actually accentuates their role as global integration and cultural unification continues (Agnew 1999, pp. 504–507). Agnew here was building on his earlier work (Agnew 1994, p. 55) where he saw space as invariably territorial and as a series of blocks defined by state boundaries. The state was so dominant that all other jurisdictions (for example global, supra-national, regional and local) could be disregarded. This could be assumed for a number of reasons that have led to a 'privileging of the territorialization of the state':

- State territory has been 'reified' which in turn has led to a process of dehistoricization and decontextualization rendering the state invioble (Weber 1978; Walker 1993; Ferguson and Gupta 2002, p. 982).
- The fixation upon links between national and international activity has obscured the processes of globalization going on and between other levels (for example, international impacts upon regional activity).
- Nation-states are seen as societally untouchable and essential. To quote Virilio (1992, p. 284): 'That an impersonal structure of domination called the state is the core of politics is an idea so deeply imbedded in our ways of thinking that any other conception of it appears counter-intuitive and implausible'.

This discussion suggests two things in the context of our wider remit. Firstly that maritime governance is no exception to this rule of the social sciences—it too has been driven by the assumption of the spatiality of the state. And secondly, perhaps the rise of globalization has created tension within a social science system that is rooted in state space; a tension that manifests itself practically in governance inconsistency and policy failure.

Taylor (1994, 1996) also suggested that getting rid of the state was going to be difficult regardless of its inconsistency in the context of globalization as states have become 'naturalized'. This term refers to the fact that states are considered to have a natural existence, similar to mountains, rivers and lakes, and as a result their existence is unquestionable precluding all other 'social worlds' and also taking precedence over all other spatial scales—notably international and local. However, Taylor suggested that from about 1970 changes in the social world order have made escape from state-centric thinking a necessity if understanding social science processes is to be possible.

Brenner (1999) considered at some length the role of the state and its relationship to the issue of geographical scale. Taking the social sciences as a whole he considered (along with others) that much of the social sciences had long been locked into a 'state-centric territorial trap' in which all social, political and economic relationships were viewed. Globalization was seen as creating problems for these issues making the permanently defined, self-enclosed and self-defined state units less than helpful. He emphasized that geographical research from the 1980s had began to question this situation and to 'challenge the iron grip of the nation-state on the social imagination'. States however, were viewed not as disappearing but as reinventing themselves. Their earlier existence as structures essential for capital formulation had now passed but the continued 'artificial' emphasis upon the state clouded our understanding of the nation-state's new role as an agent of globalization and how it was being reformulated during this process. Therefore:

> the effort to escape the 'territorial trap' of state-centrism does not entail a denial of the state's continued relevance as a major geographical locus of social power, but rather a rethinking of both state territoriality and political space in an era of intensified globalization. (Brenner 1999, p. 41)

States have thus become stabilized forms of territory for 'successive rounds of capital accumulation' since the nineteenth century. National political regimes have constructed:

> large-scale territorial infrastructures for industrial production, collective consumption, transportation and communication.

Lefebvre (1977) termed this the 'state mode of production' deliberately aimed at intensifying nationally specific capitalist production and institutional regulation of uneven forms of geographical development induced through the earliest stages of globalization of the nineteenth century to the benefit of those states which regulated the whole process. Globalization thus initially consolidated state-centrism and it is only in its later stages that the state can now be considered approaching redundancy. But much more of this later.

Brenner went on to discuss more fully how state-centrism has two main features that endure:

- It is a static platform of social action that is not constituted or modified socially.
- It assumes a preconstituted, naturalized and unchanging scale of analysis.

4.2 State Survival

The first resulted in 'spatial fetishism' in which space is timeless and immune to historical change. The second required 'methodological territorialism' which analyses all spatial forms and scales as 'self-enclosed and territorially bounded geographical units'. Thus the assumption is made of a society comprising naturally defined territorial units (states) with a permanent existence which define activities and decisions (and in our case maritime policies) and which cannot take account of changes around them (globalization) which is destroying the foundations upon which they have been built (and therefore making the policies derived through it redundant as well). To quote Brenner (1999, p. 48):

> the interstate system came to operate increasingly like a 'vortex sucking in social relations to mould them through its territoriality'. (Taylor 1994, p. 152)

Or as Lefebvre (1974) suggested [and Giddens (1985, p. 120) re-emphasized] the modern state is no less than a form of:

> violence directed towards space. Each state claims to produce a space wherein something is accomplished, a space, even, where something is brought to perfection: namely a unified and hence homogenous society. [But] the space that homogenizes... has nothing homogenous about it.

This national territorialization is represented as a 'natural precondition of social and political existence rather than... a product of historically determinate strategies of parcelization, centralization, enclosure and encaging. To quote Lefebvre (1991, p. 287):

> Abstract space is not homogenous: it simply *has* homogeneity as its goal, its orientation, its 'lens'. And, indeed, it renders homogenous. But in itself it is multiform... Thus to look upon abstract space as homogenous is to embrace a representation that takes the effect for the cause, and the goal for the reason why the goal was pursued. A representation which passes itself off as a *concept*, when it is merely an image, a mirror and a mirage; and which instead of challenging, instead of refusing, merely *reflects*. And what does such a spectacular representation reflect? It reflects the results sought. (*emphasis in the original*)

Brenner (2004, p. 43), quoting Lefebvre (1974), took it up suggesting that:

> territorialization – whether on national, sub-national or supra-national scales – must be viewed as an historically specific, contradictory, and conflictual *process* rather than as a pregiven, fixed or natural condition. By contrast, state-centric epistemologies freeze the image of state territoriality into a generalized ontological feature of social life, and thereby neglect the ways in which the latter has been continually produced, reconfigured, and transformed as a key geographical infrastructure for capital's developmental dynamic.

This may well be the key to why the state still persists despite the clear need for some sort of revision at least of its role and relationships.

Waltz (1999) emphasized that most economies, although globalized to a certain extent, still conduct the majority of their activities domestically, citing the USA as an example where Americans produce 88 % of all goods that they buy. In terms of why this should be so he suggests two main factors: states perform essential sociopolitical functions, and they foster institutions that make internal peace and prosperity possible.

Dearlove (2000) commented on how some scholars point to the continuing power and relevance of the modern state (Weiss 1998) whilst Cohen (2001, p. 77) noted the anticipated death of the nation-state which had been increasingly discussed over the previous 40 years. The usual culprit was globalization and interdependent production and consumption, international flows of capital and increasingly sophisticated information gathering and processing. He quoted Greider (1997, p. 11):

> Old verities about the rank ordering of nations are revised and a new map of the world is gradually being drawn. These great changes sweep over the affairs of mere governments and destabilize the established political orders in both advanced and primitive societies.

The decline of the state which this suggests, losing its relevance as either 'facilitators or obstacles to the function of the global economy', contrasts with the continued role that the state plays. Here perhaps is part of the key to the policy problems that besiege the maritime sector. A globalized economy that suggests little role for the nation-state; a globalized industry that seemingly needs nations less than most commercial activities; and a commercial activity with a governance structure that centres on the anachronism of the nation-state.

To counter this Cohen (2001) went on to show how the nation-state retains a role even in a globalized society emphasizing the central place that states take in promoting globalization and in regulating the finance, trade and people that cross their borders.

However, de Senarclens (1998) summed it up considering that 'social issues' have taken on a significantly more important international dimension while 'political institutions to deal with it are not up to the task'. The state remains the focus of political governance whilst the international arena has come to dominate social issues. In maritime governance terms the hierarchical state-centric structure characterized by the activities of the IMO, EU and nation-state administrations contrasts with the international focus of the shipping industry itself.

Dicken et al. (2001, p. 96) commented on the relationship between networks and states in the context of territories and scales. They saw networks that cross international borders representing a 'qualitative disjuncture between different regulatory and socio-cultural environments' (Dicken 1998; Sassen 2000; Smith et al. 1999; Taylor 1994). They saw national regulation creating 'bounded regions' and nations an 'immensely formative influence'. In their view the nation-state retained considerable influence and significance. This is evidenced in the fact that even companies operating in intensely international markets (and what better example is there than shipping) retain distinct national, organizational forms that reflect the regulatory environment of their home state, citing the work of Doremus et al. (1998) and Yeung (1998a, 2000). Using networks again as their example:

> A network methodology forces us to address the direct and indirect connectivities between economic activities stretched across geographical space but embedded in particular places. Thus we have a mutually constitutive process: while networks are embedded within territories, territories are, the same time, embedded into networks. (Dicken et al. 2001, p. 97)

This is a view that was supported by Whitley (1996, 1998, 1999) who argued that national institutional differences continue to exert a major influence on international economic activity, something reinforced by Kapstein (1991) who saw the 'very firms that are seen as being anational often act in explicit support of state functions and needs', and Ruggie (1993) who quoting Thomson and Krasner (1989) stressed that the assumption that globalization implies the 'growing irrelevance of states' is fundamentally misplaced.

Zurn (2003, p. 359) suggested that globalization has made national interpretations of politics fruitless but at the same time stressed that this does not mean the end of the nation-state, which:

> will persist and will remain of central importance. Relaxing methodological nationalism rather means that the nation, as the dominant organizing principle of politics, can no longer be taken for granted but must be given an empirical status. As governance beyond the nation-state increases in significance, the separation of political issues into nationally defined territorial units must be conceptualized as a variable – dependent and independent – rather than as a conceptual premise.

This reflected the thoughts of Drucker (1997) who had suggested much earlier that no other institution is capable of 'political integration and effective membership in the world's political economy... the nation-state will survive the globalization of the economy and the information revolution that accompanies it. But it will be a greatly changed nation-state'.

Alderton and Winchester (2002, p. 36) considered the nation-state central to shipping and in this they conflicted with Sletmo (2001) and his view that the state was irrelevant to maritime policy. They rightly state that the ship's nation of registration—its flag—has exclusive dominion over a ship and no other state can exercise dominion regardless of the vessel's location. 'Freedom is thus the guiding principle of the law of the sea, but it is a principle strongly mediated by nationality'.

Jessop (2003a, p. 40) also saw the state taking on a new form following the processes of denationalization and the reordering of political hierarchies, enabling it to 'mediate between an increasing number of significant scales of action... (playing) a central role in interscalar action'. This new form he termed a 'Schumpeterian post-national workforce regime', replacing the existing 'Keynesian national welfare state'. The important change here is in the post-national condition which Jessop viewed as reflecting a reduction in the significance of national territory and the transfer of economic and social policy-making functions upward (for example internationally and supranationally), downward (for example locally, regionally and to the individual) and sideways (for example from the public to the private sector). States remain important but significantly different in how they exercise influence. Stubbs (2005, p. 77) agreed, citing Ferguson and Gupta (2002, pp. 994, 996) who felt that it was increasingly difficult for states to present themselves as 'legitimate' or even 'natural' authorities as other jurisdictions take on responsibilities and become involved in both horizontal and vertical policy relationships.

Scholte (2004) emphasized the decline of the state but also suggested how it retains a role in modern governance albeit in a different form. Thus states could affect the impact of globalization with monetary, fiscal, consumer, labour, environmental, data protection policies and many more. Scholte actually considered that stronger states can use globalization to strengthen their impact whilst the weaker lose out. Death notices for the state are therefore premature—its role, function and even territory (note recent fragmentation in Czechoslovakia, the Yemen, Yugoslavia, the USSR and Ethiopia) may change but it has yet to disappear; we have simply entered a post-Westphalia period.

King and Kendall (2004) strongly defended the role of the nation-state despite globalization citing Hirst and Thompson's (1995) view that states were not so much seeing their authority being 'hollowed out' as retaining (if not increasing) important regulatory powers. Hirst (1997a) went on to claim that globalization was more a myth than a reality and that most multi-national companies are 'predominantly national and still dependant on national social and political solutions'. He saw the nation-state retaining a crucial role in international governance legitimizing international decisions and regulation. Nation-states and international bodies can best be seen as:

> partners seeking a larger market share of investment and economic activity and who engage in mutual 'diplomacy' to achieve the best terms. (Brown 1997)

This persistence is maintained by the strength of territoriality, the need for national defence and security and popular support for national representation. Perhaps more significantly they see international institutions maintained because they 'convey advantages for the states that make up the world community.' To quote King and Kendall (2004, p. 171) once more:

> linkages upwards towards trans-national governance institutions and downwards towards sub-national government should be more thought of as state strategies to reassert control and not as proof of states surrendering to competing models of governance.

Aalberts (2004) also retained faith in the nation-state despite the inexorable progress of multi-level governance structures. The latter were normally associated with pyramidal models of decision and policy-making but the traditional hourglass nested arenas still exist with the state acting as gatekeeper between jurisdictions both above and below. Although a clear pyramidal model may eventually emerge, currently the situation is more akin to Waltz's (1979) complex anarchy of cross-cutting relationships as the hierarchical governance structures still have to develop more clearly. As she says—'to date, sovereignty has not disappeared to make way for a European sovereign state'.

This sort of positive view of the nation-state's future is repeated in Mangat (2001, p. 6) who considered that globalization had far from eroded the nation-state. The latter was losing its autonomy of decision-making but remained almost the universal conduit for international relations. Meanwhile Braithwaite and Drahos (2000, p. 475) discussed in some detail the role of the international shipping industry and its global regulation where they concluded that although there was a

4.2 State Survival

plurality of actors, the one still with the greatest influence was the nation-state. They quoted Stopford and Strange (1991, p. 1):

> States are now competing more for the means to create wealth within their territory than for power over more territory. Where they used to compete for power as a means to wealth, they now compete more for wealth as a means to power.

However, Braithwaite and Drahos (2000, p. 484) also suggested that the nation-state was not quite as in control as it might first seem with states best seen as agents of other actors and in particular major international corporations. In similar fashion, Brown (1997, p. 120) viewed the nation-state as the key player in major UN functional agencies such as the IMO rather than as agents in themselves for policy change.

Chowdhury (2006, p. 141) went even further in her analysis of the relationship of globalization to postcolonial theory. She saw that capitalism always needs a nation-state as they are mutually beneficial to each other. Citing Wood (2001, p. 36), she viewed appropriation through the processes of capitalism as always separate from but also requiring enforcement by 'legal, political and military instruments external to the "economy", as well as support from extra-economic social institutions'. Whatever the extent of globalization, it will rely upon 'spatially limited constituent units with a political and even an economic logic of their own'. The nation-state, globalization and shipping are inseparable but not unchanging.

Finally we can consider the difficulty of the Postmodern state that Cooper (2000, p. 23) identified. He saw that a Postmodern state was developing but that the concept of democracy, and in particular the democratic institutions that continue to be developed, are 'firmly wedded to the territorial state'. He suggested that there is a package of national identity, territory, army, economy and institutions that has been 'immensely successful'. Consequently, although 'economy, law-making and defence may be increasingly embedded in international frameworks and the borders of territory may be less important' both identity and institutions remain primarily national. The traditional state will remain the fundamental unit of international relations even though it may exercise this in traditional ways. Or in the words of Cable (1995, p. 38):

> It is tempting to conclude that economic globalization has made the traditional nation-state redundant. The truth is much messier.

Borders still remain of course. Christiansen and Jorgensen (2000, p. 62) suggested that they remain very significant but reflecting the continuing changed status of the nation-state. There is border chaos with different and changing definitions dependent upon sector, issue and context and there are clear contradictions with state borders fundamental to the state system whilst conflicting with changing political reality. The concept of border still defines the modern nation-state system and the 'principle of territorial sovereignty on which states base their legitimacy and power is unthinkable without the presence of a boundary' (Christiansen and Jorgensen 2000, p. 63). However, borders have changed in the light of globalization—Cable's (1995, pp. 23–24) complex iterative process with the nation-state losing sovereignty to regional and global institutions but also acquiring new areas of influence. Policy-making and the governance of the maritime sector cannot avoid these trends.

4.3 Time for a Review!

An issue of policy failure in the maritime sector is apparent. Repeated policy initiatives which aim to overcome the failings in the sector have done little to improve the situation. We see continued inadequacies in proposals put forward, their repeated rejection and no change in the approach adopted—only the policy detail. This suggests something more than the generation of inadequate solutions to maritime problems. Rather a systemic failure of the policy-making process, something that needs to be addressed in the approach to developing and implementing policies rather than the detail of the policies themselves.

Our analysis of the existing maritime policy-making process has suggested that it has a number of significant characteristics. These centre in particular around its hierarchical construction with the concomitant requirements for jurisdictional coordination and conformity and structural rigidity; and perhaps more importantly, its state-centric characteristics stemming from its origins and focusing upon the central role of the nation-state in all jurisdictions, and the key role that the nation-state plays in the development, design, acceptance and application of all maritime policies.

Although so far we have only hinted at it, the significance of globalization as a movement in society to both these key aspects of maritime policy-making—hierarchical integrity and state-centrism—is hard to over-estimate. Given the apparent systemic failure of the maritime policy-making process, the rigid structure that characterizes it, and the emergence of a fundamental change in society based around globalization with obvious implications for both jurisdictions and states, perhaps we have a clue here to where things may be going wrong.

Consequently there is a need to look more closely at globalization and its relationship to the maritime sector and policy-making, to assess its impact upon the whole issue of governance and to see whether we can identify a relationship between its development and governance frameworks in the maritime sector which may help to explain the problems faced by policy-makers and guide us in what can be suggested to improve the process. However, before this, it is clear that the suggestion that policy problems are related to some underlying globalization issue which impacts upon the jurisdictional and state-centric characteristics of the process requires more evidence that the position of the nation-state is significantly influenced by current and recent societal globalization trends.

4.4 The Nation-State Under Global Pressure

Our discussion of the nation-state has suggested strongly that things are changing. The purpose of this section is to assess the significance of these changes and to establish whether they are significant enough to imply that globalization has fundamentally altered the context for governance in the maritime sector and consequently the need for change in the policy-making process.

We noted earlier that the nation-state has been in predicted decline for some time. In fact Drucker (1997) suggested that its fate has been anticipated by Kant (1795), Marx (1992), Russell (1916) and Davidson and Rees-Mogg (1999) amongst very many others, and yet it showed 'amazing resilience'. Drucker's view was that there was no other institution of political integration but that it will change in its 'domestic fiscal and monetary policies, foreign economic policies (and) control of international business'. Meanwhile Beck (1992, 1994) suggested that what he saw as the rising 'risk society'—characterized by ever increasing appreciation and over-recognition of environmental, social and technical risks— presented increasing threats to the nation-state as globalization makes such risks ever more apparent (and actually significant in many cases).

Ruggie (1993) questioned whether states were needed at all as territoriality is not a requirement of social organization. Myoshi (1993) concurred and Hsu (2000) cited Dicken (1998), Ohmae (1995), Lipietz (1995) and Storper (1997) as suggesting that the nation-state has ceased to be the only legitimate or even the most productive analytical unit for economic development, its effectiveness varying with industry and in the context of other institutions. Anderson (1996a) commented on how many saw the nation-state as an anachronism, eroded from below by regionalism and above by globalization although he did countenance the view that it was simply a new form of state emerging. His criticism of the nation-state coincided with that of Ruggie (1998, pp. 184–186) who saw:

> political space (come) to be defined as it appeared from a single fixed viewpoint. The concept of sovereignty then, was merely the doctrinal counterpart of the application of single-point perspectival forms to the spatial organization of politics... territorially defined, territorially fixed, and mutually exclusive state formations.

Harvey (1990, p. 108) also saw the state as struggling:

> to impose its will upon a fluid and spatially open process of capital circulation – geopolitics and economic nationalism, localism and the politics of place, are all fighting it out with a new internationalism in the most contradictory of ways.

Here he touched on a Marxist interpretation of the substantial changes taking place in society with ramifications for the nation-state and its governance. Strange (1994) supported this notion of changes in the position of the state suggesting that its power had moved upwards, sideways and downwards rendering the role of the traditional hierarchical arrangement redundant.

Tilly (1994, p. 292) further took up the idea combining the issues of time, globalization and the nation-state seeing the effect of the reduction in impact of time on the movement of capital, labour, products and information as fundamental to reducing the impact or necessity of a territorially bound nation-state. This liquidification of resources undermines the capacity of nation-states to pursue effective policies, and it encourages all parties to operate outside and across national boundaries. The effect is a decline in the significance of the nation-state, a reduction in its policy-making capabilities, and the need for new forms of governance to accommodate these changes. And how has the maritime sector adapted?

Ohmae (1993, p. 78) was particularly pessimistic about the future of the state:

> The nation state has become an unnatural, even dysfunctional unit for organizing human activity and managing economic endeavour in a borderless world. It represents no genuine, shared community of economic interests; it defines no meaningful flows of economic activity. In fact it overlooks the true linkages and synergies that exist among often disparate populations by combining important measures of human activity at the wrong level of analysis.

He continued emphasizing the role of globalization in writing the death sentence for the nation:

> On the global economic map the lines that now matter are those defining what may be called 'region states'. The boundaries of the region state are not imposed by political fiat. They are drawn by the deft but invisible hand of the global market for goods and services. They follow rather than precede, real flows of human activity, creating nothing new but ratifying existing patterns manifest in countless individual decisions.

Meanwhile Moses (1994) questioned the survival of the nation-state as financial markets grew increasingly international. Cerny (1995, p. 597) viewed the changes occurring in the scale at which goods and assets were being produced—increasingly to the international and regional and decreasibly corresponding with the national—then the:

> more the authority, legitimacy, policymaking capacity and policy-implementing effectiveness of states will be challenged from both within and without.

Holloway (1994), along with Taylor (1994) concurred identifying increasing pressures on states from international organizations and the emergence of new centres of power, challenging the nation-state but also fragmenting it and cutting across existing boundaries represented by conventional territoriality.

Jachtenfuchs (1995) outlined the contrasting international relations theories of neo-functionalism (Haas 1964) and systems theory with regards to the survival of the nation-state. Neo-functionalism anticipates the increasing transfer of decision-making competence from the state to supranational (more explicitly the EU) institutions resulting in a gradual erosion of statehood whilst systems theory suggests that the state is irrelevant and has been replaced by a political system whereby the traditional jurisdictional hierarchy is superseded by a series of separate and parallel operating systems—with clear ramifications for a maritime governance structure based upon a strictly hierarchical framework.

Brenner (1997, p. 150) also further emphasized Harvey's interpretation of events suggesting that the 'apparent territorial fixity and stability' that is

4.4 The Nation-State Under Global Pressure

commonly associated with the nation-state is in fact deeply unstable and no more than a 'precarious equilibrium'. Hinting at some of the reasons that may lie behind this fragility of the state he suggested that states try to acquire a spatial fix for capital (thus establishing both a reason for their continued existence and a means of supporting it) but citing Lefebrvre (1974):

> the Hegelian vision of this intensely contradictory process as an atemporal 'rational unity' amounts to a 'fetishization of space in the service of the state'. (Brenner 1997, p. 140)

To clarify, both Harvey and Brenner (amongst others) saw the state as a once temporary convenience, there to facilitate the accumulation of capital, and doomed once globalization made the fixity of territorial boundaries an anachronism. States would then survive only as historical oddities for however, long they proved convenient.

Brenner (1999, p. 40) further noted that the 'container-like qualities of states, this inherited model of territorially self-enclosed, state-defined societies, economies or cultures has become highly problematic'. And most interestingly for our discussion of the need for change in maritime policy-making, he felt that there was an urgent need for:

> new modes of analysis that do not naturalize state territoriality and its associated, Cartesian image of space as a static, bounded block... (challenging) the 'iron grip of the nation-state on the social imagination'. (Taylor 1996, p. 1923)

He also stressed that this emphasis upon the state and its territory is a consequence of an earlier, now largely superseded, state-centric configuration of capitalist development, hinting again at Harvey's understanding of capital accumulation at the heart of the nation-state. With changing circumstances and in particular the growth of globalization, capital accumulation now may find state-centrism a constraint rather than an opportunity.

March and Olsen (1998) felt that the condition of the nation-state had never been static, identifying the emergence of a distinctly different 'post-Westphalian order' characterized by changing national borders, increasing fragmentation associated with regionalism and sub-state identity and ever-encroaching internationalism making 'mono-centric, hierarchical and unitary states' increasingly irrelevant. They commented in particular on the disintegration of the Westphalian principle of non-intervention in internal affairs, eroded in the name of 'dispute resolution, economic stability and human rights'. This view was disputed by Blatter (2001, pp. 177–178) who characterized the modern Westphalian system as possessing a 'clear hierarchy of political authority/loyalty, with the nation-state taking centre stage'. All other identities are subordinate to the national where state territory is 'bundled' so that all state functions and responsibilities are congruent. Deleuze and Guattari (1988, p. 433) agreed seeing the central power of the state as hierarchical.

Dearlove (2000, p. 112) suggested that the decline of the state was a consequence almost entirely of the move towards 'footloose capital' giving states very limited capacity to shape any policies as they cannot risk the flight of capital—and shipping provides a classic example of this (Holloway 1994; Cerny 1996; Cox 1993). He quoted Walters and Blake (1992, p. 206):

> The primary role of the state... becomes that of accommodating the structure of the domestic economy to the imperatives of international economic development.

In addition he cited The Economist (1995): 'states can no longer set exchange rates and interest rates as 'globalization has restricted governments' ability to increase taxes, particularly on business'; and Esping-Anderson (1996): states can no longer manage 'their' economies because national economies have ceased to exist; and economic openness has 'compelled governments to cut back social expenditures'.

Cohen (2001, p. 77) emphasized how the 'sovereign state' has been the 'dominant form of political organization' in the Europeanized world for over 400 years and linked to nationalism, has become the norm for political communities. However, he also stressed the many accounts of its impending doom including commentary that suggests where maritime governance may locate its problems, quoting Greider (1997, p. 11):

> The logic of commerce and capital has overpowered the inertia of politics and launched an epoch of great social transformation. Settled facts of material life are being revised for rich and poor nations alike. Social understandings that were formed by the hard political struggles of the twentieth century are put in doubt. Old verities about the rank ordering of nations are revised and a new map of the world is gradually being drawn. These great changes sweep over the affairs of mere government and destabilize the established political orders in both advanced and primitive societies.

The common thread in almost all the commentaries is that territorial location is less and less important to the global economic system and that as a result, nation-states are less important as facilitators or obstacles to the functioning of the global economy. In the context of maritime governance, where we have seen that a state-centric framework remains dominant, such changes are inevitably fundamental. Cohen (2001, p. 77) may be a little too emphatic when he suggested that states are:

> losing control over their borders and are destined to recede into some sort of oblivion as the global economic system comes to dominate the evolution of society around the world...

...but the point made is clear and is supported by the work of Rosenau (2000) emphasizing the 'porosity of domestic-foreign boundaries'. Cohen agreed with Ruggie (1993) that the powers, authorities and capacities of the sovereign state can be 'unbundled' and repackaged so that they no longer track traditional territorial borders even though they may remain in existence, reconfigure what they do and that this may be important if different from what they have done in the past. Meanwhile Rosenau (2000, p. 7) went on:

> key dimensions of the power of the modern state have undergone considerable diminution. (Quoting Evans 1997, p. 65) ...'as wealth and power are increasingly generated by private transactions that take place across the borders of states rather than within them, it has become harder to sustain the image of states as the pre-eminent actors at the global level'.

4.4 The Nation-State Under Global Pressure

Zurn (2003, pp. 342–343) noted the decline in significance of state borders creating a challenge for nations to 'unilaterally reach... governance targets.' Four main specific problems were created for nation-states by this decline:

- *Efficiency pressures*—stemming from increasing demands to open borders and harmonize trans-national regulation to facilitate expanding markets.
- *Externality problems*—stemming from expanded social space which crosses international boundaries at will. The internet is an excellent example.
- *Competitiveness problems*—national policies may be effective but costly compared with other competing nations.
- *Representational deficits*—the difficulties of representation across national boundaries.

These problems stem from the process of globalization and present intensely difficult governance issues that need to be resolved—or at least recognized if policies for a sector are to have a chance of being effective. The response might be to fight them, to attempt to extend national jurisdiction at an international level; to coordinate national policies—or other responses. But something is needed. The position in the maritime sector today is one described by Zurn (2003, pp. 358–359) as 'methodological nationalism', locating and restricting the maritime political sphere to the national level. As Zurn commented, 'it is no longer... of any use... in the age of globalization'. He also stressed that nation-states will persist and remain of central importance.

Blatter (2003) observed the increasing dispersal of authority from the state to divergent actors whilst Axtmann (2004) identified a series of factors that had placed pressure on sovereign nation-states suggesting that tension had arisen between political problem-solving increasingly at the international level and domestic political arrangements which failed to match up with this. Aalberts (2004) stressed the move away from the traditional nation-state, international dichotomy to more transnational arrangements which have encouraged the need to focus on new forms of governance to accommodate the changes taking place. Although his focus was on multi-level governance frameworks for the EU, his comments were very relevant to maritime policy-making, governance and the changing nation-state.

Taking up this theme, Hajer (2003) stressed that the context for policy-making has changed with solutions to problems found less and less within the boundaries of sovereign politics and instead the development of transnational, polycentric networks of governance that more accurately reflect the patterns of communication and decision-making that are taking place in the real world.

Scholte (2004) as we saw earlier defined the traditional nation-state role in governance as one of 'statism' and although he stressed the continued role of the state as sites of regulation, he did feel that statism itself in the context of globalization, was no longer relevant even if states themselves would not disappear. Thus governance by the state has gone rather than the state itself. Somerville (2004) following the work of Rhodes (1997) and Jessop (1997, 1998, 2003a, b) viewed it as a process of hollowing out with the state remaining indirectly significant but with direct power flowing upwards and outwards to other institutions. Swyngedouw (1996) described this as 'rescaling' state activities whilst emphasizing the close relationship that remains between the local and the global even with the development of globalization. The improvement in communications that globalization has both encouraged and benefitted from not only allows markets, products, experiences and people to invade other communities but also for their communities to invade others. Thus the local becomes global as much as the global becomes local. This Postmodern scenario has been widely recognized in consideration of the societal impact of emails, blogs, pods and texts (for example see Short 1993, p. 171; Morgan 1997, p. 180) and forms part of our later discussion of the Postmodern era, shipping and governance failure.

Jessop (2003a, p. 46) further took up the idea of scalar mutation stressing how the nation-state was changing and required policy-makers to look beyond the state, with it embedded in an ever-widening political system and a new institutional order. Central to this was the process of rescaling and understanding that globalization as the main cause, is 'multi-centric, multi-scalar, multi-temporal, multi-form and multi-causal'. There is not a single set of pressures uniformly applied to all nation-states for globalization is a 'hyper-complex, continuously evolving product of many processes'. However, Jessop also insisted that it is not a choice between globalization and the nation-state as both will exist together albeit with the nation-state taking forms that may be almost unrecognizable from before.

Jessop also suggested that there were six trends identifiable in the restructuring of the traditional nation-state in the context of increasing global pressure, which have transformed what he terms the Keynesian National Welfare State that dominated post-war developments—a period significant for being when the major maritime policy-making institutions (e.g the IMO, UNCTAD, OECD, GATT/WTO, EU, etc.) were established. These trends were:

- Reallocation of state power—upwards and downwards to international/supranational and regional institutions respectively (hollowing out).
- Redrawing boundaries between traditional state and non-state activities.
- De-statization—including the rise of international institutions taking on responsibilities of global stabilization and the increasing role of the territorially orphaned cyberspace.
- Greater emphasis placed on a wider range of social contexts for the state to accommodate.
- The weakening of traditional jurisdictional hierarchies.
- Changes in the nature of political communities that no longer focus on national definitions.

Many have clear ramifications for the derivation and application of maritime policy and the governance framework that defines these processes. Jessop remained convinced that states still have a significant role to play especially in what he termed 'inter-scalar articulation' between the regional, supranational and international boundaries.

4.4 The Nation-State Under Global Pressure

So where does that leave us? Hirst (1997b) was significant in emphasizing the danger of 'clinging to the model of the national sovereign state' as this would simply speed up the loss of national influence. Kjaer (2004) was confident in stressing that the role of the state remains important albeit thoroughly different. In fact she suggested that its potential for action may have been enhanced by globalization using their powers through the UN, EU or WTO. She saw globalization as not necessarily a 'zero-sum game in which global governance means less national governance and vice versa' and citing Weiss (1999) suggested that the organizational forces of the global and national can be both complimentary as well as competitive.

King and Kendall (2004, p. 146) questioned whether the increasingly common view that the nation-state is about to end its 300 year history is rational given the increasing numbers of states that have emerged only recently. At the same time they stressed the continued shift of responsibilities from the state and into the marketplace, reducing the state's role. They cited Beck (1997, p. 11) describing globalization as:

> the process through which sovereign national states are criss-crossed and undermined by transnational actors with varying prospects of power, orientations, identities and networks.

They also referred back to Scholte (2000) and indirectly to Harvey (1990) in discussing space-time compression and the decline of territoriality, highly significant concepts in the consideration of globalization, governance, statehood and policy-making. This de-territorialization means that the nation-state has lost its assumed superiority over other territorial realms and governance frameworks need to adapt to these changed circumstances. Albrow (1996) saw the influence of globalization on the nation-state as a positive effect providing 'vastly increased potential for societal groups to constitute collective action outside the frame of nation-state society'.

King and Kendall referred to Hobson (2000) in suggesting that in some ways the state appears to be strong and retain autonomy over 'domestic social forces' but increasingly subject to anarchy and the compelling requirements of the international system. States as a result appear to have 'little scope to change much; rather, the needs of survival compel them to follow the herd'.

Hirst (1997b, p. 29) assessed the position of the nation-state and his comments are relevant:

> Faced with an internationalizing and increasingly volatile economy, in which competition between nations has intensified and growth rates are uncertain, national governments have proved unable to ensure the conditions for prosperity either by traditional Keynesian means or by the monetarist methods that became fashionable in the 1980s.

Taylor (2005, p. 704) takes us back to where we begun this discussion on the nation-state with the Treaty of Westphalia suggesting that globalization 'heralds a different world spatiality to replace the Westphalian mosaic' and also stressing that governance needs to be viewed as the governance of processes rather than that of space. Westphalian states could be understood through processes operating within states; globalization requires these processes to be reinterpreted as operating in a trans-state manner, issues which are central to our understanding of maritime policy-making and to which we return much later. Bennett (2000, p. 894) concurred by stressing the importance of governance beyond the state in the deliberations on policy-making in the maritime sector at the EU and their difficult relationships with the IMO. Although it may be a very strong expression of the situation, Rudd (2001, p. 5) provided a neat summary:

> The nation-state reflects a seventeenth century Westphalian concept, a territorial mode of political organization, which according to democratic theory, is supposed to provide accountability to the governed. Yet by condensing the time-space aspects of social relations, economic globalization transcends territorial states and is unaccountable to elected political officials, or is accountable to unelected market forces. The tensions between the territorial based system of political organization and economic globalization creates a disjuncture of governance at both the regional and world levels.

This disjuncture is a manifestation of the Postmodern era we shall discuss further in a later chapter and which Duncan and Savage (1989, pp. 182–183) suggested is a feature of the modern state where boundaries can now be conceived as not fixed but variable according to context, issue and consumer. This manifests itself in the uneven development of modern capitalism, itself a feature reflecting change and disorganization and an indication of societal shift. This disjuncture is revealed at the maritime level in the policy-making problems which it exhibits. Rosenau (2000, p. 7) suggested that most states are in deep crisis, with paralyzed 'policy-making processes' resulting in 'stalemate and stasis, in the avoidance of decisions that would… address the challenges posed by a world undergoing vast and continuous changes'. Governance for the maritime sector remains firmly seated in its state-centric past as the new age of globalization continues to move onwards.

4.5 Conclusions

De Vivero et al. (2009, p. 629) provided us with some hope in their discussion of the geopolitical implications of maritime policy-making suggesting that 'the mosaic of territorial and institutional pieces that can be shaped is ample and varied, which in theory translates into greater opportunities for participation'. Galgano (2009, p. 19) was less optimistic in that he saw the definition of borders and space specifically in the maritime sector as unhelpful to creating effective policies, especially the existing definitions of territorial waters. It is with this in mind that we move on to begin to try to understand the forces at play in maritime governance, their underlying social drivers and the moves that need to be made for the future.

This discussion of the nation-state has been important because the significance to maritime governance of jurisdictional relationships and particularly of the role of the nation-state is hard to over-estimate. What is remarkable is the dearth of

4.5 Conclusions

discussion about these jurisdictional issues, the apparent blind acceptance of the institutional framework that exists, and the failure to link maritime policy inadequacies to the significant place that nations retain despite globalization. De Senarclens (2001, pp. 514, 518) provided a welcome exception in considering the relationships between nation-states and international institutions. Taking the IMF and World Bank as his examples (but he could well have taken the IMO), he suggested that all the major decisions are effectively taken by a 'handful of decision-makers… (distributing) loans and recommend(ing) stabilization or structural adjustment policies to countries that have hardly any representation'. Nation-states remain desperate to retain their position at the United Nations not wanting to 'abandon their (albeit symbolic) participation in proceedings… and its main specialized agencies, and are suspicious of projects aiming to centralize the decision-making process'. It is this combination of structural inadequacy which infects maritime jurisdiction combined with the intense desire of nations to remain central to the policy-making process regardless of the forces of globalization that allows the shipping industry to take advantage of the chaos that has ensued.

Coicaud (2001, pp. 523–524) concurred. He saw the legitimacy of all international institutions coming from the nation-state members. Nation-states gave them their mandates, legitimize their work and receive their agendas. However, the very existence of such independent states forming such a vital part of international organizations leads to a 'lack of convergence and consistency' with policies within the same organization commonly revealing disagreements over values. The shipping industry again finds this type of inconsistency easy to exploit in a globalizing environment. In Coicaud's (2001, p. 533) words:

> International organizations will have to revisit their relations of cooperation with the actors that are both their partners and competitors, especially states… States will have to become less protective of their sovereign powers; more willing, for instance, to share power with international organizations, and more open to claims from individuals.

Meanwhile Hatzaras (2005, p. 4) provided a fitting conclusion to this chapter when he commented that by 1995 it was clear that there was a new gradient for European integration from the point where:

> state-centric conceptions were able to yield explanatory power and onto a new context where 'states are melded gently into a multi-level policy by their leaders and the actions of numerous sub-national and supranational actors. (Marks et al. 1996, p. 371)

Schrier et al. (1984, p. 87) claimed that national governments were increasingly involved in international maritime affairs and in turn politicizing the industry with inter-governmental debate rather than market forces dictating policy. Their rather naïve view of the significance of the nation-state was questioned by Zacher and Sutton (1996, p. 38) who placed the position of the nation-state in the context of maritime governance in an increasingly globalizing world:

> A significant aspect of international maritime transport, which is found in all international transportation and communications industries, is the legal characteristic of joint sovereignty. By its very nature international shipping requires that two or more states allow exit

and entry to a vessel before it can engage in shipping services. However, this 'bilaterally shared sovereignty can give rise to policy conflict between different national regulatory regimes – the more so when some countries more aggressively develop or protect their maritime transport capabilities'. (Schrier et al. 1984, p. 88)

In February 2009, the Director of the Cypriot Department of Merchant Shipping Serghios Serghiou commented:

> The term 'flag of convenience' is obsolete in today's globalized economy. What is of essence now is whether a particular country has a responsible maritime administration which can exercise effective jurisdiction and control on ships flying its flag… today the nationality of the shipowner has no significance whatsoever. (Lloyd's List 2009)

Strange (1976, pp. 358, 361) meanwhile had much earlier stressed how 'the authority of states over the operators and the market is generally rather weak… and dispersed' a situation arising from 'the inconsistency and inconstancy of national… policies towards shipping… The result is that there are a large number of international agencies concerned with shipping but none has much real authority'. Woods and Narlikar (2001, p. 569) did not agree with the contention that these international agencies are ineffective suggesting that they commonly deal with issues which were previously the responsibility of national governments but they certainly did agree on their proliferation and the significance that they have had for the nation-state, intruding 'deeply into national politics' (Woods and Narlikar 2001, p. 570).

As a result the:

> world shipping business seems to be headed for decreased efficiency, and for increased inequity and continued instability. It is in a condition of relative anarchy dangerous to the environment and to human life and potentially very disruptive both to the rest of the world economy and to the political relations between governments – and even perhaps to politics between states. (Strange 1976, p. 364)

It is in the context of comments like this that the design of maritime governance needs to be considered. In the next section we shall begin to look at the process of globalization and the forces which underlie its development through society. This will provide guidance for proposals of change in maritime governance which will have the potential to accommodate the new societal trends.

References

Aalberts, T. E. (2004). The future of sovereignty in multilevel governance Europe—a constructivist reading. *Journal of Common Market Studies, 42*(1), 23–46.

Agnew, J. (1988). *Geo-politics: Re-visioning world politics*. London: Routledge.

Agnew, J. (1994). The territorial trap: The geographical assumptions of international relations theory. *Review of International Political Economy, 1*(1), 53–80.

Agnew, J. (1999). Mapping political power beyond state boundaries: Territory, identity, and movement in world politics. *Millennium: Journal of International Studies, 28*(3), 499–521.

Alamgir, J. (1999). Globalization and state capacity. *International Studies Review, 1*(1), 110–113.

References

Albrow, M. (1996). *The global age*. Cambridge: Polity Press.
Alderton, T., & Winchester, N. (2002). Globalization and de-regulation in the maritime industry. *Marine Policy, 26*, 35–43.
Alesina, A., & Spolaore, E. (1997). On the number and size of nations. *The Quarterly Journal of Economics, 112*(4), 1027–1056.
Althusser, L. (1971). *Lenin and philosophy and other essays*. London: New Left Books.
Amin, A. (1997). Placing globalization. *Theory, Culture and Society, 14*, 123–137.
Amin, A. (1998). Globalization and regional development: A relational perspective. *Competition and Change, 3*, 145–166.
Amin, A. (2002). Spatialities of globalization. *Environment and Planning A, 34*, 385–399.
Anderson, B. (1991). *Imagined communities*. London: Verso.
Anderson, J. (1996a). The shifting stage of politics: New medieval and postmodern territorialities? *Environment and Planning D, 14*, 133–153.
Anderson, M. (1996b). *Frontiers: Territory and state formation in the modern world*. Cambridge: Polity Press.
Appadurai, A. (1996). *Modernity at large: Cultural dimensions of globalization*. Minneapolis: University of Minnesota Press.
Aronowitz, S. (1988). Postmodernism and politics. In A. Ross (Ed.), *Universal abandon? The politics of postmodernism* (pp. 46–62). Minneapolis: University of Minnesota Press.
Arrighi, G. (1994). *The long twentieth century*. London: Allen and Unwin.
Atkinson, M. M., & Coleman, W. D. (1992). Policy networks, policy communities and the problems of governance. *Governance, 5*(2), 154–180.
Axtmann, R. (2004). The state of the state: The model of the modern state and its contemporary transformation. *International Political Science Review, 25*(3), 259–279.
Bagwell, K., & Staiger, R. W. (2001). Domestic policies, national sovereignty and international economic institutions. *The Quarterly Journal of Economics, 116*(2), 519–562.
Bairoch, P., & Kozul-Wright, R. (1996). *Globalization myths: Some historical reflections on integration, industrialization and growth in the world economy*. UNCTAD discussion paper no. 113, UNCTAD, Geneva.
Barkin, J., & Cronin, B. (1994). The state and the nation: changing norms and rules of sovereignty in international relations. *International Organization, 48*, 107–130.
Bartelson, J. (1995). *A genealogy of sovereignty*. Cambridge: Cambridge University Press.
Beck, U. (1992). *Risk society*. London: Sage.
Beck, U. (1994). The reinvention of politics. In U. Beck, A. Giddens, & S. Lash (Eds.), *Reflexive modernization politics, tradition and aesthetics in the modern social order* (pp. 1–55). Cambridge: Polity Press.
Beck, U. (1997). *What is globalization?*. Cambridge: Polity Press.
Bell, D. (1988). Previewing planet earth in 2013. *Washington Post, 3*(6), 48 (January 3).
Bennett, P. (2000). Environmental governance and private actors: enrolling insurers in international maritime regulation. *Political Geography, 19*, 875–899.
Benz, A., & Eberlein, B. (1999). The Europeanization of regional policies: patterns of multi-level governance. *Journal of European Public Policy, 6*(2), 329–348.
Berger, S., & Dore, R. (Eds.). (1996). *National diversity and global capitalism*. Ithaca: Cornell University Press.
Biersteker, T., & Weber, C. (Eds.). (1996). *State sovereignty as social construct*. Cambridge: Cambridge University Press.
Biger, G. (Ed.). (1995). *The encyclopaedia of international boundaries*. Durham: International Boundaries Research Unit, University of Durham.
Blake, G. H., & Schofield, R. N. (Eds.). (1987). *Boundaries and state territory in the Middle East and North Africa*. Wisbech: Menas Press.
Blaney, D. L., & Inayatullah, N. (2000). The Westphalian deferral. *International Studies Review, 2*(2), 29–64.

Blatter, J. K. (2001). Debordering the world of states: Towards a multi-level system in Europe and a multi-polity system in North America? Insights from Border Regions. *European Journal of International Relations, 7*(2), 175–209.

Blatter, J. (2003). Beyond hierarchies and networks: Institutional logics and change in transboundary spaces. *Governance, 16*(4), 503–526.

Boutros-Ghali, B. (1992). *An agenda for peace*. New York: United Nations.

Braithwaite, J., & Drahos, P. (2000). *Global business regulation*. Cambridge: Cambridge University Press.

Braudel, F. (1982). *The wheels of commerce*. New York: Harper and Row.

Brenner, N. (1997). Global, fragmented, hierarchical: Henri Lefebvre's geographies of globalization. *Public Culture, 10*(1), 135–167.

Brenner, N. (1998). Between fixity and motion: Accumulation, territorial organization and the historical geography of spatial scales. *Environment and Planning D, 16*(4), 459–481.

Brenner, N. (1999). Beyond state-centrism? Territoriality and geographical scale in globalization studies. *Theory and Society, 28*(1), 39–78.

Brenner, N. (2004). *New state spaces: Urban governance and the rescaling of statehood*. Oxford: Oxford University Press.

Brown, L. R. (1973). *World without borders*. New York: Vintage.

Brown, R. (1997). *Globalization and the state*. Oxford: Polity Press.

Brown, C., & Ainley, K. (1997). *Understanding international relations*. London: Palgrave.

Bull, H. (1977). *The anarchical society: A study of order in world politics*. New York: Columbia University Press.

Bulmer, S. (1983). Domestic politics and European community policy-making. *Journal of Common Market Studies, XXI*(4), 349–363.

Burch, K. (1994). The properties of the state system and global capitalism. In S. Rosow, N. Inayatullah, & M. Rupert (Eds.), *The global economy as political space* (pp. 37–59). London: Lynne Rienner.

Burch, K. (2000). Changing the rules: Reconceiving change in the Westphalian system. *International Studies Review, 2*(2), 181–210.

Cable, V. (1995). The diminished nation-state: A study in the loss of economic power. *Daedalus, 124*, 23–53.

Camilleri, J. A. (1990). Rethinking sovereignty in a shrinking, fragmented world. In R. B. J. Walker & S. Mendlovitz (Eds.), *Contending sovereignties: Redefining political community* (pp. 13–44). Boulder: Lynne Reiner.

Camilleri, J. A., & Falk, J. (1992). *The end of sovereignty? The politics of a shrinking and fragmenting world*. Aldershot: Edward Elgar.

Caporaso, J. A. (1996). The European Union and forms of state: Westphalian, regulatory or postmodern? *Journal of Common Market Studies, 34*(1), 29–52.

Caporaso, J. A. (2000). Changes in the Westphalian order: Territory, public authority and sovereignty. *International Studies Review, 2*(2), 1–28.

Castells, M. (1996). *The rise of the network society*. Cambridge: Blackwell.

Cerny, P. G. (1995). Globalization and the changing logic of collective action. *International Organization, 49*(4), 595–625.

Cerny, P. G. (1996). International finance and the erosion of state policy capacity. In P. Gummett (Ed.), *Globalization and public policy capacity* (pp. 83–104). Cheltenham: Edward Elgar.

Chowdhury, K. (2006). Interrogating "newness". Globalization and postcolonial theory in the age of endless war. *Cultural Critique, 62*, 126–161.

Christiansen, T., & Jorgensen, K. E. (2000). Transnational governance 'above' and 'below' the state: The changing nature of borders in the new Europe. *Regional and Federal Studies, 10*(2), 62–77.

Clark, G. L., & Dear, M. (1984). *State apparatus, structures and language of legitimacy*. Boston: George Allen and Unwin.

Cohen, E. S. (2001). Globalization and the boundaries of the state: A framework for analysing the changing practice of sovereignty. *Governance, 14*(1), 75–97.

Coicaud, J.-M. (2001). Reflections on international organizations and international legitimacy: Constraints, pathologies, and possibilities. *International Social Science Journal, 170*, 523–536.
Connor, W. (1978). A nation is a nation, is a state, is an ethnic group, is a.... *Ethnic and Racial Studies, 1*, 378–400.
Cooper, R. (2000). *The post-modern state and the world order*. London: Demos.
Cox, K. R. (1993). The local and the global in the new urban politics. *Environment and Planning D, 11*(4), 433–448.
Croucher, S. L. (2003). Perpetual imagining. Nationhood in a global era. *International Studies Review, 5*, 1–24.
Davidson, J. D., & Rees-Mogg, W. (1999). *The sovereign individual: Mastering the transition to the information age*. New York: Touchstone.
De Senarclens, P. (1998). Governance and the crisis in the international mechanisms of regulation. *International Social Science Journal, 50*(55), 91–104.
De Senarclens, P. (2001). International organizations and the challenges of globalization. *International Social Science Journal, 170*, 509–522.
De Vivero, J. L. S., Mateos, J. C. R., & del Corral, D. F. (2009). Geopolitical factors of maritime policies and marine spatial planning: state, regions and geographical planning scope. *Marine Policy, 13*, 624–634.
Dear, M. J., & Clark, G. (1978). The state and geographic process: a critical review. *Environment and Planning A, 10*, 173–183.
Dearlove, J. (2000). Globalization and the study of British politics. *Politics, 20*(2), 111–118.
Decker, F. (2002). Governance beyond the nation-state. Reflections on the democratic deficit of the European Union. *Journal of European Public Policy, 9*(2), 256–272.
Deleuze, G., & Guattari, F. (1988). *A thousand plateaus: Capitalism and schizophrenia*. London: The Athlone Press.
Del Rosso, Jr., S. J. (1995). The insecure state (What future for the state?). *Daedalus, 124*(2).
Deutsch, K. W. (1966). *Nationalism and social communication*. New York: MIT Press.
Deutsch, K. W. (1981). The crisis of the state. *Government and Opposition, 16*, 331–343.
Dicken, P. (1994). Global-local tensions: Firms and states in the global space-economy. *Economic Geography, 70*(2), 101–128.
Dicken, P. (1998). *Global shift: Transforming the world economy*. London: Paul Chapman.
Dicken, P., Kelly, P. F., Olds, K., & Yeung, W. (2001). Chains and networks, territories and scales: towards a relational framework for analysing the global economy. *Global Networks, 1*(2), 89–112.
Dodds, K.-J. (1994). Geopolitics and foreign policy: Recent developments in Anglo-American political geography and international relations. *Progress in Human Geography, 18*, 186–208.
Donnan, H., & Wilson, T.-M. (1994). An anthropology of frontiers. In H. Donnan & T.-M. Wilson (Eds.), *Border approaches: Anthropological perspectives on frontiers*. Lanham: University Press of America.
Doremus, P. N., Keller, W. W., Pauly, L. W., & Reich, S. (1998). *The myth of the global corporation*. Princeton: Princeton University Press.
Drache, D. (1996). From Keynes to K-mart: Competitiveness in a corporate age. In R. Boyer & D. Drache (Eds.), *States against markets: The limits of globalization* (pp. 31–61). London: Routledge.
Drucker, P. F. (1997). The global economy and the nation-state. *Foreign Affairs, 76*(5), 159–171. (September/October).
Duchacek, I. D. (1988). Multi-communal and bi-communal polities. In I. D. Duchachek, D. Latouche, & G. Stevenson (Eds.), *Perforated sovereignties and international relations* (pp. 3–28). New York: Greenwood Press.
Duncan, S., & Savage, M. (1989). Space, scale and locality. *Antipode, 21*(3), 207–231.
Economist. (1995, October 7). *The Myth of the Powerless State*.
Einstein, A. (1971). *The evolution of physics*. Cambridge: Cambridge University Press.

Esping-Anderson, G. (Ed.). (1996). *Welfare states in transition: National adaptations in global economies.* London: Sage.

Evans, P. (1997). The eclipse of the state? Reflections on stateness in an era of globalization. *World Politics, 50*(1), 62–87.

Falk, R. (1996). An enquiry into the political economy of world order. *New Political Economy, 1*, 13–26.

Falk, R. (1997). State of siege: Will globalization win out? *International Affairs, 73*(1), 121–136.

Ferguson, J., & Gupta, A. (2002). Spatializing states: Towards an ethnography of neo-liberal governmentality. *American Ethnologist, 29*(4), 981–1002.

Ferguson, Y., & Mansbach, R. (1996). *Politics: Authority identities and change.* Columbia: University of South Carolina Press.

Forsberg, T. (Ed.). (1995). *Contested territory: Border disputes at the edge of the former Soviet empire studies of communism in transition.* Aldershot: Edward Elgar.

Franklin, J. (1998). *The politics of risk society.* Cambridge: Polity Press.

Galgano, F. A. (2009). The borderless dilemma of contemporary maritime piracy: Its geography and trends. *Pennsylvania Geographer, 47*(1), 3–33.

Galluser, W. A. (Ed.). (1994). *Political boundaries and co-existence.* Bern: Peter Lang.

Gellner, E. (1983). *Nations and nationalism.* Oxford: Blackwell.

Giddens, A. (1985). *A contemporary critique of historical materialism* (Vol. 2)., The nation-state and violence Cambridge: Polity Press.

Girot, P. (Ed.). (1994). *World boundaries* (Vol. 4)., The Americas London: Routledge.

Gordenker, L., & Weiss, T. G. (1995). Pluralising global governance: Analytical approaches and dimensions. *Third World Quarterly, 16*(3), 357–387.

Gottdiener, M. (1987). Space as a force of production: Contribution to the debate on realism, capitalism and space. *International Journal of Urban and Regional Research, 11*, 405–416.

Gottmann, J. (1973). *The significance of territory.* Charlottesville: University of Virginia Press.

Gradus, Y., & Lithwick, H. (Eds.). (1996). *Frontiers in regional development.* Lanham: Rowman and Littlefield.

Greider, W. (1997). *One world ready or not.* New York: Simon and Schuster.

Grimm, D. (1992). Der Mangel an europaischer Demokratie. *Der Spiegel, 43*, 57–59. (19th October).

Gritsch, M. (2005). The nation-state and economic globalization: Soft geo-politics and increased state autonomy? *Review of International Political Economy, 12*(1), 1–25.

Gross, L. (1948). The peace of Westphalia, 1648–1948. *The American Journal of International Law, 42*(1), 20–41.

Gross, L. (1968). The peace of Westphalia, 1648–1948. In R. A. Falk & W. Hanreider (Eds.), *International law and organization* (pp. 45–67). New York: J.B. Lippincott.

Grundy-Warr, C. (Ed.). (1990). *International boundaries and boundary conflict resolution.* Durham: International Boundaries Research Unit, University of Durham.

Gupta, A., & Ferguson, J. (1992). Beyond 'culture': Space, identity and the politics of difference. *Current Anthropology, 7*(1), 6–23.

Haas, E. (1964). *Beyond the nation state. Functionalism and international organization.* Pan Alto: Stanford University Press.

Habermas, J. (1999). The European nation-state and the pressures of globalization. *New Left Review, 235*, 46–59. (May–June).

Hajer, M. (2003). Policy without polity? Policy analysis and the institutional void. *Policy Sciences, 36*, 175–195.

Hanreider, W. (1967). *West German foreign policy 1949–1963.* Stanford: Stanford University Press.

Harding, A. (1997). Urban regimes in a Europe of the cities? *European Urban and Regional Studies, 4*, 291–314.

Hardt, M., & Negri, A. (2004). *Multitude.* London: Hamish Hamilton.

Harvey, D. (1976). The Marxian theory of the state. *Antipode, 8*, 80–98.

Harvey, D. (1989). *The urban experience.* Oxford: Blackwell.

References

Harvey, D. (1990). *The condition of postmodernity*. Cambridge: Blackwell.
Harvey, D. (2000) Reinventing geography. *New Left Review, 4*, 75–97.
Harvey, D. (2001). Globalization and the "spatial fix". *Geographische Review, 2*, 23–30.
Hatzaras, K. (2005). *Multi-level governance, Europeanization and the poorest EU region*. Epirus and the 2nd Hellenic Community Support Framework, 2nd LSE Symposium on Modern Greece.
Held, D. (1991). Democracy, the nation-state and the global system. *Economy and Society, 20*(2), 138–172.
Held, D., & McGrew, A. (Eds.). (2002). *Governing globalization. Power authority and global governance*. Cambridge: Polity Press.
Helleiner, E. (1994). *States and the re-emergence of global finance* (pp. 138–156). Ithaca: Cornell University Press.
Helleiner, E. (1999). Sovereignty, territoriality and the globalization of finance. In D. A. Smith, D. J. Solinger, & S. C. Topik (Eds.), *States and sovereignty in the global economy* (pp. 138–157). London: Routledge.
Herz, J. H. (1957). Rise and demise of the territorial state. *World Politics, 9*(4), 473–493.
Herz, J. H. (1959). *International politics in the atomic age*. New York: Columbia University Press.
Hinsley, F. H. (1986). *Sovereignty*. Cambridge: Cambridge University Press.
Hirst, P. (1995). Globalization and the future of the nation-state. *Economy and Society, 24*(3), 408–442.
Hirst, P. (1997a). *From statism to pluralism*. London: UCL Press.
Hirst, P. (1997b). *Democracy, civil society and global politics*. London: UCL Press.
Hirst, P. Q., & Thompson, G. (1995). Globalization and the future of the nation state. *Economy and Society, 24*(3), 408–422.
Hirst, P. Q., & Thompson, G. (1999). *Globalization in question*. Cambridge: Polity Press.
Hobsbawm, E. (1990). *Nations and nationalism since 1780*. Cambridge: Cambridge University Press.
Hobsbawm, E. (1998). The nation and globalization. *Constellations, 5*(1), 1–9.
Hobson, J. M. (2000). *The state and international relations*. Cambridge: Cambridge University Press.
Hoffmann, S. (1966). Obstinate or obsolete? The fate of the nation-state and the case of Western Europe. *Daedalus, 95*, 892–908.
Hoffmann, S. (1982). Reflections on the nation-state in Western Europe today. *Journal of Common Market Studies, 20*, 21–37.
Holloway, J. (1994). Global capital and the national state. *Capital and Class, 52*, 23–43.
Holloway, J. (1995). Global capital and the national state. In W. Bonefield & J. Holloway (Eds.), *Global capital, national state and the politics of money* (pp. 116–140). New York: St Martins Press.
Hsu, J.-Y. (2000). *Revisiting economic development in post-war Taiwan: The dynamic process of geographical industrialization*. Alternative 21st Century Geographies: 2nd International Critical Geography Conference, Taegu, Korea.
Hudson, A. (2000). Offshoreness, globalization and sovereignty: A postmodern geo-political economy? *Transactions of the Institute of British Geographers, New Series, 25*, 269–283.
Jachtenfuchs, M. (1995). Theoretical perspectives on European governance. *European Law Journal, 1*(2), 115–133.
Jackson, R. (1999). Sovereignty in world politics, a glance at the conceptual and historical landscape. *Political Studies, 47*, 431–456.
Jessop, B. (1997). Capitalism and its future: Remarks on regulation, government and governance. *Review of International Political Economy, 4*(3), 561–581.
Jessop, B. (1998). The rise of governance and the risk of failure: The case of economic development. *International Social Science Journal, 50*(155), 29–45.

Jessop, B. (1999). Some critical reflections on globalization and Its illogic(s). In K. Olds, P. Dicken, P. K. Kelly, L. Kong, & H. W.-C. Yeung (Eds.), *Globalization and the Asia Pacific: Contested territories* (pp. 19–38). London: Routledge.
Jessop, B. (2002). *The future of the capitalist state*. Cambridge: Polity Press.
Jessop, B. (2003a). The future of the state in an era of globalization. *International Politics and Society, 3*, 30–46.
Jessop, B. (2003b). *Governance and metagovernance: On reflexivity, requisite variety and requisite irony*. Lancaster: Department of Sociology, University of Lancaster.
Johnson, D. M. (1994). Who is we? Constructing communities in US-Mexico border discourse. *Discourse and Society, 5*, 207–231.
Johnston, R. J. (1991). *A question of place: Exploring the practice of human geography*. Oxford: Blackwell.
Johnston, R. J. (1995). Territorality and the state. In G. Benko & U. Strohmayer (Eds.), *Geography, history and social science* (pp. 213–226). Dordrecht: Kluwer.
Johnston, R. J. (2001). Out of the 'moribund backwater': Territory and territoriality in political geography. *Political Geography, 20*, 677–693.
Jordan, A. G. (2001). The European union: An evolving system of multi-level governance...or government? *Policy and Politics, 29*(2), 193–208.
Kant, I. (1795). Perpetual peace. In: H. Reiss (Ed.), (1991) *Kant: Political writings*. Cambridge: Cambridge University Press.
Kapstein, E. B. (1991). We are US. *The National Interest, Winter, 1991*(2), 56–62.
Kelly, P. F. (1999). The geographies and politics of globalization. *Progress in Human Geography, 23*(3), 379–400.
Kettl, D. F. (2002). *The transformation of governance: Globalization, devolution and the role of government*. Washington: Brookings Institution.
Kielmansegg, P. G. (1992, December 2). Ein Mass für die Grösse des Staates. Was Wird aus Europa? Europa fehlt die Zustimmung der Bürger, *Franfurter Allgemeine Zeitung*, 35.
King, R., & Kendall, G. (2004). *The State, Democracy and Globalization*. Basingstoke: Palgrave Macmillan.
Kjaer, A. M. (2004). *Governance*. Cambridge: Polity Press.
Kohn, H. (1967). *The idea of nationalism*. New York: Macmillan.
Krasner, S. D. (Ed.). (1983). *International regimes*. Ithaca: Cornell University.
Krasner, S. D. (1984). Approaches to the state: Alternative conceptions and historical dynamics. *Comparative Politics, 16*(2), 223–246.
Krasner, S. D. (1999). *Sovereignty: Organized hypocrisy* (pp. 193–216). Princeton: Princeton University Press.
Kratochwil, F. (1986). Of systems, boundaries and territoriality: An inquiry into the formation of the state system. *World Politics, 39*(1), 27–52.
Krishna, S. (1996). Cartographic anxiety: Mapping the body politic in India. In M. J. Shapiro & H. R. Alker (Eds.), *Challenging boundaries* (pp. 193–214). Minnesota: University of Minnesota Press.
Kuehls, T. (1996). *Beyond sovereign territory: The space of ecopolitics*. Minnesota: University of Minnesota Press.
Lambert, J. (1991). Europe: The nation-state dies hard. *Capital and Class, Spring, 43*, 9–24.
Langford, T. (1999). Things fall apart: State failure and the politics of intervention. *International Studies Review, 1*(1), 59–79.
Lapidoth, R. (1992). Sovereignty in transition. *Journal of International Affairs, 45*, 325–346.
Latour, B. (1993). *We have never been modern*. Hemel Hempstead: Harvester Wheatsheaf.
Le Gales, P., & Harding, A. (1998). Cities and states in Europe. *West European Politics, 21*(3), 120–145.
Lefebvre, H. (1974). *La Production de L'Éspace*. Paris: Éditions Anthropos.
Lefebvre, H. (1977). *De L'État: Le Mode de Production Étatique* (Vol. 3). Paris: Union Générale d'Editions.

Lefebvre, H. (1978). *De L'État: Les Contradictionsde L'État Moderne* (Vol. 4). Paris: Union Générale d'Editions.
Lefebvre, H. (1991). *The production of space*. Oxford: Blackwell.
Linklater, A. (1998). *The transformation of political community*. Columbia: University of South Carolina Press.
Lipietz, A. (1995). Thriving locally in the global economy. *Harvard Business Review*, September–October, 151–160.
Lloyd's List. (2007, January 30). *Europe's 'maritime space' draws fire*.
Lloyd's List. (2009, February 11). *Nationality of shipowner has 'no significance' says Serghiou*.
Lomnitz, L. A. (1988). Informal exchange networks in formal systems: A theoretical model. *American Anthropologist, 90*(1), 42–55.
Luke, T. W. (1991). The discipline of security studies and the codes of containment: Learning from Kuwait. *Alternatives, 16*, 315–344.
MacLeod, G. (2001). Beyond soft institutionalism: Accumulation, regulation and their geographical fixes. *Environment and Planning A, 33*(7), 1145–1167.
MacLeod, G., & Goodwin, M. (1999). Space, scale and state strategy: Rethinking urban and regional governance. *Progress in Human Geography, 23*(4), 503–527.
Macpherson, C. B. (1977). Do we need a theory of the state? *European Journal of Sociology, 18*, 223–244.
Mandel, E. (1975). *Late capitalism*. London: New Left Books.
Mangat, R. (2001, February). The death of distance? Globalization of international relations, E-merge. *Student Journal of International Affairs, 2*.
Mann, M. (1993). Nation-states in Europe and other continents: Diversifying, developing, but not dying. *Daedalus, 122*, 115–141.
Mann, M. (1997). Has globalization ended the rise and rise of the nation-state? *Journal of International Political Economy, 4*(3), 472–496.
March, J. G., & Olsen, J. P. (1998). The institutional dynamics of international political orders. *International Organization, 52*(4), 943–969.
Marks, G., Hooghe, L., & Blank, K. (1996). European integration from the 1980s: State-centric v. multi-level governance. *Journal of Common Market Studies, 34*(3), 342–378.
Marston, S. A. (2000). The social construction of scale. *Progress in Human Geography, 24*, 219–241.
Martin, R. (2004). Editorial: Geography: Making a difference in a globalizing world. *Transactions of the Institute of British Geographers, New Series, 29*, 147–150.
Marx, K. (1992). *Das kapital* (Vol. 1). London: Penguin.
Mathews, J. T. (1997). Power shift. *Foreign Affairs, 76*(1), 50–66.
Meyer, J. W. (1980). The world polity and the authority of the nation-state. In A. J. Bergesen (Ed.), *Studies of the modern world system* (pp. 109–137). New York: Academic.
Meyer, J. W., Boli, J., Thomas, G. M., & Ramirez, F. O. (1997). World society and the nation-state. *American Journal of Sociology, 1*, 144–181.
Miliband, R. (1969). *The state in capitalist society*. New York: Basic Books.
Milnar, Z. (1992). Introduction. In Z. Milnar (Ed.), *Globalization and territorial identities*. Aldershot: Avebury.
Miyoshi, M. (1993). A borderless world? From colonialism to transnationalism and the decline of the nation-state. *Critical Inquiry, 19*, 726–751.
Morgan, G. (1997). *Images of organization*. Thousand Oaks: Sage.
Morgan, K. & Roberts, E. (1993). The democratic deficit: A guide to quangoland. Papers in Planning Research 44, Department of City and Regional Planning, University of Wales, Cardiff.
Morgenthau, H. J. (1967). *Politics among nations: The struggle for power and peace*. New York: Alfred Knopf.
Morley, D., & Robins, K. (1995). *Spaces of identity: Global media, electronic landscapes and cultural boundaries*. London: Routledge.

Moses, J. W. (1994). Abdication from national policy autonomy: What's left to leave? *Politics and Society, 22*(2), 125–148.

Myoshi, M. (1993). A borderless world? From colonialism to transnationalism and the decline of the nation-state. *Critical Inquiry, 19,* 726–751.

Nettl, J. P. (1968). The state as a conceptual variable. *World Politics, 20,* 559–592.

Newman, D. (1995). *Boundaries in flux: The 'green line' boundary between Israel and the West Bank, boundary and territory briefing* (Vol. 5). Durham: International Boundaries Research Unit, University of Durham.

Newman, P., & Paasi, A. (1998). Fences and neighbours in the postmodern world: boundary narratives in political geography. *Progress in Human Geography, 22*(2), 186–207.

Norbert, E. (1983). *Power and civility.* New York: Pantheon.

O'Tuathail, G. (1996). *Critical geopolitics: The politics of writing global space.* London: Routledge.

O'Tuathail, G. (1997). At the end of geopolitics? Reflections on a pluralizing problematic at the century's end. *Alternatives: Social Transformation and Humane Governance, 22*(1), 35–55.

Oates, W. E. (1999). An essay on fiscal federalism. *Journal of Economic Literature, 37,* 1120–1149.

Obando-Rojas, B., Welsh, I., Bloor, M., Lane, T., Bodigannavar, V., & Maguire, M. (2004). The political economy of fraud in a globalised industry. *Sociological Review, 52*(3), 295–313.

Ohmae, K. (1993). The rise of the region state. *Foreign Affairs, 72,* 78–87.

Ohmae, K. (1995). *The end of the nation state: The rise of regional economies.* New York: Free Press.

Onuf, N. (1991). Sovereignty: Outline of a contemporary history. *Alternatives, 16*(4), 425–446.

Opello, W. C., Jr, & Rosow, S. J. (2004). *The nation-state and global order: A historical introduction to contemporary politics.* Boulder: Lynne Reinner.

Osiander, A. (2001). Sovereignty, international relations and the Westphalian myth. *International Organization, 55*(2), 251–287.

Paasi, A. (1996a). *Territories boundaries and consciousness: The changing geographies of the Finnish-Russian border.* Chichester: Wiley.

Paasi, A. (1996b). Inclusion, exclusion and territorial identities. The meanings of boundaries in the globalizing geopolitical landscape. *Nordisk Samhallgeografisk Tidskrift, 23,* 6–23.

Parekh, B. (2002). *Rethinking multiculturalism: Cultural diversity and political theory.* Cambridge: Harvard University Press.

Parker, M., & Bell, D. (Eds.). (2009a). *Space travel and culture: From Apollo to space tourism.* Malden: Wiley-Blackwell.

Parker, M., & Bell, D. (2009b). Introduction: Making space. *Sociological Review, 57*(1), 1–5.

Peters, B. G., & Pierre, J. (2001). Developments in intergovernmental relations: Towards a multi-level governance. *Policy and Politics, 29*(2), 131–135.

Petit, P., & Soete, L. (1999). Globalization in search of a future. *International Social Science Journal, 160,* 165–181.

Philpott, D. (1995). Sovereignty: An introduction and brief history. *Journal of International Affairs, 48,* 353–368.

Picciotto, S. (1997). Networks in international economic integration. *Northwestern Journal of International Law and Business, 17*(2/3), 1014–1056.

Picciotto, S. (1998, April 16). *Globalization, liberalization, regulation.* Conference on Globalization, The Nation-State and Violence, University of Sussex.

Podolny, J. M., & Page, K. L. (1998). Network forms of organization. *Annual Review of Sociology, 24,* 57–76.

Polanyi, K. (1944). *The great transformation.* Boston: Beacon Press.

Poulantzas, N. (1969). The problem of the capitalist state. *New Left Review, 58,* 67–78.

Puchala, D. J. (1972). Of blind men, elephants and international integration. *Journal of Common Market Studies, 10*(3), 267–284.

Reich, R. (1992). *The work of nations.* New York: Vintage.

Relyea, S. (1998). Trans-state entities: Postmodern cracks in the great Westphalian dam. *Geopolitics, 3*(2), 30–61.
Rhodes, R. A. W. (1997). *Understanding governance. Policy networks, governance, reflexivity and accountability*. Buckingham: Open University Press.
Robertson, R. (1992). *Globalization*. London: Sage.
Rodnik, D. (1997). *Has globalization gone too far?* Washington: Institute for International Economics.
Rosenau, J. N. (1988). The state in an era of cascading politics, wavering concept, widening competence, withering colossus, or weathering change? *Comparative Political Studies, 21*(1), 13–44.
Rosenau, J. N. (2000). *The governance of fragmegration: neither a world republic nor a global interstate system*. Quebec: Congress of the International Political Sciences Association.
Royal Institute of International Affairs. (1939). *Nationalism*. London.
Rudd, K. (2001). Globalization and regional governance. In C. Sheil (Ed.), *Globalization: Australian impacts*. Sydney: UNSW Press.
Ruggie, J. G. (1983). Continuity and transformation in the world polity: Toward a neo-liberalist synthesis. *World Politics, 35*, 261–285.
Ruggie, J. G. (1993). Territoriality and beyond: Problematizing modernity in international relations. *International Organization, 47*(1), 139–174.
Ruggie, J. G. (1998). *Constructing the world polity: Essays on world institutionalization*. London: Routledge.
Rumley, D., & Minghi, J. (Eds.). (1991). *The geography of border landscapes*. London: Routledge.
Russell, B. (1916). *Justice in war time*. New York: Cosimo Classics.
Sack, R. D. (1986). *Human territoriality: Its theory and history*. Cambridge: Cambridge University Press.
Sahlins, P. (1989). *Boundaries: The making of France and Spain in the Pyrenees*. Berkeley: University of California Press.
Said, E. (1979). Zionism from the standpoint of its victims. *Social Text, 1*, 7–58.
Sakamoto, Y. (1994). *Transformation: Challenges to the state system*. Tokyo: Tokyo University Press.
Sassen, S. (1997). *Losing control? Sovereignty in an age of globalization*. New York: Columbia University Press.
Sassen, S. (2000). Territory and territoriality in the global economy. *International Sociology, 15*, 372–393.
Sassen, S. (2003). Globalization or denationalization? *Review of International Political Economy, 10*, 1–22.
Sayer, A. (1985a). The difference that space makes. In D. Gregory & J. Urry (Eds.), *Social relations and spatial structures* (pp. 49–66). London: Macmillan.
Sayer, A. (1985b). Industry and space: A sympathetic critique of radical research. *Society and Space, 3*, 3–29.
Scharpf, F. W. (1992). Europaisches Demokratiedefizit und deutscher Föderalismus. *Staatswissenschaften und Staatspraxis, 3*, 293–306.
Scharpf, F. W. (1994). Community and autonomy: Multi-level policy-making in the European Union. *Journal of European Public Policy, 1*(2), 219–242.
Schmidt, V. A. (1995). The new world order incorporated; the rise of business and the decline of the nation-state. *Daedalus, 124*, 75–106.
Schofield, C. H. (Ed.). (1994). *World boundaries* (Vol. 1)., Global boundaries London: Routledge.
Schofield, C. H., & Schofield, R. N. (Eds.). (1994). *World boundaries* (Vol. 2)., The Middle East London: Routledge.
Scholte, J. A. (2000). *Globalization*. London: Palgrave.
Scholte, J. A. (2004). Globalization and governance: From statism to polycentrism. CSGR Working Paper 130/04, Centre for the Study of Globalization and Regionalization. University of Warwick, Coventry.

Schrier, E., Nadel, E., & Rifas, B. E. (1984). Forces shaping international maritime transport. *World Economy, 7*, 87–102.

Scott, A. J. (1996). Regional motors of the global economy. *Futures, 28*(5), 391–411.

Seton-Watson, H. (1977). *Nations and states: An inquiry into the origin of nations and the politics of nationalism.* Boulder: Westview Press.

Shapiro, M. (1994). Moral geographies and the ethics of post-sovereignty. *Public Culture, 6*, 479–502.

Shapiro, M. J., & Alker, H. R. (Eds.). (1996). *Challenging boundaries: Global flows, territorial identities.* Minneapolis: University of Minneapolis Press.

Short, J. R. (1993). The 'myth' of postmodernism. *Tijdschrift voor Economische en Sociale Geografie, 84*(3), 169–171.

Sibley, D. (1995). *Geographies of exclusion: Society and difference in the west.* London: Routledge.

Skelcher, C. (2005). Jurisdictional integrity, polycentrism and the design of democratic government. *Governance, 18*(1), 89–110.

Skowronek, S. (1982). *Building a new American state: The expansion of national administrative capacities.* New York: Cambridge University Press.

Slaughter, A.-M. (1997). The real new world order. *Foreign Affairs, 76*(5), 183–197.

Sletmo, G. K. (2001). The end of national shipping policy? A historical perspective on shipping policy in a global economy. *International Journal of Maritime Economics, 3*, 333–350.

Smith, N. (1984). *Uneven development: Nature, capital and the production of space* (1st ed.). Oxford: Blackwell.

Smith, N. (1987). Dangers of the empirical turn: Some comments on the CURS initiative. *Antipode, 19*, 59–68.

Smith, N. (1993). Homeless/global: Scaling places. In J. Bird, B. Curtis, T. Putnam, G. Robertson, & L. Tickner (Eds.), *Mapping the futures: Local cultures, global change* (pp. 87–119). London: Routledge.

Smith, D. A., Solinger, D. J., & Topik, S. C. (Eds.). (1999). *States and sovereignty in the global economy.* London: Routledge.

Soja, E. W. (1989). *Postmodern geographies: The reassertion of space in critical social theory.* New York: Verso.

Soja, E. W. (2000). *Postmetropolis: Critical studies of cities and regions.* Oxford: Blackwell.

Somerville, P. (2004). *Governance and democratic transformation.* Political Studies Association Annual Conference, Lincoln.

Sorel, G. (1908a). *Les Illusions du Progrès (The Illusions of Progress)* (John & Charlotte Stanley, Trans.). Berkeley, CA: University of California Press.

Sorel, G. (1908b). *Réflexions sur la violence (Reflections on Violence)*, (T. E. Hulme, & J. Roth, Trans.). New York, NY: The Free Press.

Spruyt, H. (1994a). Institutional selection in international relations: State anarchy as order. *International Organization, 48*, 527–557.

Spruyt, H. (1994b). *The sovereign state and its competitors.* Princeton: Princeton University Press.

Stalder, F. (2006). *Manual Castells.* Cambridge: Polity Press.

Stopford, M., & Strange, S. (1991). *Rival states, rival firms: Competition for world market shares.* Cambridge: Cambridge University Press.

Storper, M. (1997). *The regional world. Territorial development in a global economy.* New York: The Guilford Press.

Strange, S. (1976). Who runs world shipping? *International Affairs, 52*(3), 346–367.

Strange, S. (1994). The power gap: Member states and the World economy. In F. Brouwer, V. Lintner, & M. Newman (Eds.), *Proceedings of a Conference on Economic Policy Making and the European Union, London European Research Centre, London Federal Trust, University of North London* (pp. 19–26).

Strange, S. (1996). *The retreat of the state. The diffusion of power in the world economy.* Cambridge: Cambridge University Press.

Strange, S. (1999). The westfailure system. *Review of International Studies, 25*, 345–354.
Stubbs, P. (2005). Stretching concepts too far? Multi-level governance, policy transfer and the politics of scale in south–east Europe. *Southeast European Politics, 6*(2), 66–87.
Sturmey, S. G. (1975). *Shipping economics: Collected papers*. London: Macmillan Press.
Swyngedouw, E. (1992). Territrial organization and the space/technology nexus. *Transactions of the Institute of British Geographers, New Series, 17*, 417–433.
Swyngedouw, E. (1996). *Globalization or glocalization? Networks, territories and re-scaling, School of Geography and the Environment*. Oxford: St. Peter's College, Oxford University.
Swyngedouw, E. (1997a). Neither global nor local: 'Glocalization' and the politics of scale. In K. R. Cox (Ed.), *Spaces of globalization: Reasserting the power of the local* (pp. 137–188). New York: Guilford.
Swyngedouw, E. (1997b). Excluding the other: The contested production of new 'gestalt of scale' and the politics of marginalization. In R. Lee & J. Wills (Eds.), *Society, place, economy: States of the art in economic geography* (pp. 167–176). London: Edward Arnold.
Swyngedouw, E. (2000). Authoritarian governance, power and the politics of rescaling. *Environment and Planning D, 18*, 63–76.
Szczerski, K. (2004). *The EU multi-level governance in post-communist countries—Challenge in governability for the new member states*. 12th NISPAcee Annual Conference, Vilnius, Lithuania, May 13–15.
Taylor, P. J. (1994). The state as container: Territoriality in the modern world system. *Progress in Human Geography, 18*(2), 151–162.
Taylor, P. J. (1995). Beyond containers: Internationality, interstateness, interterritorality. *Progress in Human Geography, 19*, 1–15.
Taylor, P. J. (1996). Embedded statism and the social sciences: Opening up to new spaces. *Environment and Planning A, 28*, 1917–1928.
Taylor, P. J. (2000). Embedded statism and the social sciences 2: Geographies (and metageographies) in globalization. *Environment and Planning A, 32*, 1105–1114.
Taylor, P. J. (2005). New political geographies: Global civil society and global governance through world city networks. *Political Geography, 24*, 703–730.
Thomson, J. (1994). *Mercenaries, pirates and sovereigns: Statebuilding and extraterritorial violence in early modern Europe*. Princeton: Princeton University Press.
Thomson, J. E., & Krasner, S. D. (1989). Global transactions and the consolidation of sovereignty. In E.-O. Czempiel & J. N. Rosenau (Eds.), *Global Change and Theoretical Challenges* (pp. 201–203). Lexington: Lexington Books.
Thrift, N. J. (1996). *Spatial formations*. London: Sage.
Thrift, N. J., & Olds, K. (1996). Refiguring the economic in economic geography. *Progress in Human Geography, 20*, 311–337.
Tilly, C. (1994). The time of states. *Social Research, 61*(2), 269–295.
Tishkov, V. A. (2000). Forget the 'nation': Post-nationalist understanding of nationalism. *Ethic and Racial Studies, 23*, 625–650.
Valaskakis, K. (1999). Globalization as theatre. *International Social Science Journal, 160*, 153–164.
Van Creveld, M. (1999). *The rise and decline of the state*. Cambridge: Cambridge University Press.
Van Hoof, L., & Van Tatenhove, J. (2009). EU marine policy on the move: The tension between fisheries and maritime policy. *Marine Policy, 33*, 726–732.
Virilio, P. (1977). *Speed and politics. An essay on dromology*. New York: Semiotext(e).
Virilio, P. (1992). *From politics to reason of state: The acquisition and transformation of the language of politics, 1250–1600*. Cambridge: Cambridge University Press.
Virilio, P. (1995). Speed and information: Cyberspace alarm! *ctheory.net*. www.ctheory.net/articles.aspx?id=72.
Vonnegut, K., Jr. (1963). *Cat's cradle*. London: Penguin.
Walker, R. B. J. (1991). State sovereignty and the articulation of political space/time. *Millennium, 20*, 445–461.

Walker, R. B. J. (1993). *Inside/outside: International relations as political theory*. Cambridge: Cambridge University Press.
Wallerstein, I. (1984). *Politics of the world-economy: The states, the movements and the civilizations*. Cambridge: Cambridge University Press.
Walters, R. S., & Blake, D. H. (1992). *The politics of global economic relations*. Englewood Cliffs: Prentice-Hall.
Waltz, K. N. (1979). *Theory of international politics*. Reading: Addison-Wesley.
Waltz, K. N. (1999). Globalization and governance. *Political Science and Politics, 32*(4), 693–700.
Weber, M. (1978). *Economy and society: An outline of interpretative sociology*. Berkeley: University of California Press. (2 Volumes).
Weber, C. (1992). Reconsidering statehood: Examining the sovereignty intervention boundary. *Review of International Studies, 18*, 199–216.
Weiss, L. (1998). *The myth of the powerless state*. Ithaca: Cornell University Press.
Weiss, L. (1999). Globalization and national governance: Autonomy or inter-dependence. *Review of International Studies, 25*(5), 59–88.
Weiss, T. G. (2000). Governance, good governance and global governance: Conceptual and actual challenges. *Third World Quarterly, 21*(5), 795–814.
Welchman, J. C. (Ed.). (1996). *Rethinking borders*. Minnesota: University of Minnesota Press.
Whitley, R. (1996). Business systems and global commodity chains: Competing or complimentary forms of economic organization. *Competition and Change, 1*, 411–425.
Whitley, R. (1998). Internationalization and varieties of capitalism: The limited effects of cross-national coordination of economic activities on the nature of business systems. *Review of International Political Economy, 5*, 445–481.
Whitley, R. (1999). *Divergent capitalisms: The social structuring and change of business systems*. New York: Oxford University Press.
Wood, E. M. (2001). Global capital. National states. In M. Rupert & H. Smith (Eds.), *Historical materialism and globalization* (pp. 17–39). New York: Routledge.
Woods, N., & Narlikar, A. (2001). Governance and the limits of accountability: The WTO, the IMF, and the world bank. *International Social Science Journal, 170*, 569–583.
Wright-Mills, C. (1959). *The sociological imagination*. New York: Oxford University Press.
Yeung, H. W. (1998a). Capital, state and space: Contesting the borderless world. *Transactions of the Institute of British Geographers, New Series, 23*, 291–309.
Yeung, H. W. (1998b). The socio-spatial constitution of business organizations: A geographical perspective. *Organization, 5*, 101–128.
Yeung, H. W. (2000). Embedding foreign affiliates in transnational business networks: The case of Hong Kong firms in south–east Asia. *Environment and Planning A, 32*, 201–222.
Yiftachel, O., & Meir, A. (Eds.). (1997). *Ethnic frontiers and peripheries*. Boulder: Westview Press.
Zacher, M. W. (1992). The decaying pillars of the Westphalian temple: Implications for international order and governance. In J. N. Rosenau & E.-O. Czempiel (Eds.), *Governance without government: Order and change in world politics* (pp. 58–101). Cambridge: Cambridge University Press.
Zacher, M. W., & Sutton, B. A. (1996). *Governing global networks*. Cambridge: Cambridge University Press.
Zhang, X. (2005). Coping with globalization through a collaborative federate mode of governance. *Policy Studies, 26*(2), 199–209.
Zurn, M. (2000). Democratic governance beyond the nation-state: The EU and other international institutions. *European Journal of International Relations, 6*(2), 183–221.
Zurn, M. (2003). Globalization and global governance: From societal to political denationalization. *European Review, 11*(3), 341–364.

Chapter 5
Globalization

Bergquist (1993, p. 83) provided an excellent link between one of our central themes—globalization—and something to which we shall turn our attention in later sections—the role of the Postmodern. He drew on Paddy Chayefsky's 1976 film *Network*, which dwells upon the significance of globalization to society as a whole. Ned Beatty (powerful international businessman) suggests to Peter Finch (renegade newscaster) that:

> national boundaries no longer exist. The governments that used to control these boundaries have been replaced by corporate boards and stockholders. National currency has been replaced by the currency of oil. The new world order will be run like a business, not like a government. This new world order has no purpose other than the provision of profits to its anonymous stockholders and the establishment of a peaceful, business-as-usual environment. People are tranquilized by their own personal consumption of commodities skillfully marketed by those who control the international business conglomerates.

This vision of a Postmodern globalized world is familiar to the maritime sector as

> The maritime industry is thus not only pivotal to world trade, but it is also the only example of a fully globalized industry. The ship and the seafarers aboard are at the centre of a complex constellation of multiple interests which situates ship owners and seafarers in fluid, and sometimes, volatile legal, political and social circumstances. (Alderton and Winchester 2002, p. 36)

It is not the intention here to provide a full debate about globalization, its development, significance, and meaning. The literature on globalization is already immense. However, we have seen the significance of globalization in broader terms (for example Caporaso 1981), and more significantly, for the nation-state. The importance of this rests with the current preeminent role of the nation-state in the process of maritime governance and policy-making, something stressed by Lorange (2008) among others and which was clear from our earlier discussion. Consequently we need to understand a little more fully what globalization means for governance so that in turn we can begin to make progress in revitalizing policy-making in the maritime sector taking account of the new forces at work.

Globalization is far from new. Harvey (2000, p. 25) quoted extensively from Marx and Engels (1967, Sect. 1) who were writing in 1848:

> The need for a constantly expanding market chases the bourgeoisie over the whole surface of the globe. It must settle everywhere, establish connections everywhere... The bourgeoisie has through its exploitation of the world market given a cosmopolitan character to production and consumption in every country... All old established national industries have been destroyed or are daily being destroyed. They are dislodged by new industries, whose introduction becomes a life and death question for all civilized nations, by industries that no longer work up indigenous raw material, but raw material drawn from the remotest zones: industries whose products are consumed, not only at home, but in every quarter of the globe. In place of the old wants, satisfied by the production of the country, we find new wants, requiring for their satisfaction the products of distant lands and climes. In place of the old local and national seclusion and self-sufficiency, we have intercourse in every direction, universal interdependence of nations. And as in material, so also in intellectual production. The intellectual creations of individual nations become common property. National one-sidedness and narrow-mindedness become more and more impossible, and from the numerous national and local literatures, there arises a world literature...

At risk of overdoing the message they added:

> The bourgeoisie... draws all, even the most barbarian nations into civilization. The cheap prices of its commodities are the heavy artillery with which it batters down all Chinese walls, with which it forces the barbarians' intensely obstinate hatred of foreigners to capitulate. It compels all nations on pain of extinction, to adopt the bourgeois mode of production; it compels them to introduce what it calls civilization into their midst, i.e. to become bourgeois themselves. In one word, it creates a world after its own image.

Overlooking the political overtones for the moment, the applicability of the comments of Marx and Engels to current descriptions of globalization is clear. Berman (1982, pp. 90, 91) took up the Marxist interpretation of globalization drawing the relationship between it and capitalism and stressing how far from being a new phenomenon, instead it is part of a recurrent theme that forced economic and political change to take place.

Held (1991, p. 144) meanwhile harked back much earlier in identifying phases of globalization over 400 years ago with the expansion of the world economy and the rise of the modern state (Wallerstein 1974). And much earlier Callières (1963, p. 11) wrote in 1716:

> we must think of the states of which Europe is composed as being joined together by all kinds of necessary commerce, in such a way that they may be regarded as member of one Republic, and that no considerable change can take place in any one of them without affecting the condition, or disturbing the peace, of all the others. The blunder of the smallest of sovereigns may indeed cast an apple of discord among all the greatest powers, because there is no state so great which does not find it useful to have relations with the lesser states and to seek friends among the different parties of which the smallest state is composed.

Indeed both Grotius (1617) and Kant (1795) understood the notion of the society of states. However, modern 'Westernized' globalization is something different again with two distinct characteristics in the opinion of Held (1991). It implies world-wide political, economic, and social activity and not just international. And secondly, it also implies an intensification of this activity far beyond anything ever seen before. As Kegley and Wittkopf (1989, p. 511) stated eloquently, politics unfolds today, with all its customary uncertainty, contingency, and indeterminateness, against the background of a world:

> permeated and transcended by the flow of goods and capital, the passage of people, communication through airways, airborne traffic, and space satellites.

Relyea (1998, p. 31) took an extreme view seeing globalization driving 'its relations for millennia' and it is only that 'the new technologies of the twentieth century and their concomitant philosophies (that) have raised the consciousness of governments and citizens to globalization's influence'. Valaskakis (1999, pp. 153, 154) agreed with a substantial history to globalization placing its origins precisely in the fifteenth century and the struggles between feudal emperors and Popes over temporal and spiritual power. This power was transferred to the early nation-state with the Thirty Years War and the Treaty of Westphalia but it was not until the increase in colonial expansion and the development of global commodity markets that modern globalization emerged. Shaw (1997, p. 498) considered that globalization accompanied the 'multi-power actor of the west' originating in Europe 'over some centuries' (Mann 1986). Meanwhile, Cable (1995, p. 24) suggested that the economic boom of the 1980s resembled the cycles of 'lending, over lending, default, rescheduling and fresh lending' of the nineteenth century, while of the 1990s 'discovery' of emerging markets merely reflected similar investment developments in the same century in the USA, Latin America, Asia, and Eastern Europe. Similarly, the investment in the sizeable, global projects of the Suez and Panama Canals is reflected by those in the Channel Tunnel and Hong Kong Airport.

5.1 A Definition of Globalization

> We are living through a transformation that will rearrange the politics and economics of the coming century. There will be no national products or technologies, no national corporations, no national industries. There will no longer be national economies, at least as we have come to understand that concept... As almost every factor of production – money, technology, factories and equipment – moves effortlessly across borders, the very idea of an American economy is becoming meaningless as are the notions of an American corporation, American capital, American products, and American technology. A similar transformation is affecting every other nation, some faster and more profoundly than others; witness Europe hurtling toward economic union. (Reich 1992, pp. 3, 8, quoted in Dicken 1994, p. 102)

The literature on globalization is nothing short of vast and we shall only scrape the surface where it provides guidance for a definition. Garrett (1998) summarized the issues while Walker (1991, p. 447) was confident that the phenomenon of globalization really existed although he did question its origins which in turn perhaps reflect upon its long history. Gritsch (2005, p. 9) had no problem in identifying globalization, seeing it as a process of soft geopolitics or states' zero sum pursuit of geopolitical power on non-militaristic terrain. The approach was to apply both competitive and coercive economic tactics by dominant states who then impose a mix of institutional practices and structures to ensure specific and preferential politico–economic agendas for other states and markets. This has been commonly achieved through international institutions such as the UN, WTO, OECD, etc. In this way, globalization allows capitalists and states to defuse resistance to increasing accumulation from domestic labor interests in particular (Krasner 1978, p. 11; Putterman and Rueschemeyer 1992, p. 257). It is thus a strategy that Gritsch (2005, p. 18) saw as acting 'to transform the politics of accumulation by disarticulating from relationships of contestation and reciprocity that have integrated it into the nation'. One (of many) problem that emerges from this is that globalization decreases the opportunity for representation and thus the role of the stakeholder is diminished, a process of disenfranchization that has repercussion for the generation of meaningful and representative policies—maritime or otherwise (Poulantzas 1976, p. 80). Wood (1997, p. 552) alluded to the problems that globalization created for democracy suggesting that it leads to the 'breaching, transcendence, or obliteration of national boundaries by economic agencies, and correspondingly the weakening of political authorities, whose jurisdiction is confined within those boundaries'. However, Wood also offered an interesting slant on globalization and its relevance to the debate on whether it represents a final solution to capitalism's ambitions or whether it is just the next phase in a rolling program of inevitable crises aimed at generating and consuming capital. He was confident that this is so—something we return to in considering the concept of 'disorganized capitalism' later in this chapter. In Wood's (1997, p. 553) words:

> Globalization is just another step in the geographic expansion of economic rationality and its emancipation from political jurisdiction. In the long geopolitical process that has constituted the spread of capitalism, the borders of the nation-state appear to be the last frontier, and nation-state power the final fetter to burst asunder.

Shaw (1997, p. 498) meanwhile provided a neat definition—'a complex set of distinct but related processes—economic, cultural, social and also political and military—through which social relations have developed towards a global scale and with global reach, over a long historic period'.

However, Jameson (2000) was scathing of attempts to define globalization seeing them as little more than 'ideological appropriations—discussions not of the process itself, but of its effects, good or bad'. Others are not so negative. For example, Waltz (1999, p. 694) provided an informal definition:

> Free markets, transparency, and flexibility are the watchwords. The 'electronic herd' moves vast amounts of capital in and out of countries according to their political and economic merits. Capital moves almost instantaneously into countries with stable governments, progressive economies, open accounting and honest dealing, and out of countries lacking their qualities. States can 'defy' the herd but they will pay a price as did Thailand, Malaysia, Indonesia and South Korea in the 1990s. Some countries may defy the herd inadvertently (the countries just mentioned); others, out of ideological conviction

5.1 A Definition of Globalization

(Cuba and North Korea); some because they can afford to (oil-rich countries); others because history has passed them by (many African countries).

Waltz's views were openly capitalist and his comments on progressive economies and open accounting are open to some question but his interpretation of the principles and significance of globalization is clear.

Soja (2000, p. 191) cited Robertson (1992, p. 64) in his attempt to define globalization:

> the compression of the world and the intensification of consciousness of the world as a whole (bringing with it the deepening and widening of) worldwide social relations which link distant localities in such a way that local happenings are shaped by events occurring many miles away and vice versa.

Soja saw globalization not as a new thing but as an intensification and widening in scope and scale of social, political, economic, and cultural relations. He also provided a rather useful definition of the forces that are always found lying behind the process of globalization and which drive the deep restructuring that is taking place:

- Industrialization of major segments of the old Third World and the simultaneous deindustrialization of established cities and regions of Fordist industrial production.
- Creation of new forms of globally networked manufacturing.
- Accelerated movement of people, goods, services, and information across national borders and the growth of global markets for labor and globally networked commodities.
- Reorganization of international trading systems and markets including the EU and NAFTA.
- Propulsive emergence of the transnational corporation to rationalize and coordinate global investment, production, and capital accumulation
- Space compression and network effects of the telecommunications and information revolution.
- Emergence of powerful institutions to promote global financial integration consolidating commercial, financial, and industrial interest at a global scale.
- Rise of the Pacific Rim as a competitive power bloc to the North Atlantic Alliance.
- Concentration of political and economic power in a reordered hierarchy of global cities acting as command posts for the financial operations of the world economy (Soja 2000, pp. 193, 194).

While having some doubts about the hierarchical nature of world cities, the remaining points ring true. Cable (1995, pp. 25, 26) meanwhile condensed these forces into two dominantly technological developments—in transport, resulting in plummeting costs, and in communications—resulting in easier, speedier, and higher quality facilities for transmitting and exchanging information. Globalization is thus dominated by a heady mix of the service sector and the transport industry (Cable 1995, p. 33).

Mann (1993) suggested that sovereignty had been 'outreached by transnational power networks, especially those of global capitalism and Postmodern culture'. Strange (1986) termed it 'casino capitalism', in Mann's words, with 'its funds slushing rapidly through the world in a complex web of institutions which partially elude the economic planning capacities of states and partially inoperationalize them'.

Axtmann (1995) outlined what he saw as the 12 most significant principles underlying globalization while Cerny (1995, p. 596) offered another vision of globalization suggesting that it was:

> a set of economic and political structures and processes deriving from the changing character of the goods and assets that comprise the base of the international political economy.

Taylor (2000a, p. 1,107) quoted Held et al. (1999, p. 51) in defining globalization as:

> a process (or set of processes) which embodies a transformation in the spatial organization of social relations and transactions – assessed in terms of their extensity, intensity, velocity and impact – generating transcontinental or interregional flows and networks of activity, interaction, and the exercise of power.

Globalization is thus essentially a geographic scale of activity lying on 'a continuum with the local, national and regional' implying a 'stretching of social, political and economic activities across frontiers' (Held et al. 1999, p. 67).

Zurn (2000, p. 187) redefined globalization as 'denationalization' which in turn he explained was:

> the extension of social spaces, which are constituted by dense transactions, beyond national borders without being necessarily global in scope. Even though the scope of most of these cross-border transactions is indeed not global, they still cause a problem for national governance because the social space to be governed is no longer national.

Zurn went on to stress that this process of denationalization is 'jagged', differing according to issue, country, and time and occurs wherever goods, services and capital (economy), threats (force), pollutants (environment), signs (communication), and persons (mobility) are exchanged.

The debate about globalization in the context of the nation-states was taken up by Storper (1997, p.171) who suggested that on the one hand we have those who argue that both foreign and national investment will only ever flow to those areas where appropriate factors of production exist (for example high quality and cheap labor; appropriate infrastructure etc.). Consequently, the nationality of firms is unimportant. This contrasted with those who argued that nationality is important for national security, political reasons and because even multinational firms have to have a core location somewhere and that has to be within a nation. In shipping terms, both arguments have relevance and the contradictions are clearly apparent. Globalization has clearly made national loyalty secondary to capital accumulation while at the same time shipping is by tradition associated with the nation-state through ownership, flag, and most recently tax concession and the significance of globalization to international shipping has been widely recognized (see for example Bloor and Sampson 2009, p. 712). Both global and national shipping can be seen operating within the same markets, sectors, even ships.

Storper (1997, p. 174) went on to stress how globalization also had eroded the traditional vertically integrated, business firm hierarchy to be replaced by nodes and linkages ranging from 'ownership to alliance and including cross investment, technology, and production partnerships'. The impact upon the traditional hierarchical, maritime policy-making model is obvious.

Keohane (2001, p. 1) also provided a layman's definition of globalization based on the comments of Held et al. (1999) and Keohane and Nye (2001) suggesting it refers to the shrinking of 'distance on a world scale through the emergence and thickening of networks of communications—environmental and social as well as economic'. Keohane also stressed how effective globalization needed effective governance and that the two were heavily interrelated.

Jessop (2003) emphasized the complexity of globalization with multiple causes and characteristics and how the different nation-states are affected in different ways; Kjaer (2004) provided a lengthy discussion of the economic and military images of globalization while Forman and Segaar (2006) analyzed the interrelationship of the nation-state administrations, NGOs, multinational corporations and individuals in the development of a globalized society and the governance framework that they require.

Finally, Zurn (2003, p. 344) returned to recouch his conception of globalization into societal terms viewing it as a:

> process in which the world moves as an integrated global society and in which the significance of national borders decreases. It thus calls into question the distinction between domestic and foreign relations that underlies the notion of interdependence. In this view, the living conditions of people and local communities have changed through globalization; distant events of all sorts have immediate consequences not only for states but for individuals' daily lives. This notion of globalization refers to a measurable process of *social change* which, in turn, may or may not have causal effects on *political developments*... This societally based understanding of globalization facilitates a distinction between different fields and runs counter to narrow notions of globalization that are restricted to the economic or the cultural sphere...

Zurn's interpretation of globalization as a process rather than a purely spatial event and following the work of Rosenau (1990); Holm and Sorensen (1995), Hirst and Thompson (1995), and Held et al. (1999), is particularly interesting and something we shall return to later in the context of maritime governance.

5.2 Globalization and the Nation-State

Globalization and the nation-state are therefore intrinsically involved with one another and because the rise of globalization has had such a massive impact on the nation-state, then the impact upon governance frameworks and consequential policy-making (maritime or otherwise) has to be considered. The globalization/ nation-state debate is well documented and features throughout the literature. Our discussion here will be limited simply to reflect the major concerns.

We can begin with some thoughts from Berry (1989, p. 1) quoted in Dicken (1994, p. 101) who located the nation-states in a multi-dimensional cultural–geographic space, which constituted a 'differentiating checkerboard of culture' on which 'in an increasingly tightly knit global economy, multi-national corporations play their locational games interactively with nation-states'. This global/nation-state relationship was central to Berry's core of economic geography. International firms and the nation-state are central to 'shaping the changing geography of the global economy' (Dicken 1986, 1992).

Zurn (2003, pp. 342–344) considered this relationship in some depth suggesting that the significance of national borders must decrease as globalization intensifies thus challenging the nation-state to 'unilaterally reach its governance targets'. He saw effective governance depending upon 'the spatial congruence of political regulations with socially integrated areas and the absence of significant externalities'. Globalization makes this spatial congruence less effective. Specific challenges to the nation-state by globalization were identified:

- National borders no longer encompass sufficient territory to function as self-contained markets for globalized companies. This presents challenges to domestic legislation. Shipping is no exception and the issue of its external activity has always presented difficulties further exacerbated by the intensification of globalization. Favorable national discrimination distorts the marketplace (e.g tonnage taxation); unfavorable national discrimination drives away domestic business (e.g removal of domestic subsidy).
- Political regulation will be ineffective as it covers only part of the relevant social and economic space. Externalities in shipping are a classic case as, for example, shipping generated air pollution is unhindered by domestic regulation either in its production (domestic regulations drive ships to other flags) or its consumption (there are no borders for air pollution).

Zurn (2003, p. 343) identified variable responses to these challenges. Some national governments:

> fight globalization to evade these challenges (*This is pointless in shipping where globalization is a vital ingredient to the industry's success and will not go away*). Many others push for national adjustments in order to capitalize on the opportunities of globalization and at the same time reduce its unwanted effects (*A common response in shipping – take tonnage tax for example – where shipowners hope to gain the best of both worlds at the expense of the nation-state*). Still others also see the potential gains of globalization and argue in favour of international institutions to handle the challenges (*This is uncommon in shipping which continues to hold on to the nation-state paradigm and the convenience this provides*). Governments and other political actors then endeavour to regain control by establishing new international and transnational or even supranational regimes, networks and organizations for the coordination and harmonization of their policies. That is they endeavour to establish legitimate governance beyond the nation-state. *(emphasized sections added)*

Brenner (1999, pp. 44, 52) saw states as vital 'geographical components' for globalization and not something irrelevant citing Lefebrvre's (1976a, b, 1977, 1978) globally articulated state form (*le mode de production etatique*) promoting, regulating, and financing capitalist industrial development globally. Their role is providing stable territorialization for the successive rounds of inevitable capital accumulation albeit with intensified decentralization and the generation of widened global interdependencies. Cerny (1995, p. 599) continued the global/national debate, Shaw (1997, p. 498) felt that globalization did not undermine but transformed the state, while Evans (1997, pp. 66–70) discussed their relationship in some detail. Opello and Rosow (2004, p. 269) stressed that the state had been empowered by neoliberalist globalization enabling it to exercise increased 'politico–military rule' through the increased diffusion and deterritorialization of power. Globalization presented a territory problem but one which the new neoliberalist state (characterized in particular by privatization and deregulation) could accommodate far better than the traditional Westphalian state.

Wood (1997, p. 553) considered that globalization's relationship to the nation-state was contradictory. On the one hand, by definition the position of the nation-state must be weakened but on the other it:

> not only presupposes the nation-state but relies on the state as its principle instrument. If anything the new global order is more than ever a world of nation-states; and if these states are permeable to the movements of capital, then permeability has its corollary, indeed as its condition, the existence of national boundaries and state jurisdictions.

Indeed this is precisely what the shipping industry enjoys in taking the maximum advantage from a jurisdictional framework that attempts to accommodate both globalization and nation-state sovereignty. Almost inevitably this is ineffective utilizing a jurisdictional framework that is outdated and characterized by inappropriate institutional policy-making.

Zurn (2003, p. 348) suggested that due to the intensification of globalization we now have a condition he terms 'governance beyond the state'. Under Modernist influences 'governance by government' was the norm with governance provided by a nation-state government claiming a monopoly of legitimate force and ruling by hierarchical order. This was how governance for the international shipping industry was designed and the institutional and jurisdictional framework that it embodied remains with us. The mismatch between the existing Westphalian maritime jurisdiction and the Postmodern globalized shipping industry generates the policy failures we have today. The existing international institutions, typified by the IMO, address states as the ultimate regulators; they regulate transactions that commonly take place at state borders; and they tend to consider problems which are relatively certain and which as a result the relevant actors might be able to make good. In Zurn's opinion (2003, p. 351), the actions that need to be taken by international institutions need to be directed at actors and not states (thus changing the range of policy stakeholders substantially); the issues to be dealt with are no longer focused primarily on state borders but commonly on issues within borders; and the intensity of uncertainty has increased dramatically.

Yeung (1998, p. 292) continued the debate about national borders and globalization and refuted the arguments that borders have no relevance in a globalized world. Along with Zurn he insisted that borders remained but that the governance issues to be dealt with may well be within these borders as well as outside them and that globalization would intensify rather than homogenize the contrasts across borders as it permits the extended exploitation of differences that exist. Shipping could not be more globalized yet it is characterized by severe contrasts between flag states which the process of globalization has encouraged.

5.3 Globalization and Space–Time Compression

Cohen (2001) suggested that globalization is a consequence of the interplay of changes in technology and in the structure of business enterprises combined with the movement in political choice and power that has taken place. However, Brenner (1999, p. 42) provided a more useful interpretation that helps to understand better the processes going on and provides a basis to contemplate the forces that lie behind globalization, their relationship to capital accumulation and impact upon governance. He began by suggesting that globalization was a 'highly contested term'. Some concentrate upon the:

> growing role of transnational corporations, the deregulation of finance capital, the expansion of foreign direct investment, the intensified deployment of information technologies, and the dissolution of the Bretton Woods monetary regime since the early 1970s (Boyer and Drache 1996; Dicken 1992; Ruigrok and van Tulder 1995). Others emphasize various newly emergent forms of collective identity, political consciousness, and diaspora that appear to have unsettled the principle of nationality as a locus of everyday social relations, as well as new forms of technologically mediated socio-cultural interaction that seem oblivious to national territorial boundaries (Appadurai 1996; Featherstone 1990; Marden 1997). Finally some authors have defined globalization more abstractly as a process through which interdependencies among geographically distant localities, places and territories are at once extended, deepened and intensified (Giddens 1990; McGrew 1992).

Brenner went on to provide his own interpretation of globalization. His explanation is excellent and will be quoted at some length. Globalization is:

> the most recent historical expression of continual deterritorilaization and reterritorialization that has underpinned the production of capitalist spatiality since the first industrial revolution of the early nineteenth century. On the one hand, capitalism is under the impulsion to eliminate all geographical barriers to the accumulation process in search of cheaper raw materials, fresh sources of labor-power, new markets for its products, and new investment opportunities... On the other hand... the resultant process of 'time-space compression' must be viewed as one moment within a contradictory socio-spatial dialectic that continually molds, differentiates, deconstructs and reworks capitalism's geographical landscape (Harvey 1982, 1985b, 1990). According to Harvey, it is only through the production of relatively fixed and immobile configurations of territorial organization – including urban built environments, industrial agglomerations, regional production complexes, large-scale transportation infrastructures, long-distance communications networks,

5.3 Globalization and Space–Time Compression

and state regulatory institutions – that capital's circulation process can be continually accelerated temporarily and expanded spatially. Each successive round of capitalist industrialization has therefore been premised upon socially produced geographical infrastructures that enable the accelerated circulation of capital through global space. (Brenner 1999, pp. 42, 43)

Brenner, along with Armitage and Graham (2001, p. 111), commented upon Marx's contribution to this debate which was that:

> the creation of the physical conditions of exchange – of the means of communication and transport – the annihilation of space by time – becomes an extraordinary necessity for (production). Marx (1973, pp. 349, 524)

Harvey (1981, pp. 2, 2000) saw Hegel (1967) as fundamental to these early considerations of what he terms the 'spatial fix' and considered society driven by its inner dialect to:

> push beyond its own limits and seek markets, and so its necessary means of subsistence, in other lands that are either deficient in the goods it has overproduced, or else generally backward in industry.

Hegel stressed the need to found colonies and thereby permit a part of its population to:

> return to life on the family basis in a new land… (and supply) itself with a new demand and field for its industry.

Hegel viewed processes such as imperialism and colonization as essential revolutions reacting against internal contradictions that manifest themselves in mature society. Social instability is an inevitable result of the polarization of society into those accumulating excessive wealth (through overproduction) and those under-consuming (trapped as 'penurious rabble'), creating imbalances in the distribution of income. Capitalist society is forced into looking for a new spatial fix, increasingly global, as domestic (national) contradictions make internal resolution of conflicts increasingly impossible. Thus capitalist industry looks for an 'outer–transformation' through geographic expansion, manifesting itself as increasing globalization, to sustain the flow of capital accumulation and the preservation of this social imbalance that it needs to survive.

Sack (1973) considered the significance of the role of space in relation to geography specifically and more widely from a societal perspective. Meanwhile, Harvey considered at some length Marx's interpretation of spatial fix and its relationship to globalization and subsequently the everchanging process of capitalist accumulation and the constant need for mutation (Harvey 2001a, pp. 25, 26). Hsu (2000, p. 3), along with Jessop (2002, 2005, 2006, pp. 146–148) in Castree and Gregory (2006) and Wood (1997, p. 541) reaffirmed Harvey's view while Dickens and Ormrod (2007, p. 49) summarized it as concerning 'historical and contemporary imperial projects and especially those involving a reconfiguration of time and space, as attempts to resolve or fix crises that are inherent to the capitalist economy' (Hassan 2009, pp. 5, 11) while acknowledging Harvey's contribution was also critical of his concentration upon the space element rather than time.

Bauman (2000, pp. 121, 122) placed the idea of space time compression and the fixation of space into a wider context that he termed 'capitalist liposuction'. What he called the 'disembodied labour of the software era' no longer ties down capital to a particular location, allowing it to be 'exterritorial, volatile and fickle', able to travel 'hopefully, counting on brief, profitable adventures' moving 'fast and… light and its lightness and motility have turned into the paramount source of uncertainty' in its exploitation of virtual instantaneity.

> Bulkiness and size are turning from assets into liabilities. For the capitalist who would rather exchange massive office buildings for hot-air balloon cabins, buoyancy is the most profitable and the most cherished of assets; and buoyancy can be best enhanced by throwing overboard every bit of non-vital load and leaving the non-indispensable members of the crew on the ground.

One major incentive to compress and fix space is to reduce labor costs by employing cheaper and less labor under conditions which both allow and facilitate this. As Bauman (2000, p. 122) suggested:

> The managerial equivalent of liposuction has become the paramount stratagem of managerial art; slimming, downsizing, phasing out, closing down…

All this made easier by the compression of time and distance through the process of globalization as capital seeks out its next spatial fix. To quote Bauman again, 'one does not plant a citrus-tree grove to squeeze a lemon'.

Harvey has support from many for his view on spatial fixation including Richards (2004, p. 1) and Soja (1989), the latter emphasizing that:

> Since the mid-19th century, historical factors have been privileged over spatial ones. My argument is not just that space matters, but that it is a powerful force for shaping society – that the spatial shaping of society is as significant as the social shaping of space.

Marx himself viewed the spatial fix (although not using the term) as something inevitable but always temporary in that there was no permanent spatially fixed solution to the capitalist's needs. He dismissed Hegel's idea that virgin territory on the edge of the developed world could provide a permanent solution to the need for another spatial fix in that in such territory, labor could appropriate land and control of their production by always being able to move into other undeveloped areas. The capitalist would therefore be unable to control the location of labor and its productive activity—something which forms an essential part of capitalism's definition of territory (and the nation-state as one significant example as we have seen). Capitalism would thus be unable to define a spatial fix with any meaning, something it requires with regularity. Until recently, capitalism's territorial definition focused on the nation-state but increasingly as a response to globalization, a new form of spatial fix has had to be sought.

Marx reemphasized the continuous and permanent need for spatial fixes for capitalism, manifesting itself in new territories. Any suggestion of 'virgin territory' where labor could migrate to avoid their acquisition by capitalism is swiftly undermined through colonization, imperialism and globalization. The sanctity of territory is exemplified in the UK (where all land is owned by someone) and through the gradual elimination worldwide of migrating peoples ranging from gypsies to nomads whose failure to be associated with specific land ownership fails to meet the needs of capitalist society. Spatial fixes are here to stay.

Harvey (1987, p. 268) commented on the friction of distance and its association with the fixation of space. He used the rather awkward term 'distanciation' to reflect on the barriers that distance places in the way of human interaction, imposing transaction costs, manifested as a hindrance to social interaction that capitalism needs to reduce to continue to accumulate capital and has achieved to a great extent through globalization (Giddens 1984, pp. 258, 259). Harvey (1981, p. 9) continues, seeing 'primitive accumulation at the frontier (as) just as vital as primitive accumulation and technologically induced unemployment at home'. Thus, the capitalist will aim to impose a spatial fix. In recent times, this has centered around the nation-state and during this time the governance regime that has been established for shipping has focused on the nation's prime role. However, as the need for increased capital accumulation continues, the nation-state loses its advantage, the need to globalize has developed increasingly stimulated by the needs of the capitalist, the friction of distance that needs to be overcome has increased and a new spatial fix has emerged. However, what has not emerged is a new governance structure recognizing this and the consequences are manifested in the policy disasters that shipping faces. This continued process of what Harvey terms deliberate 'crisis formation' can be attenuated in the short term by a new spatial fix, but in the long term there is no permanent spatial fix solution to 'capitalism's internal contradictions'.

Harvey (1975, pp. 9, 10) provided an indication of why capitalism requires crises to develop:

- To ensure that there is a surplus of labor which is available for the expansion of production when needed. Consequently, mechanisms are developed to expand population growth and to stimulate immigration so that there exists a reserve of the unemployed. It also includes making women and children available for employment when needed—what Harvey terms drawing on 'latent elements'— and a stimulus to innovate to save labor.
- To ensure that there is a surplus of the means of production—machinery, raw materials, buildings, etc.—to permit expansion of production as and when required.
- To ensure that there is a market readily able and willing to absorb surplus production. Without desirable goods and the ability to purchase them, then capital accumulation is impossible.

In each of these categories, serious barriers to their maintenance will be reached periodically creating crises which then need to be overcome. Since capitalism creates these barriers as well as abhors them, capitalism is also the creator of its own crises which it then resolves. In the maritime case, the constraints on cheap labor, cheap means of production, and adequate markets which developed national economies face are overcome through the emergence of Open Registries, Flags of Convenience, the extended use of labor from developing countries, and the ever-present pressure on flag states to provide extended financial, legal and regulatory priveliges in turn. This maritime globalization requires new structures of governance and new policies to match. We have the crisis, we have the solution, we lack new governance.

Harvey (1975, p. 11) even referred to transport which he saw as a special case. Accumulation, which is central to the capitalist system, makes improvements to transport both inevitable and necessary. Quoting Marx (1992, p. 384):

> The revolution in the modes of production of industry and agriculture made necessary a revolution... in the means of communication and transport (so that they) became gradually adapted to the modes of production of mechanical industry by the creation of a system of river steamers, railways, ocean steamers and telegraphs.

Just substitute computers, mobile telephones, emails etc. and you have the impact of globalization today on the transport and logistics network.

Marx continued (1973, p. 524):

> The more production comes to rest on exchange values, hence on exchange, the more important do the physical conditions of exchange - the means of communications and transport – become for the costs of circulation. Capital by its nature drives beyond every spatial barrier. Thus the creation of the physical conditions of exchange... becomes an extraordinary necessity for it.

The development of new and improved transport systems is a response to the crises in capitalism that would otherwise develop. Hence, globalization, with its impact upon space and time, has both generated and been stimulated by improvements in transport—in our case shipping and ports, but of course, all the logistical framework that goes with this. Crises—disorganization—are a necessity of capitalism and they demand both their resolution and a framework within which the problems they generate can be satisfactorily addressed. Shipping has no longer such a framework.

Dickens and Ormrod (2006, p. 11) took the concept of space–time compression a stage further and looking into the future suggested that capitalism's continual search for new spatial fixes will inevitably lead to the inclusion of outer space into contestable territory. Seeing new fixes as currently occurring in Brazil, Mexico, Chile, the Former Soviet Union, and much of East Asia and facilitated by the extensive development of globalized communications, they are characterized in particular by liquid capital with no fixed geographic location or physical form. Satellite technology has been central to this and begun the process of bringing outer space into the global spatial fix by providing the mechanism for increasing communication speed in a way recognized by Virilio (1998) who saw dromology and its associated technologies of speed as central to the spread of empire. As Dickens and Ormrod suggested, 'it is because of (the) compression of space that it is now possible to envisage spatial fixes being made in the closer parts of the cosmos'. They believed that the future will see this continuous fixation upon space will bring capitalism to reach across the universe in the search for expanded territory and hand in hand with this will come the necessary process of 'universalization' (*cf* globalization) which will erode the friction of distance between

5.3 Globalization and Space–Time Compression

Earth and other planets. These 'outer transformations' as they call them, will increasingly see fixes taking place in the cosmos and will focus more on liberating the value inherent in raw materials found across the universe rather than the accumulation of cheap labor from Mars, Venus, Neptune, and beyond (although who knows…). To a certain extent this is happening already with the extended use of space as a communication channel to regulate capital and information. And these cosmic spatial fixes will then replicate the continuous process of capitalism reformulating itself through repeated fixation over greater distances as improved technology permits more distant resources to be accumulated and exploited (Dickens and Ormrod 2007, pp. 54, 55; Dickens 2009; Parker 2009; Parker and Bell 2009a, b).

The necessary internal contradictions and crises of capitalism were also considered by Harvey (1981, p. 10) citing Marx (1973, 1992) seeing crisis formation and the internal violent contradictions which follow from this generating the continual need for a new spatial fix (something he terms 'over-accumulation—devaluation'). It is:

> the height of insanity, the intense destructive power, implicit in the capitalist mode of production. Beneath its façade of market rationality, and counterposed to its creative powers to revolutionize the productive forces, the bourgeoisie turns out to be 'the most violently destructive ruling class in history' (Berman 1982). In the depths of crises capitalists unleash the violence of primitive accumulation upon each other, destroy vast quantities of capital, cannibalize and liquidate each other in that 'war of all against all' that Hobbes had long before seen as an inherent characteristic of market capitalism. What Marx nowhere anticipates… is the conversion of the process into economic, political and military struggles between nation states. At times of savage devaluation, the search for a spatial fix is converted into inter-imperialist rivalries over who is to bear the brunt of devaluation. The export of unemployment, inflation and idle productive capacity, become the stakes in an ugly game. Trade wars, dumping, tariffs and quotas, restrictions on capital flows and foreign exchange, interest rate wars, immigration policies, colonial conquest, the subjugation and domination of tributary economies, the forced reorganization of the division of labor within economic empires, and finally the physical destruction and forced devaluation of a rival nation's capital through war, are some of the options at hand.

There is a tradition of linking the nation-state and violence stimulated by the development of capitalism (Braudel 1982, pp. 516–518; Castells 1989, 2000, p. 8; Johnston 1991, pp. 549, 553; Conversi 1999; Caparaso 2000, p. 2; Allen 2003; Stalder 2006, p. 104). Capitalism is seen as intrinsically linked to the nation-state and its violent ambitions and even though the state is changing in its role and function, it remains a vital part of the capitalist regime. Deleuze and Guattari (1988, pp. 447, 448) quote Marx's observation that violence necessarily operates through the state. It is always present. It is incorporated and structural. At the same time attempts are made to justify the state going to war, and inflict violence on other states. For example, Bakunin's comments in 1848 following the disappointment of the failed European Revolution that year:

> No more wars of conquest, nothing but the last supreme war, the war of the revolution for the emancipation of all peoples! Away with the narrow frontiers, forcibly imposed by the congress of despots, in accordance with the so-called historic, geographical, commercial,

strategic necessities! There should be no other frontiers but those which respond simultaneously to nature and to justice, in accordance with the spirit of democracy – frontiers which the people themselves in their sovereign will shall trace, founded upon their national sympathies. (Bakunin 1996, quoted in Giddens 1995, p. 178)

Allen (2003, p. 57) pointed to the work of Arendt (1969) who outlined the many authoritative voices that have stressed the significant link between the state and violence; for example war is:

the continuation of politics by other means. Clausewitz (1976, originally quoted in 1832)

However, Bodin (1955) was perhaps the earliest commentator on the relationship between power, the state and violence in 1576. Hobbes (1909, p. 109) had indicated in 1651 how important violence was by stating that 'covenants without the swords are but words'. Engels defined violence as the accelerator of economic development. Trotsky in 1918 at Brest-Litovsk commented that 'every state is based on violence'. Wright-Mills (1956, p. 171) suggested that 'all politics is a struggle for power; the ultimate kind of power is violence' following on from Weber's (1921) definition of the state as 'the rule of men over men based on the means of legitimate, that is allegedly legitimate, violence' (Weber 2004). Strange (1986, p. 81) noted the significance of violence to the nation-state even as its power reduced. Gritsch (2005, p. 2) meanwhile was still emphasizing the violent credentials of the nation-state at beginning of the twenty-first century—'through hard geopolitics states use militarism, organized violence and war to competitively acquire, control and defend territory; access strategic natural resources; promote national security; and achieve military-political hegemony within the inter-state system'—and referring the reader on to a multitude of related sources to back the claims including Block (1996; p. 230); Cerny (1997, p. 52); Krasner (1999, p. 35); Mann (1993, p. 49); Rueschemeyer and Evans (1985, p. 47); Skocpol (1979, p. 202); Strange (1994, p. 5) and Weiss (1997, p. 17).

Perhaps most dramatically, Bull (1995b, p. 1) looked at how oppression by state power can be interpreted through the Bible and used as symbolism to generate radical Christian positions:

It is something deeply rooted in the primary texts of Christianity, for in the gospels Jesus is reported as identifying his ultimate triumph with the eternal rule of the heavenly Son of Man which will take place after the destruction of the Beast, presumably representing state power.

His comments were derived from:

And I saw a beast rising out of the sea, with ten horns and seven heads… and to it the dragon gave his power and his throne and great authority… Men worshipped the dragon for he had given his authority to the beast… it was allowed to make war on the saints and conquer them. And authority was given it over every tribe and people and nation. (Revelations, 13, 1–7)

Bull went on to cite Bloch's *The Principle of Hope*, published between 1938 and 1947 (Bloch 1986), which contains commentary on a number of Christological models including one referring to imperial oppression and hierarchism. This features the 'Lord Christ who is enthroned in Heaven in a glorious state similar to the imperial oppressors of the poor' (Bull 1995, p. 4).

War, violence, capital, and state are themes that have also been widely integrated. Dickens and Ormrod (2007, pp. 88, 89) took these themes and integrated them with hegemony. Chowdhury (2006, p. 126) was emphatic and quoted Rosa Luxemburg (2003):

> Everywhere war finds material enough for imperialistic desires and conflicts; creates new material to feed the conflagration that spreads out like prairie fires.

He went on to assess Lenin's theory of imperialism and how it related war, capital, nation-states, and violence to globalization, the latter acting as a 'haunting spectre' within its fabric. He claimed that capitalist industry thrives on war and the prospect of war and cited examples from the Middle East and Africa as evidence including the sale of a US$40 million air defense system to Tanzania in 2002, when the average annual income in that country was only US$250. In Chowdhury's words, there is an 'unvarnished economic logic of this violent process of endless war'.

Harvey (1981) stressed the role of Lenin, Bukharin and Luxembourg in connecting political and economic development and the geography of capitalist imperialism (and by implication violence) and this accompanied other work by him stressing similar themes (Harvey 1982 [especially pp. 439–451], 2001b, 2003; Jessop 2006) in which he emphasized the 'depredations wrought in the name of human progress by a rapacious capitalism'. However, the most extensive discussion of the relationship between violence and the nation-state was undertaken by Giddens (1985, p. 20). He defined the state as:

> a political organization whose rule is territorially ordered and which is able to mobilize the means of violence to sustain that rule.

He also stressed the relationship between the nation-state, violence, capitalism, and territoriality. Territoriality is something we will turn to in a later section. Meanwhile, the nation-state was defined as a 'bordered power container of the modern era'. He used Weber's (1949) identification of the features that a state always possesses and which again stress the importance of violence summarized as (1) the existence of a regularized staff able (2) to sustain the claim to the legitimate monopoly of violence and (3) to uphold that monopoly within a given territory.

The nation-state can be seen as an organization linked to violence and war which in turn is one major manifestation of capitalism's rush to accumulate or to unburden capital. Burrell (1988) took these issues even further. In relating the role of the nation-state and capitalism in generating violence for the purposes of capital accumulation and absorption he compared two descriptions of very different forms of historical domination. The first is the execution of the regicide Damiens on March 2nd, 1757 (quoted by Foucault 1977 in Mills 2003, p. 42):

> The flesh will be torn from his breasts, arms, thighs and calves with red-hot pincers, his right hand, holding the knife with which he committed the said patricide, burnt with sulphur and on those places where the flesh will be torn away, poured molten lead, boiling

oil, burning resin, wax and sulphur melted together and then his body drawn and quartered by four horses and his limbs and body consumed by fire.

The second refers to the rules for 'House of Young Prisoners in Paris' set out in 1837:

> at half past seven in summer, half past eight in winter, the prisoners must be back in their cells after the washing of hands and the inspection of clothes in the courtyard; at the first drum-roll they must undress and at the second get into bed. (Mills 2003, p. 83)

These two contrasting modes of discipline Burrell termed traditional and disciplinary, respectively. The more overt and public spectacle was replaced by a more subtle punishment directed toward the 'soul, the mind, the will'. Extremes of violence were replaced by 'complex, subtle forms of correction and training'. Foucault (1977) along with Sunesson (1985) saw these new forms of violence mirrored in all state institutions—prisons, hospitals, state companies, ministries, and also in semi and non state organizations—schools, factories, housing estates, barracks, etc. State violence is central to capitalist society and an intrinsic part of overcoming the difficulties of capital accumulation and disposal of capital overaccumulation.

> Within the whole range of organizations found in contemporary society, one finds not a plurality of powers but a unified power field encapsulated within the bureaucratic, military and administrative apparatus. For Foucault, power does not reside in things, but in a network of relationships which are systematically interconnected... Power should be seen in a positive sense as actively directed towards the body and its possibilities, converting it into something both useful and docile... Disciplinary power is invested in, transmitted by and reproduced through all human beings in their day-to-day existence. It is discrete, regular generalized and uninterrupted. It does not come from outside the organization but it is built into the very process of education, therapy, house building and manufacture (Donzelot, 1980). Thus the body of the individual is directly involved in a political field; power relations have an immediate hold upon it; they invest it, mark it, train it, force it to carry out tasks, to perform ceremonies, to emit signs. (Foucault 1977, p. 25)

Violence, power and the state are thus inherently tied together through capitalism and its search for capital solutions and spatial fixation. As Burrell (1988, p. 228) summarized Foucault:

> since all of us belong to organizations and all organizations are alike and take the prison as their model, we are all imprisoned within a field of bio-power, even as we sit alone.

The application of violence by the state (or by capitalist institutions in any form) has changed with Postmodernism. Rouleau and Clegg (1992, p. 10) suggested that under Modernism power could be exercised overtly violently, brutal, and frank. These forms of expression have now been largely superseded by seduction and consumption in the market and are now confined primarily to the dispossessed, the non-citizens, and the non-subjects. However, violence in Postmodern times and through the Postmodern state has not gone away.

As we have seen, capitalism continually seeks crises to achieve a temporary spatial fix to ensure the continued maintenance of the accumulation of capital or to use up overaccumulation. Globalization is the manifestation of one such crisis

5.3 Globalization and Space–Time Compression

(of which others are commonly associated with war and violence exercised through the nation-state). Globalization however, represents the emergence of a significant spatial fix from roots deep within the growth of capitalism and which springs forth from the very existence of the nation-state. This places new pressures upon governance of all sectors but perhaps shipping more than any other with its unrivaled international and global credentials. And to reinforce the significance of war, conflict and violence as part of nation-state history (and therefore that of capitalism) we need go no further than Camus (1957) in his speech of acceptance for the Nobel Literary Prize:

> Born at the outset of World War (they) became twenty at the time both of Hitler's ascent to power and of the first revolutionary trial. Then to complete their education, they were confronted in turn by the Spanish Civil war and World War Two – the universal concentration camp, a Europe of torture and prisons. Today they must raise their children and produce their work in a world threatened by nuclear destruction. Nobody, surely, can expect them to be optimists. (Giddens 1985, p. 294)

Writing later in 1981 and reaffirmed in 1982, Harvey continued to see the spatial fix as an essential element of capitalist society. Taking advantage of increasing geographic expansion, changes in spatial organization and the ever-present uneven development (later emphasized by Soja [1989, pp. 104–109] stressing 'the knife-edge path between preservation and destruction'), capitalism makes use of these tensions through its continuous migration between spatial fixes. Thus improved communications facilitate market extension for labor, raw materials, and products (expansion). The decline of the nation-state and the rise of the supranational/global authority (the EU, IMO, OECD etc.) encourages this process and the continued contrasts between the developed and developing world (for example between the UK and Vietnam; USA and Cambodia) together provide opportunities of migrating from a nation-based spatial fix that dominated through most of the twentieth century and in particular through the period of maritime institutional formation from 1920 to 1970, to a new global spatial fix that has a new agenda, new processes, new forces and a desperate need for new governance.

Anderson (1996) summarized Harvey's views on globalization and its relationship to spatial fixation as providing the opportunity to use up an overaccumulation of capital that would otherwise stultify capitalist development. Coupled with a selection of carefully located wars, the engendering of a sense of panic at the prospect of the slowdown of accumulation and the development of technological improvements to facilitate change, a new global spatial fix is acquired but one which fails to match the existing governance structures from which policy emerges. Harvey (2000, p. 54) summarized:

> something akin to 'globalization' has a long presence within the history of capitalism… Capitalism cannot do without its 'spatial fixes'. Time and time again it has turned to geographical reorganization (both expansion and intensification) as a partial solution to its crises and impasses. Capitalism thereby builds and rebuilds a geography in its own image. It constructs a distinctive geographical landscape, a produced space of transport and communications, of infrastructures and territorial organizations, that facilitates capital accumulation during one phase of its history only to have to be torn down and

reconfigured to make way for further accumulation at a later stage. If, therefore, the word 'globalization' signifies anything about our recent historical geography, it is most likely to be a new phase of exactly this same underlying process of the capitalist production of space.

Anderson (1996, p. 135) suggested that this new global fix requires policymakers to think again about policies, states and territories and emerging from all this, about governance. He emphasized how space–time compression that accompanies the development of the new global fix, has shifted the ground 'under established political arrangements and concepts'. He went on to consider how the new fix is characterized by flexibility, unpredictability, and variability, something we shall return to later when we consider the nature of the new fix in a Postmodern age. He suggested:

> One consequence even in traditional nation-states, is that it is becoming increasingly difficult to find one fixed viewpoint or perspective from which to make sense of territorial sovereignty.

Meanwhile, Harvey gave examples of how this process of spatial fixation (and progress toward the current manifestation of globalization) have emerged since Marx and rest on Marxist theories. He cited:

- Lenin's theory of imperialism (Galtung 1971).
- Luxembourg's positioning of imperialism as the savior of capitalist accumulation (Luxembourg 2003).
- Mao's depiction of primary and secondary contradictions in class struggle.
- Accumulation on a world scale (Amin 1974).
- The production of a capitalist world system (Wallerstein 1974; Arrighi 1994).
- The development of underdevelopment (Frank 1969; Rodney 1981).
- Unequal exchange (Emmanuel 1972).
- Dependency theory (Cardoso and Faletto 1979).

Harvey's views on capital accumulation, spatial fixes, flexible accumulation, crisis formation through overaccumulation and the relationship of all these factors to significant changes in the location and characteristics of the fix adopted have been supported by many commentators including Johnston (1982, 1991), Anderson (1996), Brenner (1997, 1998, 1999), and Schoenberger (2004, p. 428). For example, Schoenberger saw the spatial fix as a way to:

> productively soak up capital by transforming the geography of capitalism. Over-accumulation… is the great problem of capitalism that threatens its very existence. Because it is inherent in the system, it is unavoidable over the long run, and the system is thrown into more or less violent crises during which the excess is savagely devalued at great cost to people and places. But although there is no permanent solution to the crisis tendencies of capitalism, the system does generate some important ways of delaying them or diverting them into reasonably productive pathways. This is the spatial fix.

Johnston's (1991, p. 201) view was that:

> In order to accumulate wealth, those investing capital in production and/or distribution have to 'freeze' it in a place, where they create fixed resources (such as factories) and

5.3 Globalization and Space–Time Compression

employ others (such as labour). Their basic 'materials' capital and labour – are inherently mobile, but for accumulation strategies to succeed that mobility must be restrained by investment in what Harvey (1982) terms 'fixed and immobile infrastructures'. Those structures must then be protected, until such time as investors decide to replace them (probably by withdrawing their investment – or declining to renew it – and instead placing it elsewhere). The result is that 'capitalist investment must negotiate a knife-edge between preserving the values of past commitments made at a particular place and time, and devaluing them to open up fresh room for accumulation'. There is a continual state of tension, therefore.

...but one that manifests itself rather like a severe and cataclysmic earthquake where two plates suddenly decide to move against each other after a prolonged period of stability. The sudden shock (manifesting itself in our case through globalization and in the past through earlier phases of major economic, social, and political change) causes enormous disruption as a new spatial fix is found. Globalization represents the most recent change and as we shall see in the following sections, is mirrored by economic, social, and political changes equally as fundamental.

Meanwhile Brenner (1997) considered Lefebvre's (1974, 1976a, b, 1977, 1978, 1991) spatialized approach to state theory and globalization and saw it reaffirmed in Harvey's (1982, p. 150) view of the impermanence of spatial fixes suggesting that the 'apparent territorial fixity and stability of state institutions' (representing the most recent of spatial fixes), is 'deeply unstable' and in Lefebvre's words no more than a 'precarious equilibrium'. Brenner characterized Lefebvre's attitude to spatial fixation as 'the globalization of capital and the rescaling of state territorial power... as two intrinsically related processes within the same dynamic of global socio–spatial restructuring'. He also saw capitalism's increasing global dynamism threatening the 'Hegelian conception of state spatiality as the terminus of history'. The nation-state spatial fix is just one stage along an unending road of spatial fixes adopted by capitalism in its incessant search for more capital characterized by severe societal, political and economic disruptions as each fix is adopted and absorbed. The state strives to secure a spatial fix for capital—and for some time has achieved this—but do not be deluded. This state oriented spatial fix is as temporary as all those before and all those to follow—something Hegel claimed was an 'atemporal rational unity' but which Lefebvre saw as 'fetishization of space in the service of the state'. Increasing demands for capital from a worn and weary spatial fix always creates demands for a new fix which emerges from the old in ways that disturb and shock. The result is often a dislocation of governance and fixation exemplified in the late twentieth and early twenty-first centuries by a succession of policy failures in the maritime sector perhaps more than elsewhere because of its intense relationship to globalization and the new globalized spatial fix that has emerged. It is this dislocation that lies behind the current problems in maritime policy-making. It is these issues therefore that need addressing.

Brenner (1997, p. 141), referring to Lefebvre (1974, 1976a, b), saw spatial configurations thus 'continually constructed, deconstructed and reconstructed on all spatial scales as a means to accelerate the turnover time of capital'. Globalization is just the next phase of a continuous succession of spatial fixes and the one creating policy havoc in the maritime sector. Dale and Burrell (2007, p. 7) pursued the notion of spatial space as emphasized by Lefebrvre (1991, p. 15) who having argued that all space is a product of society continued to recognize that this may be a difficult concept to grasp as 'the idea of producing space... sounds bizarre, so great is the sway still held by the idea that empty space is prior to whatever ends-up filling it'. Space is thus constructed from two elements—'thingness'; its materiality and physicality; and its 'imaginary' existence which constitutes its 'social, cultural and historical meaning' (Dale and Burrell 2007, p. 7; Natter and Jones III 1997, p. 149).

5.4 The Relationship Between Disorganization, Globalization, and Capitalism

> Order and disorder, the one and the multiple, systems and distributions, islands and sea, noises and harmony, are subjective as well as objective. Now I am a multiplicity of thoughts, the world is now as orderly as a diamond. What fluctuates are the order and disorder themselves, what fluctuates is their proximity, what fluctuate are their relationship to and penetration of one another. (Serres 1995, p. 131)

> Development inexorably entails the destruction of old forms so that 'change and 'progress' inevitably bring with them human suffering, the obliteration of neighbourhoods (from Haussmann's Paris to *fin de siècle* Vienna to Robert Moses' much cited 'meat-ax' approach in New York) and on a global scale, even the eradication of entire societies. These phenomena were, and are, familiar outcomes of the industrial revolution and its recent extensions under the service sector and flexible accumulation. Warf (1990, p. 587)

The relationship between globalization, capitalism, and the need to periodically reinvent the spatial landscape is fundamental to the activities of the maritime sector and to its governance. Consequently, we need to consider it further. Anderson (1996) saw the process of globalization representing the development of a new spatial fix as capitalism moves through a Modern–Postmodern transformation with radical consequences for territoriality, nation-states and political, economic, and social governance. This is reinforced by Lash and Urry (1987, p. 307) who showed how the spatial fix widely adopted by the 1980s was unraveling across major industrial areas to be replaced by a Postmodern upgrade. This mirrors a similar process in space–time experience, conceptualization, and representation that was undergone in the transformation from medieval to modern society. Each is a consequence of the need for capital to search out a new spatial fix to ensure a continued accumulation. Each has had significant consequences for social, economic, and political activity. One current victim is maritime governance and policy, a consequence of the changes imposed on the nation-states by the search for a new globalized spatial fix:

> struggling to impose its will upon a fluid and spatially open process of capital circulation - geopolitics and economic nationalism, localism and the politics of place are all fighting it out with a new internationalism in the most contradictory of ways. (Harvey 1990, p. 358)

Or again to quote Harvey (1990, pp. 237, 238) (in turn quoting Deleuze and Guattari 1984):

5.4 The Relationship Between Disorganization, Globalization, and Capitalism

If space is indeed to be thought of as a system of 'containers' of social power... then it follows that the accumulation of capital is perpetually deconstructing that social power by re-shaping its geographical bases. Put the other way round, any struggle to reconstitute power relations is a struggle to reorganize their spatial bases. It is in this light that we can better understand why 'capitalism is continually reterritorializing with one hand what it was deterritorializing with the other'.

Agnew (1999, p. 513) suggested that disorganization was closely related to globalization because the nation-states traditionally exercise power monopolies over their territory representing a fixity and permanence that globalization destroys. This change from permanence to one of transience affects the stability of the nation-state creating a crisis or disorganization that is manifested through and as a consequence of globalization and its impact on space and time.

Space–time compression was a notion that had been emerging for some time. Berg (1993, p. 495) cited Gregory (1991, p. 17), who in turn cited Jameson (1984) who suggested that by the beginning of the twentieth century, new relationships between space and time had developed where 'the truth of daily experience no longer coincided with the place in which it takes place'. Luke (1991, p. 320) took the discussion of the relationship of space to time further by introducing the notion of chronopolitics. The increase in pace of flows of information that characterizes globalization means that the significant issue in policy-making and governance is no longer space (epitomized by the nation-state, borders, boundaries, territories and the like) but time and how rapidly flows can travel before meeting resistant barriers. The result is that 'dominating the pace of progress, setting the tempos of interaction, or managing the speed of exchange are the critical points of power in these informational systems of order'.

Meanwhile, the Marxist interpretation of globalization, its relationship to time-space compression and the notion of spatial fix is also taken up by Watts (1991, p. 10) with his 'recursive wave-like pattern of space-time compressions... Flexibility, New Times, industrial divides... connotes a speeding up, a hypermobility, a fast capitalism'. Further to this, King and Kendall (2004) cited work by Scholte (2000) in particular. Dale and Burrell (2007, p. 5) discussed the relationship that exists between space and time and the unsubstantiated dichotomy that has arisen. This they mirror with a similar problem between space and place illustrated by a story of indigenous people who knew that they were at the center of their God's universe because they could see the same distance in each direction. The complex relationships between time, place, and space have become even more relevant as the process of globalization has focused attention upon how one is determined by the other. Harvey's vision of space–time compression and spatial fixation is the manifestation of the impact of globalization upon these concepts with serious ramifications for governance and ultimately policy-making in jurisdictionally promiscuous activities such as shipping. Sletmo (2002, p. 4) summed it up:

> Globalization is said to be the collapse of time and space. Today's communication systems allow instant contact between businesses around the world and intermodal transport has so reduced the importance of freight charges for manufactured goods as to practically eliminate the effect of distance.

Virilio (1995, p. 2) saw time superseding space as distances and surfaces become irrelevant in the face of time spans. This is accompanied by what Virilio calls global time which is taking over local time creating a fusion of temporal jurisdictions. Virilio (1999, pp. 11, 12) put forward his vision of 'Polar Inertia' where in the post-industrial age of the 'absolute speed of light' there is no longer any need for anyone to make a journey since one has already arrived:

> We're heading towards a situation in which every city will be in the same place – in time. There will be a kind of co-existence, and probably not a very peaceful one, between these cities which have kept their distance in space but which will be telescoped in time. When we can go to the Antipodes in a second or a minute, what will remain of the city? What will remain of us? The difference of sedentariness in geographical space will continue but real life will be led in a polar inertia. (Virilio and Lotringer 1997, p. 64)

The difference between 'here and there' is obliterated by speed and 'physical geographical spaces no longer have significant human content', a view that matches that of Thompson (1979, pp. 52, 53) when he considered the value of anything and how it is defined by the opinion holder's belief in the item's value. Thus, 'rubbish' is only what the individual or society views as a worthless item, and by changing an individual's views on the age or relative location of an item to one which has value, so that item becomes of value. Thus, time and space are relative to the individual and when convenient, may be merged into a single coherent whole so that they become codeterminous, instantaneous, and represented by points rather than spaces.

Interpreting both Brenner (1998) and Harvey (1985b, 1989) and also alluding to Dickens and Ormrod (2006, p. 4; 2007, pp. 54, 58) who took the notion of capital stalking the Earth for resources and then applied it in the future to the cosmos, we can arrive at an explanation for globalization and an understanding of its origins, drivers, purpose, and hence its implications for governance. Harvey (1985a) saw it as a 'restless formation and re-formation of geographical landscapes' in which configurations of capitalist organization are continuously reworked into 'provisionally stabilized spatial fixes' for the purpose of each successive regime of capital accumulation. The nation-state space therefore is one part of a capitalist 'social dimension' continually constructed, deconstructed and reconstructed through an historically specific, multiscalar dialectic of de- and reterritorialization. Arrighi (1994, p. 14) agreed:

> the expansion of capitalist power over the last five hundred years has been associated not just with inter-state competition for mobile capital, as underscored by Weber (1978), but also with the formation of political structures endowed with ever-more extensive and complex organizational capabilities to control the social and political environment of capital accumulation on a world scale. Over the last five hundred years these two underlying conditions of capitalist expansion have been continually recreated in parallel with one another. Whenever world-scale processes of capital accumulation as instituted at any given time attained their limits, long periods of inter-state struggle ensued, during which the state that controlled or came to control the most abundant sources of surplus capital tended also to acquire the organizational capabilities needed to promote, organize and regulate a new phase of capitalist expansion of greater scale and scope than the preceding one.

5.4 The Relationship Between Disorganization, Globalization, and Capitalism

Globalization is the major manifestation of this particular and current spatial fixation. Brenner (1999, p. 435) saw it as a:

> double edged, dialectical process through which: 1) the movement of commodities, capital, money, people, images, and information through geographical space is continually expanded and accelerated: and: 2) relatively fixed and immobile socio-territorial infrastructures are produced, reconfigured, re-differentiated and transformed to enable such expanded, accelerated movement. Globalization therefore entails a dialectical interplay between the endemic drive towards space–time compression under capitalism (the moment of de-territorialization) and the continual production of relatively fixed, provisionally stabilized configurations of territorial organization on multiple geographical scales (the moment of re-territorialization)… globalization is an on-going, conflictual and dialectical process rather than a static situation or a terminal condition.

Marx and Engels (1967) had already taken up the broad theme in the Communist Manifesto:

> The bourgeoisie cannot exist without constantly revolutionizing the instruments of production, and with them the relations of production, and with them all the relations of society… Constant revolutionizing of production, uninterrupted disturbance of all social relations, everlasting uncertainty and agitation, distinguish the bourgeois epoch from all earlier times. All fixed, fast-frozen relations, with their train of ancient and venerable prejudices and opinions, are swept away, all new-formed ones become antiquated before they can ossify. All that is solid melts into air, all that is holy is profaned, and men at last are forced to face… the real conditions of their lives and their relations with their fellow men.

However, the subsequent history of capitalism has 'demonstrated an enormous resiliency in its ability to self-organize into elaborate new forms to the extent that, in terms of his detailed vision and prediction of how the future of society would unfold, Marx got it almost all wrong' (Morgan 1997, p. 290), something that Engels alluded to with his discussion of the 'negation of the negation' and the constant contradictions evident in capitalism's search for new forms of accumulation (Engels 2007). The system as a whole is constantly on the move toward another state of crisis (Morgan 1997, pp. 287, 288).

The constant need to revolutionize the 'instruments of production' is something taken up by a number of commentators in the twentieth century who have linked it to what Offe (1985) and Soja (1989, p. 171; 2000, p. 193) termed a process of 'disorganized capitalism', what Gregory (2006) labeled 'creative destruction' (after Harvey 1990) arising out of the crises within the 'circuits of capital accumulation', what Tyler (2005, p. 567) saw as the need for 'crisis management' to accommodate Postmodern communication under modern capitalism, and what Arrighi (1994) recognized as a period of restructuring and reorganization, of discontinuous change but change which was far from unprecedented. It has even been compared to the process of 'punctuated equilibria' which characterizes biological evolution and is the process of evolutionary change taking place in catastrophes rather than as a process of gradualism (Gould and Eldredge 1977, p. 115). Tyler (2005) linked the disorganization characteristic of increasing organizational disputes to Postmodern change and the need to understand the process of disorganization if policy-making is to be profitable. Stalder (2006,

p. 109) saw the continuous crisis that bedevils the nation-state as a similar issue. Meanwhile, Harvey linked it to space–time compression (or globalization in layman's terms) suggesting that capitalism has the prime objective of (in Marx's terms) attempting to 'annihilate space with time' but that to do this it needs to produce successions of 'new, fixed and relatively immobile spatial configurations'. Harvey (1982) suggested that 'spatial organization is needed to overcome space'. Some of these new configurations are significant—and globalization in the late twentieth century is undoubtedly one such significant change—and result in the 'constant revolutionizing of the spatial constraints of production'.

That disorganization in capitalist society takes place is indisputable. The most recent examples of such crises include the late 1960s and early 1970s (Armstrong et al. 1991; Webber and Rigby 1996), and of course that from about 2007–2008. Urry (1986) discussed disorganized capitalism extensively in relation to issues of class and space. A summary of his understanding of both organized and disorganized capitalism is given in Table 5.1.

Hirst and Zeitlin (1991, pp. 1, 2) suggested that there is widespread agreement that 'something dramatic has been happening in the international economy over the past two decades' characterized by:

- Radical changes in production technology;
- Changes in industrial organization;
- Restructuring of world markets;
- Large-scale changes in policies of economic management at all jurisdictions.

They saw these changes as having serious policy implications in the movement from a Fordist to a Post-Fordist environment typified by flexible specialization which is an essential part of globalization. Schoenberger (1988, p. 248) suggested that globalization inevitably spells the end of Fordism—and is not just an indication that it has ended—as Fordism requires an hegemony with only one or two core countries at its center. The inevitable diffusion of production associated with globalization indicates the demise of hegemonies as previously known and with this a new era—the Postmodern–emerges. Capitalism has moved on as part of its recurrent disorganization which is one of its central features. Schoenberger (1988, p. 245) agreed as 'we appear to be in the midst of a transition from one form of social and economic organization to another'.

Gibbon et al. (2008, p. 317) saw globalization as the cyclical consequence of such disorganization with new phases identifiable over hundreds of years and each one originating from economic depression and the chaos that ensues (Hopkins and Wallerstein 1994). The consequent unevenness of prosperity, both spatially and historically creates disorganization, the consequences of which have stimulated policy failings in the maritime sector and the need for new governance initiatives to accommodate the changes that have occurred. Lash and Urry (1994, p. 323) identified a number of processes and flows which characterized disorganized capitalism including:

5.4 The Relationship Between Disorganization, Globalization, and Capitalism

Table 5.1 Organized and disorganized capitalism

Organized capitalism	Disorganized capitalism
Concentration and centralization of industrial, banking, and commercial capital. Markets increasingly regulated	Growth of world market combined with centralized capital means that national markets are less regulated by national based corporations. Increased competition comes from increased centralization
Growth of the separation of ownership from control. Bureaucratization and elaboration of complex managerial hierarchies	Continued increase in middle class workers especially of certain types of service class (e.g educators, scientists, managers) encourages disorganization through increased mobility, individual achievement, and social movements
Growth of new sectors of managerial/scientific/ technological elite and a bureaucratically employed middle class	Decline in the absolute and relative size of manual workers as nations deindustrialize.
Growth of collective organizations in the labor market. Trade Unions, employers' associations, etc	Decline in effectiveness and importance of national bargaining. Growth of local arrangements
Increasing articulation between the state and large monopolies and between the state and collective organizations as the former increasingly intervenes in social conflicts	Increasing independence of large monopolies from direct control and regulation by individual nation-states. Breakdown of neo-corporatist forms of state regulation of wage bargaining, planning, etc.; increasing contradiction between the state and capital
Imperialist expansion and the control of markets and production overseas	Spread of capitalism into most third world countries including competition with extractive and manufacturing sectors. Move of first world economies toward services
Changes in politics and the state. Increased number of state bureaucracies, increased representation of diverse interests in and through the state, transformation of administration from keeping order to achieving national goals	Increased alienation of nation-based political movements. Decline in class vote and rise in 'catch-all' political parties
Ideological change—technical rationality, glorification of science, significance of the national interest	Cultural fragmentation and pluralism. Commodification of leisure. Reduction of time/ space creating global villages
Increased concentration of industrial capitalist relations within a few industrial sectors and a small number of much more centrally significant nation-states	Massive increase in the nation-states involved in capitalist production. Increase in number of sectors involved in capitalist production
Extractive/manufacturing industries as the dominant sectors with a relatively large number of employees	Decline in absolute/relative numbers employed in extractive/manufacturing industries. Increased significance of service industries for structuring social relations (e.g increased feminization, smaller plants etc.)
Concentration of different industries within a region. Clearly identifiable regional economies based on a limited number of manufacturing/ extractive industries	Overlapping effect of new forms of the spatial division of labor has weakened the degree to which industries are concentrated in different regions. No more regional economies dominated by one industrial type

(continued)

Table 5.1 (continued)

Organized capitalism	Disorganized capitalism
Increasing plant size with economies of scale.	Decline in average plant size because of shifts in industrial structure, labor saving investments, subcontracting, export of labor intensive activities to the third world.
Growth and increased importance of massive industrial cities dominating regions and the provision of centralized services	Industrial cities decline in their size and influence. This is followed by a population collapse of inner cities, increased population in smaller towns. Cities less centrally implicated in the circuits of capital and progressively reduced to the status of alternative pools of labor-power.

Source Urry (1986, pp. 25–29)

- The flow of capital and technology to 170 individual, self-governing, capitalist countries each concerned with defending its territory.
- Time–space compression in financial markets and the growth of global cities; growth of internationalized producer services; generalization of risks that know no national boundaries; globalization of cultures and communication structures; proliferation of forms of reflexivity, individual and institutionalized, cognitive and esthetic.
- Huge increases of global mobility and of tourists, migrants, and refugees; development of a service class with cosmopolitan tastes especially for fashionable services; the decline in efficiency and effectiveness of the nation-states (Luke 1992).

Jessop (1997, p. 571) agreed linking these changes to the 'crisis of the postwar mode of growth in the economic space occupied by Atlantic-Fordism' creating 'serious difficulties regarding the emergence of a new and stable accumulation regime as well as important discontinuities in modes of regulation and capitalism's articulation with wider societal formation'.

Santos (1995, p. 171) related this disorganization and disruption to a heady mix of changes in what he termed 'world-time and world-space' referring back to Harvey's space–time compression and globalization and on to a discussion we will have later of the significance of speed to the governance of maritime activities and the work of Virilio in particular. Rodrigue et al. (1997, p. 87) also put it clearly. They saw all systems subject to change, carrying the seeds of their own destruction and cited the changes that took place from the 1960s from the Fordist to the Post-Fordist economy as evidence of this. Thus, a Modernist society will inevitably mutate to what we can now see as a Postmodern one (or whatever it might be called) as part of an inevitable process of societal, economic, and political change. They outlined the causes of these changes with particular reference to the transport sector—in particular globalization—and the effects upon the practice of transport, presenting three models of change all of which are interpretations of the disorganization of capitalism that this represents. Jameson (1985) was similarly minded and Croucher (2003, p. 11) agreed identifying a 'new historical epoch, different in some significant ways from the one preceding it' generated by globalization and characterizing all aspects of society.

Bonacich (2003) identified the same trends in the fields of logistics and supply chain management suggesting that capitalism was undergoing restructuring with major shifts occurring between 1970 and 2000 and having disturbing impacts upon the role of the labor force. Supply chain management in particular had emerged as a response to the desire by capitalism to identify new sources of capital and to become more flexible and responsive in return.

Thompson's (1979) contribution we have noted earlier. He provided one of the more interesting analyses of disorganization, change, and society in his consideration of 'rubbish' suggesting that we all tend to think of objects (theories; beliefs; relationships etc.) to be constant and the way that they are is because of some intrinsic physical or metaphorical properties. He disagreed. The 'qualities that objects have are conferred on them by society' and that nature's only role is to reject those qualities when they prove to be physically impossible. Thus there are numerous 'natural' conditions and constant change in these conditions with context, circumstances, attitudes, and with change in society itself.

Thompson saw the societal situation as self-perpetuating in the face of the constant threat of change. Those in possession (the lead, on top, in charge) ensure that the societal view on life is one that fits them well, thus creating a 'natural' order of things which rarely if ever threatens their advantage. To quote Thompson (1979, p. 9) those in control (read Capitalists) 'are like a football team whose center-forward also happens to be the referee; they cannot lose'.

Thompson's 'Rubbish Theory' relates closely to our discussion of capitalism. Thompson considered that there are many 'transient' objects that are gradually losing value and significance and which ultimately slip from transience into a state of 'rubbish'. These items tend then to hang around until one day some gain value and then reenter the world of transience, eventually maybe to climb to a position of what he calls 'durable', with increasing value. This process of change, reinvention and revaluation is comparable with that of capitalism and its inherent disorganization, whereby what was once a prosperous and 'durable' activity, falls from grace to become effectively 'rubbish' and society (policy-makers; policies; governance) needs to accommodate this change if it is to work. He cited the example of what he terms 'hard core body products' to drive the message home:

> We can draw up a list of body products and then distinguish between those that are rubbish and those that are not. The rubbish items would include excrement, urine, finger and toenail clippings, pus, menstrual blood, scabs and so on… The non-rubbish items would include milk, tears, babies… and sometimes sperm. If rubbishness were self-evident and derived from the intrinsic physical properties of objects then this division of body products into rubbish and non-rubbish items would be fixed and unchangeable. Yet in recent years, some body products have crossed from one side to the other. Phlegm is now clearly seen as rubbish, but until recently, it had a noble connotation. English phlegm was expectorated from splendidly stiff upper lips…

It is the same with sweat. Once it was good, honest and noble. In 1940 Churchill could rally the Dunkirk spirit with offers of 'blood, toil, tears and sweat'. Today sweat is firmly in the rubbish category." (Thompson, 1979, pp. 10, 11).

Rubbish (and policies; governance) is thus socially defined and the boundary between rubbish and non-rubbish, and thus between acceptable and workable policies and those which cannot be accepted or work, moves in response to social pressure.

Lash and Urry (1987) felt that it involved three interconnected processes:

- The tendency for capital to see-saw from place to place seeking locational advantage, resembling a plague of locusts, settling on one place, devouring it, moving on to a new place while the old restores itself for another attack.
- The tendency for capital progressively to become spatially indifferent through reducing its dependence upon particular raw materials, markets, sources of energy, areas of the city, supplies of skilled labor.
- The tendency for certain characteristics of labor power (skills, costs, organization, reliability) to become of heightened importance because labor power, unlike the physical means of production, cannot be reproduced capitalistically, and hence is not subject to the same process of geographic levelling or homogenization.

This is an excellent summary of the globalization processes that have characterized the shipping industry in the late twentieth and early twenty-first centuries and suggested that we have witnessed a revolution in the instruments and process of production that needs to be reflected in the process of policy-making and the framework of governance. The rise of globalization and the decline of the role of the nation-state alongside the growth of disorganized capitalism is emphasized by a number of contributors. Arrighi (1994, p. 3) quoted Harvey (1990, p. 165):

> There had of course always been a delicate balance between financial and state powers under capitalism, but the breakdown of Fordism-Keynesianism evidently meant a shift towards the empowerment of finance capital vis-à-vis the nation-state.

Lash and Urry (1994, p. 10) similarly commented:

> contemporary capitalism is indeed disorganized. By this we mean that the flows of subjects and objects are progressively less synchronized within national boundaries.

Offe (1985, p. 284) also offered a view on the relationship between the disorganization of capitalism and the role of the nation-state seeing the latter as becoming rather more confused and possibly less relevant.

> it is unrealistic to assume that decisions of the nation-state touch *all* citizens (and with equal intensity) as implied in the link of majority rule and the nation-state. Yet the reciprocal assumption that decisions of the nation-state affect only its citizens is also unrealistic. Hence the untenability of the not-yet-democratic formula already familiar to legal thinkers of the Middle Ages: *quod omnes similiter tangit, ab omnibus comprobetur* (what touches everyone may be approved by everyone). The practice of limiting the application of majority rule to the nation-state led to the appearance of *both* logically

5.4 The Relationship Between Disorganization, Globalization, and Capitalism

conceivable deviations from this rule that are on the agenda today; either the scope of those affected is *smaller* than those participating, or it is *larger*. (*emphasis original*)

The latest manifestation of the disorganization of capitalism centering on globalization, has made the traditional national emphasis thus less relevant and the consequences for policies and governance more apparent.

Hasse and Leiulfsrud (2002) provided one of the more extensive reviews of this revolutionary change and begin by outlining what made up organized capitalism. They took Lash and Urry's (1987) definition, whereby it is characterized by features that were present in all highly developed economies until the late twentieth century and which had emerged with the development of industrial capitalist society over the previous 200 years. They suggested that it always consisted of:

- The synchronization of industrial, banking, and commercial capital as market competitive institutions and market actors.
- The separation of ownership from control with new managerial hierarchies.
- The growing number of experts.
- The rise of the new middle classes.
- Industrial relations and wage negotiations mediated primarily by collective organizations (e.g. trades unions; professional organizations).
- The state an active economic and political actor negotiating and dictating standards of collective bargaining, encouraging the growth of major blocks of industry and politicizing welfare state legislation.
- Specialization of certain core sectors of national economies.

To quote Hasse and Leiulfsrud (2002, p. 109):

the extent to which capitalist regimes could be described as highly 'organized' was related to the formation of industry, capital and the state. Trade unions, working class parties and the welfare state were considered as crucial institutions. Organization in this respect refers to a number of key characteristics and mechanisms regulating industrial relations within the boundaries of nation-states.

By the late twentieth century, what became to be known as organized capitalism was beginning to disintegrate as capitalism's demands for continued capital accumulation made it imperative that change should take place. The result was the emergence of what we saw earlier as 'disorganized capitalism'—a Postmodern response to the situation and as a reaction to the Modernist era that paralleled that of organized capitalism. We shall return to the Modernist–Postmodernist divide in the following chapters but first what constitutes this new disorganization—and how justified are we in seeing it as a new capitalist phase that needs to be accommodated by maritime policy-makers and governance frameworks?

Hasse and Leiulfsrud (2002) interpreted it as the dispersion of what they term 'specific national models and economic–political configurations'. Lash and Urry (1994, p. 314) suggested that 'nothing is fixed, given and certain, while everything rests upon greater knowledge and information'. Certain features become apparent:

- Increased competition on a global scale.
- Separation of industrial capital from bank and finance capital.
- Deregulation of national economies.
- Undermining of loyalties and spatial boundaries of firms.
- Organized labor and political parties constrained and weakened.
- Institutional modes of wage bargaining outmoded.
- Major firms can work out their own terms for wages, work regimes, and organization.
- The breakdown of neo-corporatist models as modes of economic and political negotiation.

Disorganization is mainly at the level of the nation-state—reflecting the moves toward globalization that we have identified as fundamental to problems of governance. Lash and Urry (1994, p. 314) saw more than this in that it also reflected a breakdown of traditional class politics. Work relations are no longer seen to have the same relevance within society:

> Cultural forms, images, sound and impulses that blur standard sociological interpretations and boundaries between individuals and society replace everyday life as a coherent social project at the junction between work and family. New emerging influences, cultural projects and a tendency towards individualization are believed to promote the constitution of individuals as economic actors. (Hasse and Lieulfsrud 2002)

Soja (1989, p. 173) identified what he called this 'flexible specialization' as 'derigidifying older hierarchical structures and creating… a significantly different order of responsibility and control'. He went on to define what he saw as this essentially Postmodern challenge to existing society:

> the contemporary period of restructuring has been accompanied by an accentuated visibility and consciousness of spatiality and spatialization, regionalization and regionalism. The instrumentality and the spatial and locational strategies of capital accumulation and social control is being revealed more clearly than at any time over the past hundred years.

Meanwhile, Low and Barnett (2000, p. 53) suggested that globalization is fundamentally multi-faceted and that these facets may all move at different times and to a different timescale. As such, there are periodic crises occurring creating a process of multi-temporal globalization with distinct historical overtones. This 'disorganization' is a critical and constant feature of capitalism that if misunderstood, leads to chaotic policy-making. Lacher (2005, p. 43) insisted that capitalism was far from a 'self-equilibrating' process and had an inherent tendency toward overaccumulation. This in turn leads to uneven spatial development or even worse as identified by Polanyi (1944). Disorganized capitalism is consequently a serious matter for policy-making.

Lash and Bagguley (1988) provided an extensive background to the idea of disorganized capitalism and although applied largely to labor relations, many of the points they made are relevant to the discussion here. As Lash and Urry (1987, pp. 312, 313) suggested:

5.4 The Relationship Between Disorganization, Globalization, and Capitalism

The world of a 'disorganized capitalism' is one in which the 'fixed, fast frozen relations' of organized capitalist relations have been swept away. Societies are being transformed from above, from below and from within. All that is solid about organized capitalism, class, industry, cities, collectivity, nation-state, even the word, melts into air.

Boje et al. (1996, pp. 3, 4) concurred with the concept of capitalism's tendency to deliberately self-destruct. Following Habermas (1973) and Offe's (1984) view of capitalist society as 'contradictory and entropic' and composed of four sub-systems—economic, political or administrative, socio–cultural and legitimation—self destruction arises from the foundations of these systems. Boje et al. provided the example that in advanced capitalism, 'the inherently exploitative mode of production tends to destroy the very preconditions on which the system depends' (Gephart and Pitter 1993; Offe 1984, p. 132). There emerges a desire to accumulate surplus but this is matched by a fall in profit which therefore works against this. Economic exploitation can create ecosystem crises as nature becomes commodified and legitimation crises occur when the state fails to sustain loyalty. Here, the state attempts to delay such crises by redistributing welfare—and thus advanced capitalism not only displays recurrent crisis but also can only attempt to avoid the consequences by altering societal class structures.

Harvey (1990) did not fully agree with the idea that capitalism progressively disintegrates although like Lawson (2007, p. 4) who recognized how 'Taylorist and "flexible" strategies in Canadian logging led to the shedding of jobs in core firms and made others more precarious', he suggested that the successive changes (which he acknowledged) are a coherent rather than an incoherent process. His definition of flexible accumulation incorporates this recognition of continuous and predictable change with less stress on the collapse of successive capitalist arrangements. Arrighi (1994, p. 3) emphasized Harvey's understanding of rigidity in the capitalism of the 1960 and 1970s—manifesting themselves in mass production, labor markets and contracts, state commitments and defense progammes.

Behind all these specific rigidities lay a rather unwieldy and seemingly fixed configuration of political power and reciprocal relations that bound big labor, big capital and big government into what increasingly appeared as a dysfunctional embrace of such narrowly defined vested interests as to undermine rather than secure capital accumulation. (Harvey 1990, p. 142)

Arrighi (1994, p. 9) also suggested that the process of constant change in capitalism which we have considered here possessed a close resemblance to the metamorphosis model of Mensch (1979, p. 73) of socio–economic development. Mensch at least in part supported Harvey's view of progression in the successive changes that capitalism demands as he abandoned:

the notion that the economy has developed in waves in favour of the theory that it has evolved through a series of intermittent innovative impulses that take the form of successive S-shaped cycles.

Stable growth alternates with 'crises, restructuring and turbulence' which lead once more to a phase of stable growth. This, he combined with Braudel's (1982) model of capitalism as the non-specialized top layer in a hierarchy of world trade.

As in all hierarchies the upper layers rely upon the lower ones to exist. Three layers were envisaged—the lowest is the layer of material life, the middle is the layer of the market economy and the upper layer is the world market economy with capitalist reformulation (revolution) successively occurring to move capitalist society further toward a globalized situation. Capitalism thus has increasingly globalized its operating arena with consequences for policies and governance which as yet have not been fully appreciated by those with responsibility for maritime policy-making and governance design.

Berman (1982, p. 78) compared this image of constant change necessitated by capitalism's persistent demands for increasing capital accumulation with Goethe's Faust:

> under the pressures of the modern world economy the process of development must itself go through perpetual development. Where it does, all people, things, institutions and environments that are innovative and avant-garde at one historical moment will become backward and obsolescent in the next. Even in the most highly developed parts of the world, all individuals, groups and communities are under constant relentless pressure to reconstruct themselves; if they stop to rest, to be what they are, they will be swept away. The climactic clause in Faust's contract with the devil – that if he ever stops and says to the moment, 'Verweile doch, du bist so schoen', (*Ah, linger on, thou art so fair!*) he will be destroyed – is played out to the bitter end in millions of lives every day.

It is important not to assume however, that the constant search for new spatial fixes makes the need for 'new, fixed and relatively immobile spatial configurations' unnecessary (Lash and Urry 1987). Marx suggested that the development of capitalist relations had the effect of 'overcoming all spatial barriers', and hence to 'annihilate space with time'. Lash and Urry emphasized that this might be the *objective* of capitalist production but in practice, in Harvey's (1990, p. 328) terms, 'spatial organization is necessary to overcome space'.

Harvey (1990, p. 239) also provided a neat summary of this discussion on globalization, space—time compression, spatial fixes, and capital accumulation. His interpretation is overtly Marxist, but the points made are clearly supported by the evidence we have discussed so far relating to globalization and the spatial fix.

> Spatial and temporal practices are never neutral in social affairs. They always express some kind of class or other social content and are more often than not the focus of intense social struggle. That this is so becomes doubly obvious when we consider the ways in which space and time connect with money, and the way that connection becomes even more tightly organized with the development of capitalism. Time and space both get defined through the organization of social practices fundamental to commodity production. But the dynamic force of capital accumulation (and overaccumulation), together with conditions of social struggle, renders the relations unstable. As a consequence, nobody quite knows what 'the right time and place for everything' might be. Part of the insecurity which bedevils capitalism as a social formation arises out of this instability in the spatial and temporal principles around which social life might be organized. During phases of maximal change, the spatial and temporal bases for reproduction of the social order are subject to the severest disruption.

These unstable relations manifest themselves in a continual search for an optimal spatial fix which once found needs to move on to the next. This ever

5.4 The Relationship Between Disorganization, Globalization, and Capitalism

present restlessness creates the severe disruption that Harvey noted, that has manifested itself most recently in globalization and the disorder of the nation-state and that is evidenced in governance inadequacy and policy-failure in the maritime sector. This latest (globalized) spatial fix transformation is part of a grander transformation that has occurred in society as a whole and can be found throughout politics, economics, society, culture and beyond in the movement from a Modernist to a Postmodernist epoch. This theme is taken up by Bergquist (1993, p. 28) who cited Gitlin (1988) and his 'ceaseless transformation in style' and Jameson (1991, p. 25) who suggested that fragmentation is 'too weak and primitive a term' as society is no longer breaking up its preexisting organic totality, but instead we have the emergence of new and unexpected events, discourses, modes of classification and compartments of reality'. Anderson (1984, p. 111) made the point further stressing capitalism brings a 'constant upheaval in our conditions of life… a permanent revolution… not to long nostalgically for the "fixed, fast-frozen relationships" of the real or fantasized past but to delight in mobility, to thrive on renewal, to look forward to future developments' (Berman 1982, p. 95, 96). Aglietta (1982, p. 1) considered the emergence of a new reorganization of the capitalist world in the early 1980s and the crisis that he identified. Gramsci's (1971, p. 276) comment is highly pertinent:

> The crisis consists precisely in the fact that the old is dying and the new cannot be born; in this interregnum a great variety of morbid symptoms appears.

Jessop (1997, pp. 571, 572) commented on how capitalism is characterized by periodic market failure (or more politely structural limits and contradictions) and consequently how it needs to shift direction and search for new opportunities for profit. This crisis has been typified in the 1990s by:

- The emergence of new core technologies.
- The shift from a Fordist to a Post-Fordist paradigm for industrial organization.
- Growing internationalization and globalization.
- Rise of regional and local economies as sites of international systemic competitiveness.

In turn this has led to what Jessop noted as 'important discontinuities in modes of regulation and capitalism's articulation within the wider societal formulation'. Capitalism is thus rooted in what Jessop (1997, p. 578) saw as a 'dependence upon an unstable balance between its economic and extra-economic forms' (see also Polanyi 1944; Jessop 1982; Offe 1984).

Burke (2000) stressed the relationship between disorganized capitalism and the continuous search for capital accumulation through change to the emergence of Postmodernism in the late twentieth century as one aspect of the new spatial fix being sought out. We shall turn to Postmodernism in some detail in a later chapter but for the moment need only to outline its relationship to disorganized capitalism, spatial fix and capital accumulation. Nooteboom (1992, p. 65) cited Lyotard (1987, p. 28) who suggested that Postmodernism is characterized by 'paralogy' or disruptive discourse, Burke suggested in a similar way to Lash and Urry (1987) that disorganized capitalism is characterized by the disintegration of state regulation, the expansion of world markets dominated by multinational corporations, the undermining of the nation-state, the growth of third world manufacturing, and the decline of manufacturing in the west. This is accompanied by a decline in trades union influence and an erosion of class based politics. Cultural life becomes fragmented and pluralistic. All this is classically Postmodern.

We have gone on a while here because this section is especially important to our discussion of governance and the maritime sector. Globalization is the essential force that is destabilizing the state-centric hierarchical governance model that dominates the sector, and the forces behind globalization clearly need to be understood if there is to be a meaningful change in approach to governance and policy-making. We have in the work of Harvey and Brenner (among others) a Marxist interpretation of globalization as a continuous process of spatial fixation to enhance capital accumulation. The shipping industry as a pillar of capitalism, understands these processes well. Unfortunately those who attempt to formulate its policies and to manage its governance seemingly do not.

5.5 The Nation-State, Territory, Governance, and the Role of Globalization

> Man... is as much a territorial animal as is a mockingbird singing in the clear California night. We act as we do for reasons of our evolutionary past, not our cultural present... If we defend the title to our land or the sovereignty of our country, we do it for reasons no different, no less innate, no less ineradicable, than we do lower animals... all of us will give everything we are for a place of our own. Territory, in the evolving world of animals, is a force perhaps older than sex. (Ardrey 1969, quoted in Johnston 1991, p. 187)

Ardrey may not be quite correct in his interpretation of the significance of territory and in dismissing the present (or future) as a stimulant, but his enthusiasm for the concept is appropriate for the discussion to follow and the relationship of territory to spatial fix, globalization, and the problems of governance. The rise of the nation-state has always been heavily entwined with the principle of territoriality and Almond (1989) indicated the relationship between the international and the national as essential constructs. Gottmann discussed the importance of territoriality emphasizing its role in societal development while Johnston (1990) along with Sack (1983, 1986) in the latter's extensive and significant contribution to the understanding of territoriality, suggested that it is a deliberate spatial strategy 'to affect, influence and control'. Territoriality is identified as:

> a more effective means of establishing differential access to people, or resources, than is non-territoriality (Sack 1983, p. 57). It can be bounded and thus readily communicated; it can be used to displace personal relationships, between controlled and controller, by relationships between people and the 'law of the place'. (Johnston 1995, p. 214)

5.5 The Nation-State, Territory, Governance, and the Role of Globalization

Consequently, territoriality is a major component of state sovereignty and of its 'exercise of power over people through its control of bounded space' (Johnston 2001, p. 684). Territoriality is viewed as a 'necessary, if not sufficient, strategy in the exercise of state power' (Johnston 2001, p. 684). However, before we go on we need to establish what we mean by the term.

Johnston (1991) suggested that territoriality has been a central concept to much of the social and natural sciences. Sack (1986, p. 19) defined it as a human strategy:

> by an individual or group to affect, influence or control people, phenomena and relationships, by delimiting and asserting control over a geographic area.

In the context of our earlier discussion on hierarchies and governance generally and more specifically in the maritime sector Sack (1983) emphasized that territoriality is something that applies at any scale from the 'room to the international arena'. Although Sack's definitions is not the one that is widely used by other commentators it is of particular relevance here in our discussion of the relationship between territory, power, capitalism, the nation-state and governance, and the use that can be made of an understanding of spatial diversity, political sovereignty and policy-making (Table 5.2).

Capitalism uses territoriality as a central part of its manipulation of space and time, to accumulate capital and to ensure that overaccumulation does not become excessive. Successive territorial based (commonly nation-state based) crises are used to achieve this through the processes of disorganizing capitalist structures (including financial crises and war) to sustain the existing organizational, political, economic, and power-based situation. The most recent manifestation of these crises can be seen having its effect in the move from a Fordist/Modernist society to one characterized by Post-Fordism and Postmodernism (Cooke 1990, p. 142). This in turn has disrupted the established processes of governance in the maritime sector resulting in successive policy failures which reveal themselves through inadequacies in maritime safety, security and the environment, through political disorganization between the IMO and the European Union, through inadequate representation of stakeholders and through successive economic, societal, and political maritime crises. As Sack (1983) suggested territoriality is a central feature of capitalism's deliberate generation of differences in its drive to accumulate capital.

Clegg (1990, p. 180) identified the response of West European nations and the USA to the end of the post-war boom as a manifestation of Lash and Urry's disorganized capitalism, also termed the 'second industrial divide' by Piore and Sabel (1984) as flexibility grew in manufacturing systems and organization.

Sack continued to stress the significance of territoriality in the definition and sustenance of hierarchies that exist in societal control, power and relationships and also how modern capitalism, in contrast to feudal and classless societies has the closest of all relationships. Meanwhile, Smith (1986, p. 620) suggested in very similar language that territoriality represents the:

Table 5.2 Internal relations of tendencies and combinations (in Territoriality)

A	B	C	D	E	F	G	H	I	J	
	x	x					x	x	x	Hierarchy and bureaucracy
	x	x					v	x	x	Divide and conquer
	x	x					x	x	x	Secession
	x	x					x		x	Territorial definition of social relations
	x						x		x	Mismatch and spillover
		x					x		x	Obfuscation by assigning wrong scale territory
x				x	v					Territoriality an end, not a means
x				x	v					Social conflict obscured by territorial conflict—horizontal displacement
		x					x			Long and short range planning and stages
		x					x			Obfuscation by stage. Clear at national level, unclear at local
							x			Efficient supervision—span of control
							x			Inequalities
					x	x				Magic—representation becomes powerful in itself
x	x	x	x	x			x		x	Conceptual separation and recombination of space and substances

Source Sack (1983, p. 61)
x: Tendency important
v: Tendency extremely important
A: Multiplication of territories
B: Conceptually empty space
C: Mold
D: Neutral space clearing
E: Impersonal relations
F: Displacement
G: Reification symbol
H: Enforcement of access
I: Communication
J: Classification

attempt by an individual or group to influence or establish control over a clearly demarcated territory which is made distinctive and considered at least partially exclusive by its inhabitants or those who define its bounds.

Significantly, Smith took his interpretation much further proposing that 'people use territoriality strategies to promote identity, defense, and stimulation' thus hinting at the role of territory in the definition of a spatial fix in the process of capital accumulation. These characteristics he saw as socially generated, 'conditioned primarily by cultural norms and values which vary in structure and function from society to society, from one time period to another, and in, accordance with the scale of social activity', suggesting again a central place in the process of spatial fixing.

Johnston (2001) suggested that territoriality is commonly a process of the political division of space into containers. He put forward two reasons why this takes place:

- The exercise of power.
- The ease of administration.

5.5 The Nation-State, Territory, Governance, and the Role of Globalization

Both can be interpreted as being present as capital exercises its importance in defining territoriality through spatial fixation with the ultimate aim of either generating more capital to accumulate or disposing of excess capital (for example through aggression) to subsequently stimulate the demand for more.

Gottmann (1973, p. 126) identified the significance of the decline of the nation-state and its relationship to territoriality, globalization, and governance. He saw the growth in supranationality as evidence of decline and as a change in territoriality which would ultimately impact upon governance, power elites, policy-making, and decision-making. Gottmann's view was that the emergence of supranational bodies was a reflection of nation-states recognizing the limitations of their own boundaries resulting in weakness and frailty. The result was that 'sovereignty has been bypassed, and a new fluidity has infiltrated the recently shaped map of multiple national states' (Gottman 1973, p. 154). This process of change introduced 'new currents in technology, economic demands, and political ideas... into the arena, disquieting forces that expand accessibility, modify operating structures and call for the revision of established concepts and principles'.

Sack (1986, p. 261) stressed the importance of territoriality in understanding spatiality, and therefore underpinning the spatial fix and its relationship to governance. He claimed territoriality to be:

> the basic geographic expression of influence and power...[It is] an essential link between society, space and time. Territoriality is the backcloth of geographical context - it is the device through which people construct and maintain spatial organizations. [It] is not an instinct or drive, but rather a complex strategy to affect, influence, and control access to people, things and relationships... it is not only a means of creating and maintaining order, but is a device to create and maintain much of the geographic context through which we experience the world and give it meaning.

In our terms, we can interpret this as territoriality relating consistently to the role of the nation-state, to the development and impact of globalization, to the repeated search for new spatial fixes, and subsequently to the governance inadequacies that have emerged in recent years. Gottmann (1973, p. 157) emphasized the globalization of territoriality in these terms and in a way that directly relates to the maritime sector:

> In the economic field, national sovereignties are somewhat affected and nations interrelated by the networks woven in specialized domains (*e.g. shipping*) by the great international corporations of modern business. Most of these are characteristically based in one territory and under the flag of one nation but their diverse operations are so scattered and managed by so many agencies located in such dispersed fashion that to disentangle the puzzle and determine the 'nationality' of each of the many components becomes a difficult task... The growing interdependence of the great hubs of population and transactional activity scattered around the world adds to the involvement of each territory with others. (*section in italics added*)

This is all brought together by Taylor (1994, p. 160) who identified what was then a triple layered territoriality which reflected the 'fracturing of society', noted in our debate of globalization and disorganized capitalism, caused he felt because the territorialities were operating at different scales:

The state as a power container tends to preserve existing boundaries; the state as a wealth container tends towards larger territories; and the state as a cultural container tends towards smaller territories.

In fact Johnston (2001, p. 690) saw the future as one of ever more territoriality and scale complexity with 'multiple layers and scales and perhaps multiple identities'. This might manifest itself in what he termed 'global apartheid'—a pattern of unequal development in which the:

> affluent, in a whole variety of ways and at a whole variety of scales, seek to insulate themselves in places from what they see as the cultural, social, political and... threats from 'undesirable others.

This global apartheid is a territorial inevitability that is a characteristic of globalization and a consequence of capitalism's search for a spatial fix, reflecting the chaotic change in society that manifests itself as well in governance failure.

Harvey (2000), quoting Marx and Engels (1967) followed this up introducing the idea of historical territorial expansion as part of the continuous spatial fixation that characterizes capitalism suggesting that the significance of the global setting should not be ignored over any time scale:

> The revolutionary changes that brought the bourgeoisie to power were connected to 'the discovery of America, the rounding of the Cape' and the opening up of trade with the colonies and with the East Indian and Chinese markets. (Harvey 1998, p. 5)

Harvey went on to stress Marx and Engel's (1967) emphasis on transport (especially international shipping) in extending capitalism's control of territory in its search for (in Harvey's terms) new spatial fixes:

> Modern industry has established the world market, for which the discovery of America paved the way. This market has given an immense development to commerce, to navigation, to communication by land. This development in turn reacted on the extension of industry; in proportion as industry, commerce, navigation, railways extended, in the same proportion the bourgeoisie developed, increased its capital, and pushed into the background every class handed down from the Middle Ages. (Marx and Engels 1967)

Harvey (2000, p. 59) continued stressing the long history of globalization within capitalism reflecting its need for territorial spatial fixes and saw capitalism as impelled to eliminate spatial barriers, and thus to use territoriality to solve the constant crisis of capital which in recent times has manifested itself through ever increasing globalization. Capitalism, therefore, produces a:

> geographical landscape (of space relations, of territorial organization, and of systems of places linked in a 'global' division of labour and of functions) appropriate to its own dynamic of accumulation at a particular moment of its history, only to have to destroy and rebuild that geographical landscape to accommodate accumulation at a later date.

Harvey (2001a, p. 26) continued his discussion of crisis, capitalism and spatial fix emphasizing the virulence of overaccumulation when surpluses of labor and capital occur together with seemingly no way of combining them in a socially useful form—something that Hegel viewed in inner dialectical terms—

5.5 The Nation-State, Territory, Governance, and the Role of Globalization

overaccumulation—and outer dialectical—the release of these overaccumulated resources through the spatial fix. The spatial fix is the way that capitalism overcomes these problems and globalization—with its compression of space and time—is the most recent manifestation of this process which actually has its roots centuries ago. Capital accumulation is the latest phase in 'its problematic trajectory through continuous and sometimes disruptive geographic adjustments and reconfigurations' (Harvey 2001a, pp. 26, 27). Thus, periodic crises caused by capitalist overaccumulation are overcome by rejigging space and fixing a new arrangement which normally involves a reduction (and near elimination recently) of space and time through technical and organizational development. Shipping is a major stimulus to globalization (and a major beneficiary) and the context in which it operates is affected considerably. However, policy-making and the global governance of the industry has yet to catch up. The result—policy failure and the need for a new arrangement, new governance, and quite possibily new institutions.

Deleuze and Guattari (1984) emphasized Harvey's conclusions stressing that the 'territorialization and reterritorialization of capital is an ongoing process'. Harvey (2000, p. 60) took their view further by illustrating the development of the nation-state's 'construction of territorial organization' for the purposes of regulating 'money, law and politics' with the ultimate aim of monopolizing the 'means of coercion and according to a sovereign territorial (and sometimes extra-territorial) will'. However, this national territorialization really only began with the Treaty of Westphalia in 1648 which recognized national autonomy and territorial integrity for the first time. Several centuries have since passed including for example, the reduction of nation states in Europe between 1500 and 1923 from over 800 to 23. The continued and contradictory process of nation creation—exemplified by the emergence of the former states of Yugoslavia and the disintegration of the Soviet Union in recent years –suggests that the diminution of the number of states alongside an increased stratification of global order is not a one-way process. However, there remain good arguments that the creation of supranational organizations such as the EU, and the continued March of globalization is an inexorable deterritorialization. The latter Harvey (2000) saw as a consequence of:

- Financial deregulation.
- Technological change.
- Communications developments.
- Reduction in cost and time of moving goods and people.

…each of which has a very close relationship to the maritime sector and its developments (and problems). Smith (1990) stressed the importance of scale and territoriality as foundations of uneven geographic developments. Harvey (2000, p. 77) continued:

> the hierarchical scales at which human activities are now being organized are different from, say, thirty years ago. Globalization in part signifies an important aspect of that shift.

That shift is a reflection of capital's needs for a new spatial fix to sustain its desire for capital. It also implies a move from the culture of Modernism to one of Postmodernism (Smith 1992) or similarly defined Fordism to Postfordism; and that in turn requires a reinterpretation of governance and policy-making. The relationship between disorganization, Postmodernism, and the development of globalization was emphasized by Parker (1992, pp. 13, 14):

> one strategy for outsider intellectuals is to appear to attempt to subvert the whole game – Postmodernism. With Postmodernism, traditional distinctions and hierarchies are collapsed, polyculturalism is acknowledged... kitch, the popular and difference are celebrated. Their cultural innovation proclaiming a beyond is really a within, a new move within the cultural game which takes into account the circumstances of production of cultural goods, which will itself in turn be greeted as eminently marketable by cultural intermediaries. (Featherstone 1987, p. 69)

Rajchman (1987, p. 51) cited Connor (1989, p. 19) in his summary of the link between the rise of globalization, the disorganization that exemplifies this and Postmodernism.

Meanwhile Held and McGrew (2002, p. 484) at a relatively early stage suggested that the impact of globalization was to threaten the very basis of democracy which is founded on the notion of the nation-state—a 'community that properly governs itself and determines its own future'. Globalization lessens the ability of the nation-state to control its own destiny and makes access to its markets and society much easier for organizations and institutions themselves commonly with little territorial allegiance. One cannot provide governance for any sector or activity without recognizing the new forces that are adrift.

Globalization poses a threat to territoriality as transnational activity erodes the significance of the border. States therefore lose their established role based upon control of space as borders become porous the result of which is that the established norms are threatened. Established processes and frameworks of governance and policies rely upon the territoriality of the state for their power—and with its erosion through globalization, so the governance mechanism ceases to work effectively and the policies produced become unmanageable. In the maritime sector, this is the position of established methods of policy-making and governance regimes based upon outdated statist principles of territoriality made irrelevant by the creeping presence of a globalized society. And for the time being, let us just hint at the processes underlying these changes—the role of globalization in the accumulation of capital; the recognition by capitalists of the failure of the sovereign state to form an effective governance framework; and the enthusiasm of the shipping industry to sustain the nation-state's role in governance as it sees opportunities to abuse a failing policy-making framework for its own benefit of capital accumulation. But more of this later.

Cohen (2001) suggested that the main significance of globalization is that 'it is working to disengage the boundaries of the state's authority from its territorial borders, and/or changing the ways in which these boundaries are governed'. Zhang (2005, p. 199) meanwhile took up the theme of nation-state and globalization suggesting that:

5.5 The Nation-State, Territory, Governance, and the Role of Globalization

The Janus-faced nature of globalization provides both opportunities for sustainable human development since world history shows us that advances in civilization largely emanate from the ability of states to learn from each other, and it poses new challenges for states to govern themselves more effectively.

Much earlier, Jachtenfuchs (1995) saw globalization as questioning the existence of the state both externally through pressures that cross borders, but also internally treating the nation-state as simply a functional subsystem of society, no more nor less important than any other and certainly not occupying any privileged, unquestionable position. Rosenau (1997, 2001, 2003) saw globalization as moving significant activity 'beyond the national seats of power that have long served as the foundations of economic, political, and social life.' This suggests change—what he termed a 'transformation of practices and norms—in particular in relation to territoriality—and that globalization no longer recognizes the existing jurisdictional barriers. Existing sovereign state authorities may attempt to reinforce barriers and to encourage what Rosenau terms 'localization' but they will inevitably, eventually fail.

Zurn (2003, p. 342) provided further support for these views suggesting that with globalization, the significance of national borders must decrease, challenging 'the nation-state to unilaterally reach its governance targets'. Significantly for our discussion of the problems facing maritime policy-making and governance Zurn suggested that:

> Effective governance depends upon the spatial congruence of political regulations with socially integrated areas and the absence of significant externalities. As societal interconnectedness increases with globalization, national governments are increasingly confronted with difficulties in the implementation of their policies.

The traditional maritime governance state-centric hierarchy fails to meet the needs of a globalized society because globalization has reduced the effectiveness of national borders to the point where nation-states no longer have the ability to implement national policies even within their own state territory.

Taylor (2000b, p. 1111) saw globalization in the context of 'embedded statism', treating nation-states as 'natural units' with society, economy and polity all covering identical areas represented by the sovereign state. Globalization necessitates an enhancement of trans-state processes that cross these natural boundaries. However, he was disappointed that there had been a tendency for the debate to be couched in terms of a choice that had to be made between globalization and state-centrism whereas consideration of their 'manifold inter-relationships' would be more productive. The state changes in the face of globalization but they are not adversaries (Brenner 1998).

Moreover Taylor (2000c, p. 159) regretted that most data are also defined from a state spatial viewpoint (including populations, land, economy, and their properties) with little if any consideration of the 'flows, connections and linkages that make our world work'. Consideration of these flows might well require a new view of governance frameworks to accommodate flows rather than spaces. Globalization forces these new views to be taken for four reasons:

- It involves a new scale of human activity with the state challenged from above and below (internationally/supranationally and regionally).
- Globalization is essentially transnational rather than international with the elimination of borders a central feature rather than negotiation between states.
- Globalization requires the incorporation of a 'space of flows'. Taylor (2000c) termed this a network society rather than a national society.
- The nation-state as a result can no longer be taken for granted as the 'basic unit of investigation'. And yet this is just what happens in maritime governance and policy-making.

5.6 The Consequences of Continued Globalization for Contemporary Maritime Governance

Not everyone sees globalization as a universal movement and there are some dissenting views. Gordon (1988, p. 63) quoted in Cox (1993, p. 443) stated:

> I would argue that we have *not* witnessed movement towards an increasingly 'open' international economy, with productive capital buzzing around the globe, but that we have moved rapidly towards an increasingly 'closed' economy for productive investment, with production and investment decisions increasingly dependent upon a range of institutional policies and activities and a pattern of differentiation and specialization among the countries in the LDCs... The international economy, by the standards of traditional neoclassical and Marxian models of competition has witnessed *declining* rather than *increasing* mobility of productive capital. (*emphasis in original*)

Mann (1997, p. 494) was even more doubtful:

> We must beware the more enthusiastic of the globalists and transnationalists. With little sense of history, they exaggerate the former strength of nation-states; with little sense of global variety, they exaggerate their current decline; with little sense of their plurality, they downplay their inter-national relations.

Meanwhile Virilio (1995, p. 2) was his usual oblique self:

> The very word 'globalization' is a fake. There is no such thing as globalization, there is only virtualization. What is being effectively globalized by instantaneity is time. Everything now happens within the perspective of real time: henceforth we are deemed to live in a 'one-time-system'.

Quite. In contrast, Harvey (1990) is central to the argument that globalization reflects a major change in society, politics and economics which in turn has a close relationship to the problems of policy failure which we have identified in the maritime sector. These problems stem from the need for a new approach to governance that globalization demands. Globalization itself is the manifestation of a new spatial fix that stems from the continuous chase for capital and the need to accommodate its overaccumulation. Only by searching for new spatial fixes can capitalism continue to acquire what it needs to feed off and dispose of its excesses of capital which otherwise would lay dormant. Harvey saw globalization as the latest of a succession of spatial fixes which can be identified over many centuries and it will not be the last. The current move toward extended globalization can be closely associated with the emergence of what he termed the 'Postmodern condition' from the earlier Modernist epoch—a process that is widely recognized throughout society, politics, economics and culture and one which has affected business and commerce no less than any other. In the coming chapters we shall turn to this process in more detail.

Harvey (1990, p. 284) discussed this transition from a Modernist to a Postmodernist period (or as he terms it at this stage, from a Fordist epoch to one characterized by flexible accumulation) focusing particularly upon the relationship to space and time and hence to the process of globalization. He suggested that since about 1970 the world had experienced an intense period of time–space compression that has had 'a disorienting and disruptive impact upon political–economic practices, the balance of class power as well as upon cultural and social life'. This transition to flexible accumulation he saw as a Postmodern phenomenon and a result of new organizational forms and the introduction of new technologies. These developments, manifested in such things as just-in-time logistics, subcontracting, outsourcing, electronic control of small batch manufacturing and all the data and information advances that have taken place represent a new epoch. A move on from Fordist rigidity and even Post-Fordism variation, from Modernist control, organization and mass production to a new Postmodernist era of flexibility, variability and choice characterizing the globalized markets that now dominate capitalism.

Harvey suggested that globalization was a function of improved communication systems and information flow, which mixed with streamlined distribution (for example improved inventory control) makes it possible to move commodities through the market speedier. This is supported by considerable evidence throughout the transport, shipping and logistics areas including for example work by Hansen (2003, p. 3) considering the Great Belt and Oresund Fixed Links.

> New traffic infrastructures are not isolated phenomenon (*sic*), which are established independently of their contemporary context – political, economic, social etc. Thereby, can also the newly established fixed links be seen in the context of underlying structures and tendencies embedded in the postmodern society. The postmodern society can be characterized by the *compression of time and space*... reflect(ing) a development, where distance and time is perceived in new ways caused by the instantaneous connectedness of information, product and transport flows (Beckmann 1999; Drewes Nielsen and Oldrup 2001). (Hansen 2003)

Electronic banking has matched this process for money. Coupled with this is a fundamental change in culture and taste which has also speeded up consumption through the associated globalization thus helping to generate the accumulation of more capital. Mass fashion in clothing, entertainment and life style has expanded markets dramatically for goods which conveniently become obsolete very quickly. Meanwhile a move from goods to services has been dramatic—the latter have much shorter life spans (sometimes almost instantaneous) thus generating even quicker flows of capital.

Globalization has helped these trends by spreading the global message, advancing the marketing image and facilitating the consequential flow of capital. In Harvey's words:

> We have, in short, witnessed another fierce round in that process of annihilation of space through time that has always lain at the centre of capitalism's dynamic. (Harvey 1990, p. 293)

However, this does not mean that space is no longer significant, overcome by the compression of time that is represented by globalization. In fact space has increased in significance for as 'spatial barriers diminish so we become more sensitized to what the world's spaces contain' and the ability to minimize space through time compression actually makes the choice between locations both more extensive but consequently more important in a competitive marketplace (Harrison and Bluestone 1988, quoted in Harvey 1990, p. 294).

Harvey also suggested that the reduction in spatial barriers as globalization progresses causes the existing national hierarchical system to take on a new structure that realigns itself with global characteristics. He went on to propose that this takes on a new hierarchical form but this can be doubted as the strictly defined and controlled jurisdictional structure needed for hierarchies to work—typified by the IMO/EU/national ministry framework that dominates maritime policy-making at present—would appear to be less relevant in a global environment focused on flexible accumulation. But more of that later. Harvey even mentioned the changing role of governance as a new society emerges in the ever intensifying global spatial fix. At last we can begin to see some rationale behind the policy problems emerging in the maritime sector and a pattern that does not assume that all can be made well through successive policy initiatives. Perhaps, addressing the governance changes that should accompany globalization would be more efficacious?

Harvey finished his assessment of globalization, spatial fix and the moves toward Postmodernism by considering the relationship between the change that has occurred and what he terms 'geopolitical dangers'. These have emerged because of the rapidity of change that has taken place:

> The transition from Fordism to flexible accumulation (*Modernism to Postmodernism*) such as it has been, ought to imply a transition in our mental maps, political attitudes, and political institutions. But political thinking does not necessarily undergo such easy transformations, and is in any case subject to the contradictory pressures that derive from spatial integration and differentiation. There is an omni-present danger that our mental maps (*and those of policy-makers*) will not match current realities. The serious diminution of the power of individual nation states over fiscal and monetary policies, for example, has not been matched by any parallel shift towards an internationalization of politics. Indeed, there are abundant signs that localism and nationalism have become stronger precisely because of the quest for the security that place always offers in the midst of all the shifting that flexible accumulation implies. The resurgence of geopolitics and of faith in charismatic politics (*both Postmodern tokens*) fits only too well with a world that is increasingly nourished intellectually and politically by a vast flux of ephemeral images (*again essentially Postmodern*). (emphasis added) (Harvey 1990, p. 385)

The last comment is highly suggestive of the situation for shipping with national flag concerns ever more a central feature within a highly globalized

5.6 The Consequences of Continued Globalization

industry and one which reflects the race toward a new spatial fix. The maritime industry displays the Janus face of globalization noted earlier, accommodating within its realm both highly protectionist, domestic, national characteristics alongside a globalized, flexible mobility that exceeds that of almost any other sector. This dichotomy sits uncomfortably for the policy-maker who must try and set up a governance framework to accommodate it. However, it sits very comfortably for those in the industry who find it convenient to flit in and out of the new spatial fix to maximize their accumulation of capital.

Anderson (1996, pp. 143, 144) provided added emphasis to the idea of space–time compression and globalization and also linked this to the Postmodern—which we shall look at in considerable detail at a later stage. He saw what he termed a 'political progression' from Premodern, through the Modern to the Postmodern and that this was a movement from 'relative to absolute and then back to relative space'. The relative dimensions of time and space are also seen to change. In our discussion the Modern times can be seen as a mobile temporality related to development and progress and it is from this that the first Postmodern concepts can be seen to be emerging. This embryonic Postmodernism was exemplified by the ideals of the Futurist movement and its emphasis on science, technology, machines, and speed (Tisdall and Bozzolla 1977; Lista 2001) which developed during the early twentieth century reflecting the revolutionary avant-garde of the time. The objectives were to 'abolish the limits between creation and action' making art an expression of Postmodernism itself. An example of Futurist output which concentrates on speed, acceleration and movement can be seen in Fig. 5.1 with Pannaggi's *Speeding Train*. The traditional view of time had been that it was a given process which existed of its own accord, with or without an observer or participant; or subjectively, present only as a 'sense experience' (Giedion 1941, p. 439). In 1908, the mathematician Minkowski was the first to suggest that 'space alone or time alone is doomed to fade into a mere shadow: only a kind of union of both will preserve their existence'. Futurism (along with Cubism) epitomized this view introducing the concept of space–time into art and pointing the way toward the Postmodern period.

However, in contrast, the conventional (Modernist) view of space (unrelated to time) remained fixed in terms of territory, sovereignty and states and the institutions and policy-making arrangements that emerged during this phase characterized modern maritime governance. Anderson saw the emergence of Postmodernism increasing fluidity of time and space as globalization frees society from its spatial limitations and allows the possibilities of reducing time constraints to increase. He suggested that this Postmodern fluidity would take time to develop especially in terms of territoriality where items such as factories, cities, states, and national interests will cling on, but that globalization will continue despite creating discordance between governance and governed, between governance jurisdictions, between maritime policies, and those for which they are created.

Der Derian (1990, p. 297) also identified a significant change in the relationship of speed to space as time (and therefore effectively space) is compressed. He described the processes going on as *chronopolitical* and *technostrategic* generating

Fig. 5.1 Ivo Pannaggi. Speeding Train 1922. Oil on canvas, 100 × 120 cm, Cassa di Riparmio, Macerata

a series of problems for the nation-states locked into traditional systems as the Postmodern has developed. Der Derian saw them as chronopolitical as they 'elevate chronology over geography, pace over space, in their political effects'; technostrategic in that they use these characteristics to promote ambitions through violence (Clausewitz 1976, pp. 128, 177; Der Derian 1987, Ch. 9; Klein 1989).

5.7 Conclusions

Anachronistic, state-centric, hierarchical governance structures are losing control to the inevitable forces of globalization which are making the traditional role of the nation-state outmoded. The state still has a role but it is markedly different and governance can only be effective if it recognizes these changes. The maritime sector displays a spectacular disregard for the changes in society of which shipping has been a major generator. Until it begins to accommodate them, policies will continue to fail.

The current governance model that forms the framework for policy-making in the maritime sector remains the one which formed the basis upon which the institutions that today attempt to direct the industry. It is hierarchical, with broad

5.7 Conclusions

policies developed at its highest level (for example form the IMO and OECD) at an international/global level, supposedly drifting down through the hierarchy and its subsequent and more localized levels of the supranational (for example the EU), national (respective ministries), regional, and local levels with each respecting and recognizing the policy decisions made by its superior(s). While theoretically pleasing this policy-making model suggests a very different governance structure to the one that currently dominates an increasingly globalized world. As a consequence, it is suggested that the failure of much of maritime policy-making is an inevitability until the fundamental governance framework is reconsidered.

Our next task is to try to understand better the forces that have encouraged globalization and which currently direct its progress because only then can we begin to suggest a more realistic approach toward policy-making. We will need to appreciate the shifts that have taken place which underlie the moves toward globalization. We will attempt to explain how these shifts in process have manifested themselves in the moves from a Modernist to a Postmodernist society and whether this reflects a long term change in capitalism akin to the disorganized capitalism of Lash and Urry (1987) and Arrighi (1994) and its approach to accumulating and disposing of capital or whether it is a merely a continuation of an 'inevitable' process in the economic development of mankind. We will also consider how the shipping sector has driven globalization, benefitted from the forces of capital accumulation that it represents and at the same time encouraged the continued role of the nation-states in maritime policy-making for its own advantage and at some cost to the success of maritime policies in particular and maritime governance more generally. Following this, we can then begin to look at ways that maritime governance might be modified to accommodate the forces of globalization, the chase for capital and the specific nature of the international shipping industry.

References

Aglietta, M. (1982). World capitalism in the eighties. *New Left Review, 136*(1), 3–41.
Agnew, J. (1999). Mapping political power beyond state boundaries territory, identity, and movement in world politics. *Millennium: Journal of International Studies, 28*(3), 499–521.
Alderton, T., & Winchester, N. (2002). Globalization and de-regulation in the maritime industry. *Marine Policy, 26*, 35–43.
Allen, J. (2003). *Lost geographies of power*. Oxford: Blackwell.
Almond, G. A. (1989). The international-national connection. *British Journal of Political Science, 19*(2), 237–259.
Amin, S. (1974). *Accumulation on a world scale*. New York: Monthly Review Press.
Anderson, M. (1996). *Frontiers: Territory and state formation in the modern world*. Cambridge: Polity Press.
Anderson, P. (1984). Modernity and revolution. *New Left Review, 144*, 96–113.
Appadurai, A. (1996). *Modernity at large: Cultural dimensions of globalization*. Minneapolis: University of Minnesota Press.
Ardrey, R. (1969). *The territorial imperative*. London: Fontana.
Arendt, H. (1969). *On violence*. London: The Penguin Press.

Armitage, J., & Graham, P. (2001). Dromoeconomics: towards a political economy of speed. *Parallax, 7*(1), 111–123.
Armstrong, P., Glyn, A., & Harrison, J. (1991). *Capitalism since 1945*. Oxford: Blackwell.
Arrighi, G. (1994). *The long twentieth century*. London: Allen and Unwin.
Axtmann, R. (1995). *Globalization and the democratic nation-state: Twelve theses*. Political Studies Association Annual Conference, New York.
Bakunin, M. (1996). *Bakunin on anarchism*. Montreal: Black Rose Books.
Bauman, Z. (2000). *Liquid modernity*. Cambridge: Polity Press.
Beckmann, J. (1999). Introduction: risks, benefits and tools of speed, in J. Beckmann (1999) *SPEED: A Workshop on Space, Time and Mobility* (pp. 7–10). Transportradets notatserie, nr. 99-05. Transportradet, Kobenhavn.
Berg, L. (1993). Between modernism and postmodernism. *Progress in Human Geography, 17*(4), 490–507.
Bergquist, W. H. (1993). *The postmodern organization*. San Francisco: Jossey-Bass.
Berman, M. (1982). *All that's solid melts into air: The experience of modernity*. Harmondsworth: Penguin.
Berry, B. J. L. (1989). Comparative geography of the global economy: cultures, corporations and the nation-state. *Economic Geography, 65*, 1–18.
Bloch, E. (1986). *The principle of hope*. Cambridge: MIT Press.
Block, F. (1996). *The vampire state: And other myths and fallacies about the US Economy*. New York: W.W. Norton and Co.
Bloor, M., & Sampson, H. (2009). Regulatory enforcement of labour standards in an outsourcing globalized industry. *Work, Employment and Society, 23*(4), 711–726.
Bodin, J. (1955). *Six books of the commonwealth*. Oxford: Basil Blackwell.
Boje, D.M., Fitzgibbon, D.E., and Steingard, D.S. (1996). Storytelling at Administrative Science Quarterly: warding off the postmodern barbarians. In D.M. Boje, R.P. Gephart Jr. and T.J. Thatchenkery (Eds.) (1996) *Postmodern management and organization theory* (pp. 60–92). Thousand Oaks, CA: Sage.
Bonacich, E. (2003). Pulling the plug: labor and the global supply chain. *New Labor Forum, Summer, 12*(2), 4–6.
Boyer, R., & Drache, D. (Eds.). (1996). *States against markets: The limits of globalization*. London: Routledge.
Braudel, F. (1982). *The wheels of commerce*. New York: Harper and Row.
Brenner, N. (1997). Global, fragmented, hierarchical: Henri Lefebvre's geographies of globalization. *Public Culture, 10*(1), 135–167.
Brenner, N. (1998). Between fixity and motion: accumulation, territorial organization and the historical geography of spatial scales. *Environment and Planning D, 16*(4), 459–481.
Brenner, N. (1999). Beyond state-centrism? Territoriality and geographical scale in globalization studies. *Theory and Society, 28*(1), 39–78.
Bull, M. (Ed.). (1995a). *Apocalypse theory and the end of the world*. Oxford: Blackwell.
Bull, M. (1995b). On making ends meet. In M. Bull (Ed.), *Apocalypse theory and the end of the world* (pp. 1–20). Oxford: Blackwell.
Burke, B. (2000). *Post-modernism and post-modernity*. The encyclopaedia of informal education. www.infed.org/biblio/b-postmd.htm.
Burrell, G. (1988). Modernism, postmodernism and organizational analysis 2: The contribution of Michel Foucault. *Organization Studies, 9/2*, 221–235.
Cable, V. (1995). The diminished nation-state: a study in the loss of economic power. *Daedalus, 124*, 23–53.
Callières, F. (1963). *On the manner of negotiating with princes*. Notre Dame IN: University of Notre Dame Press.
Camus, A. (1957). *Banquet speech, nobel prize in literature*. http://nobelprize.org/nobel_prizes/literature. In A. Giddens (1987) *A contemporary critique of historical materialism: The nation state and violence* (p. 294). Berkeley: UCLA Press.

Caporaso, J. A. (1981). Industrialization in the periphery: the evolving global division of labour. *International Studies Quarterly, 25*(3), 347–384.
Caporaso, J. A. (2000). Changes in the Westphalian order: Territory, public authority and sovereignty. *International Studies Review, 2*(2), 1–28.
Cardoso, F., & Faletto, E. (1979). *Dependency and development in Latin America*. Berkeley: University of California Press.
Castells, M. (1989). *The informational city: Information technology, economic restructuring and the urban regional process*. Oxford: Blackwell.
Castells, M. (2000). Materials for an exploratory theory of the network society. *British Journal of Sociology, 51*(1), 5–24.
Castree, N., & Gregory, D. (Eds.). (2006). *David Harvey: A critical reader*. Oxford: Blackwell.
Cerny, P. G. (1995). Globalization and the changing logic of collective action. *International Organization, 49*(4), 595–625.
Cerny, P. G. (1997). Paradoxes of the competition state: the dynamics of political globalization. *Government and Opposition, 32*(2), 251–274.
Chowdhury, K. (2006). Interrogating "newness". Globalization and postcolonial theory in the age of endless war. *Cultural Critique, 62*, 126–161.
Clausewitz, C. Von. (1976). *On war*. Princeton: Princeton University Press.
Clegg, S. R. (1990). *Modern organizations. Organization studies in the postmodern world*. London: Sage.
Cohen, E. S. (2001). Globalization and the boundaries of the state: a framework for analysing the changing practice of sovereignty. *Governance, 14*(1), 75–97.
Connor, S. (1989). *Postmodernist culture*. Oxford: Blackwell.
Conversi, D. (1999). Nationalism, boundaries and violence. *Millennium: Journal of International Studies, 28*(3), 553–584.
Cooke, P. (1990). *Back to the future. Modernity, postmodernity and locality*. London: Unwin Hyman.
Cox, K. R. (1993). The local and the global in the new urban politics. *Environment and Planning D, 11*(4), 433–448.
Croucher, S. L. (2003). Perpetual imagining. *Nationhood in a global era, International Studies Review, 5*, 1–24.
Dale, K., & Burrell, G. (2007). *The spaces of organization and the organization of space*. London: Palgrave Macmillan.
Deleuze, G., & Guattari, F. (1984). *Anti-Oedipus: Capitalism and schizophrenia*. Minneapolis: University of Minnesota Press.
Deleuze, G., & Guattari, F. (1988). *A thousand plateaus: Capitalism and schizophrenia*. London: The Athlone Press.
Der Derian, J. (1987). *On diplomacy: A genealogy of western estrangement*. Oxford: Blackwell.
Der Derian, J. (1990). The (s)pace of international relations: simulation, surveillance, and speed. *International Studies Quarterly, 34*(3), 295–310.
Dicken, P. (1986). *Global shift: Industrial change in a turbulent world*. London: Harper and Row.
Dicken, P. (1992). *Global shift: The internationalization of economic activity*. London: Paul Chapman.
Dicken, P. (1994). Global-local tensions: firms and states in the global space-economy. *Economic Geography, 70*(2), 101–128.
Dickens, P. (2009). The cosmos as capitalism's outside. *Sociological Review, 57*(1), 66–82.
Dickens, P., and Ormrod, J.S. (2006). *The outer spatial fix*. Critical approaches to outer space panel, BISA Annual Conference, Cork, Ireland.
Dickens, P., & Ormrod, J. S. (2007). *Cosmic society. Towards a sociology of the universe*. London: Routledge.
Donzelot, J. (1980). *The policing of families*. London: Hutchinson.
Emmanuel, A. (1972). *Unequal exchange: A study of the imperialism of trade*. New York: Monthly Review Press.
Engels, F. (2007). *Dialectics of nature*. London: Well Red Publications.

Evans, P. (1997). The eclipse of the state? Reflections on stateness in an era of globalization. *World Politics, 50*(1), 62–87.
Featherstone, M. (1987). Lifestyle and consumer culture. *Theory, Culture and Society, 4*(1), 55–70.
Featherstone, M. (Ed.). (1990). *Global culture*. London: Sage.
Formann, S., & Segaar, D. (2006). New coalitions for global governance: the changing dynamics of multilateralism. *Global Governance, 12,* 205–225.
Foucault, M. (1977). *Discipline and punish*. Harmondsworth: Penguin.
Frank, A. (1969). *Capitalism and underdevelopment in Latin America*. New York: Monthly Review Press.
Galtung, J. (1971). A structural theory of Imperialism. *Journal of Peace Research, 8*(2), 81–117.
Garrett, G. (1998). Global markets and national politics: collision course or virtuous circle? *International Organization, 52*(4), 787–824.
Gephart, R. P, Jr, & Pitter, R. (1993). The organizational basis of industrial accidents, *Canada. Journal of Management Inquiry, 3,* 238–252.
Gibbon, P., Bair, J., & Ponte, S. (2008). Governing global value chains: an introduction. *Economy and Society, 37*(3), 315–338.
Giddens, A. (1984). *The constitution of society: Outline of the theory of structuration*. Berkeley CA: University of California Press.
Giddens, A. (1985). *A contemporary critique of historical materialism. The Nation-state and Violence* Vol. 2. Cambridge: Polity Press.
Giddens, A. (1990). *The consequences of modernity*. Stanford: Stanford University Press.
Giddens, A. (1995). *A contemporary critique of historical materialism*. Cambridge: Polity Press.
Giedion, S. (1941). *Space, time and architecture*. Cambridge: Harvard University Press.
Gitlin, T. (1988). Hip deep in postmodernism, *New York Times,* Nov, 6, 1988.
Gordon, D. (1988). The global economy: new edifice or crumbling foundations? *New Left Review, 168,* 24–64.
Gottmann, J. (1973). *The significance of territory*. Charlottesville: University of Virginia Press.
Gould, S. J., & Eldredge, N. (1977). Punctuated equilibria: the tempo and mode of evolution reconsidered. *Paleobiology, 3,* 115–151.
Gramsci, A. (1971). *Selections from the prison notebooks*. New York: International Publishers.
Gregory, D. (1991). Interventions in the historical geography of modernity: social theory, spatiality and the politics of representation. *Geografiska Annaler, 73 B,* 1, 17–44.
Gregory, D. (2006). Introduction: troubling geographies. In N. Castree & D. Gregory (Eds.), *David Harvey: A critical reader* (pp. 1–25). Oxford: Blackwell.
Gritsch, M. (2005). The nation-state and economic globalization: soft geo-politics and increased state autonomy? *Review of International Political Economy, 12*(1), 1–25.
Grotius, H. (1617). *De satisfactione Christi adversus Faustum Socinum (On the satisfaction of Christ against [the doctrines of] Faustus Socinus),* Leiden.
Habermas, J. (1973). *Legitimation crisis*. Boston: Beacon Press.
Hansen, L.G. (2003). *Impacts of fixed links on firms' organization of logistics and transport*. Nectar Conference 7, European Transport Systems Between Efficiency, Equity and Sustainability, Umea, Sweden.
Harrison, B., & Bluestone, B. (1988). *The great U-turn: Capital restructuring and the polarization of America*. New York: Basic Books.
Harvey, D. (1975). The geography of capitalist accumulation. A reconstruction of the Marxian theory. *Antipode, 7*(2), 9–21.
Harvey, D. (1981). The spatial fix: Hegel, Von Thunen and Marx. *Antipode, 13*(3), 1–12.
Harvey, D. (1982). *The limits to capital*. Chicago: University of Chicago Press.
Harvey, D. (1985a). *The urbanization of capital: Studies in the history and theory of capitalist urbanization*. Baltimore: John Hopkins University Press.
Harvey, D. (1985b). *Consciousness and the urban experience*. Baltimore MD: John Hopkins University Press.

Harvey, D. (1987). Flexible accumulation through urbanization: reflections on 'post-modernism' in the American city. *Antipode, 19*(3), 260–286.
Harvey, D. (1989). *The urban experience*. Oxford: Blackwell.
Harvey, D. (1990). *The condition of postmodernity*. Cambridge: Blackwell.
Harvey, D. (1998). *The geography of class power*. http://socialistregister.com/index.php/srv/article/view/5700/2596.
Harvey, D. (2000). *Spaces of hope*. Edinburgh: Edinburgh University Press.
Harvey, D. (2001a). Globalization and the "spatial fix". *Geographische Review, 2*, 23–30.
Harvey, D. (2001b). *Spaces of capital: Towards a critical geography*. London: Taylor and Francis.
Harvey, D. (2003). The right to the city. *International Journal of Urban and Regional Research, 27*(4), 939–941.
Hassan, R. (2009). Crisis time: networks, acceleration and politics within late capitalism, ctheory.net, www.ctheory.net/articles.aspx?id=618.
Hasse, R., & Leiulfsrud, H. (2002). From disorganized capitalism to transnational fine tuning?: recent trends in wage development, industrial relations and 'work' as a sociological category. *British Journal of Sociology, 53*(1), 107–126.
Hegel, G. W. (1967). *Philosophy of right*. Oxford: Oxford University Press.
Held, D. (1991). Democracy, the nation-state and the global system. *Economy and Society, 20*(2), 138–172.
Held, D., & McGrew, A. (Eds.). (2002). *Governing globalization. Power authority and global governance*. Cambridge: Polity Press.
Held, D., McGrew, A., Goldblatt, D., & Perraton, J. (1999). *Global transformations: Politics*. Cambridge: Economics and Culture, Polity Press.
Hirst, P., & Zeitlin, J. (1991). Flexible specialization versus Post-Fordism: theory, evidence and policy implications. *Economy and Society, 20*(1), 1–56.
Hirst, P.Q., & Thompson, G. (1995) .Globalization and the Future of the Nation State, *Economy and Society, 24*(3) 408–422.
Hobbes, T. (1909). *Leviathan. The matter forme and power of a common wealth ecclesiastical and civil*. Oxford: Clarendon Press.
Holm, H-H., and Sorensen, G. (1995). Introduction: what has changed? In H-H. Holm and G. Sorensen (Eds.),*Whose World Order? Uneven Globalization and the End of the Cold War*. Boulder: Westview Press.
Hopkins, T., & Wallerstein, I. (1994). Commodity chains: construct and research. In G. Gereffi & M. Korzeniewicz (Eds.), *Commodity chains and global capitalism* (pp. 17–19). Westport: Praeger.
Hsu, J-Y. (2000). *Revisiting economic development in post-war Taiwan: The dynamic process of geographical industrialization*. Alternative 21st Century Geographies: 2nd International Critical Geography Conference, Taegu, Korea.
Jachtenfuchs, M. (1995). Theoretical perspectives on European governance. *European Law Journal, 1*(2), 115–133.
Jameson, F. (1984). Postmodernism, or the cultural logic of late capitalism. *New Left Review, 146*, 53–92.
Jameson, F. (1985). Postmodernism and consumer society. In H. Foster (Ed.), *Postmodern culture* (pp. 111–125). London: Pluto Press.
Jameson, F. (1991). *Postmodernism or the cultural logic of late capitalism*. Durham: Duke University Press.
Jameson, F. (2000). Globalization and political strategy. *New Left Review, 4*, 49–68.
Jessop, B. (1982). Accumulation strategies, state forms and hegemonic projects. *Kapitalistate, 10*, 89–111.
Jessop, B. (1997). Capitalism and its future: Remarks on regulation, government and governance. *Review of International Political Economy, 4*(3), 561–581.
Jessop, B. (2002). *Critical realism and semiosis, the future of the capitalist state*. Cambridge: Polity Press.

Jessop, B. (2003). The future of the state in an era of globalization. *International Politics and Society, 3*, 30–46.

Jessop, B. (2005). The governance of complexity and the complexity of governance: preliminary remarks on some problems and limits of economic guidance. In A. Amin & J. Hausner (Eds.), *Beyond market and hierarchy* (pp. 111–147). Cheltenham: Edward Elgar.

Jessop, B. (2006). Spatial fixes, temporal fixes and spatio-temporal fixes. In N. Castree & D. Gregory (Eds.), *David Harvey: A critical reader* (pp. 142–166). Oxford: Blackwell.

Johnston, R. J. (1982). *Geography and the state*. London: Macmillan.

Johnston, R. J. (1990). The territoriality of law: An exploration. *Urban Geography, 11*, 548–565.

Johnston, R. J. (1991). *A question of place exploring the practice of human geography*. Oxford: Blackwell.

Johnston, R. J. (1995). Territorality and the state. In G. Benko & U. Strohmayer (Eds.), *Geography, history and social science* (pp. 213–226). Dordrecht: Kluwer.

Johnston, R. J. (2001). Out of the 'moribund backwater': Territory and territoriality in political geography. *Political Geography, 20*, 677–693.

Kant, I. (1795). Perpetual peace. In H. Reiss (Ed.) (1991) *Kant: Political writings*. Cambridge: Cambridge University Press.

Kegley, C. W., & Wittkopf, E. R. (1989). *World politics*. London: Macmillan.

Keohane, R. O. (2001). Governance in a partially globalised world. *American Political Science Review, 95*(1), 1–13.

Keohane, R. O., & Nye, J. S, Jr. (2001). *Power and interdependence*. New York: Addison-Wesley.

King, R., & Kendall, G. (2004). *The state, democracy and globalization*. Basingstoke: Palgrave Macmillan.

Kjaer, A. M. (2004). *Governance*. Cambridge: Polity Press.

Klein, B.S. (1989). The textual strategies of military strategy: or, have you read any good defense manuals recently? In J. Der Derian and M. J. Shapiro (Eds.) *International/Intertextual relations: Postmodern readings of world politics* (pp. 97–112). Lexington: Lexington Books.

Krasner, S. D. (1978). *Defending the national interest: Raw materials investments and US Foreign Policy*. Princeton: Princeton University Press.

Krasner, S. D. (1999). Globalization and sovereignty. In D. Smith, D. Solinger, & S. Topik (Eds.), *States and sovereignty in the global economy*. New York: Routledge.

Lacher, H. (2005). International transformation and the persistence of territoriality: toward a new political geography of capitalism. *Review of International Political Economy, 12*(1), 26–52.

Lash, S., & Bagguley, P. (1988). Labour relations in disorganized capitalism: a five-nation comparison. *Environment and Planning D, 6*, 321–338.

Lash, S., & Urry, J. (1987). *The end of organized capitalism*. Cambridge: Polity Press.

Lash, S., & Urry, J. (1994). *Economies of signs and space*. London: Sage.

Lawson, J. (2007). *Pressurized timber: Occupational health and safety prevention initiatives in BC*. Canadian Political Science Association, Ottawa. http://www.cpsa-ascp.ca/papers-2007/lawson.pdf.

Lefebvre, H. (1974). *La Production de L'Éspace*. Paris: Éditions Anthropos.

Lefebvre, H. (1976a). *De L'État: L'Etat dans le Monde Moderne* (Vol. 1). Paris: Union Générale d'Editions.

Lefebvre, H. (1976b). *De L'État: De Hegel à Marx par Staline*, Vol. 2. Paris: Union Générale d'Editions.

Lefebvre, H. (1977). *De L'État: Le Mode de Production Étatique* (Vol. 3). Paris: Union Générale d'Editions.

Lefebvre, H. (1978). *De L'État: Les Contradictionsde L'État Moderne* (Vol. 4). Paris: Union Générale d'Editions.

Lefebvre, H. (1991). *The production of space*. Oxford: Blackwell.

Lista, G. (2001). *Futurism*. Paris: Terrail.

Lorange, P. (2008). *Shipping: Competing in a global market*, www.imd.ch.

Low, M., & Barnett, C. (2000). After globalization. *Environment and Planning D, 18*, 53–61.

Luke, T. W. (1991). The discipline of security studies and the codes of containment: learning from Kuwait. *Alternatives, 16*, 315–344.
Luke, T.W. (1992). New world order or neo-world orders: Power politics and ideology in the informationalizing global order. Theory Culture and Society 10th Anniversary Conference, Champion, Pennsylvania, Aug, 16–19, 1992.
Luxembourg, R. (2003). *The accumulation of capital*. New York: Routledge.
Lyotard, J.-F. (1987). *La condition postmoderne, (Dutch translation),*. Kok Agora: Kampen.
Mann, M. (1986). *The sources of social power* (Vol. I). Cambridge: Cambridge University Press.
Mann, M. (1993). *The sources of social power* (Vol. II). Cambridge: Cambridge University Press.
Mann, M. (1997). Has globalization ended the rise and rise of the nation-state? *Journal of International Political Economy, 4*(3), 472–496.
Marden, P. (1997). Geographies of dissent: globalization, identity and the nation. *Political Geography, 16*(1), 37–64.
Marx, K. (1973). *Grundisse. Foundations of the critique of political economy*. London: Penguin.
Marx, K. (1992). *Das kapital* (Vol. 1). London: Penguin.
Marx, K., & Engels, F. (1967). *The communist manifesto*. London: Penguin.
McGrew, A. (1992). A global society? In S. Hall, D. Held and A McGrew (Eds.), *Modernity and its futures* (pp. 62–102). Cambridge: Open University Press.
Mensch, G. (1979). *Stalemate in technology*. Cambridge: Ballinger.
Mills, S. (2003). *Michel Foucault*. Abingdon: Routledge.
Morgan, G. (1997). *Images of organization*. Thousand Oaks: Sage.
Natter, W., & Jones, J. P, I. I. I. (1997). Identity, space, and other uncertainties. In G. Benko & U. Stroymayer (Eds.), *Space and social theory* (pp. 144–164). Oxford: Blackwell.
Nielsen, L.D., and Oldrup, H.H. (Eds.). (2001). Mobility and transport, Transportradet, Aalborg University, Department of Development and Planning.
Nooteboom, B. (1992). A postmodern philosophy of markets. *International Studies of Management and Organization, 22*(2), 53–76.
Offe, C. (1984). *Contradictions of the welfare state*. Cambridge: MIT Press.
Offe, C. (1985). *Disorganized capitalism*. Cambridge: Polity Press.
Opello, W. C, Jr, & Rosow, S. J. (2004). *The nation-state and global order: A historical introduction to contemporary politics*. Boulder: Lynne Reinner.
Parker, M. (1992). Post-modern organizations or postmodern organization theory? *Organization Studies, 13*(1), 1–17.
Parker, M. (2009). Capitalists in space. *Sociological Review, 57*(1), 83–97.
Parker, M., & Bell, D. (Eds.). (2009a). *Space travel and culture: From Apollo to space tourism*. Malden: Wiley-Blackwell.
Parker, M., & Bell, D. (2009b). Introduction: making space. *Sociological Review, 57*(1), 1–5.
Piore, M. J., & Sabel, C. F. (1984). *The second industrial divide: Possibilities for prosperity*. New York: Basic Books.
Polanyi, K. (1944). *The great transformation*. Boston: Beacon Press.
Poulantzas, N. (1976). *Classes in contemporary capitalism*. London: New Left Books.
Putterman, L., & Rueschemeyer, D. (1992). Synergy or rivalry? In L. Putterman & D. Rueschemeyer (Eds.), *State or market in development: Synergy or rivalry?* (pp. 243–262). Boulder CO: Lynne Rienner Publishers.
Rajchman, J. (1987). Postmodernism in a nominalist frame. *Flash Art, 137*, 49–51.
Reich, R. (1992). *The work of nations*. New York: Vintage.
Relyea, S. (1998). Trans-state entities: Postmodern cracks in the great Westphalian dam. *Geopolitics, 3*(2), 30–61.
Richards, H. (2004). Has postmodernism killed social sciences? *Times Higher Education*, October, 15, 2004.
Robertson, R. (1992). *Globalization*. London: Sage.
Rodney, W. (1981). *How Europe underdeveloped Africa*. Washington: Howard University Press.
Rodrigue, J.-P., Comtois, C., & Slack, B. (1997). Transportation and spatial cycles: Evidence from maritime systems. *Journal of Transport Geography, 5*(2), 87–98.

Rosenau, J. N. (1990). *Turbulence in world politics. A theory of change and continuity.* Princeton: Princeton University Press.

Rosenau, J. N. (1997). The complexities and contradictions of globalization. *Current History, 96,* 360–364.

Rosenau, J. N. (2001). Stability, stasis, and change: A fragmegrating world. In S. J. Flanagan, E. L. Frost, & R. L. Kugler (Eds.), *Challenges of the global century.* Washington: Report of the Project on Globalization and National Security, Institute for National Strategic Studies, National Defense University.

Rosenau, J. N. (2003). Globalization and governance: bleak prospects for sustainability. *Internationale Politik und Gessellschaft, 3,* 11–29.

Rouleau, L., & Clegg, S. R. (1992). Postmodernism and postmodernity in organization analysis. *Journal of Organizational Change, 5*(1), 8–25.

Rueschemeyer, D., & Evans, P. (1985). The state and economic transformation; toward an analysis of the conditions underlying effective intervention. In P. Evans, D. Rueschemeyer, & T. Skopcol (Eds.), *Bringing the state back in* (pp. 69–92). Cambridge: Cambridge University Press.

Ruigrok, W., & van Tulder, R. (1995). *The logic of international restructuring.* New York: Routledge.

Sack, R. D. (1973). A concept of physical space in geography. *Geographical Analysis, 5,* 115–139.

Sack, R. D. (1983). Human territoriality: A theory. *Annals of the Association of American Geographers, 73,* 55–74.

Sack, R. D. (1986). *Human territoriality: Its theory and history.* Cambridge: Cambridge University Press.

Santos, M. (1995). Contemporary acceleration: World-time and world-space. In G. Benko & U. Strohmayer (Eds.), *Geography, history and social sciences* (pp. 171–176). Dordrecht: Kluwer.

Schoenberger, E. (1988). From Fordism to flexible accumulation: Technology, competitive strategies, and international location. *Environment and Planning D, 6,* 245–262.

Schoenberger, E. (2004). The spatial fix revisited. *Antipode, 36*(3), 427–433.

Scholte, J. A. (2000). *Globalization.* London: Palgrave.

Serres, M. (1995). *Genesis.* Ann Arbor: University of Michigan Press.

Shaw, M. (1997). The state of globalization: Towards a theory of state transformation. *Journal of International Political Economy, 4*(3), 497–513.

Skocpol, T. (1979). *States and social revolutions.* Cambridge: Cambridge University Press.

Sletmo, G.K. (2002). *National shipping policy and global markets: A retrospective for the future.* International Association of Maritime Economists Annual Conference (IAME), Panama City, Nov, 13–15, 2002.

Smith, G. E. (1986). Territoriality. In R. J. Johnston, D. Gregory, & D. M. Smith (Eds.), *The dictionary of human geography* (2nd ed., p. 620). Oxford: Blackwell.

Smith, N. (1990). *Uneven development; Nature, capital and the production of space* (2nd ed.). Oxford: Blackwell.

Smith, N. (1992). Geography, difference and the politics of scale. In J. Doherty, E. Graham, & M. Malek (Eds.), *Postmodernism and the social sciences* (pp. 57–79). London: Macmillan.

Soja, E. W. (1989). *Postmodern geographies; The reassertion of space in critical social theory.* New York: Verso.

Soja, E. W. (2000). *Postmetropolis: critical studies of cities and regions.* Oxford: Blackwell.

Stalder, F. (2006). *Manual Castells.* Cambridge: Polity Press.

Storper, M. (1997). *The regional world. Territorial development in a global economy.* New York: The Guilford Press.

Strange, S. (1986). *Casino capitalism.* New York: Blackwell.

Strange, S. (1994). The power gap: member states and the World economy. In F. Brouwer, V. Lintner and M. Newman (Eds.), *Proceedings of a Conference on Economic Policy Making and the European Union, London European Research Centre, London Federal Trust* (pp. 19–26). London: University of North London.

Sunesson, S. (1985). Outside the goal paradigm; power and structural patterns of non-rationality. *Organization Studies, 6*(3), 229–246.
Taylor, P. J. (1994). The state as container: territoriality in the modern world system. *Progress in Human Geography, 18*(2), 151–162.
Taylor, P. J. (2000a). Embedded statism and the social sciences 2: geographies (and metageographies) in globalization. *Environment and Planning A, 32,* 1105–1114.
Taylor, P. J. (2000b). World cities and territorial states under conditions of contemporary globalization. *Political Geography, 19,* 5–32.
Taylor, P. J. (2000c). World cities and territorial states under conditions of contemporary globalization II: looking forward, looking ahead. *GeoJournal, 52,* 157–162.
Thompson, M. (1979). *Rubbish theory.* Oxford: Oxford University Press.
Tisdall, C., & Bozzolla, A. (1977). *Futurism.* London: Thames and Hudson.
Tyler, L. (2005). Towards a postmodern understanding of crisis communication. *Public Relations Review, 31,* 566–571.
Urry, J. (1986). Class, space and disorganised capitalism. In K. Hoggart & E. Kofman (Eds.), *Politics, geography and social stratification* (pp. 16–32). London: Croom Helm.
Valaskakis, K. (1999). Globalization as theatre. *International Social Science Journal, 160,* 153–164.
Virilio, P. (1995). Speed and information: cyberspace alarm! ctheory.net, www.ctheory.net/articles.aspx?id=72.
Virilio, P. (1998). Military space. In J. Der Derian (Ed.), *The virilio reader.* Oxford: Blackwell.
Virilio, P. (1999). Indirect light: extracted from Polar inertia. *Theory, Culture and Society, 16, 1*(5–6), 57–70.
Virilio, P. and Lotringer, S. (1997). Pure war. New York: Semiotext(e).
Walker, R. B. J. (1991). State sovereignty and the articulation of political space/time. *Millennium, 20,* 445–461.
Wallerstein, I. (1974). *The modern world system.* New York: Academic.
Waltz, K. N. (1999). Globalization and governance. *Political Science and Politics, 32*(4), 693–700.
Warf, B. (1990). Can the region survive post-modernism? *Urban Geography, 11*(6), 586–593.
Watts, M. J. (1991). Mapping meaning: denoting difference, imagining density; dialectical images and postmodern geographies. *Geografiska Annaler B, 73,* 7–16.
Webber, M., & Rigby, D. L. (1996). *The golden age illusion: Rethinking postwar capitalism.* New York: The Guilford Press.
Weber, M. (1949). The Methodology of the Social Sciences. In E.A. Shils, H.A. Finch (Ed.), New York: The Free Press.
Weber, M. (1978). *Economy and society; An outline of interpretative sociology,* (2 Vols.). Berkeley, CA: University of California Press.
Weber, M. (2004). Politics as a vocation. In M. Weber (Ed.), *The vocation lectures.* Indianapolis: Hackett Publishing Company.
Weiss, L. (1997). Globalization and the myth of the powerless state. *New Left Review, 225,* 3–27.
Wood, E. M. (1997). Modernity, postmodernity or capitalism? *Review of International Political Economy, 4*(3), 539–560.
Wright-Mills, C. (1956). *The power elite.* New York: Oxford University Press.
Yeung, H. W. (1998). Capital, state and space: contesting the borderless world. *Transactions of the Institute of British Geographers NS, 23,* 291–309.
Zhang, X. (2005). Coping with globalization through a collaborative federate mode of governance. *Policy Studies, 26*(2), 199–209.
Zurn, M. (2000). Democratic governance beyond the nation-state; the EU and other international institutions. *European Journal of International Relations, 6*(2), 183–221.
Zurn, M. (2003). Globalization and global governance; from societal to political denationalization. *European Review, 11*(3), 341–364.

Chapter 6
Modernism

To understand the current problems that surround maritime governance our first step must be to assess the substantive developments that have occurred in the context for the maritime (and wider) world by examining what went before. The aim is not to present an historical analysis of cultural or economic change over many centuries but to take the more immediate developments that we have noticed affecting the maritime sector—focused upon but not exclusively restricted to globalization. We begin with Modernism which provides a framework to begin to understand the maritime policy failures and the structure upon which the existing governance of the sector rests. In looking at Modernism we will deliberately make the discussion wide to encompass issues and activities that are not normally part of the maritime scene—for example arts, architecture, design—as well as business, commerce, finance, politics, the media, and society. In this way we will be able to show that the grand forces that determine the Modernist (and subsequently Postmodernist) phases—stimulated by capitalism and its desire for capital accumulation and disposal—are not confined to the maritime sector but are significant trends in economic, cultural, social, and political life that need to be recognized and accommodated if the policy failures in any sector are to be addressed. Others have taken this approach. Thrift (1997, p. 29) suggested that we can only understand capitalism if we also understand its relationship to wider cultural changes that have 'swept across the social science and humanities'. He saw capitalism as having undergone 'its own cultural turn' founded on the idea that it has to become more 'knowledgeable' in what he termed a 'turbulent and constantly fluctuating world'—something we can clearly recognize in the disorganized capitalism of Arrighi (1994; Lash and Urry 1987; Carroll 2002) discussed in Chap. 5 and what Krasner (1984, p. 242) termed 'punctuated equilibrium' when 'short bursts of rapid institutional change (are) followed by long periods of stasis'.

Central to these changes and the characteristics of disorganization is governance. Inadequate or inappropriate governance structures result in policy failure. The underlying capitalist forces that determine a Modernist, Postmodernist, or other subsequent genre in turn need a governance framework that has to be adapted to their new and changing context. But first to Modernism as the predominant social force of recent times and which remains the major underlying determinant of the governance framework for the maritime sector today.

6.1 Modernism: Its Character and Characteristics

> There followed on the birth of mechanization and modern industry... a violent encroachment like that of an avalanche in its intensity and its extent. All bounds of morals and nature, of age and sex, of day and night, were broken down. Capital celebrated its orgies. Marx (1992, p. 21)

Implicitly Marx was commenting on Modernism and its emergence accompanied by industrialization and capitalism. Denzin (1986, p. 196) suggested that the industrial age to which Marx belonged, and which continued well into the twentieth century, was one in which:

> men were induced to believe that their labor (use value) defines their worth (exchange value). The assumption that use value (concrete value behind exchange value) structures economic production emerges in the early capitalist period. This moment is organized in terms of an ideology which links man as a producer to man as a moral being. The myth or mirror of production thus underlies Marxism's analysis of capitalism (Baudrillard 1975, p. 31) for it convinces men that that are alienated because they sell their labor power.

Our discussion will focus upon Modernism, its relationship with capitalism, and its decline and replacement by Postmodernism. There is no shortage of commentary on Modernism and we shall consider only a small part of what is available. An indication of the general issues that are felt to be significant can be found in Giddens (1981; Anderson 1984; Curry 1991, pp. 211–213), (Boje and Dennehy 1993, pp. 5–10) and Burrell (1994). Meanwhile (Habermas 1981, p. 3) takes us back to the very origin of the word 'modern', to the late fifth century when it was used for the first time to distinguish Christian modernity from the pagan past.

More specifically, Warf (1993, p. 162) provided a clear and relatively concise summary of what Modernism means—not an easy matter for an era so diffuse and complex in nature. He saw Modernism searching for a universal truth which was independent of time, place, or social circumstances. Universal laws 'unveil a single, underlying truth that exists external to human beings'—objective reality, not socially constructed conveniences. Modernism was 'objective rationality', science, organization, control, and discipline.

Boje et al. (1996, p. 24) stressed the close relationship between the emergence of the Modernist era and the growth of capitalism. Capitalism has always emphasized hard work as a moral ethic and an obligation (Bendix 1960, pp. 52–53), coupled with rational economic pursuits, seeking Weber's (1958, p. 64) 'profit rationality'. The capitalist state which developed within a Modernist framework is characterized by a strict legal order, voluntary labor, planned division within labor, a bureaucratic government, and a market economy incorporating planned elements. Problems can always be solved by state intervention, science

6.1 Modernism: Its Character and Characteristics

and social theory (Peukert 1989, p. 134). Boje et al. (1996, p. 29) laid out the rules of a Modernist state:

- Emphasis on (upward) hierarchical obedience, thereby disrupting the two-way flow of authority and responsibility.
- Technical rationality, concerned with optimizing efficiency as a function of inputs and outputs.
- Stewardship with regard to husbanding organizational resources for the interests of hierarchical elites.
- Pragmatism in decision making. A devotion to expediency guided by organizational imperative.

Parker (1992, p. 3) described Modernism as elevating faith in reason to a level where it was equated with progress. The two became almost synonymous during the period of 'high Modernism' (arguably the 1920–1970s). The world was a system increasingly under man's control and expanding knowledge. Positivism, empiricism, and science were in full control. Nature was 'there' and could be entirely understood (and by implication, controlled). Modernism was unchallengeably rational and had the ability to communicate everything to other rational beings.

Nooteboom (1992, p. 58) outlined how Parmenides considered that 'all becoming, change and plurality are illusory'. Behind this view lies Plato's 'eternal being which can be grasped by thought'. Alternatively both Nooteboom and Burke (2000) cited the Heraclitus who maintained that society always reflected constant change suggesting that 'you never step in the same river twice' and all is in flux, and the conflict of opposites is essential for all of us. The migration from Modernism to Postmodernism can be seen as a move from a Parmenidic to a Heraclitan view. This dialectical debate has a considerable history particularly associated with the writings of Hegel (see for example Soll 1969; Norman 1976) and its strained relationship with political theory, and especially through the debate around dialectical materialism characterizing Communism in the Former Soviet Union (see for example Cornforth 1952; Lefebvre 1968, 1971; Goldman 1969; Birnbaum 1969, 1971; Habermas 1970, 1971, 1973; Lukacs 1971; Markovic 1974). McLennan (1995, p. 59) suggested it has close relationships with pluralism citing the importance of Popper's (1963) 'notion of science as moving forward through a dialectic of conjecture and refutation'. A particularly significant contribution came from Benson (1977) as noted in Evans (2001, p. 544) who assessed the relationship between dialectics and organizations and in so doing hinted at the contribution that the study of dialectics could make to understanding the rise of the Postmodern. The dialectic view considered a 'general perspective on social life' (Benson 1977, p. 2), seen by Marsh (1995, p. 14) as derived from Marx's analysis of the capitalist economy with society 'committed to the concept of process and… in a continuous state of becoming' (Benson 1977, p. 3). Classically Postmodern, any social arrangements that exude permanence are actually only temporary and represent only one of the many possibilities. Dialectical analysis looks for 'fundamental principles which account for the emergence and dissolution of specific social orders' (Benson 1977, p. 3).

Four main principles can be identified each of which are clear Postmodern characteristics. These will be considered in more detail in the next chapter:

- Social construction—the social world is constructed by people through their interactions constrained by existing social structures and contexts but open to variation, flexibility, and manipulation.
- Totality—society must be studied in its totality. This holistic approach assumes that everything is interconnected and nothing exists in isolation or abstraction.
- Contradiction—'the social order produced in the process of social construction contains contradictions, ruptures, inconsistencies and incompatibilities' (Benson 1977, p. 4) These contradictions produce radical breaks—the emergence of the Postmodern era is one example and may be constructive or destructive. Godelier (1972) suggests that each economic system will pass as an inevitable consequence—for example Premodernism, Modernism, Postmodernism.
- Praxis—dialectism assumes that people can reconstruct such social arrangements on the basis of reason within the constraints that exist. It therefore undermines existing constructs, de-reifying established patterns and arrangements.

There have been many other attempts at defining dialectal analysis (see for example Evans 2001, p. 543; Marsh and Smith 2000, p. 5; Morgan 1997, pp. 287–288; Brenner 1999, p. 68; Lourenco and Glidewell 1975, p. 1) but each returns to the main principles outlined by Benson which reveal clear Postmodern tendencies.

The relationship between dialectics and many of the issues we have considered earlier in the context of governance and policy failure and with direct relevance to the maritime sector has also been a central feature of the debate. These include the antagonism between state and society (Borzel 1998, p. 260), the significance of the end of the nation-state as we know it (King and Kendall 2004, p. 161), the role of struggle and argument in policy making in the EU (Neyer 2003, p. 691), and the difficult relationship between the state and pluralism (Atkinson and Coleman 1992, p. 155). To quote the latter:

> For modern states, the problem has become one of maintaining ultimate control yet sharing the exercise of public authority. Public officials want to escape blame, but also to claim credit. They want to husband political power, but they must mobilize social forces to obtain it in the first place. As a result the range of organized groups and state forms has expanded considerably.

All very Postmodern and yet seemingly unrecognized some 18 years later by the maritime sector. In a similar vein dialecticism has been seen to relate closely to changes in the role of the nation-state. Harvey (1990, p. 109) emphasized the struggle that continued between the 'fixity (and hence stability) that state regulation imposes and the fluid notion of capital flow', presenting a serious problem for capitalism that the new Postmodern era characterized by globalization has exacerbated. Morgan (1997, p. 288) agreed and Rosenau (2000, pp. 2–3) also

6.1 Modernism: Its Character and Characteristics

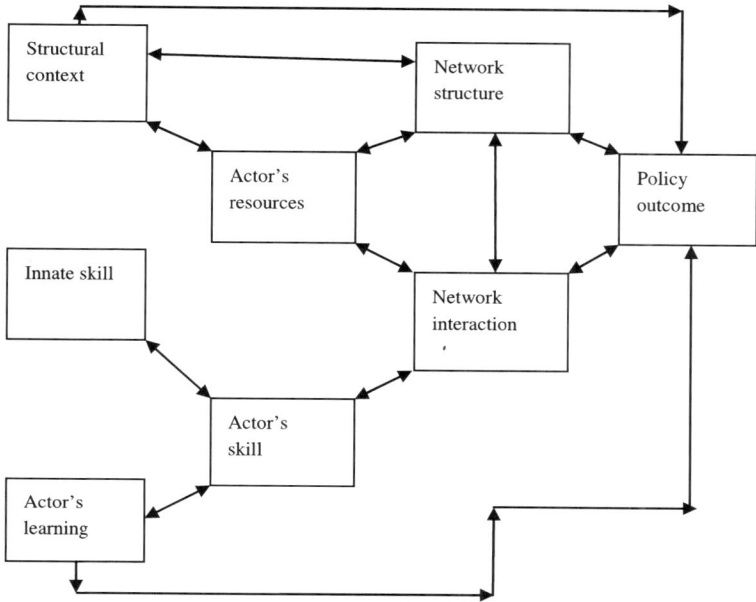

Fig. 6.1 Policy networks and policy outcomes: a dialectical approach. *Source* Derived from Marsh and Smith (2000, p. 10)

supported this view commenting on the 'emergent epoch', the 'tensions involving three central polarities'—globalization and localization, centralization and decentralization, and integration and fragmentation. He described them as opposing poles, conflictual, and, ongoing. Table 6.1 gives an indication of where he saw conflict manifesting itself as a result of the Postmodern, fragmegrating (his term) scenario. Brenner (1999, p. 50) stressed the conflicting process inherent in the new relationships between jurisdictions (national, supranational, subnational) with few if any pregiven or fixed/natural conditions. Traditional assumptions of state-centricism neglect the clear continuous transformations that actually occur and that are so fundamental to the success or otherwise of policy making and application (Table 6.1).

The challenge was to:

> conceive configurations of geographical scales at once as the territorial scaffolding within which the dialectic of de- and re-territorialization unfolds and as the historically produced, incessantly changing medium of that dialectic. (Brenner 1999, p. 69)

Or to cite Rudd (2001, p. 5) quoting Mittelman (1999, p. 34) perhaps the most significant aspect of Postmodernism for the condition of governance, the erosion of the nation-state, has occurred as a consequence of 'a dialectic of subnationalism and supra-nationalism'.

Table 6.1 Sources of Fragmegration at four levels of aggregation

	Micro	Macro	Macro–Macro	Micro–Macro
Skill revolution	Expands peoples' horizons on a global scale; sensitizes them to the relevance of distant events; facilitates a reversion to local concerns	Enlarges the capacity of government agencies to think 'out of the box', seize opportunities, and make challenges	Multiplies quantity and enhances quality of links among states; solidifies their alliances and enmities	Constrains policy making through increased capacity of individuals to know when and how to engage in collective action
Authority crises	Redirect loyalties; encourage individuals to replace traditional criteria of legitimacy with performance criteria	Weaken ability of both governments and other organizations to frame and implement policies	Enlarge the competence of some IGOs and NGOs; encourage diplomatic wariness in negotiations	Facilitate the capacity of publics to press and/or paralyze their governments, the WTO, and other organizations
Bifurcation of global structures	Adds to role conflicts, divides loyalties, and foments tensions among individuals; orients people toward local spheres of authority	Facilitates formation of new spheres of authority and consolidation of existing spheres in the multi-centric world	Generates institutional arrangements for cooperation on major global issues such as trade, human rights, the environment, etc	Empowers transnational advocacy groups and special interests to pursue influence through diverse channels
Organizational explosion	Facilitates multiple identities, subgroupism, and affiliation with transnational networks	Increase capacity of opposition groups to form and press for altered policies; divides publics from their elites	Renders the global stage ever more transnational and dense with non-governmental actors	Contributes to the pluralism and dispersion of authority; heightens the probability of authority crises

(continued)

6.1 Modernism: Its Character and Characteristics

Table 6.1 (continued)

	Micro	Macro	Macro–Macro	Micro–Macro
Mobility upheaval	Stimulates imaginations and provides more contacts with foreign cultures; heightens salience of the outsider	Enlarges the size and relevance of subcultures, diasporas, and ethnic conflicts as people seek new opportunities abroad	Heightens the need for international cooperation to control the flow of drugs, money, migrants, and terrorists	Increases movement across borders that lessen capacity of governments to control national boundaries
Microelectronic technologies	Enable like-minded people to be in touch with each other anywhere in the world	Empower governments to mobilize support; renders their secrets vulnerable to spying	Accelerate diplomatic processes; facilitates electronic surveillance and intelligence work	Constrain governments by enabling opposition groups to mobilize more effectively
Weakening of territoriality, states, and sovereignty	Undermines national loyalties and increases distrust of governments and other institutions	Adds to the porosity of national boundaries and the difficulty of framing national policies	Increases need for interstate cooperation on global issues; lessens control over cascading events	Lessens confidence in governments; renders nationwide consensus difficult to achieve and maintain
Globalization of national economies	Swells ranks of consumers; promotes uniform tastes; heightens concerns for jobs	Complicate tasks of state governments vis-à-vis markets; promotes business alliances	Intensifies trade and investment conflicts; generates incentives for building global financial institutions	Increases efforts to protect local cultures and industries; facilitates vigor of protest movements

Source Rosenau (2000, p. 3)

Zurn (2000, p. 184) also stressed the struggle between the system and the individual (a key Postmodern feature and reflected in moves toward effective stakeholder involvement in policy making that maritime governance has yet to address). Globalization and dialectics also feature strongly. Harvey (1990, pp. 278–279) for example discussed in some depth the difficult relationship between traditional nation-states, internationalism, war, and governance. Rosenau (1997, pp. 361–364) viewed the world as characterized by tensions between the global and the local while Jessop (1998, p. 33) saw the:

world economy... reshaped by a complex dialectic of globalization-regionalization. This has allegedly made it more difficult for (national) states to control economic activities within their borders – let alone global capitalist dynamics.

Meanwhile Harvey (2000, p. 194) returned to the issue of dialectical struggle emphasizing how capitalism on the one hand demands free trade and a reduction in nation-state influence while on the other does what it can to retain the nation-state to protect its interests. The shipping industry and the role of national flags is perhaps the very best example of this Postmodern manifestation of dialectical abuse. Ruggie (2003, p. 301), along much earlier with Puchala (1972, p. 279), concurred in stressing the need to control and ameliorate the continuous arguments that characterize the national–international debate.

In addition, there has been considerable attention given to the relationship between governance networks and dialectical considerations. Davies (2005, pp. 312, 314, 326, 331) emphasized how the relationships between hierarchies, markets, and networks are essentially dialectic especially in the context of local governance. Markets are always in conflict in the struggle for survival and one result is state intervention. Stoker (2004, p. 24) stressed that network governance—suggested as increasingly significant and replacing that of the traditional hierarchy—is susceptible to conflict which can only be temporarily resolved by state intervention and which will always reemerge as an inevitability of the fragmentation that characterizes modern governance.

Marsh and Smith (2000, p. 5) identified three dialectical relationships which must exist with the increased adoption of policy networks—structure and agency; network and context; and network and outcome—although their definition of a dialectical relationship centered upon interaction between components and less on the traditional 'struggle'. Their definition has generated much criticism (see for example Evans 2001) as it is a watered-down version of others that in themselves are commonly unclear. Plato defined it as the 'art of defining ideas' (Allen 2006; Fichte 1998) as 'thesis, synthesis and antithesis', or Marx's implicit definition (he never used the word dialectic) 'progressive unification through the contradiction of opposites'. Figure 6.1 gives an indication of a dialectical interpretation to network-based policy making.

However, we need to return to Modernism before we can consider the Postmodern in any detail. Societal views in more recent times have been characterized by a belief in immutable and unchanging laws, seen as a progressive process. Humankind, as a result of rational and scientific thinking was overcoming and understanding the challenges that surrounded it. This process was termed Modernism. Cooper and Burrell (1988, p. 92) provided a definition:

> with its belief in the essential capacity of humanity to perfect itself through the power of rational thought.

Thrift (1997, p. 31) provided an excellent summary of what he saw as Modernism, although he never quite uses the term. Quoting him unashamedly in full:

6.1 Modernism: Its Character and Characteristics

One of the prevalent discourses in western intellectual cultures of the last two thousand years – one which has waxed and waned, which has adjusted to historical custom whilst still holding to a series of central tenets – has been what Jowitt (1992) calls the 'Joshua discourse'. It is founded on the idea of transcendental rationality, on the notion of a single, correct, God's-eye view of reason that transcends the way human beings... think, and which imparts the idea of a world that is 'centrally organized, rigidly bounded, and hysterically concerned with impenetrable boundaries'. (Jowitt 1992). This discourse usually involves a series of linked and self-supporting tenets (Lakoff 1987): that the mind is independent of the body, reason being a dismembered phenomenon; that emotion has no conceptual content but it is a pure force; that meaning is based on truth and reference and concerns the relationship between symbols which represent things in the real world; and that categories are independent of the world, defined by the internal characteristics of their members rather than by the nature of those doing the categorizing.

It is contended here that this Modernist interpretation of the world underlies the structured, hierarchical, and deterministic framework that created the maritime institutions that generate and implement policies. Centrally organized; rigidly bounded; deterministic; based on a single truth and reference; unaffected by those forming the policy but acting as commanded by some higher (and unquestionably 'right') power; and characterized by institutions with impenetrable boundaries. A 'single, God's eye reason' behind maritime policy making which has become an anachronism as capitalism's inevitable progression from Modernism to the next phase of capital disposal and accumulation takes place.

Modernism is an historical period in Western culture but the timing of its emergence is subject to some debate. Benko (1997, p. 3) citing Habermas (1981, pp. 3–4, 1983) traced in some detail the development of the 'modern' from the time of Charlemagne in the twelfth century but suggested that it really took off from the discovery of the New World from the fifteenth century and the emergence of the Enlightenment (Gergen and Thatchenkery 2004, p. 229). Onuf (1991, p. 425) in line with Wood (1997, p. 541) felt that the main elements of Modernity only became clear between 1600 and 1800, offering an interpretation of the world which was human centered and full of meaning and experience. Emphasis was placed upon 'individuality, reason and mastery over circumstance' and progress was a central issue measured in terms of tasks and prosperity. 'Liberalism is modernity's core ideology, capitalism its paymaster and the state its highest social realization, primary agent and paramount problem'. Onuf, along with Giddens (1985) saw the sovereign state as always modern. Williams (1989) provided an extensive discussion of what he termed the 'modern quest for the foundation of truth' which mirrors the period identified by Onuf and which places the debate within the realms of science, religion, knowledge, and God. Meanwhile, although Thrift (1997) implied a much earlier origin, Burke suggested it emerged with the Enlightenment in the eighteenth century, a period characterized by three major features:

- The power of reason over ignorance.
- The power of order over disorder.
- The power of science over superstition.

Modernity was seen as revolutionary—characterized by the French Revolution of 1789 and much later the Russian Revolutions of 1905 and 1917—and it heralded among other things, the advent of capitalism. This represented a new form of production and societal order, something that Denzer (undated) saw as emerging particularly in the early twentieth century in the fields of architecture and industrial design. Virginia Woolf chose 1910 as the year in which a cultural revolution occurred coalescing as the Modernist movement. Robert Adams (1978), quoted in Reid (1992, p. 262) agreed:

> Within five years either way of that date a great sequence of new and different works appeared in Western culture, striking the tonic chords of Modernism. Ten years before that fulcrum of December 1910, Modernism is not; ten years after, it is already.

This mirrored the rise of capitalism, industrial society, and the machine, responding to 'sweeping changes in technology and society'. Capitalism provided the new basis from which progress and all aspects of culture and society could develop. The emphasis was forward looking with the application of reason taking center stage. Humanity could then be emancipated from ignorance, poverty, insecurity, and violence (Leonard 1997).

Harvey (1990) provided an alternative view on Modernism and its rise in the early twentieth century. Citing Sorel (1908), he identified a Modernist corporatist movement that became a powerful organizing tool for the fascist right in the 1930s appealing to myths of a:

> hierarchically ordered but nevertheless participatory and exclusive community, with clear identity and close social bonding, replete with its own myths of origin and omnipotence. (Harvey 1990, p. 34)

Harvey went on to view Modernism of the 1930s 'heroic' but 'fraught with disaster', and then saw the universal or high Modernism of post 1945 exhibiting a comfortable relation to the dominant power centers of the society. This was a result of the stabilization of the international power system (in our case focusing upon the IMO, OECD, WTO, EU etc.), on Fordist lines and under the watchful eye of the USA.

> High Modernist art, architecture, literature etc. became establishment arts and practices in a society where a corporate capitalist version of the Enlightenment project for development for progress and human emancipation held sway as a political-economic dominant.

Cooper and Burrell (1988) agreed with Burke that Modernism's historical roots lay in the Enlightenment when man no longer saw himself as a 'reflection of God or Nature'. Its origins lay in Reason, which Kant described as when man thinks for himself and ceases 'depending on an external authority to make up our minds for us'. Kant's comment was '*aude sapere*'—'dare to know'. In more modern times this has been translated into 'instrumental rationality' which is typified by the organization of knowledge expressed in terms of the needs of large-scale technological systems.

Williams (1989, pp. 48–49) traced Modernism from its roots in the use of the word 'modern' from the late sixteenth century but in truth really emerging between 1890 and 1940, although he was dismissive of the whole idea of defining eras as it consisted of 'identifying the machinery of selective tradition'. Adams (1978, pp. 19–21) suggested that Modernism's emergence is reflected in a series of art works that began with Picasso's *Demoiselles d'Avignon* in 1906–1907. This was followed by Ezra Pound's book *Personae* in 1909, Stravinsky's *Sacre de Printemps* in 1913, and Joyce was completing *Ulysses* in 1914. During the same period post-impressionist art was rising in the USA, D.H. Lawrence emerged as a serious poet and novelist and Marinetti was criss-crossing Europe lecturing on his esthetic dream of *Futurism*. Adams suggested that the list could be extended almost indefinitely at this time—to include Bartok, Braque, Modigliani, Epstein, Kafka, Klee, and Kandinsky—all of which suggested that something was happening. Tying many of this diverse list of artists together was their rejection of the past, typified by the Futurist movement which championed dynamism, efficiency, and mechanism and repudiated historicism, humanitarianism, and softness. Violence, organization, and stripped energy were key characteristics whether one was dealing with poetry, painting, architecture, music, or even beyond in business, management, industrialization, and society more generally.

Adams was not happy with defining Modernism except through its products and concluded (1978, p. 33):

> When it is understood to refer to distinct structural features that some artistic works of this period (1905–1925) have in common, it has real meaning... As it departs from that specific meaning, it gets fuzzier and fuzzier, and so sometimes doesn't mean much of anything at all.

Soja (1989, pp. 24–31, 2000) wrote extensively on Modernism tracing its origins from the enlightenment but seeing its development in the nineteenth century and the contribution of Marx (1992) and Comte (2005) as especially significant:

> The comprehensive deconstruction and reconstruction of modernity arising from its spatio-temporal convergence with urban-industrial capitalism and the rising power of the nationalist state changed the critical discourse in nearly every field of knowledge and action. In particular it radically redefined and refocused political theory, philosophy and socio-spatial praxis.

He cited Berman (1982, p. 16) in describing a series of 'material forces' that contributed to the construction of Modernism:

- The industrialization of production, which transforms scientific knowledge into technology, creates new environments and destroys old ones, speeds up the whole tempo of life, generates new forms of corporate power, and class struggle.
- Immense demographic upheavals, severing millions of people from their ancestral habitats, hurling them halfway across the world into new lives.
- Rapid and often cataclysmic urban growth.
- Systems of mass communications, dynamic in their development, enveloping and binding together the most diverse people and societies.
- Increasingly powerful national states, bureaucratically structured and operated, constantly striving to expand their powers.

- Mass social movements of people and peoples, challenging their political and economic rulers, striving to gain control over their lives.
- Finally, bearing and driving all these people and institutions along, an ever-expanding, drastically fluctuating capitalist world market.

Cooper (2000, p. 17), like Soja (1989) and Berman (1982) emphasized the central role of the nation-state to Modernism describing it as the great engine behind the movement, and stressing the importance placed on state sovereignty and the separation of domestic and foreign affairs with consequential Westphalian implications.

Meanwhile, Luke (1991, p. 318) suggested that Modernism was characterized by an overemphasis upon location, hierarchy and organization, something taken up by Thompson (1993, p. 325) who applied the term to the process of systemizing knowledge and the rational organization of social life, terms highly reminiscent of the governance structure that typifies the maritime sector in the early twenty-first century. Thompson (1993, pp. 328–330) went on to describe Modernism as the pursuit of the truth in defiance of the forces of illusion and dogmatic belief (Kuhn 1957). He saw Modernism as summarized in science, truth, and God. Cooper and Burrell (1988) saw Modernism reflecting the belief that 'the essential capacity of humanity (is) to perfect itself through the power of rational thought'. Modernism centers upon faith in an intrinsically logical and meaningful world based on reason. Cooper and Burrell saw this as taking two forms:

- Discourse mirrors reason and order already in the world.
- There is a thinking agent which can make itself conscious of this external order.

Depending on your standpoint, Cooper and Burrell saw the rational subject as either the system itself which works according to the cybernetic discourse of 'control and communication in the animal and the machine' (Wiener 1948) with laws that can be discovered through science and mathematics; or the subject is a network of interacting human individuals reaching a 'universal consensus of human experience'. This assumed unity provides authority for the critical position thus sustaining the establishment or status quo.

Dear (1986) saw Modernism as grouping together a range of visions and values which suggested that as Berman (1982, pp. 345–346) described it, shattered Modernism into 'a multitude of fragments with incommensurable private languages'. Berman went on to define Modernism in a much wider context:

> For many thinkers the whole point of Modernism is to clear the decks of all these entanglements so that the self and the world can be created anew… To be modern… is to experience personal and social life as a maelstrom, to find one's world and oneself in perpetual disintegration and renewal, trouble and anguish, ambiguity and contradiction; to be part of a universe in which all that is solid melts into air. To be a Modernist is to make oneself somehow at home in the maelstrom, to make its rhythms one's own, to move within its currents in search of the forms of reality, of beauty, of freedom, of justice, that its fervid and perilous flow allows.

6.1 Modernism: Its Character and Characteristics

Thrift (1997, p. 37) stressed that the immediate post Second World War period was characterized by developments that we can see as wholly Modernist even though he never used the term. In particular he associated the emergence of a new 'managerial discourse' with the 1940s and 1950s which was clearly Modernist in character and which formed the foundation of maritime policy making governance to this day. He saw this as a period where:

> striated spaces abounded: the buttoned-down personality of the company man (Whyte 1957; Sampson 1995); the enclosed, hierarchical world of the multidivisional corporation (Chandler 1977) with its monolithic goal of achieving ever-greater size and scale by means of a single corporate strategy realized through a relatively static and formal bureaucratic inner core which passed information slowly upwards from an 'external' environment and control slowly downwards from a closed-off headquarters. Then there were the rigidities that resulted from rules of nation-states, including fixed exchange rates and high tariff barriers. And finally, orchestrating the whole was the idea of management 'science' which would be able to produce the cognitive wherewithal to predict and thereby control the world. At least in the rhetoric of time, then, the world was an organized place, made of carefully closed-off spaces which could be rationally appropriated and controlled.

Despite not directly referring to Modernism, Thrift's comments are almost ludicrously appropriate. Buttoned-down company man, hierarchical, multi-divisional corporations, monolithicism, single corporate strategies, static and formal bureaucratic inner cores, with information passing slowly upwards and control downwards, all sounds horribly familiar in the context of maritime policy making and governance in the late twentieth and early twenty-first century.

Modernism was thus a social movement that affected everything. Castells (1989) outlined how Modernism followed the process of capital reconstruction which we have observed repeats itself with monotonous regularity in an attempt to dispose of and accumulate more capital. Castells saw the Modernist phase of capital reorganization emerging rather later than Adams (1978) and Denzer (undated) whom we noted earlier but did suggest that it had a rather more direct relationship to capitalism's repeated needs. He identified the Great Depression of the 1930s and the consequential World War as fundamental to the institutional, political, cultural, and social framework that emerged and which formed the basis for the system of governance to which the maritime sector remains welded.

A Modernist influence is reflected in almost all aspects of today's society and culture. Ballard (2006) saw Modernism as attempting to build a better world as it emerged from the ruins of the First World War utilizing all the benefits of science and technology and all the organizational and institutional regulation of a defined and controlled society. He noted Bertolt Brecht (Thomson and Sachs 1994) who remarked that the:

> mud, blood and carnage of the first world war trenches left its survivors longing for a future that resembled a white-tiled bathroom.

Ballard also suggested that architecture was in the 'vanguard of the new movement' with Le Corbusier and the Bauhaus Design School as major forces and 'function defin(ing) form, expressed in a pure geometry that the eye could easily grasp in its entirety' (Le Corbusier 1978). Of particular relevance to our discussion is the contribution of Cooper and Burrell (1988, p. 92) who noted the significant implications of Modernism for how we understand the role of and nature of organizations in the modern world. They saw it as a:

> shift away from the prevailing definition of organization as a circumscribed administrative-economic function ('the organization') to its formative role in the production of systems of rationality.

Thus, the whole social and cultural ethos of Modernism was seen as central to business, commerce, and management as much as art and architecture. We shall go on to see how this has moved on to a particularly Postmodern position but of significance at this stage is the holistic relevance of such grand concepts for understanding the processes of societal governance and in our case, more specifically maritime policy making (Eagleton 1985).

Steingard and Fitzgibbons (1993, p. 28) took Morgan's (1997) 'machine metaphor' to describe Total Quality Management (TQM) in organizations where progress and outcomes are commonly achieved at the expense of individuality. The assumption was that there was no 'concrete relationship between societies' and that societal interaction was definitely off the agenda. Organizations worked to achieve well-defined and controllable ends which were independent of social relationships except where objectives and ambitions either coincided or conflicted (Robertson and Lechner 1985, p. 104). Fixation and inflexibility ruled the day; Schoenberger's (1988, p. 251) Fordist stability was a central theme of the Modernist vision. Mayntz (1976, p. 119) emphasized the problems that are faced by accusing current Modernist models of organization as being 'normatively rational'—goals are supposedly set by the organization (itself) and this step is followed by a search for the best solution from competing alternatives.

> Thus action appears to be touched off by pre-conceived goals or purposes.

Mayntz (1976, p. 119), supported by Hirschman and Lindblom (1962), continued suggesting that in practice this order is reversed and that:

> organizational activity in general and policy-making in particular is primarily triggered by situational factors which constitute a pressure to act, rather than being generated by deliberations on how certain abstract values can be achieved.

In either case, the central role of the notably substantial and relatively inflexible Modernist institutional governance framework is apparent.

> (Modernism) is clearly a return to the grand concerns that Weber introduced into the study of modern social systems, in which bureaucratic organization had created the 'iron cage' of the modern economic order and whose other significant effect had been to purge the world of the auratic and magical… Weber made us see modern organization as a process which emblemized the rationalization and objectification of social life. (Mayntz 1976, p. 119)

Clegg (1990, pp. 177–178, 1996, pp. 248–249) identified what he saw as the Modernist organization, comparing it with Weber's (Weber 1978) 'bureaucratized,

6.1 Modernism: Its Character and Characteristics

mechanistic structures of control'. Weber's infamous (but very appropriate) 'iron cage' of bureaucracy was composed of 15 Modernist characteristics. These included: specialization; hierarchicization; stratification; standardization; centralization; organization; mechanization; rationalization. These characteristics typified the labor workforce and their work environment (Taylor 1911). More 'creative, mental work' was to be undertaken by the new management stratum (Clegg and Dunkerley 1980). Clegg saw this Modernist mix of bureaucracy and Fordism emerging from the earliest years of the twentieth century as a process of mass production and reflected upon in Gramsci's Prison Notebooks (1971). Weber's characteristics noted above were supplemented by high employment, high wages, centralized industrial regions, and increased urbanization. Meanwhile, for Anderson (1984, p. 97), Modernism was generated by a host of social processes—scientific discoveries, urban expansion, national states, mass movements—propelled by the 'ever-expanding, drastically fluctuating capitalist world markets' (Berman 1982, p. 16). This latter issue of fluctuation reflected the earlier discussion of disorganized capitalism.

> This atmosphere of agitation and turbulence, psychic dizziness and drunkenness, expansion of experiential possibilities and destruction of moral boundaries and personal bonds, self-enlargement and self-derangement, phantoms in the street and in the soul—is the atmosphere in which modern sensibility is born. (Berman 1982, p. 18)

Aglietta (1982, p. 6) further emphasized the relationship between Modernism and disorganization suggesting that recurrent disequilibrium was a characteristic of the world economy and stemmed from national modes of regulation. The response was hegemony which created a forced stability accompanied by a hierarchical structure which dampened change in the complex, open system of the world economy. Both hegemony and hierarchy were seen as fundamental characteristics of Modernism.

Modernism was thus significant to society in general and the maritime sector was no exception possessing an institutional and policy making structure with its hierarchical and state-centric characteristics that reflect a Modernist take on the world. Rigid, inflexible, controlled, predictable, narrowly defined, formal in its approach and institutionalized so that those forces with the power to direct and influence retain them indefinitely regardless of the implications for policies. The maritime sector also presents examples of the disorganization of capitalism which characterizes the eventual moves from Modernism and toward the Postmodern. Chlomoudis et al. (2000, pp. 2–3) commented that the port industry exhibits many examples of restructuring which have taken place over a number of decades in the second half of the twentieth century and we shall return later to the clear moves from a 'post-war industrial paradigm of mass-production' with standardized port products to one typified by substantial organizational, technical, and political changes including adaptations to (JIT) manufacturing, logistics, and supply chain management practices and increased multimodalism. Van de Loo and Van de Velde (2003, pp. 6–8, 14) also showed how the port and shipping industries have experienced massive changes in recent years akin to Lash and Urry's disorganized capitalism.

Modernism is a powerful discourse that continues to pervade our way of thinking and the institutions that direct the maritime sector whether it be the way policies are considered, developed, adopted, and applied, the design of the institutions responsible for this, the stakeholders who are permitted to have possession of these processes or the hierarchical, state-centric formulation that remains central to an increasingly non-hierarchical and state-impotent maritime industry. Within this discourse, fables are spun which are constructed to legitimize 'constructs, subject positions and affective states' over other and competing institutions, individuals, and forces. These myths of Modernism have been all pervasive in the past century and have become almost unquestionable (Thrift 1997).

- The myth of total knowledge—that it is possible to know everything although we are yet to reach this position.
- The world is ordered and homogeneous, defined by oneness, consistency, and integrity. This provides ideal territory upon which 'purified theoretical orders could operate and permeate'.
- There is a material world which can be separated from the world of imagination, symbols, and semiotics.
- Knowledge comes from a 'God-like gaze' which emanates from an individual focal point. Each human therefore has an innate endowment 'received at the point of conception; with no grasp of the individual and being a modulated effect' (Thrift 1991). There is no conception of human capacities arising out of 'emergent properties of the total developmental system constituted by virtue of an individual's situation, from the start, within a wider field of relations—including most importantly, relations with other persons' (Ingold 1995).

Thrift (1997, pp. 33–34) combined these four myths into what he described as 'how we are now'—the myth of the modern which he characterizes as combining speed and transience. Enormous organizations:

> are involved in a whirl of constant information-gathering which feeds into systems of control which produce an iron cage of surveillance and discipline. These organizations are supported by myths of instrumental rationality which allow the world to be trussed-up like a Christmas turkey, with nothing out of place. Furthermore... these organizations are able to drain sociality out of the world, leaving behind nothing but a systematized shell. Thus we arrive at a world populated by anomic and hard-bitten individuals who have had to develop all kinds of survival skills. And, of course, there is a price to pay for the hubris. Not so slowly, but certainly surely, modernity builds towards a climax, usually involving a runaway apocalypse based upon either technology, or the arms race, or mass communications (Norris 1995) in which, in one way or another, human subjectivity is annihilated.

Thrift's vision had a number of associations with our argument concerning capitalism, spatial fixation, and the problems of maritime governance. Modernism's predictable and inevitable race toward successive apocalypses reflects capitalism's reinvention of itself in its desire for both acquisition and disposal of capital through spatial fixes more often than not utilizing technical investment and war. The existing framework for maritime governance can be seen as reflecting a systematized shell driven by instrumental rationality, creating a trussed-up framework for maritime policy making. Subjectivity is annihilated, and individual policy makers claim an innate endowment which makes no allowance for modulation according to the developmental system that surrounds them. Modernist governance and the institutions that characterize it fail to reflect the globalized world in which the industry operates with its new characteristics of flexibility, instantaneity, individualism, recognition of variation, and operation in a context of interrelationship and influence. Thus, the Modernist institutional framework for maritime governance is yet to readjust to capitalism's demands for capital accumulation and disposal exercised through its latest (globalized) spatial fix. While the existing Modernist institutions continue to attempt to generate effective maritime policies (and fail), the shipping industry searches for new methods to accumulate and dispose of excess capital in a Postmodern scenario. The friction that exists between these Modernist and Postmodernist regimes results in policy failure.

The increasing irrelevance of Modernism to today's maritime governance is reflected through the failures we have identified in earlier sections and more explicitly through the work of a number of other commentators. For example, Ballard (2006) suggested that Modernism's heroic architectural period from 1920 to 1939 was now dead, and although his choice of dates might be questionable his examples with maritime relevance in particular including the blockhouse architecture of Utah Beach in Northern France certainly have credence. He commented that:

> Modernism was a vast utopian project, and perhaps the last utopian project we shall ever see, now that we are well aware that all utopias have their dark side.

Using examples from other Modernist architectural development and placing it in its wider social context (of significant relevance to the maritime sector reflecting as the latter does society's needs and desires) he suggested:

> Modernism saw off the dictators (Hitler and Stalin who both had a predilection for Modernist architecture) and among its last flings were Brasilia, the Festival of Britain and Corbusier's state capital buildings at Chandigarh in India (see examples at Figs. 6.2, 6.3 and 6.4). But it was dying on its pilotis, those load-bearing pillars with which Corbusier lifted his buildings into the sky. Its slow death can be seen, not only in the Siegfried Line and the Atlantic Wall, but in the styling of Mercedes Cars, at once paranoid and aggressive, like medieval German armour. We see its demise in 1960s kitchens and bathrooms, white tiled laboratories that are above all clean and aseptic, as if human beings were some kind of disease. We see its death in motorways and autobahns, stone dreams that will never awake, and in the turbine hall at that middle-class disco, Tate Modern – a vast totalitarian space that Albert Speer would have admired, so authoritarian that it overwhelms any work of art inside it.

Norcliffe et al. (1996) provided evidence of the existence and subsequent decline of Modernism in a port context. Although defining Modernism as Fordism the association of the two is clear. Bauman (2000, p. 56) saw Fordism in Modernist terms as described by Lipietz (1996, pp. 116–117):

Fig. 6.2 Brasilia. *Source* US Geological Survey

(a) combination of forms of adjustment of the expectations and contradictory behavior by individual agents to the collective principles of the regime of accumulation...

The industrial paradigm included the Taylorian principle of rationalization, and constant mechanization. That 'rationalization' was based on separation of the intellectual and manual aspects of labour... the social knowledge being systematized from the top and incorporated within machinery by designers. When Taylor and the Taylorian engineers first introduced those principles at the beginning of the twentieth century, their explicit aim was to enforce the control of management on the workers.

Bauman went on to view Fordism as an 'epistemological building site' on which the 'whole world-view was erected', towering majestically over 'the totality of living experience'.

The Fordist factory – with its meticulous separation between design and execution, initiative and command-following, freedom and obedience, invention and determination, with its tight inter-locking of opposites within each of such binary oppositions and the smooth transition of command from the first element of each pair to the second – was without doubt the highest achievement to date of order-aimed social engineering. No wonder it set the metaphorical frame of reference... For everyone trying to comprehend how human reality works on all levels – the global-societal as well as that of the individual life. (Bauman 2000, pp. 60–61)

Fordism, although only part of the movement, thus mirrored much of the development of Modernism in the twentieth century and the two can be seen in many ways to be to be inseparable. Bauman went on:

6.1 Modernism: Its Character and Characteristics

Fig. 6.3 The festival of Britain site, London. *Source* National Media Museum/SSPL

> Fordism was the self-consciousness of modern society in its 'heavy', 'bulky', or 'immobile' and 'rooted', 'solid' phase. At that stage in their joint history, capital, management and labour were all, for better or worse, doomed to stay in one another's company for a long time to come… tied down by the combination of huge factory buildings, heavy machinery and massive labour forces. To survive they had to 'dig in', to draw boundaries and mark them with trenches and barbed wire, while making the fortress large enough to enclose everything necessary to endure a protracted, perhaps prospectless siege. Heavy capitalism was obsessed with bulk and size, and, for that reason, also with boundaries, with making them tight and impenetrable. (Bauman 2000, p. 61)

The last comment relating to boundaries is particularly relevant to our discussion of globalization, their relaxation, and the need to understand the movement that has occurred from a Fordist (Modernist) societal model to one lighter, more flexible, and distinctly different as the Postmodern phase of capitalism cranks in necessitating a change in policy making and governance style.

Norcliffe et al. (1996, p. 127) saw Fordism as achieving its 'apotheosis in the standardization of the moving assembly line' and they argued that there were 'broadly contemporaneous parallels in the cultural realm'. In architecture:

Fig. 6.4 Palace of the assembly, India. *Source* Vikramāditya Prakāsh

> the functional use of modern materials created austere rectangular office blocks and apartment buildings of steel, glass and reinforced concrete that eschewed all ornamentation, those of Mies van der Rohe and Le Corbusier being the best known (Relph 1987; Cooke 1990). In urban planning, we find acts of public vandalism on parts of the inner city, the better to serve the port and the central business district with public buildings, expressways and other transport facilities in the name of functional efficiency (Berman 1982). Here… it is possible to discern an ideology of monotheism – the pursuit of a functional, unadorned uniformity of style, irrespective of geographical content.

Norcliffe et al.'s comments upon architecture and urban planning may seem remote from our discussion of maritime policy failure but in fact are highly pertinent as they reflect the wholly encompassing characteristics of Modernism which influenced society, economics, politics, business organization, and much more. 'Functional, unadorned uniformity' swept through physical society, business organization, the arts, and science along with all other activities and discourse— and the maritime sector and all its features was no exception.

Norcliffe et al. (1996, p. 128) went on to show how the organization of ports was altered by Modernism, becoming physically 'brutally functional' and also port operations were markedly affected:

> The dock labour force formed an internal labour market (Doeringer and Piore 1971) with a well-paid and highly job-demarcated core workforce. Operations tended to form integrated sequences of unloading, warehousing and shipping, and the reverse for loading operations. The growing size of these Fordist ports stimulated the move down river or to the outer

6.1 Modernism: Its Character and Characteristics

harbor, but without at first discarding the old port on the central urban waterfront. The result was to create vast, post-industrial spaces with dock and wharf systems stretching from the old city to the downstream suburban waterfront. In many instances these ports had a fairly captive trade and were operated by public agencies that were not very profit motivated.

Norcliffe et al. suggested that a common 'raft of features' in Modernist ports could be identified:

- Increased scale of operations.
- Specialization in sequential tasks in handling particular commodities within a port.
- Embeddedness within dominant, spatially integrated economic systems functioning increasingly on a global scale.
- Increased relocation freed from city center locations.

In our discussion of Postmodernism in the maritime sector in the next two chapters we shall go on to look how these Modernist characteristics have become increasing marginalized.

Burrell (1994, p. 6) meanwhile was concerned by the crises which regularly affected capitalism with the severest of consequences for individuals. We have discussed these crises (referring to disorganized capitalism) in more detail in Chap. 5 and Burrell noted that the (then) future implied increasing hyper-centralization of economic and political power as we moved toward a vision of the Postmodern, 'profoundly antagonistic to emancipation, and therefore deeply conservative' (Crook, et al. 1992). The Modernist era was thus one to cling to. Wood (1997, pp. 541–542) also reflected on repetitive capitalist crises suggesting that Modernism was just one phase to be superseded by others—in this case the Postmodern—characterized by what he termed a major rupture. Wood saw this disorganization and repeated capitalist disorganization as 'universal and transhistorical, almost natural tendencies… capitalism by definition means constant change and development, not to mention cyclical crises' (Wood 1997, p. 549).

Habermas (1981, p. 7), as discussed by Pusey (1987), agreed that there were problems that Modernism had to face. In particular, the erosion of tradition, morals, and standards had lost their strength, undermined by ever more powerful markets and bureaucracy. How can the principles of democracy be put back at the center of society (Floud 1992a, b)?

> Already at that time, my problem was a theory of modernity, a theory of the pathology of modernity from the viewpoint of the realization – the deformed realization of the reason in history. (Habermas 1981, p. 7)

Burrell (1994, p. 11) went on to quote Rasmussen (1990, p. 112) who praised Modernism before going on to dismiss Postmodernism:

> democratic theory can be read into modern philosophy in such a way as it can be linked with the unfinished project of modernity… The path of Nietzsche, Heidegger and Postmodernity is relatively easy to chart. By opting out of this tradition they can be understood

to have given up on the normative question. Hence Postmodernism can be appropriately dismissed.

Burrell interpreted this as the 'noble ideal' of Modernism standing for emancipation and liberation from oppression; Postmodernists threatened this 'ideal'.

Alfven (1983, p. 313) was especially critical of Modernism and its application to science quoting Weisskopf (1981) in support:

> There is a strong trend towards clear-cut, universally valid answers that exclude different approaches. Whenever one way of thinking is developed with great force and success, other ways are unduly neglected. It was aptly expressed by Marcus Fierz, the Swiss physicist–philosopher. 'The scientific insights of our age shed such glaring light on certain aspects of human experience that they leave the rest in even greater darkness'.

Wright-Mills (1959, pp. 15–18) provided further support to the rejection of Modernism and its pretensions citing C.P. Snow in identifying two cultures—scientific and humanistic—and the significance that should rest with the latter, compared with the public presence that had been acquired by the former (Kimball 1994). In recent years the criticism leveled at Modernism has increased directly proportional to the rise of Postmodernism. Anderson (1984, p. 100), for example, in discussing a future characterized by a new Communist society, suggested that Modernity would inevitably eventually destroy itself in just the same way that it would in a society dominated by Capitalism. He quoted Berman (1982, p. 114): 'Ironically, then, we can see Marx's dialect of Modernity re-enacting the fate of the society it describes, generating energies and ideas that melt it down into its own air'.

Jameson (1971, p. 105) was equally as critical of the Modernist paradigm:

> Henceforth, in what we might call post-industrial capitalism, the products with which we are furnished are utterly without depth: their plastic content is totally incapable of serving as a conductor of psychic energy. All libidinal investment in such objects is precluded from the outset, and we may well ask ourselves, if it is true that our object universe is henceforth unable to yield any 'symbol apt at stirring human sensibility', whether we are not here in a presence of a cultural transformation of signal proportions, a historical break of an unexpectedly radical kind.

Anderson (1984, pp. 112–113) was even more damning:

> Modernism as a notion is the emptiest of all cultural categories. Unlike the terms Gothic, Renaissance, Baroque, Mannerist, Romantic or Neo-Classical, it designates no describable object in its own right at all: it is completely lacking in positive content... what is concealed beneath the label is a wide variety of very diverse – indeed incompatible – aesthetic practices: symbolism, constructivism, expressionism, surrealism. These, which do spell out specific programmes, were unified post hoc in a portmanteau concept whose only referent is the blank passage of time itself. There is no other aesthetic marker so vacant or vitiated. For what was once modern, is soon obsolete.

Warf (1993, p. 162) noted mounting criticism citing ahistoricism, lack of a specific conception of social relations, neglect of issues of power, inability to theorize human consciousness, insistence on unobtainable value-free objectivity, its substitution of technique for theory, and its tendency to empiricism. By

6.1 Modernism: Its Character and Characteristics

ignoring the power of history this assumes that the historical context is irrelevant. Marx's Modernist claim to 'scientific status' and tendency to economic determinism was also viewed poorly (Duncan and Ley 1982; Gould 1988). There was an obsession with universal laws, a tendency to dismiss human consciousness, a failure to incorporate the importance of language and a privileging of time over space—the latter is especially pertinent in our governance and policy debate. As we shall see, all these issues are ones dominant in the moves toward Postmodernism.

Caporaso (1996, p. 45) considered the Modernist tradition of upholding the preeminent position of the nation-state not only outdated but also morally indefensible. He viewed the nation-state as a way of 'carving up the world into territorially exclusive enclaves—people, capital, ideas, foreign powers and so on'. Modern political and economic authority failed to conform. Cooper (2000, pp. 17–18) backed him up in discussing the role of the United Nations (and by implication the IMO). Nation-states and their international relations are still based on the assumption that, in Palmerston's (1848) words, 'Britain has no permanent friends or enemies… only its interests were eternal'. The UN attempts to combine the uncombinable—the inviolability of state sovereignty along with the use of force against states when appropriate in an attempt to throw the weight of international governance behind the status quo of state domination thus creating a dog's dinner of policy making. Van Ham (2001, p. 19) continued the theme with respect to international relations claiming that until recently they were rooted in 'realism' which 'consciously ignored non-useful questions regarding culture, economy and identity'. Structure was abstracted from 'concrete reality'. Van Ham suggested that this formal and rigid Modernist view of international relations and the role of the nation-state stemmed from the Cold War period, but has lost its claim to universal truth of state-centricism and its aura of universality and superiority. Modernist reality suppressed dissonant voices and new approaches and explanations through maintenance of an increasingly irrelevant status quo. In international relations terms, the Modernist model was designed for 'violent response and big-power coercion rather than for prevention and sensitive negotiation' (George 1996, p. 35).

Hajer (2003) suggested that the Modernist inheritance had created an institutional void for policy making produced by the discrepancy between the existing institutional order and the actual practice of policy making. In our case the situation is slightly different in that the discrepancy is between the structure of the governance of the sector and the policy-making institutions themselves, but the scenario is remarkably similar in effect. Hajer illustrated the difference using the distinction between the classical-Modernist political institutions and what he termed new political spaces on the other; in our case between the Modernist, preglobalization institutions of policy making on one hand and the Postmodernist, post globalization industry on the other.

Hajer (2003, p. 176) defined classical-Modernist institutions in terms which are eminently applicable to the current maritime governance framework as codified arrangements:

that provide the official setting of policy-making and politics in the post-war era in Western societies: representative democracy, a differentiation between politics and bureaucracy, the commitment to ministerial responsibility and the idea that policy making should be based on expert knowledge. New political spaces then refer to the ensemble of mostly unstable practices that emerge in the struggle to address problems that the established institutions are unable to resolve in a manner that is perceived to be legitimate and effective. Here we may think of the activity of consumer organizations, the role of NGOs in agenda setting and in monitoring the implementation of treaties and also the surprising role of non-political actors like designers in creating the preconditions for a good deliberation.

Hajer's argument was that the rules of the classical-Modernist politics told us little about the 'new rules of the game'. The new world is discursive and cannot be 'captured in the comfortable terms of generally accepted rules', but is created through deliberation. The polity, long considered stable in policy analysis, thus becomes a topic for empirical analysis again. Modernism was dead. Long-live Postmodernism. Maritime policy governance must follow.

The classic-Modernist hierarchical, state-centric institutional dominated mold was seen to be failing. Braudel (1982), citing Sombart (1913a, b), felt that Modernism depended on the strength of the state and with its decline, had to move on. Hajer suggested that this is a consequence of the twin features of globalization on one hand and individualization on the other, together eroding the self-evidence of classical-Modernist governance. Its consequences were governance and policy failure and a new model was needed reflecting the changes in society that have taken place. In the maritime sector these changes have certainly occurred but the new model is yet to be recognized.

6.2 Modernism to Postmodernism

The new capitalist model of accumulation is widely regarded as reflected in the Postmodernist movement that has emerged from a rejection of Modernism (Vattimo 1988). We can now look to see how this has impacted upon a wide range of maritime activities and its consequences for both policy making and governance. The discussion will be far-ranging as the Postmodern condition is one that has impacted upon all facets of society, economy, and culture creating a new scenario for the maritime sector within which the industry must operate and also within which policies need to be couched and a governance framework for the development of such policies needs to be derived. It is this failure to match a new Postmodern governance framework to the policy-making institutions and policy creation and implementation that has led to the current inadequacies that are displayed in the maritime sector. We begin by looking at how Postmodernism has emerged from Modernism before going on in the next two chapters to examine its characteristics and then examples of its condition within the maritime sector.

There is considerable support for the notion that the Postmodern has emerged from a radical societal change with Jencks (1986, p. 39) suggesting that it 'springs from a post-industrial society'. Portoghesi (1983, p. 11) cited in Albertson (1988, p. 352) agreed. Postmodern architecture is the:

> most convincing answer given thus far by architectural culture to the profound transformations of society and culture, to the growth of a 'Postmodern condition' from the development of post-industrial society.

Jameson (1984, p. 78) also supported this notion, considering that Postmodernism was 'related to Mandel's (1975) third stage in the evolution of capitalism—consumer capitalism, preceded by market capitalism and monopoly capitalism'. These three stages are a mirror of Realism, Modernism, and Postmodernism. Jameson (1985, pp. 111–112) continued emphasizing that the majority of Postmodern events are a reaction to high Modernism from 1960: 'the establishment and the enemy—dead, stifling, canonical, the reified monuments one has to destroy to do anything new'. Williams (1989, p. 51) meanwhile was not surprised by the linkages between capitalism and Modernism, the latter losing its anti-bourgeois stance and becoming integrated with the capitalist ethos. Modernism's pretensions at achieving a 'universal market, trans-frontier and trans-class' were clear with a dominant theme of obsolescence and repeated shifts in schools, styles, and fashions so essential to the capitalist market. Thus, Modernism became capitalism, and the Postmodern movement was just the next phase in capitalism's continuous metamorphosis, characterized by disorganization, mutation, and change.

However, this move from Modernism to Postmodernism has neither been universally welcomed nor accepted and remains resisted by many who cling to the safety, tradition, and security of the established condition. Virilio (1999, pp. 6, 8) may not be so fearful of Postmodernism but he has seriously resisted recognizing it as progress from Modernism, considering it a sterile development. He remained highly optimistic of Modernism and proceeded to develop his own 'hypermodernism' rather than pursuing the Postmodern route. Meanwhile, Wright-Mills (1959, p. 48) was more of a traditionalist in pointing out the disbenefits of moving from a Modernist to a Postmodernist perception. Using the (embryonic Postmodern) work of Talcott Parsons' *The Social System* (Parsons 1951) he first showed how this 555 page epic can be reduced to four paragraphs of substance. Parsons is accused not only of saying little (albeit what he does say is of good quality), but also of being possessed by the idea that one model of social order (the one he has constructed incidentally) is 'some kind of universal model'. Wright-Mills called this a 'fetishization of concepts'. What Parsons claims is 'systematic' about this concept (and the detail of the concept is not the issue, but the failure to recognize that alternatives might be more appropriate and will change with time):

> is the way it outruns any specific and empirical problem. It is not used to state more precisely or more adequately any new problem of recognizable significance. It has not been developed out of any need to fly high for a little while in order to see something in the social world more clearly, to solve some problem that can be stated in terms of the historical reality in which men and institutions have their concrete being. (Wright-Mills 1959, p. 48)

Table 6.2 Social science legitimations

John Locke (1689)	Principle of sovereignty
Georges Sorel (Meisel 1951)	Ruling myth
Thurman Arnold (1937)	Folklore
Émile Durkheim (1912)	Collective representations
Karl Marx (1992)	Dominant ideas
Jean-Jacques Rousseau (Swenson 2000)	General will
Harold Lasswell (Ascher and Ascher 2004)	Symbol of authority
Max Weber (Mommsen 1984)	Legitimations
Karl Mannheim (1936)	Ideology
Herbert Spencer (1851)	Public sentiments

Source After Wright-Mills (1959, p. 36)

Wright-Mills went on to talk about those in authority attempting to 'justify their rule over institutions by linking it, as if it were a necessary consequence, with widely believed-in moral symbols, sacred emblems, legal formulae':

> These central conceptions may refer to a god or gods, the 'vote of the majority', 'the will of the people', 'the aristocracy of talent or wealth', to the 'divine right of kings' or to the allegedly extraordinary endowment of the ruler himself. (Wright-Mills 1959, p. 36)

There are various terms used to refer to such concepts (Table 6.2) and this is precisely where Postmodernism comes in—an alternative understanding and interpretation to the established and traditional conditions and organization that characterize the Modernist view of the maritime sector which were wholly appropriate at the time that the current maritime governance structures were setup but which are now wholly inappropriate. Change is difficult. Change is resisted. But change is needed continuously to reflect societal change. The move from a Modernist, inflexible, hierarchical, state-centric maritime governance model to a Postmodern one that accommodates alternatives, cross-jurisdictional communication, universal stakeholder needs, and the widest impacts of globalization is apparent.

Hardt and Negri (2004, p. 4) observed that Modernism and Postmodernism are linked but at the same time Cooper and Burrell (1988) suggest that 'two radically different systems of thought and logic were at work in the Modernist-Postmodernist confrontation'. They saw that Modernism was essentially a reflection of formality which had worked its way into every facet of life by the late 1980s, generating an opposing reaction in Postmodernism which rejected such formality in every way possible creating an environment of informalism. Meanwhile, (Harvey 1990, p. 44), cited in Dear (1991, p. 534) suggested that the links between the two were clear and characterized by continuity.

Although commenting only upon architecture and urban planning, Bertens (1995) agreed seeing Postmodernism as reflecting a crisis in Modernism suggesting that Postmodernism takes Modernism and moves on emphasizing:

the fragmentary, the ephemeral, and the chaotic… while expressing a deep skepticism as to any particular prescriptions as to how the eternal and immutable should be conceived of, represented or expressed. (Harvey 1990, p. 116)

Lash and Urry (1987, p. 16, 88–89) provided extensive consideration of the move from Modernism to Postmodernism suggesting as well that it was a reflection of the move from organized to disorganized capitalism—which we have seen is a process generated by a desire of capitalist society to accumulate which generates the need for a periodic reformulation of the framework for governance. Table 6.3 provides some detail and incorporates an adaptation of Lash and Urry's original formulation showing the move from Premodernism through Modernism to Postmodernism that it implies suggesting a linked and direct relationship between each phase.

They borrowed from the work of Taylor and Thrift (1982) in suggesting that these three phases were aligned with three identifiable stages of capitalism—they termed these liberal capitalism (our Pre-Modernism), where firms operate at mostly the local or occasionally the regional level; organized capitalism (Modernism) where three types of firms were identifiable—small firms, a regional national company, and the multinational. The latter was especially significant as it kept tight control over its subsidiaries and branches. Capital was 'nationally owned and clearly tied into the fortunes of the country in which it was owned and to which it was indissolubly bound'. Finally, disorganized capitalism (Postmodernism) where polycentric structures were developed with subsidiaries operating independently and a multidivisional structure emerging. Attachment to a single economy becomes 'tentative, as capital expands (and contracts) on a global basis'. However, although recognizing the three stages, Van Ham (2001, p. 16) questioned their sequentiality, suggesting that they existed simultaneously with 'some regions of the globe dominated by one phase of development, other regions by others' (Falk 1996, p. 23).

The hierarchical governance structure that we see in the maritime sector (and elsewhere) is a direct product of the era of Modernism, characterized by formal structures in decision making, a recognized set of relationships between authorities and individuals, and a limited capacity for flexibility and change. Modernism characterized Western society for long periods in the twentieth century, manifesting itself in architecture, art, music, design, industrial organization, education, civil engineering, manufacturing as well as management, policy making, and governance. The formal structure to maritime governance which exists today is a direct product of the Modernist era, emerging as it did from the consequences of the disasters of the First World War and the 1930s depression. Focused upon the new international organizations of the UN, World Bank and IMF, joined later by the WTO, the OECD and the EU it inherited Westphalian state-centricism as a central focus. Postmodernism is the latest reaction by capitalism against the Modernist framework and the need for increased and renewed capital accumulation.

This shift has also been viewed as a move from a Fordist production regime to one that is Postfordist. Schoenberger (1988, p. 245, 247) considered that the late

Table 6.3 Temporal and spatial changes in liberal, organized, and disorganized capitalism

	Phase of capitalist development	Predominant temporal, spatial and organizational structures	Spatial changes within each territory	Predominant means of transmitting knowledge and executing surveillance
Premodernist	Liberal	Large-scale collapsing empires that had been built-up around dynastic rulers or world religions: emergence of weak nation-states	Growth of tiny pockets of industry. Importance of substantial commercial cities as well as the expansion of new urban centres in rural areas	Handwriting and word of mouth
Modernist	Organized	Nation-states within the ten or so major Western economies increasingly dominate large parts of the rest of the world through colonization	Development of distinct regional economies organized around growing urban centres. Major inequalities between new industrial and non-industrial regions and nations	Printing developed through 'print capitalism'
Post-modernist	Disorganized	Development of world economy, an international division of labor, and the widespread growth of capitalism in most countries	Decline of distinct national/regional economies and of industrial cities. Growth of industry in smaller cities and rural areas and the development of service industry. Separation of finance and industry	Electronically transmitted information dramatically reduces the time-space distance between people and increases the power of surveillance

Source Adapted by the Author from Lash and Urry (1987)

6.2 Modernism to Postmodernism

Table 6.4 Modernism and postmodernism

Modernism	Postmodernism	The maritime effect
Using rational, scientific, logical means to know the world. Optimism that we can understand and control an objective world	A reaction against rationalism, scientism, or objectivity of Modernism	Shipowners gambling on the markets. Dotcom revolution and failure. Irrational man (and woman)
There is an absolute, universal truth that we can understand through rationalism and logic	There is no universal truth. Rationality by itself does not help us truly understand the world	Flexibility in the market place. Understanding the irrationality of the consumer. Continuous market mutation. Seeking alternative modes, energy, ports, routes, financial solutions etc
Humans are material machines. We live in a purely physical world. Nothing exists beyond what our senses perceive	Suspicious of such dogmatic claims to knowledge	Open to new ideas. Flexibility in institutional arrangements, stakeholders, maritime power-bases, etc
Humankind is progressing by using science and reason	"Progress" is a way to justify the domination by European culture of other cultures	Recognition of the move in power to the Far East. Loss of status and new roles for the TMNs
Time, history, progress	Culture on fast forward: time and history replaced by speed, futureness, accelerated obsolescence	Obvious impacts in IT, communications, etc. Speed/time compression is central to the maritime marketplace
History as a narrative of what happened with a point of view and cultural/ideological interests	Postmodern historians and philosophers question the representation of history and cultural identities: history as what "really happened" is from one group's point of view	Recognition of new cultural dominance. Different priorities from new power bases. New rules of the game
Faith in "depth" (value content, how things work) over "surface" (appearances, the superficial, how we use things)	Attention to play of surfaces, images, things mean what we make them mean, no concern for "depth" but with how things look and respond	Recognition of the importance of image in the industry and its markets
Disenchantment with material truth and search for abstract truth	There is no universal truth, abstract, or otherwise	Values change as do rules. Nothing is permanently 'correct'. Traditional approaches, methods, contacts all change

(continued)

Table 6.4 (continued)

Modernism	Postmodernism	The maritime effect
Faith in the real beyond media and representations; authenticity of "originals"	Hyper-reality, image saturation, simulacra seem more powerful than the "real"; images and texts with no prior "original". "As seen on TV" and "as seen on MTV" are more powerful than the unmediated experience	Importance of media and image to the industry. More important than truth is belief
Time line		
(Renaissance?) Enlightenment > 1750s > 1890–1945	Post WWII—especially after 1968	Reflected in container age, decline of TMNs, new port locations, flexible finance, manning, changing markets
General		
Attempt to achieve a unified, coherent world-view from the fragmentation that defines existence	Attempt to overturn the distinction between "high" and "low" culture	No preconceived ideas of who 'should' represent the industry. Reframing of IMO dominant nations. Relegation of TMN influence
High Modernism 1920 and 1930s, following WWI—outmoded political orders and old ways of portraying the world no longer seem appropriate or applicable: reaction against existing order	Eclecticism, a tendency toward parody and self-reference, and a relativism that knows no ultimate truth: no distinctions between good and bad	Open-minded approach to maritime activities. Willing to accept new players, new flags, new sources of finance etc
Computers		
PCs/UNIX/command line environments. Stand-alone mainframe computers. Culture of calculation	Macintosh/Windows; internet/www computer networks. Culture of simulation	IT, shipping and communications. GPS. Online activities. Decline of face-to-face interaction
Hierarchy, order, centralized control	Subverted order, centralized control, fragmentation	Networking, cross-jurisdictional communications
Culture		
High culture vs. low culture—strictly divided. Only high culture deserves to be studied, analyzed	Everything is popular culture and merits studying; pluralizing, commodification of culture—everything can be bought and sold	Shipping's image part of culture—liveries, logos, paraphernalia, memorabilia, etc

(continued)

6.2 Modernism to Postmodernism

Table 6.4 (continued)

Modernism	Postmodernism	The maritime effect
Humans are self-governing and free to choose their own direction	People are the product of their culture and only imagine they are self-governing	Incorporation of new cultures into the traditional shipping structures—committees, international organizations, decision-makers, stakeholders
Reality can be discovered through science and can be expressed abstractly	The transformation of reality into images	There are no permanent rules, methodologies, institutions. In fact everything is value-laden JIT, lean manufacturing, personalized products, etc
	De-massified culture, niche products, and marketing; smaller group identities	
Style		
Strict rules and regimentation of styles	Pastiche and parody of multiple styles: old forms of content merely become styles	Rejection of tradition for tradition's sake. Pink container ships, orange cruise ships
New form of Modernist style identified	In a world in which stylistic innovation is no longer possible all that is left is to imitate dead styles. Can only remix what has been done	Everything comes back in cycles. Modern sailing ships; customized products
Serious meaning in style	Stylistic masks, image styles without present content, the meaning is in the mimicry	Marketing in shipping; recognizing that a game is being played with the consumer
No illusion of individualism. The individual is part of society and community	Postmodern attempts to provide illusions of individualism through images that define possible subject positions or create desired positions	Individualism within the shipping sector. Value attached to new approaches and ideas that break tradition. EasyCruise use of media, and politicians
Symbolism		
Symbol and meaning—e.g hammer and sickle = world communism	Symbols drained of meaning; hammer and sickle in advertising	Use of image consultancies in shipping
Architecture		
Form follows function; Le Corbusier; machine aesthetic; international style; straight, clean lines	Multiple historical references; playful mix of styles; past and present; Las Vegas; Pompidou Centre; Venturi; Robert Stirling	Container houses and shopping; Guggenheim Museum

(continued)

Table 6.4 (continued)

Modernism	Postmodernism	The maritime effect
Body		
Clear dichotomy between organic and inorganic, human and machine	Cyborgian mixing of organic and inorganic, human and machine and electronic	The seafarer/officer as the major element in machine failure and misuse
Politics		
Big ideas/big, centralized political parties rule	Fragmented ideas, decentralized power; micro-politics; interest groups rule; Foucault—everyone has a little power; TV politics; You Tube; blogs; how will it play on the Six o'Clock News?	Stakeholders and shipping. Seafarers, port users, voters, media, all having their say in maritime policy
Door to door politics; big rallies	Late capitalism rules	Sophisticated market manipulation by flag states, shipowners—the rise of tonnage tax
Capitalism vs. Communism. Clash of ideologies—Dr Strangelove, The Making of the President; Orwell's *Animal Farm*	The "Selling of the President"; pastiche; tail wags the dog	Big shipowners, morphing with banking interests to manipulate the capitalist market place and to satisfy fragmented stakeholders—or seemingly so
Identity		
Sense of unified, centred self; individualism unified identity	Sense of fragmentation and decentred self; multiple, conflicting identities	Traditional maritime institutions take on new, untraditional roles and symbolism—green, efficient, caring shipping
Arts		
Artist is creator rather than preserver of culture Impressionism, Cubism, abstract impressionism, suprematism (Malevich's *Black Square*). Photograph never lies. Photos and video are windows/mirrors of reality	Artist plays with different styles; aesthetics; pastiche all-important. Pop Art, Dada, montage	Morley and Postmodern realist ships. Moves from traditional maritime art to symbolist, maritime art
Art fights capitalism. Seriousness of intention and purpose, middle-class earnestness	Photoshop; photos and videos can be altered completely; montage; art is consumed by capitalism. Digital media; there is no distinction between an original and a copy. CGI	The use of IT art to promote the maritime sector

(continued)

6.2 Modernism to Postmodernism 287

Table 6.4 (continued)

Modernism	Postmodernism	The maritime effect
Novel is the dominant form; movies; author determines meaning; canon of great works; Shakespeare, Kafka, Joyce. Art critics. Can tell good from bad	TV, www; meaning is indeterminate. Thomas Pynchon, Kathy Acker, William Gibson. Rise in importance of popular culture. Impossible to tell good from bad—it is all relative	www central to maritime art and communication. Meaning is unclear and malleable. Relativity is central to the interpretation of the maritime image
Media		
Knowledge mastery, attempts to embrace totality	Navigation, information management, just-in-time knowledge	The use of PR and media consultants for the maritime sector. Media first, lawyers second; accountants third; the public convinced
The encyclopedia	The web	Maritime www
Broadcast media, centralized one-to-many communications	Interactive, client server, distributed, many-to-many media	Blogs, pods, iPads, always in touch. Never out of sight
Centering/centredness, centralized knowledge	Dispersal, dissemination, networked, distributed knowledge	Fundamental and central to the maritime marketplace for everything
Business		
Small, limited in scope, vertical structures	Size, international, footloose	Shipping mergers; international port management companies, outsourcing, horizontal rather than vertical structures
Organization		
Hierarchies; institutions; highly structured; inflexible	Electronic and instantaneous communication. Private–public partnerships. Networks. Cross-jurisdictional discussion	The role of the blog, email, etc. Significance of maritime networking. Decline in the efficiency of traditional institutional structures
Financial		
Traditional financial sources and centres	Footloose	New centres in Far East. New financial sources for shipping
Violence		
Total, global, unconstrained war. Land-based attrition. Located in 'developed' countries	From total war to terrorism. Footloose. War takes place on 'other' countries' territories	A change from land-based to air and sea-based combat. New maritime theatres of war often remote from the conflict but taking part using air support

1980s was a period characterized by a 'transition from one form of social and economic organization to another' and that Fordism—as a specific period of technology, organization and consumption'—had been replaced by the Postfordism of flexible accumulation and globalization and all this brings. This is reflective of the move from Modernism to Postmodernism that has occurred over a similar period and with similar implications for flexible production, flexible consumption, and flexible markets.

6.3 Conclusions

Figure 6.8 provides a summary of the movement from the Modern to the Postmodern and the substantial change that has taken place. These changes are societal, wide-ranging, almost universal and make no exception for any activity—cultural, scientific, commercial—including the maritime sector. Clearly, not all the changes are directly relevant, nor all the time but there is enough here to suggest that Modernist institutions, infrastructure, and attitudes are inadequate for a Postmodern industry acting in a Postmodern world. Shipping is just that. There is no wonder that such a globalized sector has struggled to adapt to policies generated through a process of inadequacy and what can be readily viewed as deliberate neglect (Table 6.4).

In the words of Negri (2005, pp. 28–29) 'This (Modernist) era is now over; the shift to a Postmodern global governance has been fully achieved. This shift moreover has been of such intensity as to dissolve not only the modern but also its memory'. Opello Jr. and Rosow (2004, p. 267) agreed, suggesting that globalization had destroyed the nation-state and since Modernity and the territorial state were synonymous, then Modernity itself was dead. It is to the Postmodern we now turn, recognizing that substantive societal change has taken place and that this new era requires change in maritime governance as existing and outmoded institutions and their policy making structures need to change as well as to meet the Postmodern agenda (Cooke 1990, p. 140).

References

Adams, R. M. (1978). What was modernism? *Hudson Review, 31*, 19–33.
Aglietta, M. (1982). World capitalism in the eighties. *New Left Review, 136*(1), 3–41.
Albertson, N. (1988). Postmodernism, post-Fordism and critical social theory. *Environment and Planning D, 6*, 339–365.
Alfven, H. (1983). On hierarchical cosmology. *Astrophysics and Space Science, 89*, 313–324.
Allen, R. E. (2006). *Plato: The republic*. New Haven: Yale University Press.
Anderson, P. (1984). Modernity and revolution. *New Left Review, 144*, 96–113.
Arrighi, G. (1994). *The long Twentieth Century*. London: Allen and Unwin.

References

Ascher, W., & Ascher, B. H. (2004). Linking Lasswell's political psychology and the policy sciences. *Policy Sciences, 37*(1), 23–36.
Atkinson, M. M., & Coleman, W. D. (1992). Policy networks, policy communities and the problems of governance. *Governance, 5*(2), 154–180.
Ballard, J. G. (2006). A handful of dust, *Guardian*, March 20th.
Baudrillard, J. (1975). *The mirror of production.* St Louis MO: Telos Press.
Bauman, Z. (2000). *Liquid modernity.* Cambridge: Polity Press.
Bendix, R. (1960). *Max Weber: An intellectual portrait.* Garden City: Anchor Books.
Benko, G. (1997). Introduction: Modernity, postmodernity and the social sciences. In G. Benko & U. Strohmayer (Eds.), *Geography, history and social science* (pp. 1–48). Dordrecht: Kluwer.
Benson, J. K. (1977). Organizations: A dialectical view. *Administrative Science Quarterly, 22*, 1–21.
Berman, M. (1982). *All that's solid melts into air: The experience of modernity.* Harmondsworth: Penguin.
Bertens, H. (1995). *The idea of the postmodern, a history.* London: Routledge.
Birnbaum, N. (1969). *The crisis of industrial society.* NY: Oxford University Press.
Birnbaum, N. (1971). *Towards a critical sociology.* NY: Oxford University Press.
Boje, D. M., & Dennehy, R. F. (1993). *Managing in the postmodern world; America's revolution against exploitation.* Kendall/Hunt: Dubuque IA.
Boje, D. M., Fitzgibbon, D. E., & Steingard, D. S. (1996). Storytelling at Administrative Science Quarterly: Warding off the postmodern barbarians. In D. M. Boje, R. P. Gephart Jr, & T. J. Thatchenkery (Eds.), *Postmodern management and organization theory* (pp. 60–92). Thousand Oaks: Sage.
Borzel, T. A. (1998). Organizing Babylon—On the different conceptions of policy networks. *Public Administration, 76*, 253–273.
Braudel, F. (1982). *The wheels of commerce.* NY: Harper and Row.
Brenner, N. (1999). Beyond state-centrism? Territoriality and geographical scale in globalization studies. *Theory and Society, 28*(1), 39–78.
Burke, B. (2000). Post-modernism and Post-Modernity. The Encyclopaedia of Informal Education. www.infed.org/biblio/b-postmd.htm.
Burrell, G. (1994). Modernism, postmodernism and organizational analysis 4: The contribution of Jurgen Habermas. *Organization Studies, 15*(1), 1–45.
Caporaso, J. A. (1996). The European Union and forms of state: Westphalian, regulatory or post-modern? *Journal of Common Market Studies, 34*(1), 29–52.
Carroll, W. K. (2002). Does disorganized capitalism disorganize corporate networks? *Canadian Journal of Sociology, 27*(3), 339–371.
Castells, M. (1989). *The informational city: Information technology, economic restructuring and the urban regional process.* Oxford: Blackwell.
Chandler, A. (1977). *The visible hand.* Cambridge: Belknap Press.
Chlomoudis, C.I., Karalis, A.V., Pallis, A.A. (2000). *Transition to a New Reality: Theorising the Organizational Restructuring of Ports.* Infomare, International Workshop, Special Interest Group on Maritime Transport and Ports, Genoa, June 8–10.
Clegg, S. R. (1990). *Modern organizations, organization studies in the postmodern world.* London: Sage.
Clegg, S. R. (1996). Postmodern management. In G. Palmer & S. Clegg (Eds.), *Constituting management. Markets, meanings and identities* (pp. 235–265). Berlin: Walter de Gruyter.
Clegg, S. R., & Dunkerley, D. (1980). *Organization, class and control.* London: Routledge and Kegan Paul.
Comte, A. (2005). *Positive philosophy of Auguste Comte Part I.* Whitefish: Kessinger Publishing.
Cooke, P. (1990). *Back to the future. Modernity, postmodernity and locality.* London: Unwin Hyman.
Cooper, R. (2000). *The post-modern state and the world order.* London: Demos.
Cooper, R., & Burrell, G. (1988). Modernism, postmodernism and organizational analysis. *Organization Studies, 9*(1), 91–112.

Cornforth, M. (1952). *Dialectical materialism. An introduction. Volume 1: Materialism and the dialectical method*, Lawrence and Wishart: London.
Crook, S., Paluski, J., & Waters, M. (1992). *Postmodernization*. London: Sage.
Curry, M. R. (1991). Postmodernism, language and the strains of modernism. *Annals of the Association of American Geographers, 81*, 210–228.
Davies, J. S. (2005). Local governance and the dialectics of hierarchy, market and network. *Policy Studies, 26*(3/4), 311–335.
Dear, M. J. (1986). Postmodernism and planning. *Environment and Planning D: Society and Space, 4*, 367–384.
Dear, M. J. (1991). Book review: The condition of postmodernity: An inquiry into the origins of cultural change, David Harvey. *Annals of the Association of American Geographers, 81*, 533–539.
Denzer, A.S. (undated). Masters of modernism. www.mastersofmodernism.com.
Denzin, N. K. (1986). Postmodern social theory. *Sociological Theory, 4*, 194–204.
Doeringer, P. B., & Piore, M. J. (1971). *Internal labour markets and manpower analysis*. Lexington: Heath.
Duncan, J., & Ley, D. (1982). Structural Marxism and human geography. *Annals of the Association of American Geographers, 81*, 533–539.
Durkheim, E. (1912). *The elementary forms of the religious life*. Oxford: Oxford Classics.
Eagleton, T. (1985). Capitalism, Modernism and Postmodernism. *New Left Review, I*(152), 60–73.
Evans, M. (2001). Understanding dialectics in policy network analysis. *Political Studies, 49*, 542–550.
Falk, R. (1996). An enquiry into the political economy of world order. *New Political Economy, 1*, 13–26.
Fichte, J. (1998). *Early philosophical writings*. Ithaca: Cornell University Press.
Floud, J. (1992a). *Jurgen Habermas; Social Structure and the Limits of the Law*, Radcliffe Lecture, University of Warwick.
Floud, J. (1992b). *Law and Society in a Welfare State*, Radcliffe Lecture, University of Warwick.
George, J. (1996). Understanding international relations after the Cold War: Probing beyond the British legacy. In M. J. Shapiro & H. R. Alker (Eds.), *Challenging boundaries: Global flows, territorial identities*. Minneapolis: University of Minnesota Press.
Gergen, K. J., & Thatchenkerry, T. J. (2004). Organization science as social construction: Postmodern potentials. *The Journal of Applied Behavioral Science, 40*(2), 228–249.
Giddens, A. (1981). Modernism and postmodernism. *New German Critique, 22*, 15–18.
Giddens, A. (1985). *A contemporary critique of historical materialism. Volume 2. The Nation-state and Violence*. Cambridge: Polity Press.
Godelier, M. (1972). Structure and contradiction in capital. In R. Blackburn (Ed.), *Ideology in social science* (pp. 334–368). NY: Vintage Books.
Goldman, L. (1969). *The human sciences and philosophy*. London: Jonathan Cape.
Gould, P. (1988). The only perspective: A critique of Marxist claims to exclusiveness in geographical enquiry. In R. Golledge, H. Couclelis, & P. Gould (Eds.), *A ground for common search* (pp. 1–10). Santa Barbara: Santa Barbara Geographical Press.
Gramsci, A. (1971). *Selections from the prison notebooks*. NY: International Publishers.
Habermas, J. (1970). *Toward a rational society*. Boston: Beacon Press.
Habermas, J. (1971). *Knowledge and human interests*. Boston: Beacon Press.
Habermas, J. (1973). *Legitimation crisis*. Boston: Beacon Press.
Habermas, J. (1981). Modernity versus postmodernity. *New German Critique, 22*, 3–14.
Habermas, J. (1983). Modernity—An incomplete project. In T. Docherty (Ed.), *Postmodernism: A reader* (pp. 98–109). Brighton: Harvester Wheatsheaf.
Hajer, M. (2003). Policy without polity? Policy analysis and the institutional void. *Policy Sciences, 36*, 175–195.
Hardt, M., & Negri, A. (2004). *Multitude*. London: Hamish Hamilton.
Harvey, D. (1990). *The condition of postmodernity*. Cambridge: Blackwell.

References

Harvey, D. (2000). *Spaces of hope*. Edinburgh: Edinburgh University Press.
Hirschman, A. O., & Lindblom, C. E. (1962). Economic development, research and development, policy-making, some emergent views. *Behavioral Science, 7*, 211–222.
Ingold, T. (1995). Man, the story so far, *Times Higher Education Supplement*, June 2nd, pp. 16–17.
Jameson, F. (1971). *Marxism and form: Twentieth Century dialectical theories of literature*. Princeton: Princeton University Press.
Jameson, F. (1984). Postmodernism, or the cultural logic of late capitalism. *New Left Review, 146*, 53–92.
Jameson, F. (1985). Postmodernism and consumer society. In H. Foster (Ed.), *Postmodern culture* (pp. 111–125). London: Pluto Press.
Jencks, C. (1986). *What is post-modernism?* London: Academy Editions.
Jessop, B. (1998). The rise of governance and the risk of failure: The case of economic development. *International Social Science Journal, 50*(155), 29–45.
Jowitt, K. (1992). *New world disorder. The Leninist extinction*. Berkeley: University of California Press.
Kimball, R. (1994). The *Two Cultures* today, on the C. P. Snow–F. R. Leavis controversy. *The New Criterion*, 12 Feb.
King, R., & Kendall, G. (2004). *The state, democracy and globalization*. Basingstoke: Palgrave Macmillan.
Krasner, S. D. (1984). Approaches to the state: Alternative conceptions and historical dynamics. *Comparative Politics, 16*(2), 223–246.
Kuhn, T. S. (1957). *The Copernican revolution*. Cambridge: Harvard University Press.
Lakoff, G. (1987). *Women, fire and dangerous things*. Chicago: Chicago University Press.
Lash, S., & Urry, J. (1987). *The end of organized capitalism*. Cambridge: Polity Press.
Le Corbusier. (1978). *The city of tomorrow*. London: Architectural Press.
Lefebvre, H. (1968). *Dialectical materialism*. London: Jonathan Cape.
Lefebvre, H. (1971). *Everyday life in the modern world*. NY: Harper and Row.
Leonard, P. (1997). *Postmodern welfare: Reconstructing an emancipatory project*. London: Sage.
Lipietz, A. (1996). The next transformation. In M. Cangiani (Ed.), *The Milano papers: Essays in societal alternatives*. Montreal: Black Rose Books.
Locke, J. (1689). *Two Treatises of Government*, Anonymous Publisher.
Lourenco, S. A., & Glidewell, J. C. (1975). A dialectical analysis of organizational conflict. *Administrative Science Quarterly, 20*, 489–492.
Lukacs, G. (1971). *History and class consciousness: Studies in Marxist dialectics*. Cambridge: MIT Press.
Luke, T. W. (1991). The discipline of security studies and the codes of containment: Learning from Kuwait. *Alternatives, 16*, 315–344.
Mandel, E. (1975). *Late capitalism*. London: New Left Books.
Mannheim, K. (1936). *Ideology and Utopia*. London: Routledge.
Markovic, M. (1974). *From affluence to praxis*. Ann Arbor: University of Michigan Press.
Marsh, D. (1995). State theory and the policy network model, Paper 102, University of Strathclyde Papers on Government and Politics, Glasgow.
Marsh, D., & Smith, M. (2000). Understanding policy networks: Towards a dialectical approach. *Political Studies, 48*, 4–21.
Marx, K. (1992). *Das Kapital* (Vol. 1). London: Penguin.
Mayntz, R. (1976). Conceptual models of organizational decision-making and their application to the policy process. In G. Hofstede & S. Kassem (Eds.), *European contribution to organization theory* (pp. 114–125). Amsterdam: Van Gorcum.
McLennan, G. (1995). *Pluralism*. Buckingham: Open University Press.
Meisel, J. H. (1951). *The genesis of Georges Sorel*. Ann Arbor: Wahr.
Mittelman, J. H. (1999). Rethinking the 'New Regionalism' in the context of globalization. In J. Hettne, A. Inotai, & O. Sunkel (Eds.), *Globalism and the new regionalism, new regionalism series* (Vol. 1, pp. 25–53). London: Macmillan.

Mommsen, W. J. (1984). *Max Weber and German politics 1890–1920*. Chicago: University of Chicago Press.
Morgan, G. (1997). *Images of organization*. Thousand Oaks: Sage.
Negri, A. (2005). Postmodern global governance and the critical legal project. *Law and Critique, 16*, 27–46.
Neyer, J. (2003). Discourse and order in the EU: A deliberative approach to multi-level governance. *Journal of Common Market Studies, 41*(4), 687–706.
Nooteboom, B. (1992). A postmodern philosophy of markets. *International Studies of Management and Organization, 22*(2), 53–76.
Norcliffe, G., Bassett, K., & Hoare, T. (1996). The emergence of postmodernism on the urban waterfront. *Journal of Transport Geography, 4*(2), 123–134.
Norman, R. (1976). *Hegel's phenomenology. A philosophical introduction*. Brighton: Sussex University Press.
Norris, C. (1995). Versions of apocalypse: Kant, Derida, Foucault. In M. Bull (Ed.), *Apocalypse theory and the ends of the world* (pp. 227–249). Oxford: Blackwell.
Onuf, N. (1991). Sovereignty: Outline of a contemporary history. *Alternatives, 16*(4), 425–446.
Opello, W. C., Jr, & Rosow, S. J. (2004). *The nation-state and global order: A historical introduction to contemporary politics*. Lynne Reinner: Boulder CO.
Palmerston, 3rd Viscount (Henry Temple) (1848). Speech to the House of Commons, 1st March, *Hansard's Parliamentary Debates*. 3rd series, vol. 97, col. 122.
Parker, M. (1992). Post-modern organizations or postmodern organization theory? *Organization Studies, 13*(1), 1–17.
Parsons, T. (1951). *The social system*. NY: Free Press.
Peukert, D. (1989). *The Weimar Republic*. NY: Hill and Wang.
Popper, K. (1963). *Conjectures and refutations*. London: Routledge and Kegan Paul.
Portoghesi, P. (1983). *Postmodern: The architecture of the Postindustrial society*. NY: Rizzoli.
Puchala, D. J. (1972). Of blind men, elephants and international integration. *Journal of Common Market Studies, 10*(3), 267–284.
Pusey, M. (Ed.). (1987). *Jurgen Habermas*. London: Ellis Horwood/Tavistock/Routledge.
Rasmussen, D. (Ed.). (1990). *Reading Habermas*. Oxford: Basil Blackwell.
Reid, D. (1992). *Sex, death and God in LA, death and God in LA*. Berkeley: University of California Los Angeles Press.
Relph, E. (1987). *The modern urban landscape*. Baltimore: John Hopkins University Press.
Robertson, R., & Lechner, F. (1985). Modernization, globalization and the problem of culture in World Systems Theory. *Theory Culture and Society, 2*(3), 103–118.
Rosenau, J. N. (1997). The complexities and contradictions of globalization. *Current History, 96*, 360–364.
Rosenau, J. N. (2000). *The governance of Fragmegration: Neither a world republic nor a global interstate system*. Quebec: Congress of the International Political Sciences Association.
Rudd, K. (2001). Globalization and regional governance. In C. Sheil (Ed.), *Globalization: Australian impacts*. Sydney: UNSW Press.
Ruggie, J. G. (2003). The United Nations and globalization: Patterns and limits of institutional adaptation. *Global Governance, 9*, 301–321.
Sampson, A. (1995). *Company man: The rise and fall of company life*. London: Harper Collins.
Schoenberger, E. (1988). From Fordism to flexible accumulation: Technology, competitive strategies, and international location. *Environment and Planning D, 6*, 245–262.
Soja, E. W. (1989). *Postmodern geographies; the reassertion of space in critical social theory*. NY: Verso.
Soja, E. W. (2000). *Postmetropolis: Critical studies of cities and regions*. Oxford: Blackwell.
Soll, I. (1969). *An introduction to Hegel's Metaphysics*. Chicago: University of Chicago Press.
Sombart, W. (1913a). *Krieg und Kapitalismus*. Duncker und Humblot: Munchen.
Sombart, W. (1913b). *Luxus und Kapitalismus*. Duncker und Humblot: Munchen.
Sorel, G. (1908). *Les Illusions du Progress, translated as The Illusions of Progress by John and Charlotte Stanley*. Berkeley: University of California Press.

Spencer, H. (1851). *Social statics*. London: Chapman.
Steingard, D. S., & Fitzgibbons, D. E. (1993). A postmodern deconstruction of total quality management (TQM). *Journal of Organizational Change Management, 6*(5), 27–42.
Stoker, G. (2004). *Transforming local governance: From Thatcherism to new labour*. Palgrave: Basingstoke.
Swenson, J. (2000). *On Jean-Jacques Rousseau*. Palo Alto: Stanford University Press.
Taylor, F. W. (1911). *Principles of scientific management*. NY: Harper and Row.
Taylor, M. J., & Thrift, N. J. (1982). Models of corporate development and the multinational corporation. In M. J. Taylor & N. J. Thrift (Eds.), *The geography of multinationals* (pp. 14–32). London: Croom Helm.
Thompson, C. J. (1993). Modern truth and postmodern incredulity: A hermeneutic deconstruction of the metanarrative of "scientific truth" in marketing research. *International Journal of Research in Marketing, 10*, 325–338.
Thomson, P., & Sachs, G. (Eds.). (1994). *The Cambridge companion to Brecht, Cambridge companions to literature series*. Cambridge: Cambridge University Press.
Thrift, N. J. (1991). For a new regional geography 2. *Progress in Human Geography, 15*, 456–466.
Thrift, N. J. (1997). The rise of soft capitalism. *Cultural Values, 1*(1), 29–57.
Thurman, A. (1937). *The folklore of capitalism*. New Haven: Yale University Press.
Van de Loo, B., &Van de Velde, S. (2003). *Key Success Factors for Positioning Small-Island Sea Ports for Competitive Advantage*, Faculteit der Bedrifskunde/Rotterdam School of Management, Erasmus University, Rotterdam.
Van Ham, P. (2001). *European integration and the postmodern condition*. London: Routledge.
Vattimo, G. (1988). *The end of modernity*. Cambridge: Polity Press.
Virilio, P. (1999) Indirect light: extracted from Polar inertia. *Theory, Culture and Society, 16* (5–6), 57–70.
Warf, B. (1993). Postmodernism and the localities debate: Ontological questions and epistemological implications. *Tijds. voor Econ. En Soc. Geog., 84*(3), 162–168.
Weber, M. (1958). *The protestant ethic and the spirit of capitalism*. NY: Charles Scribner's Sons.
Weber, M. (1978). *Economy and society; an outline of interpretative sociology* (Vol. 2). Berkeley: University of California Press.
Weisskopf, V. F. (1981). Frontiers and limits of physical sciences. *Bulletin of the American Academy of Arts and Sciences, 35*, 2.
Whyte, W. H. (1957). *The organization man*. London: Jonathan Cape.
Wiener, N. (1948). *Cybernetics, or control and communication in the Animal and the Machine*. NY: Wiley.
Williams, R. (1989). When was modernism? *New Left Review, 175*(1), 48–53.
Wood, E. M. (1997). Modernity, postmodernity or capitalism? *Review of International Political Economy, 4*(3), 539–560.
Wright-Mills, C. (1959). *The sociological imagination*. NY: Oxford University Press.
Zurn, M. (2000). Democratic governance beyond the nation-state; the EU and other international institutions. *European Journal of International Relations, 6*(2), 183–221.

Chapter 7
Postmodernism

I don't know what you mean by 'glory' Alice said.
  Humpty Dumpty smiled contemptuously. "Of course you don't – till I tell you. I meant 'there's a knock-down argument for you!'"
  "But 'glory' doesn't mean 'a nice knock-down argument'" Alice objected.
  "When *I* use a word," Humpty Dumpty said in a rather scornful tone, "it means just what I choose it to mean – neither more nor less."
  "The question is," said Alice, "whether you can make words mean so many different things."
  "The question is," said Humpty Dumpty, "which is to be master - that's all."
  Alice was too much puzzled to say anything, so after a minute Humpty Dumpty began again. "They've a temper, some of them – particularly verbs, they're the proudest – adjectives you can do anything with, but not verbs – however I can manage the whole lot of them! Impenetrability! That's what I say!". (Carroll 1982, p. 124)

Postmodernism is summed up by Humpty Dumpty's recognition of the use of language and its possibilities of endless interpretation. This inherent flexibility and multitudinous characterization can be applied across the whole gamut of society including in our case the maritime sector and the policy-making institutions that make it up. It is this difficult move from the regimentation and clarity of Modernism to the confusion and choice of Postmodernism that forms the center of the policy-making debate. Whether Postmodernism has a relationship to policy-making and governance will be considered throughout this section. Gaudin (1998, p. 48) for example, had no doubt seeing that modern governance needs to make reference to 'Post-modernity' and to the 'homogenization of action which currently runs counter to the Weber-style specializations of classic management administration.' Read Modernist for Weber, and the existing maritime policy-making institutional structures that are characteristically Modern and you have the problem in one.

The sheer breadth of Postmodern territory is hard to comprehend and this is only matched by its diversity. For example take breadth:

> During the last weeks of his life Vaughan thought of nothing else but her death, a coronation of wounds he had staged with the devotion of Earl Marshal.
>
> In his vision of a car crash… Vaughan was obsessed by many wounds and impacts – by the dying chromium and collapsing bulkheads of their two cars meeting head-on in complex collisions endlessly repeated in slow-motion films, by the identical wounds inflicted on their bodies, by the image of the windshield glass frosting around her face as she broke its tinted surface like a death-born Aphrodite, by the compound fractures of her thighs impacted against their handbrake mountings, and above all by the wounds to their genitalia, her uterus pierced by the heraldic beak of the manufacturer's medallion, his semen emptying across the luminescent dials that registered for ever the last temperature and fuel levels of the engine.

Both taken from Ballard (1973, pp. 7–8).

Ballard's marriage of sex and car crashes is classic Postmodern literature in this case with a transport theme. In this section we hope to show how maritime policy failure is also a classic Postmodern reaction to the failure of the governance framework to react to the moves that have taken place in society from a Modernist to a Postmodernist scenario.

And then take diversity:

> Postmodernism is a hydra-headed monster and a chameleon, impossible to characterize without entering into life-threatening contradictions… Postmodernism could be summed up as a belief that large-scale ideas and political philosophies are inherently dangerous… In its place we need to celebrate the fragmentation of subjectivity. (Graham 1988, p. 61)

There is an enormous literature on Postmodernism which covers a vast range of topics from engineering to politics, arts to science, social sciences to architecture and much more. We can only touch on the most relevant in the coming sections but indications of significant contributions to the broader debate can be found particularly at Benko (1997, pp. 9–10) but also from Antin (1980), Greenberg (1979), Kristeva (1980), Habermas (1981), Foster (1983, 1985), Jameson (1985), Lash (1988), Arrington and Francis (1989), Cooper (1989), Bailey et al. (1993), Boje and Dennehy (2008), Dijkink (1993), Ley (1983), Short (1993), Warf (1993a, b), Radhakrishnan (1994), Dicken (1994), Kumar (1995), Clark (1996), Rhodes (1997) and Chan (2000).

7.1 The Origins of Postmodernism

> Firstly… nothing exists; secondly… even if anything exists, it is inapprehensible by man; thirdly… even if anything is inapprehensible, yet of a surety it is inexpressible and incommunicable to one's neighbor. (Gorgias, fifth Century BC; taken from Sextus Empiricus 1935, p. 179)

Christopher Jencks (1986a, p. 9) dates the symbolic end of modernist architecture and the passage to the post-modern as 3.32 pm on July 15th, 1972 when the Pruitt-Igoe Housing

7.1 The Origins of Postmodernism

Fig. 7.1 Pruitt-Igoe housing development, St Louis, USA. *Source* US Geological Survey

development (a version of Le Corbusier's 'machine for modern living') was dynamited as an unlivable environment for the low-income people it housed Harvey (1987, p. 260) (Fig. 7.1).

Boje et al. (1996, pp. 2, 33) suggested that the origins of Postmodernism were underlain by three themes. It was a new and distinct social order (Giddens 1990, p. 46); it was a clear break with the past forming a new era—something to which we will return later; and it was a social era that followed on from twentieth century Modernism (Rosenau 1992a; Newman 1995, p. 257), the latter despite the clear Postmodern suggestions within Gorgias's fifth century BC comments. Allmendinger (2001, p. 27) agreed although he did point out that not everyone sees Postmodernism as a distinctly different phase from Modernism citing in particular Berger (1998, p. 12). Boje et al. went on to consider that Postmodernism emerged as a consequence of the bankruptcy of traditional authority based on 'rational/legal grounds' and organizational fiefdoms'. A similar process can be seen going on in the maritime sector where traditional organizations such as the IMO, national ministries, and the EU have developed fiefdoms which no longer serve those for whom policies have been derived.

Etzioni (1968, p. vii) claimed that we had entered a Postmodern era but proceeded in the following 670 pages to never define what Postmodern meant (Bell 1974, p. 53). His only clue to his definition is located in the preface:

A central characteristic of the modern period has been continued increase in the efficacy of the technology of production which poses a growing challenge to the primacy of the values they are supposed to serve. The Post-modern period, the onset of which may be set at 1945, will witness either a greater threat to the status of these values by the surging technologies or a reassertion of their normative priority. Which alternative prevails will determine whether society is to be the servant or the master of the instrument it creates.

Bell's comment is that the Postmodern era is thus only a question. Meanwhile Porter (1968, pp. 12–13) also commented the same year that a new stage of major industrial societies—which he termed the Postmodern—faced a shortage of trained labor because of new trends in the demands by young people to go on to college in the USA.

Meanwhile Benko (1997, p. 10) noted the use of the word Postmodern by the English painter Chapman in 1880, and Welsch (1987) identified the same usage by the German writer Rudolf Pannwitz in 1917, who as a 'nietzscheist' proclaimed in *Die Krisis der europaischen Kultur*:

> Postmodern man, toughened by sports, trained militarily, and fired by metaphysics, is a hard-shelled creature, an equal mix of decadent and barbarian, swept along on the outpourings from the fertile vortex of the final decadence of the radical revolution of European nihilism. (Quoted in Breisach 2011)

Hassan (1980, pp. 117–118, 1985, p. 120) identified de Onis's use of the word in his writings in 1934 (de Onis 1961) and also Fitts (1942), both referring to 'a minor reaction to Modernism'. This is continued in Toynbee's 'A Study of History' (Toynbee 1946). Denzin (1986, p. 195) considered its first use was in the USA to describe changes in dance and architecture in 1949, again in 1974 and once more in 1985 (Jencks 1986a). In the 1970s Postmodernism moved to Europe through the work of Lyotard, Foucault, Kristeva, and Habermas although Wright-Mills had used the terminology in 1959 (Wright-Mills 1959, p. 166; Huyssen 1984) to describe the significant changes that were occurring at the time which he attributed to a 'new epoch'. Porter (1968, pp. 5–6) associated the Postmodern with the rise of a cybernated society, representing the second industrial revolution and characterized by information technology and the increase in technical speed (Michael 1962, p. 185ff). Burke (2000) stressed its Modernist roots, the latter emerging from the Enlightenment of the eighteenth century which in turn possessed three basic values:

- Intellectually: power of reason over ignorance.
- Power of order over disorder.
- Power of science over superstition (and religion).

Modernity was thus revolutionary (evidenced through the French [1789] and Russian [1917] revolutions amongst others). Capitalism and the nation-state were major strands that emerged through the Enlightenment and which changed the social order providing the basis for permanent social progress (Simon 1996). Essential to this was the application of reason: humanity could be emancipated from ignorance, poverty, insecurity, and violence (Leonard 1997).

Burke suggested that in the 1970s a new movement emerged amongst French intellectuals which questioned this view of society and some interpreted the changes taking place around them as a new phase in capitalism. This was Postmodernism, viewing the horrors of two world wars, Stalinism, neo-colonialism,

7.1 The Origins of Postmodernism

Eurocentrism, racism, and third world hunger as representing the true face of the twentieth century. Modernism was therefore now an outdated concept—albeit one which clearly had emerged from the Enlightenment—and was being replaced by Postmodernism—itself a creation of the capitalist society in which we live.

Carroll and Henry (1975) suggested that Postmodernism was a period of 'major historical transition' but one which was easier to define as the end of something identifiable rather than a clear beginning of something new although they were unwilling to date its emergence even roughly. Dijkink (1993) provided a geographical slant on its origins remarking that the 'roots of Postmodernism are to be found in certain shifts in postindustrial society' combined with the development of mass communication, a failure of social studies to fully encompass its contradictions and complexities and disappointment with the success achieved in social control. Parker (1992), along with Gergen and Thatchenkerry (2004, p. 229), viewed the origins of Postmodernism in an organizational context. Such organizations are what Lash (1990) termed 'de-differentiated', flexible, serving niche markets and with multi-skilled workforces. The whole organization is held together by a mix of information technology and specialist subcontractors. Rouleau and Clegg (1992) felt that Postmodernism emerged in the 1960s beginning in esthetics and architecture and moving quickly on to the humanities and culture more generally (Galinsky 1992). By the 1980s it could be found throughout the social sciences including management, finance, accounting, economics, business, and even logistics and maritime affairs. Meanwhile, Jencks (1986a) saw Postmodernism as emerging from around 1960 but with considerably longer roots than most other commentators. For example his definition of Modernism which was an essential precursor to Postmodernism, dates back to 1450 with the pre-modern phase, an essential forerunner of this, apparent from 10,000BC (Table 7.1).

Postmodernism sees all previous relations of production changed and a new value system introduced. This centers on information which is seen as ever increasing in size as it reproduces itself (rather than consumes itself) with use. Traditional divisions of power are totally altered and the polarities of society are revised to form a complex mélange of interests founded upon knowledge and skills with power dispersed into new and sometimes surprising locations acting alongside traditional seats. The maritime sector reflects all these trends.

Walker (1991, p. 445) felt that although the Postmodern had only been recently recognized in fact its characteristics of fracturing and dispersal that had been noted by Benko (1997, p. 7) blurring existing norms, could be identified over 'several centuries'. Warf (1993a, p. 163) considered that the emergence of Postmodernism reflected several parallel developments from a number of disciplines. These included Soja's (1989) 'reassertion of time and space'; the 'revival of regionalism' and 'local uniqueness' (Warf 1988); 'explorations of the questions of language, knowledge and ideology' (Giddens 1984); the 'flowering of textual analysis, literary theory and deconstructionism' with their 'multiple, contradictory meanings'; the 'maturity of ethnomethodology' (Geertz 1983); the 'demise of structuralism and teleological explanation'; and the 'shift away from class based

Table 7.1 Three types of society based on their major form of production

	Production	Society	Time	Orientation	Culture
Pre Modern 10,000 BC –1450 AD	*Neolithic revolution*	*Tribal/feudal*		*Local/city*	*Aristocratic*
	Agriculture	Ruling class of kings, priests, and military	Slow-changing	Agrarian	Integrated style
	Handwork Dispersed	Peasants	Reversible		
Modern 1450–1960	*Industrial revolution*	*Capitalist*		*Nationalist*	*Bourgeois*
	Factory Mass-production Centralized	Owning class of bourgeoisie workers	Linear	Rationalization of business Exclusive	Mass-culture reigning styles
Post-modern 1960–?	*Information revolution*	*Global*		*World/local*	*Taste-Cultures*
	Office Segmented production	Para-class of cognitant	Fast-changing Cyclical	Multinational Pluralist Eclectic	Many genres
	Decentralized			Inclusive	

Source Jencks (1986b, p. 47)

models of social analysis to other forms of social determination such as gender and ethnicity' (Kellner 1988). Warf stressed how these all exhibited a:

> ... deep appreciation for the complexity of social determination, the openness of social systems, the heterogeneity of social life, and the importance of consciousness and subjectivity.

Bertens (1995) suggested that the origins of Postmodernism were much earlier with citations starting from 1870 (Welsch 1983, 1987) and featuring throughout the 1920s, 1930s, and 1940s. However, the use of the term Postmodernism was fragmentary and clearly discontinuous. It was in literary criticism that the first coherent strands of the term became apparent with the work of Charles Olson (Olson 1967). Through the 1950s, 1960s, and 1970s Postmodernism continued to infiltrate new areas of the arts and architecture in particular with the work of Hassan (1967, 1970, 1971, 1973, 1978) especially influential. Developments in and extensions to Postmodern influence followed throughout the 1980s and into the twenty first century where today it remains a central interpretation of society and culture and has spread its interests and application to all areas beyond its traditional focus on arts and architecture to organization, business, and society in general as we shall see later. The maritime sector in all its facets—economy, politics, culture, technology, organization, and the like—is no exception, thus suggesting the need for an appreciation of this new context for governance.

7.2 The Characteristics of Postmodernism

Before we go on it might be helpful to attempt to define what we mean by Postmodernism—not an easy task as the very nature of Postmodernism and its application to all aspects of society makes its definition variable, flexible, and commonly redundant as soon as it is agreed.

Harvey (1990, p. 116) provided an extensive discussion of Postmodernism and what he, along with others (for example see Jencks 1992b, p. 13), defined as the Postmodern condition:

> ... with its emphasis upon the ephimerality of *jouissance*, its insistence upon the impenetratability of the other, its concentration on the text rather than the work, its penchant for deconstruction bordering on nihilism, its preference for aesthetics over ethics.

Harvey saw Postmodernism as taking matters (and particularly relevant to our consideration of maritime policy-making):

> ... beyond the point where any coherent politics are left, while that wing of it that seeks a shameless accommodation with the market puts it firmly in the tracks of an entrepreneurial culture that is the hallmark of reactionary neo-conservatism.

He considered Postmodernism as rejecting all meta-theories that are capable of grasping conventional (i.e. Modernist and traditional institutional-bound) political-economic processes (and here he included elements that are central to the maritime policy-making framework—money flows, international divisions of labour, financial markets) and as such it plays directly into the hands of what he termed the reactionary enemy. In this he meant the market which then can extend its power over the whole range of culture—arts, architecture, society, business, politics, in fact everything—generating an entirely new scenario for the renewal of capital accumulation, distinctly different from the Modernist scene that was the background for the institutional framework that continues to generate and implement maritime policy-making today. What hope is there with a policy-making culture of institutional, state-led formalism attempting to administer governance for an industry steeped in consumerism, informality, and state confusion?

In truth, very few people are clear about what the term Postmodernism means. It is easier to define when Postmodernism is apparent than to describe what it represents or what makes it up. However, we define it—and we shall make some attempt in this section, it has become widely used as a term and accepted that a Postmodern condition in society clearly exists. Dear (1988, p. 265) attempted to guide those inexperienced with Postmodernism beginning with a quote which in some ways summed up much of what we are about to consider:

> A mix-in, for those who have not yet followed aerobic eating into its Postmodern era, may be butterscotch chips and walnuts, pulverized Reese's peanut butter cups, crushed Oreos, M&Ms or – in some temples of asceticism – granola (Notice on carton of Steve's Ice Cream, USA).

While unhelpful for those still struggling with understanding the Postmodern condition, this quote from a real-life ice cream carton gives an idea of the chaotic and undisciplined nature of the Postmodern.

Webster (2002) provided an extensive discussion on the meaning of Postmodern while Dear (1988) suggested that it represented an attack on contemporary philosophy and saw it as directly opposed to Modernism. It therefore is a 'philosophic culture freed from the search for ultimate foundations or the final justification'. In practice, it has commonly taken on the form of a revolt against all parts of society and culture questioning the dominance of any one discourse over any other. The result is that all prior paradigms and theoretical frameworks, including those that characterize and reflect maritime policy-making during Modernist times, lose their 'privileged status'. In the context of our work, capitalism's desire for accumulation has challenged the Modernist institutional framework of the maritime sector and created the need for a Postmodern response. Unfortunately no such response has occurred. The result is governance and policy failure.

Cooper and Burrell (1988) saw Postmodernism as centered around the concept of difference and accommodating differences within everything—culture, society, business, organization, policy, arts, technology, and so on. Derrida (1973) suggested that difference is at the center of all social discourse and this 'irreducible indeterminacy' is placed at the 'vanguard' of the endeavors of Postmodernism. This instability is found everywhere today—Lyotard (1979) cited modern scientific thought as an example where more knowledge creates less predictability, more data simply raise more questions—and this interpretation can be applied just as well to policy making and governance. The concept of dialectics which was discussed in the previous chapter raises its head again at this point, focused on the perpetual struggle that goes on in all discourses.

Thrift (1997, 32) provided an extensive consideration of Postmodernism seeing it as a challenge once again to Modernism and to the idea that 'a God's eye view of reason is possible'. He suggested that:

> There are, instead, many rationalities and these rationalities are all: embodied, relying on our bodily natures; able to engage the emotions since feeling is conceptualized and conceptualization always involves feeling; based on a notion of meaning involving symbols which are constitutive of the world and not just mirrors of it; and reliant on categories that are not independent of the world but are defined by mental processes… which mean that there can be no objectively correct description of reality.

At this stage it may seem that we have come a long way from maritime policy-making and governance. However, in time we will show the linkages that exist and also propose changes that will help to draw maritime policy-making into the Postmodern society that surrounds, contains, and influences it. For the time being the issues of human interpretation, flexibility, and indetermination, all inherently Postmodern and all largely missing from Modernist maritime policy-making and governance, can help us to light the way.

7.2 The Characteristics of Postmodernism

Rhodes (1997) assessed what he saw as Postmodernism in an organizational setting, quoting Rosenau (1992a). He saw modern social science as requiring:

> simplification. It surrenders richness of description and feel for complexity in return for approximate answers to the questions, often limited in scope, that are eligible for consideration within its own terms.

This deliberate constriction of what a Modernist interpretation of priorities includes is close to some of the problems faced within maritime governance where there is the tendency to focus on traditional preconceived issues by a limited range of conventional and narrowly focused institutions. Simply think of the public consultation which took place in 2007 for the Blue Paper of Maritime Policy of the EU.

Meanwhile Rhodes considered that in a Postmodern environment:

> Only an absence of knowledge claims, an affirmation of multiple realities, and an acceptance of divergent interpretations remain.

The variability, flexibility, all-embracing nature of Postmodernism is stressed again reflecting a need for maritime governance to be similarly defined.

Gill (2000, p. 131) provided an alternative definition which implies the degree of debate that has taken place about the Postmodern condition generally, suggesting that Postmodern:

> refers to a set of conditions, particularly political, material, and ecological that are giving rise to new forms of political agency whose defining myths are associated with the quest to ensure human and intergenerational security on and for the planet, as well as democratic human development and human rights.

This all-embracing view suggests that Postmodernism cannot be confined solely to the arts and architecture—a common misconception—and in fact belongs to all facets of society—not least the maritime sector, seeking in Gill's terms to 'develop a global and universal politics of radical (re)construction'.

The breadth of Postmodern interpretation is continued through Rosenau (1989, 2004) stressing that the concept has been applied in public administration (Caldwell 1975; French 1992), planning and management (Carter and Jackson 1987), organization theory (Burrell 1988; Clegg 1990; Cooper and Burrell 1988; Cooper 1989), total quality management (Steingard and Fitzgibbons 1993); accounting (Arrington and Francis 1989; Loft 1986; Hoskin and Macve 1986) and systems analysis (Pollack 1993; Nodoushani 1987) amongst others. Both Kreiner (1992) and Ruggie (1993) emphasized Jameson (1985) and his interpretation of Postmodernity as central to political economy, a Marxist view of affairs that was subsequently taken on by Harvey (for example 1990), Brenner (1997, 1998, 1999) and others. Quoting Jameson (1984, p. 88), Postmodernism depicts:

> the third great original expansion of capitalism around the globe (after the earlier expansions of the national market and the older imperialist system)... global capitalism today internalize(s) within its own institutional forms relationships that previously took place among distinct national capitals.

In this we can see the whole syndrome of maritime governance failure come together within a Postmodern scenario. The process of capitalism's third expansion has centered on globalization and the desperate search for a spatial fix to ensure continued demand and supply for the market. This in turn has changed the role of the nation-state substantially facilitating the growth and liberation of the shipping sector to the extent that it now can operate almost outside nation-state regulation. At the same time the nation-state remains the building block of all maritime policy-making institutions. And the result is inconsistency, incoherence, and failure in the policies generated.

Postmodernism therefore has a close relationship to globalization and together they need to be reflected in systems of governance (maritime included) if policy making is to be effective. Waterman emphasized the relationship clearly citing a number of sociological reference points (Beck 1992; Giddens 1990; Hall et al. 1992; Harvey 1990, 1996; Melucci 1989; Poster 1984, 1990; de Sousa Santos 1995). Amis (2003, p. 12) contrasted the new Postmodern with its immediate predecessor in the context of 'left' politics:

> There were demonstrations, riots, torching, street battles in England, Germany, Italy, Japan and the USA. And remember the Paris of 1968: barricades, street theatre, youth worship ('The young make love, the old make obscene gestures'), the resurgence of Marcuse (the wintry dialectician), and Sartre standing on street corners handing out Maoist pamphlets… The death throes of the New Left took the form of vanguard terrorism (the Red Brigades, the Baader-Meinhof gang, the Weatherman). And its afterlife is antichristic, opposing itself to the latest mutation of capital: after imperialism, after fascism, it now faces globalization.

Meanwhile Lacher (2005, p. 27) stressed in discussing international relations, how there was a very close relationship among the demise of sovereignty, the emergence of the Postmodern, and the development of a modern system of global politics. Jencks (1992a) suggested that a Postmodern society was only possible because of technological development and the information society which has followed based upon a network of world producers and consumers and which in turn relies upon an integrated system of communication that is quick and effective.

Radhakrishnan (1994, p. 307–308) took a different view although accepting that the Postmodern has essential global characteristics, requirements, and impacts. He also identified a simultaneous regional impact in that more local developments through global communication have the ability to remain regional and spread their influence.

Rosenau (1992b, p. 1) quoting Edelman (1977, 1988) saw Postmodernism as a 'philosophy, an approach, a world view, a form of critique, a method, a guide to political (in)action' all of which apply directly to maritime policy-making. Extreme views suggest that to summarize any Postmodern activity is to 'do violence' to it (Ashley and Walker 1990). Lyotard (1988) even claimed that because Postmodernism has a 'special logic' it defies efforts at clarification.

Norcliffe et al. (1996) presented one of the few Postmodern interpretations of the maritime sector in looking at the urban waterfront and stressed how Harvey (1990) emphasized how Postmodernism has strong links to the production side of

7.2 The Characteristics of Postmodernism

the economy. Meanwhile Simon (1996) took transport development in the third world and applied a Postmodern perspective focusing on four main characteristics:

- The downplaying of universal theories and methodologies.
- The rise of pluralism with a range of more local and sensitive perspectives.
- Downgrading of the primary role of the nation-state in favor of more local and responsive forms of organization.
- Increase in diversity to reflect the increasingly mixed culture and society of today.

Simon applied these principles to transport and the development of the third world suggesting that the old, Modernist model of north versus south with the former facilitating the development of the latter at the same time as abusing the dominant position held, is no longer tenable. Parallels can be seen that despite maritime power remining in TMNs, there has been a rise of third world flags, of fleets from beyond the traditional maritime nations, and a dramatic growth in third world piracy. There has been a concomitant need in a fully globalized world to take account of changing maritime pressures and events in all parts of the world.

Radhakrishnan (1994) following Lyotard, continued this theme and suggested that Postmodernism is merely a condition rather than an 'item', founded on knowledge and was taking place almost exclusively within the 'first world' while its impacts were global.

Meanwhile Hebdige (1988, pp. 181–182) quoted at length in Woods (1999, pp. 2–3) made it clear that there is some debate over whether Postmodernism as a concept has any intellectual rigor as its foundation. The quote is repeated here as a good example of the debate that continues about the validity of the Postmodern condition:

> When it becomes possible for people to describe as 'Postmodern' the décor of a room, the design of a building, the diagesis (narrative) of a film, the construction of a record, or a 'scratch' video, a television commercial, or an arts documentary, or the 'intertextual' relations between them, the layout of a page in a fashion magazine or critical journal, an anti-teleological... tendency within epistemology, the attack on the 'metaphysical of presence... a general attenuation of feeling, the collective chagrin and morbid projections of a post-war generation of baby-boomers confronting disillusioned middle age, the 'predicament' of reflexivity, a group of rhetorical tropes, a proliferation of surfaces, a new phase in commodity fetishism, a fascination for images, codes and styles, a process of cultural, political or existential fragmentation and/or crisis, the 'de-centering' of the subject, an 'incredulity towards meta-narratives', the replacement of unitary power axes by a plurality of power/discourse formations, the 'implosion of meaning', the collapse of cultural hierarchies, the dread engendered by the threat of nuclear self-destruction, the decline of the university, the functioning and effects of the new miniaturized technologies, broad societal and economic shifts into a 'media', 'consumer' or 'multi-national' phase... or... a generalized substitution of spatial for temporal coordinates – when it becomes possible to describe all these things as 'Postmodern'... then it's clear we are in the presence of a buzzword.

He went on (Hebdige 1988, p. 181):

it becomes more and more difficult... to specify exactly what it is that 'Postmodernism' is supposed to refer to as the term gets stretched in all directions across different debate and different disciplinary and discursive boundaries, as different factions seek to make it their own, using it to designate plethora of incommensurable objects, tendencies, emergencies.

Despite these doubts, the sheer volume of debate suggests that something has (or is) happening in society that is widespread, cross-cultural, multifaceted and as relevant to commerce and industry as it is to the arts and literature. This something may as well be termed Postmodernism as anything else and manifestly affects the maritime sector as much as anywhere else. To ignore its effects is to ignore a major societal influence upon policy effectiveness and to continue to channel policy making through a governance structure that bears no resemblance to a Postmodern society is to be simply negligent.

There are many other definitions of Postmodernism including attempts by Lash and Urry (1987), Burrell (1988), Harvey (1990), Fox et al. (1992) and Schultz (1992, p. 16) who saw Postmodernism focusing on multiplicity, difference, and discontinuity and as such contrasting with the traditional formalism of Modernity, 'peeling off the layers of Modernist illusions' acting as a kaleidoscope in which conceptual fragments interrelate in different ways whenever (it) is shaken'. Tyler (2005) discussed Postmodernism in the context of crisis communication citing alternative views by Derrida, Lyotard, Baudrillard, Nietzsche, and Foucault (Boje et al. 1996) and Denzin (1986) who referred explicitly to its emergence from the criticism of scientific knowledge and the realism of the late capitalist era, and who quoted Baudrillard (1983) who saw the Postmodern experience as a form of social organization in which polity, economy, culture, and the mass media endlessly reproduce one another. Arrington and Francis (1989) suggested it centered upon the challenges that can now be made of any authority's claim to a method of knowledge production that is privileged over others—in our case the structured and established process of policy making and the associated institutions. It rejects the Modernist traditions of 'rationality, theory, facticity and history' as evidence of progress.

Parker (1992, p. 3) cited Heydebrand (1989, p. 345) in suggesting that we are witnessing in organizations a:

thoroughly intentional, conscious Postmodern strategy of increasing the flexibility of social structures and making them amenable to new forms of indirect and internalized control, including cultural and ideological control.

Boje et al. (1996) spent considerable time and effort trying to settle on a definition of Postmodernism. Suggesting initially that it is the social era immediately after Modernism (Rosenau 1992b), involving something 'different, a break with the past' and distinct social order (Giddens 1990). They took Offe's (1984) interpretation that Postmodernism centers upon how 'phenomena outside the market are brought into the economy and acquire economic use values, something he termed 'commodification'. Postmodernism is central to the commodification of everything possible, a new 'privatized society' and where consumption is an end in itself. This societal shift in turn generates a need for new governance frameworks for all aspects of life. Quoting Jameson (1991):

> Postmodernism thus involves a profound transformation in the logic of modern, industrial society, based on changes in the processes of commodification... A form of society wherein culture has greatly expanded and Nature has been replaced with technology as the 'other' in our society.

Of particular relevance to our discussion is the notion that Postmodernism is characterized by transnational business, a new international division of labor, computerization and automation. Organizational hierarchies are seen to have disappeared through 'corporate downsizing, restructuring and reengineering' (Keidel 1994; O'Neill 1994) to be replaced by market networks.

7.3 The Relationship of Postmodernism to Modernism and Post-Fordism

> From the Modernism you choose you get the Postmodernism you deserve (Antin 1972).

Earlier, we suggested that Postmodernism was a product of a sequence of societal change that has affected all that surrounds us as it has progressed from Pre-Modernism through Modernism to the current Postmodern phase, each representing capitalist societal forces generating consumption of capital. The current problems facing maritime sector policy-makers is a manifestation of the emergence of the relatively new Postmodern era, a consequence of the lag between the maritime institutional framework and policy-making processes and the globalization of society—itself a Postmodern effect. In this section we shall try and show how Postmodernism is linked to Modernism and how this inseparability is central to the current Postmodern condition that creates the policy failures in the maritime sector.

The relationship between Modernism and Postmodernism is one that has been considered by many but there has also been some discussion on the process of change emerging from the Post-Fordist era.

Fordism is a widely understood concept that refers to 'the centrality of industrial production' (Soja 2000) and characterizes the post-war economic boom of the late Modernist era—'controlled, hierarchized, stable' (Stalder 2006, p. 46). It is not the same as Modernism but has many common elements and features that can make discussion of both confusing at times. As a termm, it has its roots in the writings of Gramsci (1971) and has been widely associated with the subsequent emergence of a Post-Fordist and Postmodern era including the work by Harvey and Scott (1989).

Post-Fordism is again commonly confused with the more generic process of Postmodernism and there has been much debate about whether it represents a clean break from Fordism, whether it is simply a development or whether it even exists at all. Soja (2000, p. 170) concluded that it represented a:

> movement beyond its established regime of accumulation and mode of regulation to a significantly different economic order… a deconstruction (not destruction or erasure) and reconstitution (still only partial, ongoing, and incorporating selective components of the older order) of Fordist and Keynesian political economics.

Whereas Fordism was a specific 'composite of economic practices that are ideally typified in the production-processes and labor-management relations initiated by Henry Ford', Post-Fordism implies the development of 'an alternative 'emergent' configuration of propulsive economic practices, significantly different but not completely disconnected from Fordism' (Soja 2000, pp. 170–171). It is vaguer than Fordism and more open-ended and as such mirrors that of Postmodernism. However, its range of application is more constrained and typically applied to urban-industrial capitalism more than anywhere else. Its links with the Modernist–Postmodernist transition are clear in particular in its association with flexibility in an economy, in turn emerging from successive economic and financial crises, and the move from a heavily structured form of production to Harvey's 'flexible accumulation', a central feature of Postmodernism in the urban-industrial-economic scene.

Rhodes and Marsh (1992) identified a movement in the UK economy from mass to small batch production and from hierarchical to network relationships, both reflecting a Post-Fordist trend occurring at the same time, as others were suggesting a Postmodernist movement.

Burke (2000) saw Fordism characterized by mass production and consumption, sustained by mass marketing, the protection of national markets, and intervention by the state to guarantee demand. Post-Fordism has been facilitated as much as anything by the development of IT which has globalized markets, destroyed national priorities and facilitated the splitting of mass markets and the targeting of specific groups and demands. All these latter characteristics feature in a Postmodern society. Flexibility in the market is again a key feature in the move from a Fordist to a Post-Fordist society and disorganization characterizes this change, something reiterated by Hess (2004). This view is supported by Arrighi (1994, p. 2) who saw a crisis in Fordism and its 'vertically integrated, bureaucratically managed, giant corporations' (think IMO) creating opportunities for 'flexible specialization' coordinated by a network of market exchanges (Piore and Sabel 1984; Sabel and Zeitlin 1985, 2008; Hirst and Zeitlin 1991).

Boyer and Durand (1993) provided two visual expressions of the decline in Fordism and its problems in the context of an emerging flexible society (Figs. 7.2, 7.3).

In both cases the transition from Modernism to Postmodernism is clearly apparent in the move from Fordism to Post-Fordism and this can be applied to the maritime sector as much as elsewhere. For example, in Fig. 7.2, we find the new Post-Fordist model exhibiting the Postmodern trends of flexibility, differentiation, and market reaction each of which are far more apparent in the Postmodern vision

7.3 The Relationship of Postmodernism to Modernism and Post-Fordism

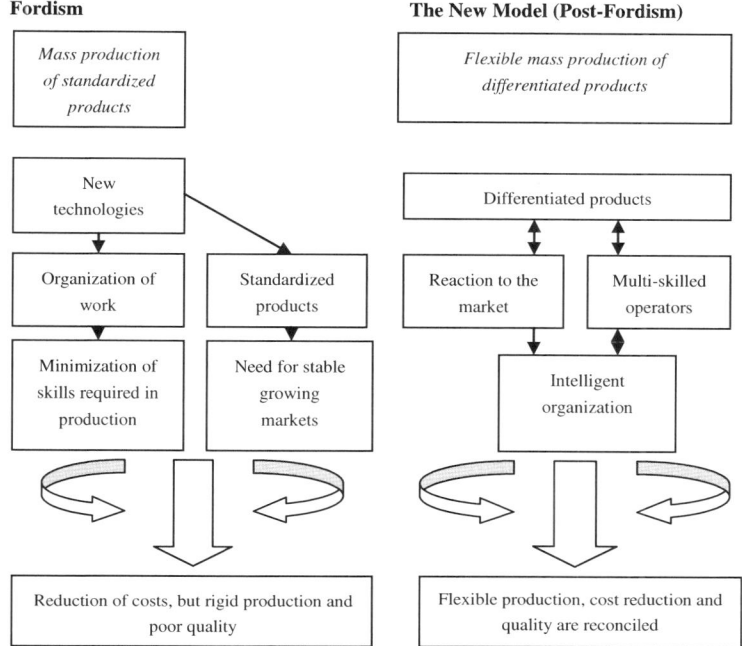

Fig. 7.2 The two production models of Fordism and Post-Fordism. *Source* Derived from Boyer and Durand (1993, p. 32)

of the maritime sector than that of the Modern. Meanwhile in Fig. 7.3, the deterioration in the Fordist model of the maritime sector is also evident with the increasing irrelevance of centralized decision-making, the destabilizing effects of new technology and new companies, and the problems of control and organization of the policy-making process. There is considerable evidence from these figures that the process of Fordism–Post-Fordism partially mirrors that of Modernism to Postmodernism and is reflected in the widest way in the maritime sector. Policy-making institutions remain locked in their Modernist/Fordist origins resulting in ineffective governance.

The significance in our case of the Fordist/Post-Fordist debate is summed up by Harvey (1990, p. 109) who suggested that:

> The tension between the fixity (and hence stability) that state regulation imposes, and the fluid motion of capital flow, remains a crucial problem for the social and political organization of capitalism.

In other words the fixity of the Fordist/Modernist model and its inherent state contrasts with the fluidity of capital that is an intrinsic formulator and a consequence of globalization and the creation of the Post-Fordist/Postmodern market society that characterizes the late twentieth and early twenty-first centuries. The Fordist/Modernist institutional framework for maritime policy-making, epitomized

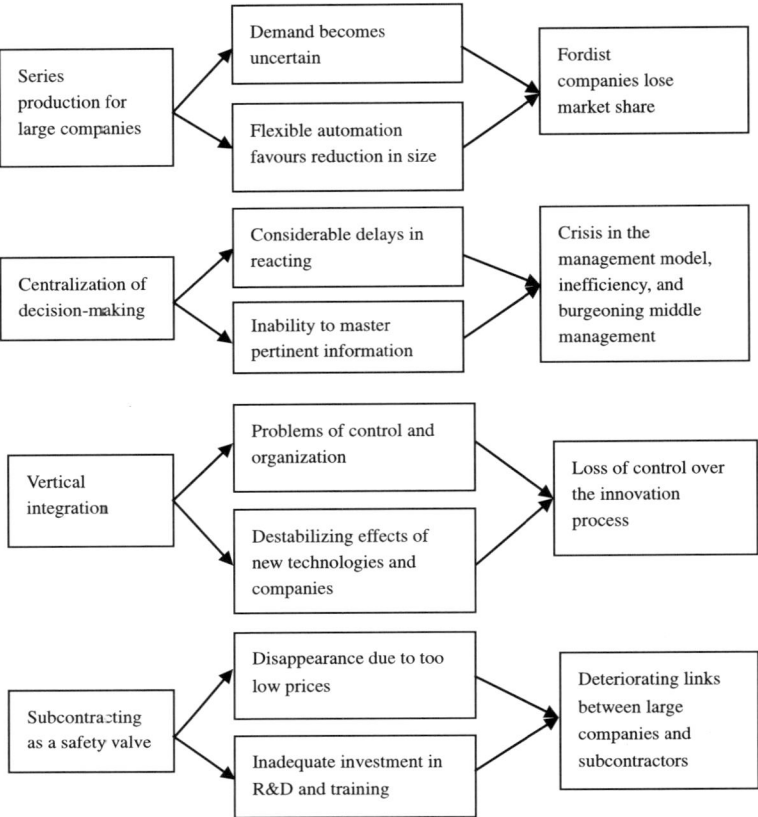

Fig. 7.3 Fordist organization proves ineffective following macroeconomic changes. *Source* Boyer and Durand (1993, p. 14)

by the IMO, OECD, WTO, EU, national ministries, and the like accompanied by a vast range of other nation-state orientated institutions with significant if indirect influence upon maritime policy—the International Monetary Fund (IMF), World Bank, etc.—remains central and has shown little enthusiasm for change to mirror that of the globally intense maritime sector. This governance mismatch has predictable (and tragic) results. To quote Harvey again:

> the more flexible motion of capital *(which Harvey stresses is a result of the transition from a Fordist model)* emphasizes the new, the fleeting, the ephemeral, the fugitive, and the contingent in modern life, rather than the more solid values implanted under Fordism. *(section in italics added)*

Hajer (2003) suggested that Postmodernism is a direct reaction to the instability that emerged in the Modernist institutions that dominated broader society—in maritime terms the IMO, EU, national ministries, etc. These institutions he saw as

7.3 The Relationship of Postmodernism to Modernism and Post-Fordism

representing the 'codified arrangements that provide the official setting of policy making and politics in the post-war era of Western societies'. In turn, he saw these arrangements as characterized by:

- Representative democracy.
- Differentiation between politics and bureaucracy.
- Commitment to ministerial responsibility.
- Policy making based on expert knowledge.

Each can be viewed as laudable but their principles and operational effectiveness eroded with the spread of globalization making the arrangements noted above anachronistic. Hence policy failure. New 'political spaces' have emerged that reflect the instability of the new system caused by the institutional/globalization incompatibilities. Existing maritime policy-making institutions are unable to resolve problems legitimately or effectively as they are rooted in a Modernist framework yet operating in a Postmodern environment. The emergence of NGOs, consumer organizations, and previously unimagined stakeholders are evidence of the friction between the Modernist and Postmodern eras. To quote Hajer (2003, p. 176):

> the constitutional rules of the well-established classical-Modernist polities do not tell us about the new rules of the game. In our world the polity has become discursive; it cannot be captured in the comfortable terms of generally accepted rules... the polity, long considered stable in policy analysis, thus becomes a topic for empirical analysis again. As politics is conducted in an institutional void, both policy and polity are dependent on the outcome of discursive interactions.

Harvey (1990) concurred citing Toffler (1970) in identifying a manifest societal change from Modernist to Postmodern with the development of a 'temporariness in the structure of both public and personal value systems' providing the context for the 'crack-up of consensus' within a 'fragmenting society', this in turn reflecting Simmel's (1971) identification of a crisis emerging in Modernist living at the turn of the nineteenth–twentieth centuries. Harvey emphasized the time-space compression which we focused on earlier as central to this Modernist–Postmodernist transition.

However, not everyone stresses the distinctness of the move from Modern to Postmodern and Hassan (1980, p. 120) was convinced that 'Modernism and Postmodernism are not separated by an Iron Curtain or a Chinese Wall, for history is palimpsest, and culture is permeable to time past, time present, and time future'. Thus, although a new Postmodern epoch may be identified it rests squarely on its inheritance of institutions, relationships, physical features, and tradition that is clearly the case with maritime governance. A new era has emerged but from an identifiable history and with clear links to the future.

Harvey, quoted in Dear (1991, p. 534), agreed in that he saw a new Postmodern era that emerged from around 1972 but one which fundamentally had changed nothing in the underlying organization of society albeit with significant impacts on the way that society operates:

> There has been a sea-change in cultural as well as in political-economic practices since around 1972.
> This sea-change is bound up with the emergence of new dominant ways in which we experience space and time.
> While simultaneity in the shifting dimensions of time and space is no proof of necessary or causal connection, strong a priori grounds can be adduced for the proposition that there is some kind of necessary relation between the rise of Postmodern cultural forms, the emergence of more flexible forms of capital accumulation, and a new round of 'time-space compression' in the organization of capitalism.
> But these changes when set against the basic rules of capitalist accumulation, appear more as shifts in surface appearance rather than as signs of the emergence of some entirely new postcapitalist or even postindustrial society. Harvey (1990, p. vii)

Soja (1996) provided an interpretation of the move from Modernist to Postmodernist as one reflecting Harvey's (1973) transition from liberal to socialist formulations and with some slight adaptations, quotes himself in Soja (2000, p. 109):

> The evolution which occurs between our approaches to the old (Modernist) and the new (Postmodernist) cultural politics naturally gives rise to contradictions and inconsistencies. The general approach to the new cultural politics is substantially different (and I believe substantially more enlightening). Yet the new approach takes on more meaning if it is understood how the viewpoints it espouses were arrived at… It is also important to note that the material content and action strategies of the old cultural politics are not being rejected but incorporated and given additional meaning by the evolving framework of the new.

Bergquist (1993) considered the Postmodern to be a move from the modern in that the latter focused on 'objectivism' assuming that there is a reality that we can get to know and to understand. This reality is characterized by universal truths which can be 'applied to the improvement of the human condition, resolution of human conflicts, restoration of human rights, or (*and in our case particularly pertinent to effective maritime policy-making*) even to the construction of a global order and community'. This formalized view of society contrasts with that of the Postmodern whereby the world can be seen to be becoming more 'segmented and differentiated' despite (or perhaps because of) increasing globalization. Rosenau (1992b, p. 137) cited in Rhodes (1997) concurred. Modern social science:

> requires simplification. It surrenders richness of description and feel for complexity in return for approximate answers to the questions, often limited in scope, that are eligible for consideration within its own terms.

Whereas in Postmodernism:

> Only an absence of knowledge claims, an affirmation of multiple realities, and an acceptance of divergent interpretations remain.

Gephart (1996, p. 31) also saw that Modernism had transformed into Postmodernism repeating Scott's (1988) claim that:

> instead of an era of growth and prosperity, society faces a period of social decay, resource scarcity and social conflict because of declining reserves of natural resources, resistance to

7.3 The Relationship of Postmodernism to Modernism and Post-Fordism

environmental pollution, changing work expectations, and declining confidence in established institutions.

All of which are pertinent to any debate on the effectiveness of maritime policy-making. Scott saw a rising nihilism challenging the 'assumption that organizational leadership will provide solutions to social problems or will enhance individuals' goodwill' (Scott 1974, p. 250).

Gephart (1996, pp. 31–32) went on to suggest that at the organizational level:

> the transformation from classical to late Modernism is evident in the displacement of generalized technical rationality as a basis for organizational actions and the substitution of 'local' institutional and organizational logics and rationalities as bases for action and legitimation. These local logics are evidenced in the partial reversion to traditional aspects of authority that occur in fiefdom-style organizations. Clearly specified bureaucratic positions with a restricted range of competence and authority are being eroded by the need to meet the performance commands of organizational elites which require members to do whatever is necessary to achieve management's operative goals, even if this means undertaking activities outside of their formal position or authority. Technical training as the basis for promotion, enhancement and success is thus eroded and supplanted with alternative criteria such as loyalty, style and fit with organizational logics and style. Fixed salaries are supplemented with performance pay, stock bonuses, and other rewards that reinforce the fealty structure of the organization. Corporate officials are no longer independent but are (part) owners of the corporation; hence they have a vested interest in its success. The control of behaviour by rules is supplanted by discretionary behavior which is allowed for fiefdom rulers or leaders.

These changes in Modernism are the precursor to the emergence of the Postmodern era and reflect a crisis of late capitalism which was touched upon earlier in our discussion on disorganized capitalism. This disorganization occurs with the steady disintegration of Modernist, capitalist society and includes 'economic, political, motivational and legitimation crises' (Habermas 1973; Offe 1984). Dear and Flusty (1998, p. 53) commented on Soja's (1989, p. 246) view of Los Angeles as a 'decentered, decentralized, metropolis powered by the insistent fragmentation of post-Fordism, that is an increasingly flexible, disorganized regime of capitalist accumulation'.

Gephart (1996, p. 33) saw this stage as the 'brink of Postmodernism' where:

> Authority based on rational/legal grounds is supplemented by traditional authority based on organizational fiefdoms. The myths of the organizational imperative are becoming difficult to legitimate because individual good is often obtained through organizational activities that undermine general organizational and social welfare. And the changing moral commitments of management threaten social totalitarianism and encourage the emergence of simulacral management meritocracy, based not on true merit but on representations or simulations of superiority that arise from the fiefdom-based status frameworks.

The relevance to the difficulties of maritime governance are clearly apparent. The progression from Modernism through to Postmodernism is reflected in organizational changes which have destroyed the benefits of the Modernist model of formalism, control, predictability, security, and stability and replaced them with 'management meritocracy' undermining social welfare. The Postmodern reaction

Table 7.2 Empirical Modernism and Postmodernism in management and organization

Premodern era	Modern era	Postmodern era
1800–1870	1900–1970s	2000–?

Source Clegg (1996), derived from Drucker (1957, 1992) and Boje and Dennehy (2008)

Table 7.3 A first axis for the analysis of empirical modernity and postmodernity

Empirical modernity	Empirical postmodernity
Formal bureaucracy; high degree of differentiation	Networks and contracts; tendency to de-differentiation

Source Derived from Clegg (1996) in Palmer and Clegg (1996, p. 239)

to this presents difficulties to those very same Modernist organizations that dominate the policy-making process and which together form the governance of the maritime sector. Maritime policy-making remains essentially Modernist, a process undertaken in a society increasingly Postmodern creating a disjuncture that leads to policy inadequacy and failure.

Clegg (1996) cited Giddens (1992, p. 12) in suggesting that the Postmodern is directly linked to the Modern. Giddens discussed:

> the drift from Modernity, a set of ideas, values and institutions firmly anchored by an organizing reason, to Postmodernity, which implies a dissolution of these scientific and cultural forms. The reflexivity demanded by this transformation may come more easily to the social sciences with their less determinate and more ambiguous knowledge than to the natural sciences more deeply attached to objective, that is context-free, truths, or even the humanities with their instinctive commitment to notions of culture, worth and discrimination.

Policy making and governance are inherently social science disciplines regardless of the issues which they consider. Hence a Postmodern interpretation in Giddens' eyes at least, is wholly appropriate. Clegg went on to suggest that we can think of an epochal periodization in which what is termed the Pre-modern 'slowly dissolves into the Modern and the Modern slowly dissolves into the Postmodern, each leaving residues and traces in the new history in which it is inscribed' (Table 7.2). Even if the process is agreed, the precise dates might be argued about (and have been) endlessly.

Clegg also presented schematic interpretations of the Modernist–Postmodernist dichotomy and relations. Table 7.3 suggests a starting point for differentiating the two. The characteristics of maritime policy institutions are reflected in the Modernist camp.

> The characteristic enterprise of Modernity was a bureaucracy, involved in the mass production of a relatively undifferentiated product for a mass market. In the epoch of Postmodernity size is no longer everything: small enterprises access large resources through

7.3 The Relationship of Postmodernism to Modernism and Post-Fordism 315

Table 7.4 Thematic discontinuities in Premodern, Modern, and Postmodern

	Premodern era	Modern era	Postmodern era
Organization principle	Craft	Control	Commitment
Value source	Tradition	Exploitation	Knowledge
Social relations	Custom	Hierarchy	Networks
Management theory	Local and patriarchal	Contingently bureaucratic	Institutionally embedded
Key tendency	Task continuity through mastery	Task-continuity through differentiation	De-differentiation within complex segmentation
Social mobility	Spatially restricted	Occupationally restricted	Organizationally restricted

Source Clegg (1996, p. 239)

the density of their networks and ties and can compete in small markets; de-differentiated production maps on to market differentiation. The flexibility of the former services the complexity of the latter.

He went on to suggest that there were clear dangers in 'applying a Modernist frame to a Postmodernist reality', something clearly happening in maritime governance. Thus maritime policy-making institutions reflect Weber's Modernist 'iron-cage' of bureaucracy. We shall see how Weber's concept relates to Fordism in the next section but for now if we take just some of his key processes the empathy with maritime policy-making institutions is clear. For example, Weber cited:

- An increase in specialization.
- Hierarchization.
- Stratification.
- Formalization.
- Standardization.
- Centralization of organizational activity.

Each is apparent in the nation-state dominated hierarchy of maritime policy-making that is apparent today. Yet by all accounts we have entered a Postmodern era where such qualities are less valued and appropriate in times of ever-increasing globalization.

In addition, Palmer and Clegg present contrasts among the Premodern, Modern, and Postmodern ages which are again reflected in the context for maritime policy-making (Table 7.4).

Boje and Dennehy (2008) also contrasted the Premodern, Modern, and Postmodern introducing a more detailed chronological division and indicating the widest social manifestations (Table 7.5) which helps to clarify the contrast between the phases and also the links between them.

Table 7.5 Alternative formulations of premodern, modern and postmodern eras

	Pre-modern	Early modernism	Mid-modernism	Late modernism	Neo-modernism	Post-modernism
Dates	???–1300	1450–???	1630s–???	1750s–1920s???	1960s–???	1970s–???
Class	Feudal	Gutenberg and renaissance humanists	Descartes' philosophy; Newton science; Hobbes politics	Late industrial revolution	Computer revolution	Individual control
	Noble class. Knights carry swords	Apprentice to journeyman to master. Knights decline	Restrictive church and state repression	Managerial class restrictive of everything	Information workers and core control	
Labor	Farm vassalage for basic survival	Child, serf, and slave labor. Fraternal guilds	Child, serf, and slave labor in city slums	Factory employs child labor. Worker is cog in the machine. De-skilling of labor specialities	Worker as digitized input. Third world cheap labor. Seek low eco-restrictions	Worker as skilled. Ecologically responsible. Social audits
Leisure	Not much	Festivals	Time off for nobility and clergy	High stress work cultures	Computer work wherever you go	Time off for leisure and family
Rationality	Work slaves to death. All "others" are beast labour	Tolerance of rationality and humanism	Descartes' philosophy. Rational enquiry of Galileo, Kepler, Newton. Political theories of Hobbes	Hire slaves, children and third world at sub-wages. 1750s steam engine. Taylorism and Fordism	Re-intro of "lean" Taylorism internalized time and motion. TQM, JIT, Kaizen idea control	Raise wages of skilled workers. Fall of TQM. Cult of performavity and stress
Humanism	Humanism for nobility	Renaissance humanists. Desiderius Erasmus, Francois Rebelias, William Shakespeare	Decline of renaissance humanism. Rise of nation states	Technological fix to human condition. Humanism reserved for nobility and managerial class	System is more important than person. Myth of information society	Re-assertion of individualistic humanism of Montaigne and Bacon

Source Boje (1996) in Palmer and Clegg (1996, p. 332)

7.3 The Relationship of Postmodernism to Modernism and Post-Fordism

Boje (1996) went on to develop this model further providing more detail of the Modernist/Postmodern transition (Table 7.6). Meanwhile Boje and Dennehy (2008) also developed a model of the Modern/Postmodern dimensions as applied to management which again is particularly applicable to the issues of governance in the maritime sector and helps to show the change in conditions from one era to the next (Table 7.7). Clegg (1990) provided an alternative visualization of Modern/Postmodern synergy and antagonism as applied to organizations (policy-making or otherwise) (Fig. 7.4). The close linkage between the two phases is clear as is also the fundamental change that has taken place in organizations of all types with the growth of globalization.

Finally in our review of Modern/Postmodern models we can consider Hassan (1985, pp. 123–124) who provided a very neat and simple summary of the issues that characterize both eras and how they have changed (Table 7.8). Clearly, policies developed in a Modernist framework can never be appropriate for a Postmodern industry. Cooper and Burrell (1988, p. 23) observed that 'two radically different systems of thought and logic are at work in the Modernist–Postmodernist confrontation' which they saw as:

> fundamentally irreconcilable because they derive from that basic split in the structure of human logic associated with the 'formal-informal' distinction which has been exacerbated by the extension of formal organization into so many facets of modern life.

Thus the Postmodern is a formidable social rupture from the Modern but one derived from Modernist origins. In the maritime sector this rupture is yet to be recognized in the way that its governance is developed, maintained, and applied.

Not everyone is happy with this interpretation of a Postmodern movement emerging from Modernist origins. The suggestion that the Postmodern is the 'alternative to the metanarrative of scientific truth (Thompson 1993), the latter a major building block of Modernism, has been associated with charges of irrationality, incoherence (and) nihilism (Rorty 1989) a view supported by Foucault (1987). He saw this as a challenge to the established order of logo-centric rationality, and its procedures for establishing truth'. The result of accommodating the move from the Modern to the Postmodern would be the 'demise of meaning, knowledge, science and morality'.

Woods (1999) suggested that there are many others who do not accept the viability of the Postmodern citing Habermas (1983), Jameson (1984), and Callinicos (1989). Sim (1992), Soper (1986), and Dews (1995a, b) each separately suggested it is philosophically contradictory while Norris (1993, 1995) argued that Modernism is not representative of a narrow, unitary, monolithic movement (typified by characterizations of the maritime institutions of the IMO, EU, etc.) which inevitably resulted in 'authoritarianism and false progress'. Thompson (1993) went on to discuss at some length the heritage of the Modernist movement and its relationship to the Postmodern which is centred on the latter's incorporation of sociocultural influences into all interpretations of life. The Postmodern interpretation thus implies that there is no fixed structure to society, no absolute truth or single best way of doing or organizing anything but that society and its activities

Table 7.6 The Postmodern model

	Pre modern	Modern	Post modern
Plan	CRAFT	PYRAMID	NETWORK
	Craftsmen	Police	Needs
	Rituals	Yoke	Expectations
	Apprenticeship	Reports	Team
	Fraternal	Atomize	6 Ws
	Tales	Monitor Distribute	Organize Responsiveness KISS
Organize	CREWS	DISCIPLINE	FLAT
	Crews	Discipline	Flat
	Rough	Inspection	Latticed
	Entrepreneurial	Surveillance	Autonomous
	Well-knit community	Centralization	Team based
	Social	Impersonal	
		Penal mechanisms	
		Layers	
		Individual cells	
		Neural	
		Elites	
Influence	SOLACE	COMPLY	INDIVIDUAL
	Solace (fines)	Conformity	Independent
	Order	Obedience	Narcissist
	Lazy	Motivate	De-centred Individual
	Attitude	Performativity	Voices
	Culture	Logical	Irrational
	Entrepreneurial	Yielding	Diversity
			Unconforming
			Affirms the self
			Linguistic
Lead	MASTER	PANOPTIC	SERVANT
	Master	Panoptic	Servant
	Authoritarian	Authoritarian	Empowers
	Slave driver	Network of	Recounts stories
	Tyrant	Mechanisms	Visionary
	Elite	Organizational layers	Androgenous
	Ruler	Pyramid	Networker
		Inspector	Team-builder
		Centralist	
Control	SLAVE	INSPECT	CHOICE
	Slavery	Impersonal	Choices
	Levels-classes	Normative	Heterogeneity
	Arbitrariness	Short-term goals	Oppositional
	Venal Elders	Pyramid-surveillance	Individualism
			Co-responsibility
		Externally-driven	Environmental audit
		Conform to standards Technical gaze	

Source Boje and Dennehy (2008)

7.3 The Relationship of Postmodernism to Modernism and Post-Fordism

Table 7.7 Modern versus Postmodern principles of management

	Modern	Postmodern
Planning	Short-term profit goals	Long-term profit goals
	Mass-production	Flexible production
	Worker is a cost	Worker is an investment
	Vertical planning	Horizontal planning
	Top down focus	Internal and external customer focus
	Planning leads to order	Planning leads to disorder and confusion
Organizing	One man, one job and de-skilled jobs	Work teams, multi-skilled workers
	Labour-management confrontation	Labor-management cooperation
	Division of departments	
	Tall is better	
	Homogeneity is strength	Flexible networks with permeable boundaries
	Top has voice and diversity is tolerated	
	Efficiency increases with specialization, formalization, routinization, fragmentation, division of labour	Flat is better
		Diversity is strength
		Many voices and diversity is an asset
		Efficiency decreases with specialization, formalization, routinization, fragmentation and division of labor
Influencing	Authority vested in superior	Authority delegated to leaders by team
	Extrinsic rewards and punishment	Intrinsic, empowered, ownership over work process
	Surveillance mechanisms everywhere	
	Women paid 68 % of men; minorities paid less	
	Discourse is white male-based	People are self-disciplined
	Individual incentives	Women and minorities equally paid
		Polyvocal/polylogic discourse
		Team incentives
Leading	Theory X or Y	Theory S (servant leadership)
	Centralized with many layers and rules	Decentralized with few layers and wide spans
	Boss centered	People centered
	White male career tracks	Tracks for women and minorities
	Tell them what to do	Visionary
Controlling	Centralized control	Decentralized control
	End of line inspection	Quality control is everyone's job
	Micro surveillance	Two-way surveillance
	Red tape	Cut red tape
		Dump procedures
	Lots of procedures, rules, MBO, and computers for surveillance	Train people
		Measure process criteria
		Information is given to all
	Train top of pyramid	Self-control
	Measure result criteria	
	Hoard information	
	Fear– and violence– based controls	

Source Boje and Dennehy (2008, p. xxix)

Modernity	⬅➡	Postmodernity
Missions, Goals, Strategies and Main Functions		
Specialization	⬅➡	Diffusion
Functional Alignments		
Bureaucracy	⬅➡	Democracy
Hierarchy	⬅➡	Market
Coordination and Control		
	In Organizations	
Disempowerment	⬅➡	Empowerment
	Around Organizations	
Laissez-faire	⬅➡	Industry Policy
Accountability and Role Relationships		
Extra-organizational	⬅➡	Intra-Organizational
	Skill formation	
Inflexible	⬅➡	Flexible
Planning and Communication		
Short-term techniques	⬅➡	Long-term Techniques
Relation of Performance and Reward		
Individualized	⬅➡	Collectivized
Leadership		
Mistrust	⬅➡	Trust

Fig. 7.4 Organizational dimensions of Modernity and Postmodernity. *Source* Clegg (1990, p. 203)

(including shipping) have to be 'interpreted'. This in turn suggests that a highly flexible and sensitive approach to governance and policy making is essential—one that is far from apparent in the constrained and restricted, formalized, nation-state dominated approach that characterizes the maritime sector at present. However, to re-emphasize that there remains much disagreement in this area let us quote Hunt (1989, p. 188):

> it is the case that societal consensus alone constitutes the reason why there are astronomy departments in universities but not astrology departments, that there are medical science departments but not palmistry departments. I and many others believe that it is not just societal consensus but societal consensus backed by very good reasons.

Woods (1999, p. 7) stressed that Postmodernism implies that it comes from Modernism either as a replacement or following it. However, he saw it as a continuous engagement between the two, in that each needs the other in some form of social symbiosis. As a consequence, it is necessary to arrive at an understanding of Modernism if one is to understand Postmodernism. Woods saw one fundamental characteristic of Modernism as the use of 'quasi-scientific modes of conceptualization and organization… as the expression of a rationalistic, progressive society'. This is reflected in the formal structure of maritime policy-making institutions. Postmodern policy-making would understand these structures and then go on to adapt processes and formalities to meet a Postmodern, globalized world.

Jameson (1983) considered Postmodernism as a 'specific reaction against the established forms of high Modernism' which means that there are as many

7.3 The Relationship of Postmodernism to Modernism and Post-Fordism

Table 7.8 Contrasting characteristics of Modernism and Postmodernism

Modernism	Postmodernism
Romanticism/symbolism	Paraphysics/dadaism
Form (conjunctive/closed)	Antiform (disjunctive/open)
Purpose	Play
Design	Chance
Hierarchy	Anarchy
Mastery/logos	Exhaustion/silence
Art object/finished work	Process/performance/happening
Distance	Participation
Creation/totalization/synthesis	Decreation/deconstruction/antithesis
Presence	Absence
Centering	Dispersal
Genre/boundary	Text/intertext
Semantics	Rhetoric
Paradigm	Syntagm
Hypotaxis	Parataxis
Metaphor	Metonymy
Selection	Combination
Root/depth	Rhizome/surface
Interpretation/reading	Against interpretation/misreading
Signified	Signifier
Lisible (readerly)	Scriptable (writerly)
Narrative/grand histoire	Anti-narrative/petite histoire
Master code	Idiolect
Symptom	Desire
Type	Mutant
Genital/phallic	Polymorphous/androgynous
Paranoia	Schizophrenia
Origin/cause	Difference–difference/trace
God the father	The Holy Ghost
Metaphysics	Irony
Determinacy	Indeterminacy
Transcendence	Immanence

Source Hassan (1985, p. 123)

Postmodernisms as there were Modernisms since the former are reactions against the latter. The 'unity of the new impulse—if it has one—is given not in itself but in the very Modernism it seeks to displace'.

The Postmodern era also is one that has a close relationship with the nation-state in a way that reflects the changes we have observed occurring as a response to globalization (Caparaso 1996, p. 45; Shaw 1997, p. 511). Aalberts (2004) suggested that the Westphalian state was a unitary center of authority (and as such suitable for the organization and focus of shipping policy) but the Postmodern state is very different. Quoting Wallace (1999, p. 506) he saw the Postmodern state operating:

within a much more complex, cross-cutting network of governance, based upon the breakdown of the distinction between domestic and foreign affairs, on mutual interference in each other's domestic affairs.

This matched Caparaso's description of the Postmodern state and its contrast with the traditional Westphalian:

> disjointed, increasingly fragmented, not based on stable and coherent coalitions of issues or constituencies, and lacking in a clear public space within which competitive visions of the good life and pursuit of the self-interested legislation are discussed and debated. (Caparaso 1996, p. 45)

This resonates with much going on in maritime policy-making in recent years and further comment by Wallace (1999, p. 519) reaffirmed this.

> One element of Postmodern statehood is that sovereignty is considered to be increasingly 'held in common', pooled among governments, negotiated by thousands of officials through hundreds of multilateral committees, compromised through acceptance of regulations and court judgments.

Aalberts (2004, p. 34) continued:

> The once pivotal rule of non-intervention has been replaced by more or less legitimized (non-military) interference in each other's domestic affairs. The sacrosanct Westphalian principle of sovereign rule (*that underlies maritime policy-making and the institutional framework that supports it,*) based upon jurisdictional exclusivity, has been abandoned. (*italicized section added*)

A Postmodern state reflects the new globalized world and has moved on significantly from that envisaged at the time of Westphalia. Cooper (2000, p. 19) described the 'state system of the world' as collapsing into an ordered Postmodern condition—more pluralist, more complex, less centralized. Meanwhile, the policy-making institutions and governance process and framework for the maritime sector (and quite possibly many others) have not.

However, Anderson (1996, p. 133) questioned this Postmodern vision of the nation-state citing the claim that the member states of the EU find themselves crushed between emerging regional and supranational jurisdictions, and yet continue to survive and even thrive. As a result:

> These 'Postmodern'... visions of a 'Europe of regions' replacing a 'Europe of states', although currently fashionable, lack plausibility, not least because of the continuing force of nationalism.

He went on to suggest that actually the EU may represent a new form of Postmodern governance whereby nation-states in the EU were not located in some sort of transitional form between national dominance and supranational superstatism but in a permanent intermediate form that Ruggie (1993) considered is characterized by an unbundling of territory. The EU thus typifies the new Postmodernism within which maritime governance must operate effectively (Anderson 1996, p. 143):

7.3 The Relationship of Postmodernism to Modernism and Post-Fordism

Postmodernity in political space may mean that the single-point perspective and singular sovereignty of independent statehood are being displaced by multiple and overlapping sovereignties.

In considering the relationship of Postmodernity to the nation-state this inevitably draws us into a discussion of territory and space, and ultimately mobility, absence, and speed.

Luke (1991) discussed these issues in great depth relating the changes in Postmodern society to the issues of nation-states, globalization, space and territory, governance, and authority. He contrasted the traditional (now defunct) society characterized by location, hierarchy, and organization—something widely termed Modernist. This society was centered around places in a way that he compared with power drawing boundaries which erect:

> monetary, military, and managerial borders around space; and exercise a monopolistic writ of sovereignty within these delimited expanses by exerting, guiding, or directing its effects from point to point or place to place within space. The stability, security and sovereignty of state power then, most often, have been stated and comprehended in essentially spatial terms through geopolitical discourses of expansion, military defense or economic development. Panoptic surveillance from the center and top of this space by state agencies works to normalize activities within it to suit the monetary, military and managerial agendas of its state structures' leaderships. (Luke 1991, p. 318)

We shall return to these issues and the relationship of institutional power to space and territory and the contrast with a Postmodern society at a later stage but for now note the significance of these observations in terms of the institutions that dominate maritime policy-making throughout the jurisdictional hierarchy that exists—and its co-existence with the range of policy failures.

Luke, along with Campbell (1990), went on to stress the anarchy that this concentration upon space rather than process created and the resistance to change that is a consequence. He saw an almost mythical quality of sovereignty 'writing and drawing lines of identity and antagonism on the Earth':

> States... are those legitimate monopolies charged with inscribing ... writs of difference – in money, religion, markets, ideology, and militaries – from what transpires within and without the geographical spaces framed by international borders... By endogenizing various disciplines of monopolistic order inside, and exogenizing diverse practices of free-for-all anarchistic conflict outside, those borders defining each nation-state's place on the planet's terrain, the fictive practices of political self-rule, or national sovereignty, define themselves spatially against the landscapes of the Earth in a conjunctive, centralizing hierarchical order. (Luke 1991, p. 318)

Such is the current organizational foundation for maritime policy-making, contrasting with the new power dynamics identified by Luke which he saw 'nested in flexible accumulation's rapid and intense flows of ideas, goods, symbols, people, images and money on a global scale'. These he described as 'disjunctive and fragmenting, anarchical and disordered'. These changes he also saw as intensely related to speed, for the flows noted above actually predate the Modernist nation-state but it is recent times that the speed of such flows has dramatically increased transmuting it into something quite different; quite Postmodern; and quite in need of a new framework for global governance of the maritime sector.

> Power today... often flows more placelessly beneath, behind, between, and beyond boundaries set into space as new senses of artificial location become very fluid or more mobile... Geopolitical barriers are articulated as cartographic traces, memberships in military pacts, and diverse denominational codes in national monetary currencies. Informational flows rarely are stymied for long by such barriers: indeed cross-border flows of money, influence and knowledge are heavily eroding such notions of geopolitical borders. (Luke 1991, p. 319)

This creates something Luke termed 'chronopolitics', and what Der Derian (1990) saw replacing the politics of space (the Modern) with that of time (the Postmodern); the acceleration that is at the basis of globalization has created a new Postmodern era where the anachronisms of the nation-state are retained solely where they can be used to manipulate the new global market place and a new governance is needed based around flows rather than territory. Westphalian states thus have become outdated and yet retain significant presence in institutions of governance where it is convenient to be able to hide behind the protection of sovereignty. Coupled with institutional inertia and the reluctance of individuals and institutions to relinquish power, we have a recipe for the current policy-making disaster that is the maritime sector.

To quote Luke (1991, p. 320) again:

> The ethnogeographic settings of self-rule defined by the classical Westphalian universe of borders, shorelines and airspaces in spatially construed grids of/for sovereignty increasingly collide in the transnational multiverse of techno-regions generated out of global monetary transactions, commodity exchanges, technical commerce, telecommunication links and media markets.

How much closer can we get to the current situation for the governance scenario that characterizes the maritime sector today?

Dear (1986) also saw space and time as fundamental elements of the new Postmodern order. Earlier, we have seen how Harvey (1981, 1990, 2001) stressed the vital significance of the compression of space and time in creating the Postmodern, globalized marketplace in which shipping operates. Dear cited Jameson (1984, pp. 83–84) in that old systems of organization and perception are destroyed to be replaced by what he terms 'Postmodern hyperspace' where time and space are stretched to accommodate the 'multinational global space of advanced capitalism'.

The significance of speed in Postmodernism is also stressed by Beckmann (2004) who although discussing mobility and safety in its cultural setting, used the term Postmodern in his consideration of Haraway's (1985) myth of the cyborg and how this resembles the car driver and the endless desire for speed. The notion of acceleration (in contrast to simply speed) is also raised with a Postmodern interpretation of time being wasted unless it is reduced to being negligible or replaced by an alternative pursuit whilst (for example) transport takes place. This emphasis on time reduction mirrors the processes going on in the Postmodern world.

7.3 The Relationship of Postmodernism to Modernism and Post-Fordism

Boundaries in organizations are also seen as undergoing a Postmodern revolution—with implications for those institutions active in maritime policy-making (Morgan 1997). Boundaries within organizations refer to the interfaces between different elements of an organization, between the various work groups and departments and between the organization itself and its professional and interactive environment. Here the new flexibility inherent in a Postmodern society changes these boundaries and in particular makes them more porous. This has major implications for governance and in the maritime sector the ability for new stakeholders to gain access to policy making and the policy-making process, and perhaps more importantly the need to provide access for these groups where they remain excluded, is highly significant.

Malcolm Anderson (1996) compared Postmodernity with the Cubists of the early twentieth century where objects were presented from several different perspectives in the same painting. Postmodernity in political space was replacing a 'single-point perspective and singular sovereignty of independent statehood' by multiple and overlapping sovereignties. Soja (1987, p. 292) also suggested that the Postmodern was a move toward making Modernity spatial, representing in the process a 'flexible assertiveness'.

> No longer were geographers alone in inserting space into the central problems of social theory, into a contemporary critique of historical materialism and an appreciation for the 'limits' to capital, into the cultural logic of late capitalist architecture, music, art, literature, and film into the very 'constitution' of society and social life.

This spatial assertiveness of the Postmodern is reflected in Harvey's view of space/time compression which has been the central feature of globalization and made possible (at least in part) by advances in the maritime and logistics sector. Postmodernism is tied up with boundaries, territories, and space and the relentless drive toward the elimination of the former and the diffusion of the latter two. Governance has to recognize this to be effective.

We shall turn again to flows, speed, and the placelessness of globalization in our concluding sections but for now the significance of these Postmodern developments in governance is notable.

Zukin (1992) stressed the significance of 'absences' in the Postmodern approach and although applied to urban political economy and more specifically to the work of Berman (1982), Harvey (1985, 1990), and Jameson (1984), this idea that by focusing on what traditionally has 'not' featured in traditional policy-making is helpful in identifying what has gone wrong. Consequently, attention is placed upon the process rather than the location of policy debate and construction; on the flow of information rather than the placement of it; and upon those normally excluded from the process and objectives rather than the traditional stakeholders. This in turn has become an attack on all that is fundamental and essential to existing governance including all 'foundational philosophies of knowledge, objectivities of observers and integrity of subject and author'. Postmodernism questions it all.

Postmodernism has long had a historical dimension to it, something pointed out by Ruggie (1993, p. 168) and his consideration of the seminal work by Mackinder (1904)—a geographer who identified a 'new global epoch' as he entered the twentieth century following the demise of Europe's global supremacy which had taken place since around 1500AD. This new epoch, commencing at what he terms 'the geographical pivot of history', had two dimensions—the former was actively considered until the post 1945 theory of containment raised its head and centered upon the strategic consequences of the unity of the world's oceans for geopolitics. However, his lesser known contribution focused on the:

> spatial and temporal implosion of the globe, featuring the integration of separate and co-existing world systems, each enjoying a relatively autonomous social facticity and expressing its own laws of historicity, into a singular post-Columbian world system.

Ruggie viewed this as a very early Postmodern interpretation of events by Mackinder and took it as a first indication of the 'unbundling of territoriality' that globalization, Postmodernism, and space–time compression suggests.

Caporaso (1996, p. 45) considered that the Postmodern state contrasted distinctly with the traditional Westphalian state which forms the basis of the maritime policy-making process. It is 'abstract, disjointed, increasingly fragmented, not based on stable conditions of issues or constituencies, and lacking a clear public space within which lie competitive visions'. Caporaso, referring in particular to the EU, claimed that the Postmodern state faced endless contradictions in that:

> politics and governance occupy different sites (Basle, Brussels, the national capitals, Luxembourg, bilateral meetings among economic and finance ministries), and these sites can change. Process and activity become more important than structure and fixed institutions. The state becomes not so much a thing… as a set of spatially detached activities, diffused across Member States but reflecting no principled – let alone constitutional – considerations.

The contradictions present in maritime policy-making are reflected in this view of the Postmodern state.

7.4 Postmodernism and Governance

This discussion on Postmodernism and its relationship to Post-Fordism and the Modern has reflected the variety of definitions and origins, its significance for the nation-state, and the integration of these ideas and concepts into our recognition of the importance of time, space and territory in considering the problems facing policy making in the maritime sector. We will now turn more closely to the Postmodern view of governance and introduce some issues which are fundamental to maritime policy-making.

Looking back at our earlier discussion on hierarchies, networks, and markets we concluded that there was a distinct rupture between the existing framework for governance and policy making in the maritime sector—characterized by a state-

7.4 Postmodernism and Governance

centred hierarchical model—and the changes in society that had been taking place—which we have now clearly identified as Postmodern—and which appear, in Castells' (Castells 2000, p. 16) terms, to demand a more flexible, network-based approach with greater consideration of markets (and as we shall see later stakeholders).

Podolny and Page (1998, p. 59) suggested that the classic hierarchy, network and market, forms of organization were false, quoting Laumann (1991) in considering that 'markets and hierarchies are simply two pure types of organization that can be represented with the basic analytical constructs of nodes and ties. However, when extended more specifically to the governance of organizations or activities then networks can be more distinctly characterized.

Podolny and Page saw hierarchies as characterized by relationships that endure for long periods of time, are easily and clearly recognized and exercise what they term 'legitimate authority'. These are clearly the characteristics of the Modernist interpretation of governance rather than the Postmodern. Networks in particular were seen as being flexible and varied—far more Postmodern—whereby policy makers might include 'joint ventures, strategic alliances, business groups, franchises, research consortia, relational contracts and outsourcing agreements' (1998, p. 59). 'No clear mapping of formal organizational arrangements on to the network form' (1998, p. 60) is possible. This placed a big question mark against the formal institutionalism of current maritime policy-making governance.

Fox et al. (1992, p. 7) in their preface to Nooteboom's (1992) paper on Postmodern markets, commented on the latter's consideration of industrial organization, economics, and marketing. Nooteboom saw firms (and shipping is no exception to this):

> embedded in a constellation of firms and other institutions... Consumers express and distinguish themselves in consumption. Entrepreneurs do not try to make a given product most efficiently with a given technology, but to distinguish themselves with new products and technologies. (Nooteboom 1992, p. 61)

This is a Postmodern view of the market with emphasis upon communication, exchange, and dialog rather than just satisfying needs and wants. The marketeer is not seen as having unilateral influence over the consumer but there is a dialog between them, and this discussion creates not only consensus but also conflict (Lyotard 1979). From this springs new ideas, products, and services. Hierarchies are to be viewed in decline as market relations replace them in the business world as Postmodernism progresses. Fox et al. (1992, p. 8) suggested that Nooteboom saw lateral, horizontal relationships 'in vogue, joint ventures necessary... while a premium is being placed on networking and influencing skills and maintaining trustworthy relationships'.

Nooteboom (1992) considered the relationships between Postmodernism and the market in some depth providing an understanding of how Postmodern thought can contribute to the interplay between firms and their environment (Bagozzi 1975, 1979, 1984; Houston and Gassenheimer 1987) and also shed light on the notion of exchange and the symbiotic relationship between buyer and seller

(Adler 1966; Varadarajan and Rajaratnam 1987). These relationships can help in understanding the new Postmodern environment in which policy-making organizations must operate and processes must take place. Placing itself outside the existing safe policy-making environment:

> Postmodernism inspires a radical departure from methodological individualism and from the concept of equilibrium in traditional economics; from the notion that economic agents have an identity, with given perceptions, interpretations, and preferences, prior to exchange; from the idea that supply settles down to equal a given demand, on the basis of given, common knowledge and technology. (Nooteboom 1992, pp. 54–55)

Nothing is as simple as this. In fact:

> there is no universal and permanent, but at best only a local and temporary unity of meaning and consensus on the rules of the game, which is continually broken and shifted in an on going process of differentiation and change. (Nooteboom 1992, p. 54)

He stressed that Postmodern markets feature on-going disequilibrium, learning by doing, interaction, changes in perception and technology, inter-subject difference and product differentiation, all characteristics that can be found in the shipping marketplace. He cited Hakansson (1982, 1987), Johanson and Mattson (1987) and Hellgren and Stjernberg (1987) in identifying the move from the structured, Modernist marketing environment to one that is flexible, variable, and network based.

Parker (1992) quoting Featherstone (1987, p. 69) indicated how the Postmodern sounded the end for hierarchies:

> one strategy for outsider intellectuals is to appear to attempt to subvert the whole game – Postmodernism. With Postmodernism, traditional distinctions are collapsed, polyculturalism is acknowledged... kitsch, the popular and difference are celebrated. Their cultural innovation proclaiming a beyond is really a within, a new move within the cultural game which takes into account the circumstances of production of cultural goods, which will itself in turn be greeted as eminently marketable by the cultural intermediaries.

The language may be a little pompous but we have to remember this is Postmodernism and in spite of this the condemnation of the hierarchical is clear.

Finally, Luke (1991, p. 319) provided a detailed discussion on the moves that Postmodernism suggests away from hierarchies and what he termed 'the flow of power in space' to 'the power of flows'. We have noted these moves and Luke's comments earlier. They are only repeated here to re-emphasize the 'conjunctive, centralizing hierarchical order' that existed along with the sovereign state, which has now been replaced by the flows of information which have reduced the power of the nation-state and at the same time moved the process of governance from one of hierarchy to networks and flows of information, 'goods, symbols, people, images and money on a global scale' creating disorder, disjuncture, fragmentation, and anarchy. However, the response should not be what it has been up to now in maritime policy-making—to ignore and deny—but the need is to absorb these changes and build a new governance framework which accommodates difference, change, and impermanence.

7.4 Postmodernism and Governance

Earlier we also discussed the relationship among capitalism, its continuous search for new ways of accumulating, and disposing of surplus capital and the rise of the Postmodern reflecting the latest major shift in society. Disorganized capitalism was one way of viewing these changes which have taken place as globalization progresses and society looks for ways of reducing the friction of time and space in the process opening up new markets for old products and inventing markets for products and services which have only recently emerged.

Evidence of disorganization in the capitalist marketplace is not difficult to find but at the same time it is important to realize that this is more a reflection of an organized (or at least controlled disorganized) revolution in society—what we describe the Postmodern—rather than anarchy and chaos reflecting the end of all that we know. The maritime sector is one of the most affected economic and business activities with its long held and intensely strong international linkages and corporate flexibility which have made it central to the process of globalization and the disorganization of capitalism that accompanies it and which in turn have made it most susceptible to the Postmodern changes that have taken place.

We will not enter into another extensive discussion about disorganized capitalism but we do need to make some connections between Postmodernism and the disorganization that has been widely recognized accompanying the emergence of globalization. As we saw earlier, Lash and Urry (1987) are the major proponents of the idea of disorganized capitalism and they stressed its emergence as part of a Postmodern phenomenon and not just a random disintegration of an established economic and social system. They emphasized the development of a 'disorganization thesis' of which Postmodernism was only a part, but a significant part at that. They described it as:

> The appearance and mass-destruction of a cultural-ideological configuration... this affects high culture and the symbols and discourse of everyday life. (Lash and Urry 1987, p. 7)

Their view was disorganized capitalism manifested itself in a transition from Modernist to Postmodernist forms of relationship in contemporary capitalism. Modernism they saw as not just limited to high culture but pervading everyday discourse including nationalism, knowledge, and morals amongst much else. The process of disorganization stemmed from this transition. They suggested that Postmodernism reflects this disorganization, in fact is an inherent part of it and although they use rather obtuse and obscure language, the message is largely clear:

- Postmodernism is about the transgression of boundaries—between what is inside and what is outside of a cultural 'text', between reality and representation, between the cultural and the social, and between high culture and popular culture.
- Whereas Modernism and Postmodernism (though much more ambiguously) can be said to break with an esthetics of representation, the more Apollonian Modernism stands in an affinity with the conscious mind than the Dionysian Postmodernism with the Freudian id.

- If communications in liberal capitalism are largely through conversation, and in organized capitalism (Modernism) through the printed word, disorganized capitalism's (Postmodern) communications are through images, sounds, and impulses. That is, to draw on some concepts of J.F. Lyotard, Modernism's *discours* is replaced by Postmodernism's and disorganized capitalism's *figure* (Lash and Urry 1987, p. 14).

They went on to stress that not all disorganized capitalist characteristics are Postmodern, but that the two have close affinity and that the Postmodern is clearly opposed to hierarchy with its close links to modernity and organized capitalism, and also to 'traditional and patriarchal superego', in the process advocating breaking down boundaries wherever possible.

Boje et al. (1996) indicated that Postmodernism emerged from the crises in late Modernism which was one of the recurring themes of disorganization in capitalism that happen with some predictability. They see Postmodernism being a product of changes in era, movement, and style. These changes mirror to an extent those identified by Habermas (1973) in what he termed the late capitalist system which was contradictory and entropic. These self-destructive tendencies arise from the very bases of the system. For example:

> in advanced capitalism, the inherently exploitative mode of production tends to destroy the very preconditions on which the system depends (Offe 1984, p. 132; Gephart and Pitter 1993). In particular there is a desire to accumulate economic surplus and yet a tendency of profit to fall and thereby limit capital accumulation. In this context a crisis cycle emerges as deficits in rationality in one sector accumulate and spill over into other sectors. Boje et al. (1996, p. 3)

Gephart et al. continued with an extensive discussion of the inevitability of crises in the capitalist state which simply is the process whereby capitalist business reorganizes itself (through a process of disorganization) to generate increased and renewed possibilities of further capital accumulation. The Postmodern era is simply the latest of a succession of such crises which are largely predictable, well understood, and welcomed (by the accumulators at least). What is particularly significant for us is that maritime policy-makers have seemingly failed to recognize this process (while the shipping industry itself is a central part and takes advantage of the crisis) and as a consequence there is a governance schism between the old and the new, the Modern and the Postmodern, the inflexible and the flexible, the national and the global.

Nooteboom (1992) confirmed his views that disorganization and crises of capitalism were similar citing Lyotard (1987, p. 28) in that Postmodernism finds its ground not in 'homology' (unified discourse) of experts, but in the 'paralogy' (disruptive discourse) of the discoverers. There is an essential, ongoing difference between agents which drives novelty.

Clegg (1990, p. 180) saw the changes that have manifested themselves within organizations in the past 30 years as just part of a 'global tendency' during the period of the post-war boom in particular across the USA and Western Europe. Soja (1989, p. 171) cited Piore and Sabel (1984) noting a 'second industrial divide'

7.4 Postmodernism and Governance

as a consequence of flexible manufacturing systems. This new global tendency and second industrial divide are also manifestations of the Postmodern in an organizational context and represent further evidence of a profound (if predictable and understandable) change in society that policy makers should not ignore or deny. Meanwhil,e Short (1993, p. 170) noted that there was a 'condition of change in the world', what he termed a 'fundamental restructuring', and that 'Postmodernism is as good as a descriptive term as any to indicate the uncertainty' suggesting new forms of analysis and explanation'.

Burke (2000) kept up the theme of disorganized capitalism and suggested it is characterized by:

> the disintegration of state regulation, the expansion of world markets dominated by international corporations, the undermining of the nation-state, the growth of manufacturing in the third world, and the decline of manufacturing in the west. Accompanying this is the growth of the 'service class' that undermines labour power, with the consequent erosion of class-based politics, and cultural life becomes more fragmented and pluralistic.

Thus the Postmodern era is entered through a process of disorganization, destroying the organization of the Modernist era and representing a broad, societal movement that is all embracing. Bauman (1992, pp. vii–ix) suggested it is:

> all-deriding, all-eroding, all-dissolving destructiveness... (and) does not seek to substitute one truth for another... one life ideal for another... It braces itself for a life without truths, standards and ideals.

Unlike the Modernist conception of the world within which the global policy institutions that dominate today were designed and constructed, under Postmodernism 'identity is not unitary or essential, it is fluid or shifting, fed by multiple sources and taking multiple forms' (Kumar 1997, p. 98).

Havel (1992, pp. 1–2) provided a broader view on the disorganization inherent in capitalism and the need for change, written in the context of the revolutions in Eastern Europe where he was a major protagonist of the decline of the influence of Communism and the old Soviet Union. His comments on the transition from the Modern to the Postmodern were made in the context of the broader political changes of the time but nevertheless were very relevant to what we are saying here about the role of capitalism, its recurrent and predictable disorganization, the relationship among business, decision-making and disorganization, and the implications this has for society as a whole and policy making in particular.

> The fall of Communism can be regarded as a sign that modern thought – based on the premise that the world is objectively knowable, and that the knowledge so obtained can be absolutely generalized – has come to a final crisis. This era has created the first global, or planetary, technical civilization, *but it has reached the limit of its potential, the point beyond which the abyss begins.* The end of Communism is a serious warning to all mankind. *It is a signal that the era of arrogant, absolutist reason is drawing to a close and that it is high time to draw conclusions from that fact.* (emphasis added)

Havel was a visionary, both in the lead up to the fall of Communism in which he played a significant part, and also in his ability to understand the disintegration of capitalism as it existed along with Communism, and the inevitability of its replacement by the next version. In this he suggested that:

> We are looking for new scientific recipes, new ideologies, new control systems, new institutions, new instruments to eliminate the dreadful consequences of our previous recipes, ideologies, control systems, institutions and instruments.

He believed the world needed something he described as 'larger', and the need to:

> abandon the arrogant belief that the world is merely a puzzle to be solved, a machine with instructions for use waiting to be discovered, a body of information to be fed into a computer in the hope that, sooner or later, it will spit out a universal solution.

And turning to the Postmodern he stressed the need to:

> release from the sphere of private whim such forces as a natural, unique and unrepeatable experience of the world, an elementary sense of justice, the ability to see things as others do, a sense of transcendental responsibility, archetypal wisdom, good taste, courage, compassion and faith in the importance of particular measures that do not seek to be the universal key to salvation. Things must once more be given the chance to present themselves as they are, to be perceived in their individuality. We must see the pluralism of the world, and not bind it by seeking common denominators or reducing everything to a single common equation.

He went on to describe politics as having a 'technocratic, utilitarian approach to Being and therefore to political power as well'. A new Postmodern face was needed which he describes in forceful but sensitive language and which sums up the qualities of politicians that are needed and the problems which face maritime policy-makers in a new Postmodern society:

> Soul, individual spirituality, first-hand personal insight into things; the courage to be himself and go the way his conscience points, humility in the face of the mysterious order of Being, confidence in its natural direction and above all, trust in his own subjectivity as his principle link with the subjectivity of the world.

Denzin (1986, pp. 198–199) summed up for us the notion of the Postmodern. Using concepts derived from Lyotard (1987) he suggested that Postmodernism is a 'different moment in the socioeconomic organization of society'. He believed that consumerism had now taken over as the most recent phase of capitalism following a period of disorganization—surpassing earlier stages where markets and monopolies dominated—in our terms where structures and choice were more limited and scientific truth and objectivity paramount. This latest capitalist phase has produced:

> a media society, a society of the spectacle, a bureaucratic society controlled by consumption and the computerization of knowledge. This has created a crisis in the legitimation of science, technology and society.

Denzin suggested that the two 'myths' that underlie the 'narrative, legitimating structures' upon which the formality of Modernist society rested—belief that science could liberate humanity and that there is unity in all knowledge which

7.4 Postmodernism and Governance

culminates in rational understanding of man, nature and society—have both collapsed. The old, positivist version of society argued that 'science could reproduce reality objectively' and focused upon stability and epistemological realism. The new Postmodern is characterized by 'paralogy', whereby knowledge is used to undermine previous understandings and seeks legitimation through pragmatism. New knowledge works for a time before being overturned by new de-stabilizing discoveries. It centres on loosening and revising prior assumptions held under the earlier Modernist conceptions of formality, normality, realism, fixity, predictability, and stability. The context for policy making (or any other societal pursuit for that matter) is no different, and a new attitude to governance, its purpose, application, and responsibility is needed.

So does Postmodernism represent a new epoch, a new era for society, distinctly separable from the existing and the past, and one with which we can then associate globalization and the problems facing policy makers and those charged with governance?

Kreiner (1992, p. 41) saw a number of features that suggested a new Postmodern epoch for organizations and these were emphasized further by Hajer's (2003, pp. 178–180) analysis of society's reaction to the 'institutional void' which he claimed had emerged. These include evidence of rejection of Weber's 'iron cage' and of a technically rational machine with 'economic man' central to institutional organization. Others are represented by clear growth in idiosyncrasy, pastiches of ideas and perspectives, reduction in the significance of rules and regulations and scientific order, and a decline in the belief that there is a single approach to solving society's problems, a scientific truth to be discovered and applied.

Østerud (1996, p. 386) was more emphatic suggesting that an 'epochal cultural shift' had occurred—from the 'traditional via the Modern to the Postmodern stage'. This was characterized by mobility, fragmentation, incoherence, fluctuating life-styles':

> If Modernity by itself was a more closed and solid conception of progress, control and emancipation, then the present scene is Postmodern.

If truth is what many mean by convention, tradition and the present, then the implication of the creeping Postmodern consciousness is that 'Truth is in Trouble' (Gergen 1991, p. 81).

Mann (1997, p. 472) identified a revolutionary trend toward a new Postmodern society. Berg (1993, p. 492) cited Curry's (1991) description of a new Postmodern epoch although he went on to question its validity. Benko (1997, p. 7) noted a 'crisis of representation, a blurring of existing norms, a fracturing within society' with the oncoming of Postmodernism. Van Ham (2001, p. 15) interpreted the change as a serious move toward what he called a new Postmodern state. He suggested that Postmodernism stands for a scientific approach and to an era that parallels contemporary modernity. These new Postmodern states look superficially like classical modern states but emphasize 'welfare rather than warfare'. This assumption is debatable but Van Ham's subsequent elaboration is less contentious:

In a Postmodern environment, traditional concerns like borders, national identity and state sovereignty are of less concern than the pursuit of prosperity, democratic governance and individual well-being. This reflects the strong pluralist and individualist streak of Postmodern society, which is tolerant to cultural and political dissent, stresses multiculturalism and legitimizes multiple identities and lifestyles.

Jarvis (1998, p. 101) took this even further. He emphasized the difficulty of identifying with a specific Postmodern 'period' while its interpretation simply as a process (for example characterized by deconstruction) is clearer. However, he also emphasized that things are changing:

we no longer inhabit an era understood simply as Modernist, but where hyperactivity in communications, transportations, trade and electronic images presuppose a 'new' set of political, social, economic and transnational realities. 'That we live in Postmodern times' notes Wendy Brown (1991, p. 63), 'is nearly inarguable' albeit that there is no agreement 'about the configuration of this condition, its most striking marker, implications and portents'.

Toynbee (1946) concurred (Jarvis 2000, p. 58):

The world had just entered the last phase of Western history – 'the post modern' era an age that would be marked by anxiety and despair.

Jarvis went on (1998, p. 101):

This sense of millennial anxiety, of absolute historical breakage and rupture, of 'new' ages and 'new' associations have been endemic themes in the social sciences and humanities, reflecting, perhaps, not only a fascination with change, science, technology and the speed of innovation but a sense of new horizons as conceptions of 'locality' and 'space' have been obliterated with inter-planetary travel, jet-setting tourists, indeed of inter-state commuters who jet from London to Brussels to work and home again in time for dinner. Dazzled by such 'transformations' it is easy to speculate that we have entered a 'new' historical phase, or at the very least that we are approaching the 'end of [modern] history. (see for example Fukuyama 1992)

He continued to speak of new, changed, transformed, and reordered realities. A world composed of new economic, political and spatial configurations; the restructuring of global industry (Caporaso 1981); the rise of transnational finance capital; the new international division of labor (Frobel et al. 1978); and the reordering of world capitalism (Peet 1983; Thrift 1986). These views were further emphasized by Caldwell (1975, p. 567) who identified a 'widely held view that we are entering a period of major historical transition'. He used the term Postmodern to characterize this new phase as a movement on from the Modern: a time of significant transition in contrast to the constant but more gradual change that occurs as time passes.

During the new Postmodern phase, territorial political forms have ceased to dominate in policy making, implementation, and organization, replaced by nonterritorial elements deriving their power through technical sophistication. These new policy-making power bases constitute a new logic whether they have yet to become part of the logic of governance. They have to in time and it is only institutional inertia rather than their optimal configuration that sustains the old

7.4 Postmodernism and Governance

regimes. Maritime policy-makers take note. As Ashley (1991, p. 48) stated (quoted in Jarvis 1998, p. 104), European people and places:

> long certain of their absolute presence as a centre of meaning and origin of authority [have] had to accommodate their situation in a wider world of contesting cultures that at once effectively resist and effectively penetrate the European territory of truth.

This represents a new Postmodern sensibility, a 'kind of relativistic-plural world full of competing interpretations with no sovereign centre'.

Dear (1986, p. 373) was quite confident of the emergence of a new 'Postmodern epoch' calling it its 'grandest dimension' representing a 'niche in time-space'. He saw it as an epoch of transition and a radical break with the past as envisaged by Jameson (1984, p. 53). Jameson (1985, p. 113) also emphasized (quoted in Dear 1986, p. 374), that Postmodernism was:

> a periodizing concept whose function is to correlate the emergence of new formal features in culture with the emergence of a new type of social life and a new economic order.

He went on to suggest that it may not be easy to recognize when the new order emerges. Radical breaks between periods:

> do not generally involve complete changes of content but rather the restructuration of a certain number of elements already given; features that in an earlier period or system were subordinate now become dominant, and features that had been dominant again become secondary. (Jameson 1985, p. 123)

Despite these problems in identifying the emergence of a new era, Jameson is confident that we have a 'Postmodern cultural dominant'. Dear (1986, p. 374), quoting Jameson (1984, pp. 55–65), saw it as:

> generated by the 'late capitalistic' era of commodity production, which has imposed its own peculiar stamp on society. The imprint thus left is characterized by a new depthlessness, in which reality is visible only through a multitude of superficial reflections: by a dominant mode of pastiche, a 'stupendous proliferation of social codes'.

Dear (1988, p. 273) emphasized that identifying this new epoch means:

> grappling with one of the most fundamental problems in human knowledge: that of theorizing contemporaneity (Davis 1985, p. 107). How do we begin to make sense of an infinity of overlapping realities? Obviously, the contemporaneous appearance of two objects need not imply a causal relation: any landscape over time and space is more likely to consist of an anachronistic mixture of the obsolete, current and newborn artifacts. How can we make sense of such variety? There can be no easy answer to these questions. However, there is little doubt about their significance, especially given the scale and extent of current change in the world.

In a later paper Dear (1997, p. 50) also stressed the significance of the work of Jacques Derrida who emphasized that only by 'assuming a radical break had occurred would our capacity to recognize it be released' (Derrida 1973). Along with Dear and Flusty (1998, p. 50) he also noted the work of Wright-Mills (1959) who remarked upon the new era that was emerging as early as the late 1950s:

> We are at the ending of what is called the Modern Age. Just as Antiquity was followed by several centuries of Oriental ascendancy, which Westerners provincially called the Dark Ages, so now The Modern Age is being succeeded by a Post-modern period.

Curry (1991, p. 214) outlined the new epoch while Denzin (1986, p. 198) stressed that Postmodernism marked a:

> different moment in the socioeconomic organization of society. Capitalism has moved to a third stage, consumerism, having surpassed prior stages where market and monopoly capitalism dominated. This third stage has produced... a bureaucratic society controlled by consumption and the computerization of knowledge. This has created a crisis in the legitimation of science, technology and society. The grand narrative legitimating structures of the past turned on two myths: the belief that science could liberate humanity... and the belief that there is a unity to all knowledge, producing cumulative rational understandings of man, nature and society... Lyotard (1979) contends that these myths have collapsed.

The effect of the Postmodern upon maritime governance is not confined to changes in the European context but the relevance of this view of societal change for maritime policy-making internationally is significant. Ashley's European comments (noted earlier) are essentially a Modernist, organizational, and inflexible view of a stable and 'correct' society which no longer holds good. Policy making must adapt to recognize this. It also needs to recognize the plurality of modern governance, something that Hutton (2004, p. 1956) emphasized in relation to a diverse, urban society and Postmodern planning principles. Ley (1989, p. 53) concurred commenting on the:

> transition from a Cubist grid to what geographers have called a sense of place (which) reveals a more complex attention to theories of space in the plural styles of what has become known as Post-modern architecture.

This is supported by Butler (2002, p. 15) who suggested that Postmodernism was a rejection of master narratives and any kind of overall, totalizing explanation. It recognized the value of the minority and of those who do not fit. It was essentially a pluralist concept where all viewpoints are what he termed 'quasi-narratives'. Distinctly anti-Modernist, anti-master narrative, anti the whole idea of nation-state governance. 'All is fiction'. Postmodernism calls for an irreducible pluralism cut off from any unifying framework of belief that might lead to common political action. Enlightenment (Modernist) ideals that underlie the legal structures of most Western democracies are aimed at 'universalizable' ideals of equity and justice. In fact it is a system of repression and control. 'Reason itself, particularly in alliance with science and technology, is incipiently totalitarian'. Although maybe a little over the top, the point is clear.

Bergquist (1993, p. 68) suggested that Postmodern organizations tend toward the 'less bounded and more open' implying a distinct pluralist agenda and reflecting Jameson's (1991, p. 115) comments on Postmodern buildings. He contrasted the traditional view of society with Postmodern ideas of 'at home', 'on the road' and 'at work', none of which are quite as straightforward as they sound and each of which suggests an openness; a plurality. Bergquist went on to stress

7.4 Postmodernism and Governance

the importance of globalization in creating these ambiguities about boundaries and the moves toward plurality:

> The world we now live in has witnessed the simultaneous destruction of large multinational states and the resurrection of smaller nation-states that had been engulfed by their larger and more powerful neighbouring states during most of the twentieth century. We have seen the demise of large conglomerate states – the British Empire and the Soviet Union – within the past fifty years. We have also seen a growing push toward division and dissolution in many other nation-states such as Canada, Yugoslavia, Czechoslovakia, the United Kingdom and India. Even within the United States we find this tendency towards 'Balkanized' communities. (Bergquist 1993, p. 82)

Bauman (1992, pp. vii–ix), cited in Burke (2000), pulled together some of the more significant Postmodern characteristics when he described its 'all deriding, all-eroding, all dissolving destructiveness'. Postmodernity 'does not seek to substitute one truth for another, one life ideal for another... It braces itself for a life without truths, standards and ideals'. Kumar (1997, p. 98) suggested that in Postmodernism 'identity is not unitary or essential. It is fluid or shifting, fed by multiple sources and taking multiple forms'. Meanwhile Schwartz (1995), cited in Van Ham (2001, p. 22), suggested that Postmodernism is in many ways like Toto, the small dog in *The Wizard of Oz*... who pulls back the curtain of the Holy of the Holies and sees the all-too human wizard from Kansas generating his own *mysterium tremendum* at a microphone'. It is some of these issues that give Postmodernity its strength and fuels many of its critics.

This has been a long discussion on the nature of Postmodernity but even after this can we be sure that it has a relationship to governance that has seriously undermined maritime policy-making? Or are we just experiencing trends in society which may be significant (for example globalization) but which are all a part of what has gone on for many years and which will continue satisfactorily supported by the existing hierarchical, state-centric policy-making infrastructure?

Østerud (1996, p. 385) is convinced that the Postmodern interpretation of international policy-making is highly relevant. Warf (1993a, p. 164) commented on how Postmodernity has serious political and policy implications in that it exposes the interests of all those involved—the widest range of stakeholders in the maritime sector. We shall return to stakeholders in a later chapter but for the moment Warf's words ring true:

> every theory, every model, whether self-consciously or not, legitimates some interpretations of reality and not others.

In other words, Postmodernism flags up the fact that any chosen policy is only one interpretation of a given situation and that the institutional structure and governance framework that lies behind it is a chosen one that may vary; perhaps should vary, according to circumstances, issue, context, and the like. Thus, Postmodern policy-making cannot rely upon a vision of governance that is determined, inflexible, unchanging and (supposedly) optimal. New flexible, innovative and responsive frameworks are needed. It is to this we turn in the closing chapter.

This is all down to the troubled relationship between discourse and power. Butler (2002, p. 44) considered a discourse as a historically evolved set of interlocking and mutually supporting statements which are used to define and describe a subject matter—commonly law, medicine, and the police but just as easily could refer to shipping. They imply a 'dominating theory' that guides and justifies them. They are used to categorize people, places, things, and activities which then self-determine the justification for dealing with these items in the way that the dominant authority does—immigrant, tourist, passenger, tax free zone, flag state,, etc.—and the rights each has. Thus established, the existing shipping authorities (policy-makers) sustain their position through such discourses against the tide of Postmodern change (itself reflecting the new spatial fix) creating a governance void manifesting itself in practical failure.

Or as Butler (2002) puts it:
Prisoner: As God is my judge, my Lord, I am not guilty.
Judge: He is not. I am. You are. 6 months.

Lash and Urry (1987, pp. 312–313) concluded that 'transformations are occurring in the very structuring of western societies' and (although perhaps taking it a little too far), that as a result the governance of society must transform as well:

> the 'fixed, fast-frozen relations' of organized capitalist relations have been swept away. Societies are being transformed from above, from below and from within. All that is solid about organized capitalism, class, industry, cities, collectivity, nation-state, even the world, melts into air.

Rosenau (1992a, pp. 1–2) suggested that Postmodernism is central to policy making in that it questions what is 'taken for granted by policy-makers and the policy process'. As such it is fundamental to understanding the failure of modern maritime policies. If Postmodernism is with us—and certainly something new has happened whatever it is called, manifesting itself most obviously as globalization—then a new approach to governance is needed to accommodate its demands. Rosenau rejected the Modernist in governance heritance suggesting that Postmodernism:

> undermines unqualified confidence in modern technology, rational organization, reasoned consideration, scientific assessment, and all analysis grounded on Enlightenment logic (Dear 1986; Cooper and Burrell 1988). It rejects any suggestion that policy be based on efficiency, integration, coordination and revision in the light of feedback… It sets aside whatever may drive the modern policy process such as a preoccupation with meeting objectives, a concern that policy is feasible, or the belief that specialists or experts, technicians or generalists, have a privileged voice. The Postmodern questions policy dependence on statistics and the possibility that data can arbitrate between policy positions or allow us to conclude that one policy is superior to another.

Rosenau suggested that Modernist policy making assumes that time can be controlled, distance measured and space predicted. Globalization is the manifestation of space–time compression, both its cause and consequence, and it represents the rejection (the 'deconstruction') of these previously stable and predictable dimensions. Neither 'linear time' nor meaning of space can be relied upon. In the

7.4 Postmodernism and Governance

words of Peters (1993), Fish (1984) and Edelman (1988), the tools of policy making have been deconstructed. Universal standards and impartial criteria are to be treated with intense suspicion. Thus, established policy institutions, vehicles, representation, and relationships need to be questioned and probably rejected.

Rosenau (1992a, p. 2) provided some advice for Postmodern policy-making:

> to look beyond the center, to the margins, the forgotten, the left out, and the uncommon. It tells us to be alert for diversity, fragmentation, discontinuity, uncertainty, and indeterminacy rather than to always expect order, unity and consensus. All knowledge is community specific and relative in a Post-modern context and consciously unsystematic, decentred and heterological.

Hutton (2004, p. 1955), in considering the emergence of Vancouver's Central Area, saw that the influential theories of urban development—namely post-industrialism and Postmodernism have been significant for politics and governance. Caldwell (1975, p. 569) commented on the transition in policy making from the Modern to the Postmodern as traumatic particularly in terms of government institutions and the relationship that should exist with the governed (e.g the IMO and shipowners). This 'unstabilizing stress' during transition is what current policy failure represents.

> When the ideas of some significant portion of the governed are no longer consistent with the assumptions upon which the institutions have been premised, a variety of tensions in human relations occurs.

Precisely. He went on:

> When the stress between agency and citizen reaches a point of mutual alienation the legitimacy of the institution may become questionable and its credibility impaired with its informed critics. To the extent that these critics are influential in the communications media, the reputation of the agency may suffer over a much wider sector of the public than that immediately opposing agency programs. A self-image of agency personnel may differ dramatically from their image as perceived by those who no longer share their assumptions. The self-image of dedication to mission objectives and the public service may appear from outside the agency to be routine-bound and self-serving.

Maritime policy-makers take note as the stakeholders become increasingly 'contemptuous' of an:

> administrative system that... seems unable to cope not only with what its leaders allege to be society's problems but what the more enlightened and forward-looking members of society perceive to be the more fundamental problems of a transitional age.

Arts et al. (2006, p. 97) considered the Postmodern impact upon sociology, political modernization, and policy making concluding that 'Western societies have reached a new qualitatively different form of Modernity' (Albrow 1996; Beck 1986; Giddens 1990; Beck et al. 1994; Inglehart 1995). Finally, Nooteboom (1992, p. 65) indicated how he saw policy making having to change in a Postmodern environment, reflecting the clear move from a Modernist, state-centric, hierarchical model that then existed. Taking the example from marketing, he proposed a network approach that was characterized by:

- A shift from dyadic relations (supplier-customer) to networks of multiple participants or stakeholders, with indirect as well as direct links. This compares to the Postmodern notion of a subject having an identity only by virtue of his or her position as a nexus in a communication network.
- A shift from incidental transactions based on present competencies and preferences to investments in ongoing relations, with learning and mutual adaptation of competence and preference.

We shall return to these ideas in the concluding sections.

There is considerable agreement about the emergence of some sort of a new era during the late twentieth century. Writing anonymously in The Economist (1991, p. 51), and quoting Cooper (2000):

> in the new post cold-war, post balance-of-power world, countries fell broadly into one of three categories. Some, characterized chiefly by chaos (Somalia, Afghanistan, Liberia), might be considered 'pre-modern'; some, more familiar in appearance and behavior (Brazil, China), were 'modern'. Others, in which international concerns about sovereignty had yielded to mutual inspection and interference, were 'Post-modern'.

The Pre-modern was considered to be where the state no longer had Weber's 'legitimate monopoly of force'. Cooper suggested that these states, once ordered through empire, now are left to rot with the exception of occasional humanitarian aid, punitive measures or where oil is discovered. In shipping terms much is the same. States such as Cambodia, North Korea, and the Philippines are politically and economically ignored except when they provide convenient locations for dubious ship registration and cheap labor and the subsequent avoidance of international shipping policy. The modern state meanwhile, is the foundation of current shipping governance and is characterized by state sovereignty and non-interference by one nation into another's affairs. Cooper saw Postmodern states as different, encouraging mutual interference in each other's domestic affairs and inviting constraints and surveillance in military affairs.

Whereas the Treaty of Rome is an excellent example of the Postmodern in international affairs, the Treaty of Westphalia was a Modern approach to sovereignty and encapsulated the relationship between nations that lay behind the institutional development of maritime policy. In this Postmodern era, national borders mean little and the emphasis is on the transnational. Maritime governance fails to recognize this in its approach to policy making and the three characteristics that Cooper describes as fundamental to it—openness, monitoring, and interference across and between nation-states.

Dear and Flusty (1998, pp. 61–62)] in discussing urban Postmodernism summarized much of what we have seen happening to the world of shipping and how maritime policy-making and the structure of maritime governance needed to change. This change they termed 'flexism':

> a pattern of econo-cultural production and consumption characterized by near-instantaneous delivery and rapid redirectability of resources flows. Flexism's fluidity results from cheaper and faster systems of transportation and telecommunications, globalization of capital markets, and concomitant flexibly specialized, just-in-time production processes

enabling short product- and production-cycles. These result in highly mobile capital and commodity flows, able to out-maneuver geographically fixed labour markets, communities and bounded nation-states. Globalization and rapidity permit capital to evade long-term commitment to place-based socioeconomies, (thus enabling a crucial social dynamic of flexism whereby it requires) little or no labour at all from a given locale. Simultaneously, local down-waging and capital concentration operate synergistically to supplant locally owned enterprises with national and supranational chains, thereby transferring consumer capital and inventory selection even farther away from direct local control.

Apart from the reference to 'national chains', Dear and Flusty's summary of the Postmodern in the urban context is entirely appropriate for the maritime scenario.

7.5 Postmodernism: The Case Against

With such a contentious concept and one that purports to be so wide-ranging there are those, many in fact, who disbelieve. We cannot provide a thorough review of all those who have doubted the existence and/or the significance of the Postmodernist movement but it is only right that some indication of the debate is outlined here.

Some of the fiercest criticisms of the Postmodern stance came from Sokal and Bricmont (1998) in a widely publicized attempt to discredit the movement through the publication of a 'spoof' Postmodern article considering the role of science. Its acceptance by a Postmodern journal for publication despite the absence of any real academic credibility was used as a way of showing how feeble and thin Postmodernism was as a concept. Front page coverage of the parody by the *New York Times* (October 3, 1986) and UK-based *The Observer* amongst others did little to help the Postmodern cause. They followed this by an extensive critique of the major Postmodern philosophers (including Virilio, Lacan, Kristeva, Boudrillard, and Deleuze) and with some justification showed that some of the Postmodern claims were unsubstantiated and at times wholly pretentious.

Meanwhile Cohen (1997, pp. 390–391), summarized many of the criticisms of the Postmodern school when he stated:

> in the wrong hands it can quickly degenerate into collage and pastiche in which everything is rendered equivalent in the cultural supermarket of ideas.

Thompson (1993) commented upon many Modernist narratives that 'interpret the Postmodern mood as an irrational one that threatens the order of scientific knowledge'. In terms of this Modernist framing, 'meaningless', 'conceptual anarchy', and 'nihilism' are the outcomes consistent with the implications of Postmodernism' (Rorty 1989). Postmodernism is seen as challenging the 'established order of logo-centred rationality, and its procedures for establishing truth, will somehow precipitate the demise of meaning, knowledge, science and morality' (Thompson 1993, p. 328).

John Rajchman, Professor of Art History and Architecture at Columbia University was quoted in Connor (1989, p. 19) adding to the cynicism that often surrounds the Postmodern concept and suggesting that it:

> is like the Toyota of thought: produced and assembled in several different places and then sold everywhere.

Meanwhile Brown (1994, p. 142), cited in Jarvis (1998, p. 6), commenting on Postmodernism, was convinced that 'those that like this sort of thing will find this the sort of thing they like – those who do not, will not'. Gephart and Pitter (1993) in their discussion of the Postmodern organization emphasized that there had been much skepticism of the approach citing Rosenau (1992a, pp. 14–17) who questioned its legitimacy. Bull (1995, p. 10)] was also highly critical in his assessment of apocalypse theory:

> The Postmodernists take refuge in irony and a sort of juvenile frivolity, when they do not simply express boredom with the world. The mood is... profoundly negative. The story has ended; there is nothing more to be said.

Parker (1992, p. 15) suggested that Postmodernists often 'retreat into an intellectual ghetto which has little relation to the problems and politics of the real world and take a comic view of the whole process:

> The Enlightenment is dead, Marxism is dead, the working class movement is dead... and the author does not feel very well either. (Smith 1984, in Harvey 1990, p. 325)

Lash (1988, p. 311) presented a substantial case against the Postmodern although mainly in the context of literature and philosophy. He cited Anderson (1984), Frisby (1985) and Callinicos (1985) who each disputed Lyotard's announcement that:

> the contemporary skepticism before 'metanarratives' had been midwife to the birth of Postmodern condition. (Lash 1988, p. 311)

Robertson (1991, p. 289) provided an extensive dismissal of the Postmodern in his assessment of change that has occurred throughout religion in the twentieth century. He saw that globalization has made religions globally available as a 'source of collective identity declarations..., as well as something to be consumed and spent by individuals. And it is here that he saw that 'light can be cast upon the relationship between globalization and Postmodernization'. His view was that as with Modernity, Postmodernity is a Western society issue, and a reaction to the failure in Western values (perhaps to capitalism?) which claims to have global origins—but the very fact that Postmodernism is restricted to western society (albeit with implications for elsewhere where societies interact)—suggests that its global pretensions are just that.

Not everyone is convinced that the Postmodern represents a new epoch even if they are willing to recognize that something different has happened. Calhoun (1993, p. 75) for example agreed that real and major changes have taken place but that they 'do not yet amount to an epochal break'. He felt that to describe a

7.5 Postmodernism: The Case Against

Postmodern epochal change is problematic and that really we are seeing the Modern era reconfigured and nothing more. Other doubters included Pred (1992, p. 305) who rather than identifying a new Postmodern era, saw it as an:

> inaccurate, uncritical, deceptive and thereby politically dangerous 'epochal' labeling of the contemporary world, of everyday life under contemporary capitalisms.

Virilio (1999, pp. 2–9) was fiercely critical of the idea of a Postmodern epoch; Allmendinger (2001, pp. 61–62) cited Kumar's (1995, p. 154) undermining of the whole idea of a new cultural or social world—'the imperatives of power, profit and control seem as predominant now as they have ever been in the history of capitalist industrialism'; Lovering (1989) dismissed the whole idea in the context of geographical research; Entrikin's (1976) comment on humanism in that Postmodernism 'suffices as a critique but not as alternative'; Hannah and Strohmayer (1992, p. 308) along with Onuf (1991, p. 438) were highly critical of the notion of Postmodernism as representing anything different from a Modernist perspective; Dear (1997, p. 66) suggested that identifying a new epoch is a 'challenge'; Opello and Rosow (2004, p. 267) questioned whether Postmodernism really represented a new order; and Bourriard (2009), cited in Hatherley (2009, p. 155) asserted that 'Postmodernism is dead and buried' replaced by a rejuvenated Modernism.

Much more criticism of the nature and characteristics of Postmodernism can be found but the acceptance that something fundamental has changed in society is widespread, even if it is not quite universal. It is the contention here that this new era (call it Postmodern or something else) has massive implications for all aspects of society and must be accommodated in the way that society works. This includes industry, economy, culture, politics, and of course policy and governance—not least in that most globalized of all sectors, maritime.

References

Aalberts, T. E. (2004). The future of sovereignty in multilevel governance Europe—A constructivist reading. *Journal of Common Market Studies, 42*(1), 23–46.
Adler, L. (1966). Symbiotic marketing. *Harvard Business Review, 44*, 59–71.
Albrow, M. (1996). *The global age*. Cambridge: Polity Press.
Allmendinger, P. (2001). *Planning in postmodern times*. London: Routledge.
Amis, M. (2003). *Koba the dread: Laughter and the twenty million*. London: Vintage.
Anderson, P. (1984). Modernity and revolution. *New Left Review, 144*, 96–113.
Anderson, M. (1996). *Frontiers: Territory and state formation in the modern world*. Cambridge: Polity Press.
Antin, D. (1972). Modernism and postmodernism: approaching the present in American poetry. *Boundary, 2*(1), 98–113.
Antin, D. (1980). Is There a Postmodernism? In, H.R. Garvin, (Ed.) (1980a). *Romanticism, Modernism, Postmodernism, Bucknell Review, 25*(2), 127–135.
Arrighi, G. (1994). *The long twentieth century*. London: Allen and Unwin.
Arrington, C. E., & Francis, J. R. (1989). Letting the chat out of the bag: Deconstruction, privilege and accounting research. *Accounting, Organizations and Society, 14*(1/2), 1–28.

Arts, B., Leroy, P., & van Tatenhove, J. (2006). Political modernization and policy arrangements. A framework for understanding environmental policy change. *Public Organization Review, 6,* 93–106.

Ashley, R. K. (1991). The state of the discipline: Realism under challenge. In R. Higgott & J. L. Richardson (Eds.), *International relations: Global and Australian perspectives on an evolving discipline*. Canberra: Department of International Relations, Research School of Pacific Studies, The Australian National University.

Ashley, R., & Walker, R. B. J. (1990). Speaking the language of exile: Dissident thought in international studies. *International Studies Quarterly, 34*(3), 259–268.

Bagozzi, R. P. (1975). Marketing as exchange. *Journal of Marketing, 39,* 32–39.

Bagozzi, R. P. (1979). Toward a formal theory of marketing exchanges. In O. C. Ferrell, S. W. Brown, & C. W. Lamb (Eds.), *Conceptual and theoretical developments in marketing* (pp. 431–447). Chicago: American Marketing Association.

Bagozzi, R. P. (1984). A prospectus for theory construction in marketing. *Journal of Marketing, 48,* 11–29.

Bailey, D., Heon, F., & Steingard, D. (1993). Post-modern international development: Interdevelopment and global interbeing. *Journal of Organizational Change Management, 6*(3), 43–63.

Ballard, J. G. (1973). *Crash*. London: Jonathan Cape.

Baudrillard, J. (1983). *Simulations*. New York: Semiotext(e) Foreign Agents Press.

Bauman, Z. (1992). *Intimations of postmodernity*. London: Routledge.

Beck, U. (1986). *Risikogesellschaft Auf Dem Weg in eine andere Moderne*. Frankfurt am Main: Suhrkamp.

Beck, U. (1992). *Risk society*. London: Sage.

Beck, U., Giddens, A., & Lash, S. (1994). *Reflexive modernization politics tradition and aesthetics in the modern social order*. Cambridge: Polity Press.

Beckmann, J. (2004). Mobility and safety. *Theory Culture and Society, 21*(4/5), 81–100.

Bell, D. (1974). *The coming of post-industrial society*. London: Heinemann.

Benko, G. (1997). Introduction: Modernity, postmodernity and the social sciences. In G. Benko & U. Strohmayer (Eds.), *Geography, history and social science* (pp. 1–48). Dordrecht: Kluwer.

Berg, L. (1993). Between modernism and postmodernism. *Progress in Human Geography, 17*(4), 490–507.

Berger, A. A. (1998). *The postmodern presence*. London: Sage.

Bergquist, W. H. (1993). *The postmodern organization*. San Francisco: Jossey-Bass.

Berman, M. (1982). *All that's solid melts into air: The experience of modernity*. Harmondsworth: Penguin.

Bertens, H. (1995). *The idea of the postmodern: A history*. London: Routledge.

Boje, D. (1996). Lessons from premodern and modern for postmodern management. In G. Palmer & S. Clegg (Eds.), *Constituting management. Markets, meanings and identities* (pp. 329–345). Berlin: Walter de Gruyter.

Boje, D. M., & Dennehy, R. F. (2008). *Managing in the postmodern world; America's revolution against exploitation* (2nd ed.). Dubuque: Information Age: Kendall Hunt.

Boje, D. M., Fitzgibbon, D. E., & Steingard, D. S. (1996a). Storytelling at administrative science quarterly: warding off the postmodern barbarians. In D. M. Boje, R. P. Gephart Jr., & T. J. Thatchenkery (Eds.), *Postmodern management and organization theory* (pp. 60–92). Thousand Oaks: Sage.

Boje, D. M., Gephart, R. P., Jr., & Thatchenkery, T. J. (Eds.). (1996b). *Postmodern management and organization theory*. Thousand Oaks: Sage.

Bourriard, N. (2009). *The radicant*. New York: Lukas and Sternberg.

Boyer, R., & Durand, J. (1993). *After Fordism*. Basingstoke: Macmillan.

Breisach, E. (2011). *On the future of history: The postmodernist challenge and its aftermath*. Chicago: University of Chicago Press.

Brenner, N. (1997). Global, fragmented, hierarchical: Henri Lefebvre's geographies of globalization. *Public Culture, 10*(1), 135–167.

Brenner, N. (1998). Between fixity and motion: accumulation, territorial organization and the historical geography of spatial scales. *Environment and Planning D, 16*(4), 459–481.
Brenner, N. (1999). Beyond state-centrism? Territoriality and geographical scale in globalization studies. *Theory and Society, 28*(1), 39–78.
Brown, W. (1991). Feminist hesitations, postmodern exposures. *Differences: Journal of Feminist Cultural Studies, 3*(1), 63–84.
Brown, C. (1994). Review of 'the political subject of violence'. *Millennium: Journal of International Studies, 23*(1), 142–144.
Bull, M. (1995). On making ends meet. In M. Bull (Ed.), *Apocalypse theory and the end of the world* (pp. 1–20). Oxford: Blackwell.
Burke, B. (2000). *Post-modernism and Post-Modernity,* The Encyclopaedia of Informal Education, www.infed.org/biblio/b-postmd.htm.
Burrell, G. (1988). Modernism, postmodernism and organizational analysis 2: the contribution of Michel Foucault. *Organization Studies, 9*(2), 221–235.
Butler, C. (2002). *Postmodernism: A very short introduction.* Oxford: Oxford University Press.
Caldwell, L. K. (1975). Managing the transition to post-modern society. *Public Administration Review, 35*(6), 567–572.
Calhoun, C. (1993). Postmodernism as pseudohistory. *Theory, Culture and Society, 10,* 75–96.
Callinicos, A. (1985). Poststructuralism, postmodernism, postmarxism. *Theory, Culture and Society, 2*(3), 85–102.
Callinicos, A. (1989). *Against postmodernism.* Cambridge: Polity Press.
Campbell, D. (1990). Global inscription: How foreign policy constitutes the United States. *Alternatives, 15,* 280.
Caporaso, J. A. (1981). Industrialization in the periphery: The evolving global division of labour. *International Studies Quarterly, 25*(3), 347–384.
Caporaso, J. A. (1996). The European Union and forms of state: Westphalian, regulatory or postmodern? *Journal of Common Market Studies, 34*(1), 29–52.
Carroll, L. (1982). Through the looking glass. In L. Carroll (Ed.), *Journeys in wonderland.* Glasgow: Galley Press.
Carroll, J. D., & Henry, N. (1975). A symposium: Knowledge management. *Public Administration Review, 35*(6), 567–572.
Carter, P., & Jackson, N. (1987). Management, myth and metatheory—from scarcity to postscarcity. *International Studies of Management and Organization, 17*(3), 64–89.
Castells, M. (2000). Materials for an exploratory theory of the network society. *British Journal of Sociology, 51*(1), 5–24.
Chan, A. (2000). Redirecting critique in postmodern organization studies: The perspective of Foucault. *Organization Studies, 21*(6), 1059–1075.
Clark, J. (1996). Fredric Jameson's postmodern marxism. *Cogdito, 4*(2), www.mun.ca/phil/codgito/vol4/v4doc2.html.
Clegg, S. R. (1990). *Modern organizations. Organization studies in the postmodern world.* London: Sage.
Clegg, S. R. (1996). Postmodern management. In G. Palmer & S. Clegg (Eds.), *Constituting management. Markets, meanings and identities* (pp. 235–265). Berlin: Walter de Gruyter.
Cohen, P. (1997). *Rethinking the youth question: Education, labour and cultural studies.* London: Macmillan.
Connor, S. (1989). *Postmodernist culture.* Oxford: Blackwell.
Cooper, R. (1989). Modernism, post modernism and organizational analysis 3: the contribution of Jacques Derrida. *Organization Studies, 10*(4), 479–502.
Cooper, R. (2000). *The post-modern state and the world order.* London: Demos.
Cooper, R., & Burrell, G. (1988). Modernism, postmodernism and organizational analysis. *Organization Studies, 9*(1), 91–112.
Curry, M. R. (1991). Postmodernism, language and the strains of modernism. *Annals of the Association of American Geographers, 81,* 210–228.

Davis, M. (1985). Urban renaissance and the spirit of post-modernism. *New Left Review, 151,* 106–113.
De Onis, F. (1961). *Antologia de le Poesia Espanola e Hispanoamericana (1882–1932).* New York: Las Américas Publishing.
De Sousa Santos, B. (1995). Globalization, nation-states and the legal field, from legal diaspora to legal ecumene? In B. de Sousa Santos (Ed.), *Towards a new common sense: Law, science and politics in the paradigmatic transition* (pp. 250–377). New York: Routledge.
Dear, M. J. (1986). Postmodernism and planning. *Environment and Planning D: Society and Space, 4,* 367–384.
Dear, M. J. (1988). The postmodern challenge: Reconstructing human geography. *Transactions of the Institute of British Geographers, NS 13,* 262–274.
Dear, M. J. (1991). Book review: The condition of postmodernity: An inquiry into the origins of cultural change, David Harvey. *Annals of the Association of American Geographers, 81,* 533–539.
Dear, M. J. (1997). Postmodern bloodlines. In G. Benko & U. Strohmayer (Eds.), *Geography, history and social science* (pp. 49–71). Dordrecht: Kluwer.
Dear, M. J., & Flusty, S. (1998). Postmodern urbanism. *Annals of the Association of American Geographers, 88*(1), 50–72.
Denzin, N. K. (1986). Postmodern social theory. *Sociological Theory, 4,* 194–204.
Der Derian, J. (1990). The (s)pace of international relations: Simulation, surveillance, and speed. *International Studies Quarterly, 34*(3), 295–310.
Derrida, J. (1973). *Speech and phenomena.* Evanstown: Northwestern University Press.
Dews, P. (1995a). *Logics of disintegration.* London: Verso.
Dews, P. (1995b). *The limits of disenchantment: Essays on contemporary European philosophy.* London: Verso.
Dicken, P. (1994). Global-local tensions: Firms and states in the global space-economy. *Economic Geography, 70*(2), 101–128.
Dijkink, G. (1993). Postmodernism, power, synergy. *Tijdschrift voor Economische en Sociale Geograpfie, 84*(3), 178–180.
Drucker, P. F. (1957). *Landmarks of tomorrow.* New York: Harper.
Drucker, P. F. (1992). *Managing for the future.* New York: Dutton.
Economist (1991) *Inner Space,* 18 May.
Edelman, J. M. (1977). *Political language: Words that succeed and policies that fail.* New York: Academic.
Edelman, J. M. (1988). *Constructing the political spectacle.* Chicago: University of Chicago Press.
Entrikin, N. (1976). Contemporary humanism in geography. *Annals of the Association of American Geographers, 66*(4), 615–632.
Etzioni, A. (1968). *The active society.* New York: Free Press.
Featherstone, M. (1987). Lifestyle and consumer culture. *Theory, Culture and Society, 4*(1), 55–70.
Fish, S. (1984). Fish v fiss. *Stanford Law Review, 36,* 1325–1347.
Fitts, D. (Ed.). (1942). *Anthology of contemporary Latin–American poetry.* Washington: Pan-American Union.
Foster, H. (1983). *The anti-aesthetic: Essays on postmodern culture* (pp. 111–125). Port Townsend: Bay Press.
Foster, H. (Ed.). (1985). *Postmodern culture.* London: Pluto Press.
Foucault, M. (1987). What is enlightenment? In P. Rainbow & P. Sullivan (Eds.), *Interpretive social sciences: A second look* (pp. 157–176). Berkeley: University of California Press.
Fox, S., Cooper, R., & Martinez, L. T. (1992). Preface. *International Studies of Management and Organization, 22*(2), 3–14.
French, R. D. (1992). Le gouvernement a l'ere du postmodernisme. *Optimum, 23*(1), 46–56.
Frisby, D. (1985). *Fragments of modernity.* Cambridge: Polity Press.

Frobel, F., Heinrichs, J., & Kreye, O. (1978). The new international division of labour. *Social Science Information, 17*(1), 123–142.
Fukuyama, F. (1992). *The end of history and the last man.* New York: Free Press.
Galinsky, K. (1992). *Classical and modern interactions: Postmodern architecture, multiculturalism, decline, and other issues.* Austin: University of Texas Press.
Gaudin, J. (1998). Modern governance, yesterday and today: Some clarifications to be gained from French government policies. *International Social Science Journal, 50*(155), 47–56.
Geertz, C. (1983). *Local knowledge.* New York: Basic Books.
Gephart, R. P., Jr. (1996). Management, social issues and the postmodern era. In D. M. Boje, R. P. Gephart Jr, & T. J. Thatchenkery (Eds.), *Postmodern management and organization theory* (pp. 21–44). Thousand Oaks: Sage.
Gephart, R. P., Jr, & Pitter, R. (1993). The organizational basis of industrial accidents. *Canada Journal of Management Inquiry, 3*, 238–252.
Gergen, K. J. (1991). *The saturated self-dilemmas of identity in contemporary life.* New York: Basic Books.
Gergen, K. J., & Thatchenkerry, T. J. (2004). Organization science as social construction: Postmodern potentials. *The Journal of Applied Behavioral Science, 40*(2), 228–249.
Giddens, A. (1984). *The constitution of society: Outline of the theory of structuration.* Berkeley: University of California Press.
Giddens, A. (1990). *The consequences of modernity.* Stanford: Stanford University Press.
Giddens, A. (1992, January 17). Uprooted signposts at century's end. *Times Higher Education Supplement,* p. 21–22.
Gill, S. (2000). Towards a postmodern prince? The battle in Seattle as a moment in the new politics of globalization. *Millennium: Journal of International Studies, 29*(1), 131–140.
Graham, J. (1988). Post-modernism and marxism. *Antipode, 20*(1), 60–66.
Gramsci, A. (1971). *Selections from the prison notebooks.* New York: International Publishers.
Greenberg, C. (1979, October 31). *Modern and Postmodern,* William Dobell Memorial Lecture, Sydney, Australia.
Habermas, J. (1973). *Legitimation crisis.* Boston: Beacon Press.
Habermas, J. (1981). Modernity versus postmodernity. *New German Critique, 22,* 3–14.
Habermas, J. (1983). Modernity—an incomplete project. In T. Docherty (Ed.), *Postmodernism: A reader* (pp. 98–109). Brighton: Harvester Wheatsheaf.
Hajer, M. (2003). Policy without polity? Policy analysis and the institutional void. *Policy Sciences, 36,* 175–195.
Hakansson, H. (Ed.). (1982). *International marketing and purchase of industrial goods—An interaction approach.* Chichester: Wiley.
Hakansson, H. (1987). *Industrial technological development. A network approach.* London: Croom Helm.
Hall, S., Held, D., & McGrew, T. (Eds.). (1992). *Modernity and its futures.* Cambridge: Polity Press.
Hannah, M., & Strohmayer, U. (1992). Postmodernism (s)trained. *Annals of the American Association of Geographers, 82*(2), 308–312.
Haraway, D. (1985). Manifesto for cyborgs—science, technology and socialist feminism in the 1980s. *Socialist Review, 80,* 65–108.
Harvey, D. (1973). *Social justice and the city.* Baltimore: The John Hopkins University Press.
Harvey, D. (1981). The spatial fix—Hegel, Von Thunen and Marx. *Antipode, 13*(3), 1–12.
Harvey, D. (1985). *Consciousness and the urban experience.* Baltimore: John Hopkins University Press.
Harvey, D. (1987). Flexible accumulation through urbanization: Reflections on 'post-modernism' in the American city. *Antipode, 19*(3), 260–286.
Harvey, D. (1990). *The condition of postmodernity.* Cambridge: Blackwell.
Harvey, D. (1996). *Justice, nature and the geography of difference.* Oxford: Blackwell.
Harvey, D. (2001). Globalization and the "spatial fix". *Geographische Review, 2,* 23–30.

Harvey, D., & Scott, A. (1989). The practice of human geography: theory and empirical specificity in the transition from Fordism to flexible accumulation. In B. Macmillan (Ed.), *Remodelling geography* (pp. 217–229). Oxford: Blackwell.
Hassan, I. (1967). The literature of silence: From Henry Miller to Beckett and Burroughs. *Encounter, 28*(1), 74–82.
Hassan, I. (1970). Frontiers of criticism: Metaphors of silence. *Virginia Quarterly Review, 46*(1), 81–95.
Hassan, I. (1971). Postmodernism: A paracritical bibliography. *New Literary History, 3*(1), 5–30.
Hassan, I. (1973). The new gnosticism: Speculations on the aspect of a postmodern mind. *Boundary, 21*(3), 547–569.
Hassan, I. (1978). Culture, indeterminacy and immanence: margins of the (postmodern) age. *Theory, Culture and Society, 1*(1), 51–85.
Hassan, I. (1980). The question of postmodernism. *Bucknell Review, 26*(2), 117–126.
Hassan, I. (1985). The culture of postmodernism. *Theory, Culture and Society, 2*(3), 119–131.
Hatherley, O. (2009). Post-postmodernism? Review of the radicant, by Nicolas Bourriard. *New Left Review, 59*, 153–160.
Havel, V. (1992, March 1). *The end of the modern era*, New York Times.
Hebdige, D. (1988). *Hiding in the light: On images and things*. London: Routledge.
Hellgren, B., & Stjernberg, T. (1987). Networks: an analytical tool for understanding complex decision processes. *International Studies of Management and Organization, 17*, 88–102.
Hess, M. (2004). 'Spatial' relationships? Towards a reconceptualization of embededness. *Progress in Human Geography, 28*(2), 165–186.
Heydebrand, W. (1989). New organizational forms. *Work and Occupations, 16*(3), 323–367.
Hirst, P., & Zeitlin, J. (1991). Flexible specialization versus post-fordism: Theory, evidence and policy implications. *Economy and Society, 20*(1), 1–56.
Hoskin, K., & Macve, R. (1986). Accounting and examination; genealogy of disciplinary power. *Accounting, Organizations and Society, 11*(2), 105–136.
Houston, F. S., & Gassenheimer, J. B. (1987). Marketing and exchange. *Journal of Marketing, 47*, 3–18.
Hunt, S. D. (1989). Naturalistic, humanistic, and interpretive enquiry: Challenges and ultimate potential. In E. C. Hirschman (Ed.), *Interpretive consumer research* (pp. 185–208). Provo: Association for Consumer Research.
Hutton, T. A. (2004). Post-industrialism, post-modernism and the reproduction of Vancouver's central area: re-theorising the 21st century city. *Urban Studies, 41*(10), 1953–1982.
Huyssen, A. (1984). Mapping the postmodern. *New German Critique, 33*, 5–51.
Inglehart, R. (1995). *Cultural shift in advanced industrialized societies*. Princeton: Princeton University Press.
Jameson, F. (1983). Postmodernism and the consumer society. In H. Foster (Ed.), *The anti-aesthetic: Essays on postmodern culture* (pp. 111–125). Port Townsend: Bay Press.
Jameson, F. (1984). Postmodernism, or the cultural logic of late capitalism. *New Left Review, 146*, 53–92.
Jameson, F. (1985). Postmodernism and consumer society. In H. Foster (Ed.), *Postmodern culture* (pp. 111–125). London: Pluto Press.
Jameson, F. (1991). *Postmodernism or the cultural logic of late capitalism*. Durham: Duke University Press.
Jarvis, D. S. L. (1998). Postmodernism: A critical typology. *Politics and Society, 26*(1), 95–142.
Jarvis, D. S. L. (2000). *International relations and the challenge of postmodernism: Defending the discipline*. Columbia SC: University of South Carolina Press.
Jencks, C. (1986a). *The language of postmodern architecture*. New York: Basic Books.
Jencks, C. (1986b). *What is post-modernism?*. London: Academy Editions.
Jencks, C. (Ed.). (1992a). *The post-modern reader*. London: Academy Editions.
Jencks, C. (1992b). The post-modern agenda. In C. Jencks (Ed.) (1992a) The post-modern reader, London: Academy Editions, pp. 10–39.

Johanson, J., & Mattson, L. G. (1987). Interorganizational relations in industrial systems—a network approach compared with the transaction cost approach. *International Studies of Management and Organization, 17*, 34–48.

Keidel, R. W. (1994). Rethinking organizational design. *Academy of Management Executive, 4*(4), 12–27.

Kellner, D. (1988). Postmodernism as social theory: Some challenges and problems. *Theory, Culture and Society, 5*, 239–270.

Kreiner, K. (1992). The postmodern epoch of organization theory. *International Studies of Management and Organization, 22*(2), 37–52.

Kristeva, J. (1980). Postmodernism? *Bucknell Review, 25*, 136–141.

Kumar, K. (1995). *From post-industrial to post-modern society*. Oxford: Blackwell.

Kumar, K. (1997). The post-modern condition. In A. H. Halsey, H. Lauder, P. Brown, & A. S. Wells (Eds.), *Education: Culture, economy and society*. Oxford: Oxford University Press.

Lacher, H. (2005). International transformation and the persistence of territoriality: Toward a new political geography of capitalism. *Review of International Political Economy, 12*(1), 26–52.

Lash, S. (1988). Discourse or figure? Postmodernism as a 'regime of signification'. *Theory, Culture and Society, 5*, 311–336.

Lash, S. (1990). *Sociology of postmodernism*. London: Routledge.

Lash, S., & Urry, J. (1987). *The end of organized capitalism*. Cambridge: Polity Press.

Laumann, E. O. (1991). Comment on "the future of bureaucracy and hierarchy in organizational theory: A report from the field". In P. Bourdieu & J. S. Coleman (Eds.), *Social theory for a changing society* (pp. 90–93). Boulder: Westview.

Leonard, P. (1997). *Postmodern welfare: Reconstructing an emancipatory project*. London: Sage.

Ley, D. (1983). Postmodernism, or the cultural logic of advanced intellectual capital. *Tijdschrift voor Economische en Sociale Geografie, 84*(3), 171–174.

Ley, D. (1989). Modernism, post-modernism and the struggle for place. In J. Agnew & J. S. Duncan (Eds.), *The power of place* (pp. 44–65). Boston: Unwin Hyman.

Loft, A. (1986). Towards a critical understanding of accounting; the case of cost in the UK 1914–1925. *Accounting, Organizations and Society, 11*(2), 137–169.

Lovering, J. (1989). Postmodernism, Marxism and locality research: The contribution of critical realism to the debate. *Antipode, 21*, 1–12.

Luke, T. W. (1991). The discipline of security studies and the codes of containment: Learning from Kuwait. *Alternatives, 16*, 315–344.

Lyotard, J.-F. (1979). *The postmodern condition: A report on knowledge*. Manchester: Manchester University Press.

Lyotard, J.-F. (1987). *La condition postmoderne*. Kampen: Kok Agora. (Dutch translation).

Lyotard, J.-F. (1988). *The differend: Phrases in dispute*. Minneapolis: University of Minneapolis Press.

Mackinder, H. J. (1904). The geographical pivot of history. *The Geographical Journal, 23*(4), 421–437.

Mann, M. (1997). Has globalization ended the rise and rise of the nation-state? *Journal of International Political Economy, 4*(3), 472–496.

Melucci, A. (1989). *Nomads of the present: Social movements and individual needs in contemporary society*. London: Hutchinson.

Michael, D. N. (1962). *Cybernation: The silent conquest*. Santa Barbara: Center for the Study of Democratic Institutions.

Morgan, G. (1997). *Images of organization*. Thousand Oaks: Sage.

Newman, D. (1995). *Boundaries in flux: The 'green line' boundary between Israel and the West Bank, boundary and territory briefing* (Vol. 5). Durham: International Boundaries Research Unit, University of Durham.

Nodoushani, O. (1987). A note on progress: Postmodern transformation in the systems age. *Systems Research, 4*(1), 59–64.

Nooteboom, B. (1992). A postmodern philosophy of markets. *International Studies of Management and Organization, 22*(2), 53–76.

Norcliffe, G., Bassett, K., & Hoare, T. (1996). The emergence of postmodernism on the urban waterfront. *Journal of Transport Geography, 4*(2), 123–134.

Norris, C. (1993). *The truth about postmodernism*. Oxford: Blackwell.

Norris, C. (1995). Versions of apocalypse: Kant, Derida, Foucault. In M. Bull (Ed.), *Apocalypse theory and the ends of the world* (pp. 227–249). Oxford: Blackwell.

O'Neill, H. M. (1994). Restructuring, re-engineering and rightsizing: do the metaphors make sense? *Academy of Management Executive, 4*(4), 9–11.

Offe, C. (1984). *Contradictions of the welfare state*. Cambridge: MIT Press.

Olson, C. (1967). *Human universe and other essays*. New York: Grove Press.

Onuf, N. (1991). Sovereignty: Outline of a contemporary history. *Alternatives, 16*(4), 425–446.

Opello, W. C., Jr., & Rosow, S. J. (2004). *The nation-state and global order: A historical introduction to contemporary politics*. Boulder: Lynne Reinner.

Østerud, Ø. (1996). Antinomies of postmodernism in international studies. *Journal of Peace Research, 33*(4), 385–390.

Palmer, G., & Clegg, S. (Eds.). (1996). *Constituting management. Markets, meanings and identities*. Berlin: Walter de Gruyter.

Parker, M. (1992) Post-modern organizations or postmodern organization theory? *Organization Studies, 13*(1), 1–17.

Peet, R. (1983). Introduction: The global geography of contemporary capitalism. *Economic Geography, 59*(2), 105–111.

Peters, B. G. (1993). *American public policy: Promise and performance*. Chatham: Chatham House.

Piore, M. J., & Sabel, C. F. (1984). *The second industrial divide: Possibilities for prosperity*. New York: Basic Books.

Podolny, J. M., & Page, K. L. (1998). Network forms of organization. *Annual Review of Sociology, 24*, 57–76.

Pollack, A. (1993, January 27). Forget all the theorizing, Japan puts chaos to work. *New York Times*.

Porter, J. (1968). The future of upward mobility, *American Sociological Review, 33*(1) 5–19.

Poster, M. (1984). *Foucault, marxism and history: Mode of production versus mode of information*. Cambridge: Polity Press.

Poster, M. (1990) *The mode of information: Poststructuralism and social context*. Cambridge: Polity Press.

Pred, A. (1992). Straw men build straw houses? *Annals of the American Association of Geographers, 82*(2), 305–319.

Radhakrishnan, R. (1994). Postmodernism and the rest of the world. *Organization, 1*(2), 305–340.

Rhodes, R. A. W. (1997). *Understanding governance. Policy networks, governance, reflexivity and accountability*. Buckingham: Open University Press.

Rhodes, R. A. W., & Marsh, D. (1992). New directions in the study of policy networks. *European Journal of Political Research, 21*, 181–205.

Robertson, R. (1991). Globalization, modernization and postmodernization. The ambiguous position of religion. In R. Robertson & W. R. Garrett (Eds.), *Religion and global order* (pp. 281–291). New York: Paragon House.

Rorty, R. (1989). *Contingency, irony and solidarity*. Cambridge: Cambridge University Press.

Rosenau, J. N. (1989). The state in an era of cascading politics. In J. A. Caporaso (Ed.), *The elusive state* (pp. 17–48). Beverly Hills: Sage.

Rosenau, J. N. (1992a). Governance, order and change in world politics. In J. N. Rosenau & E. O. Czempiel (Eds.), *Governance without government: Order and change in world politics* (pp. 1–29). Cambridge: Cambridge University Press.

Rosenau, J. N. (1992b). *Post-modernism and the social sciences: Insights, inroads and intrusions*. Princeton: Princeton University Press.

Rosenau, J. N. (2004). Strong demand, huge supply: Governance in an emerging epoch. In I. Bache & M. Flinders (Eds.), *Multilevel governance* (pp. 31–48). Oxford: Oxford University Press.

Rouleau, L., & Clegg, S. R. (1992). Postmodernism and postmodernity in organization analysis. *Journal of Organizational Change, 5*(1), 8–25.

Ruggie, J. G. (1993). Territoriality and beyond: Problematizing modernity in international relations. *International Organization, 47*(1), 139–174.

Sabel, C., & Zeitlin, J. (1985). Historical alternatives to mass production: Politics, markets and technology in nineteenth-century industrialization. *Past and Present, 108*, 133–174.

Sabel, C. F., & Zeitlin, J. (2008). Learning from difference: The new architecture of experimentalist governance in the EU. *European Law Journal, 14*(3), 271–327.

Schulz, M. (1992). Postmodern pictures of culture. *International Studies of Management and Organization, 22*(2), 15–35.

Schwartz, R. (1995). Montheism and the violence of identities. *Raritan, 14*, 3.

Scott, A. J. (1988). *New industrial spaces: Flexible production, organization and regional development in North America and Western Europe.* London: Pion.

Scott, W. G. (1974). Organization theory: A reassessment. *Academy of Management Journal, 17*, 242–254.

Sextus Empiricus (1935) *Against the logicians* (R. G. Bury Trans.), (Loeb edn). Heinemann: London, p. 179.

Sextus Empiricus (1935) *Against the logicians* (R. G. Bury Trans.), (Loeb edn). Heinemann: London, p. 179.

Shaw, M. (1997). The state of globalization: Towards a theory of state transformation. *Journal of International Political Economy, 4*(3), 497–513.

Short, J. R. (1993). The 'myth' of postmodernism. *Tijdschrift voor Economische en Sociale Geografie, 84*(3), 169–171.

Sim, S. (1992). *Beyond aesthetics.* Brighton: Harvester Wheatsheaf.

Simmel, G. (1971). The metropolis and mental life. In D. Levine (Ed.), *On individuality and social form* (pp. 324–339). Chicago: University of Chicago Press.

Simon, D. (1996). *Transport and development in the third world.* London: Routledge.

Smith, N. (1984). *Uneven development: Nature, capital and the production of space.* Oxford: Blackwell.

Soja, E. W. (1987). The postmodernization of geography: A review. *Annals of the American Association of Geographers, 77*(2), 289–294.

Soja, E. W. (1989). *Postmodern geographies: The reassertion of space in critical social theory.* New York: Verso.

Soja, E. W. (1996). Margin/alia: Social justice and the new cultural politics. In A. Merrifield & E. Swyngedouw (Eds.), *The urbanization of injustice.* London: Lawrence and Wishart.

Soja, E. W. (2000). *Postmetropolis: Critical studies of cities and regions.* Oxford: Blackwell.

Sokal, A., & Bricmont, J. (1998). *Intellectual impostures.* London: Profile Books.

Soper, K. (1986). *Humanism and anti-humanism.* London: Hutchinson.

Stalder, F. (2006). *Manuel Castells.* Cambridge: Polity Press.

Steingard, D. S., & Fitzgibbons, D. E. (1993). A postmodern deconstruction of total quality management (TQM). *Journal of Organizational Change Management, 6*(5), 27–42.

Thompson, C. J. (1993). Modern truth and postmodern incredulity: A hermeneutic deconstruction of the metanarrative of "scientific truth" in marketing research. *International Journal of Research in Marketing, 10*, 325–338.

Thrift, N. J. (1986). The geography of international economic disorder. In R. J. Johnston & P. J. Taylor (Eds.), *A world in crisis?* (pp. 12–67). Oxford: Blackwell.

Thrift, N. J. (1997). The rise of soft capitalism. *Cultural Values, 1*(1), 29–57.

Toffler, A. (1970). *Future shock.* New York: Random House.

Toynbee, A. (1946). *A study of history.* New York: Oxford University Press.

Tyler, L. (2005). Towards a postmodern understanding of crisis communication. *Public Relations Review, 31*, 566–571.

Van Ham, P. (2001). *European integration and the postmodern condition*. London: Routledge.
Varadarajan, P., & Rajaratnam, D. (1987). Symbiotic marketing revisited. *Journal of Marketing, 50*, 7–17.
Virilio, P. (1999). Indirect light: Extracted from polar inertia. *Theory, Culture and Society, 16*(1), 5–6–57–70.
Walker, R. B. J. (1991). State sovereignty and the articulation of political space/time. *Millennium, 20*, 445–461.
Wallace, W. (1999). The sharing of sovereignty: The European paradox. *Political Studies, 47*(3), 503–521.
Warf, B. (1988). The resurrection of local uniqueness. In B. Golledge, H. Couclelis, & P. Gould (Eds.), *A ground for common search* (pp. 51–62). Santa Barbara: Santa Barbara Geographical Press.
Warf, B. (1993a). Postmodernism and the localities debate: Ontological questions and epistemological implications. *Tijdschrift voor Economische en Sociale Geografie, 84*(3), 162–168.
Warf, B. (1993b). Embalming the detritus of modernism: reply to Short, Ley, De Pater, and Dijkink. *Tijdschrift voor Economische en Sociale Geografie, 84*(3), 181–184.
Webster, F. (2002). *Theories of the information society*. London: Routledge.
Welsch, W. (1983). Modernite et postmodernite. *Les Cahiers de Philosophie, 6*, 21–31.
Welsch, W. (1987). *Unsere postmoderne moderne*. Weinheim: Acta Humanoria.
Woods, T. (1999). *Beginning postmodernism*. Manchester: Manchester University Press.
Wright-Mills, C. (1959). *The sociological imagination*. New York: Oxford University Press.
Zukin, S. (1992). The postmodern invasion. *International Journal of Urban and Regional Research, 16*(3), 489–496.

Chapter 8
Maritime Postmodernism in Practice

Eagleton (1986, p. 80), cited in Soja (1987, p. 289) suggested that:

> To 'deconstruct', then, is to reinscribe and resituate meanings, events and objects within broader movements and structures; it is, so to speak, to reverse the imposing tapestry in order to expose in all its unglamorously disheveled tangle the threads constituting the well-heeled image it presents to the world.

Postmodernism has the potential to provide a mechanism to expose the complex structures that drive maritime policy-making. Arthur (1994), cited in Waldrup (1993, p. 329) and quoted in Thrift (1999, p. 32) emphasized the difficulties of understanding real life. He was considering the value of complexity theory and its application in geography but his comments are equally as relevant to the issues that surround maritime governance. Complexity theory understands that:

> logic and philosophy are messy, that language is messy, that chemical kinetics is messy, that physics is messy and finally that the economy is naturally messy. And it's not that this is a mess created by the dirt that's on the microscope glass. It's that this mess is inherent in the systems themselves. You can't capture any of them and confine them to a neat box of logic.

So it is with maritime policy-making where the neat and organized Modernist vision of maritime governance just does not work.

Postmodernism is a concept that we have identified as central to developments in maritime policy-making but it will already be apparent that it can be, and has been, widely applied across innumerable other disciplines, scenarios, and situations. This variety of applications is not irrelevant to our discussion of maritime policy failure as a narrow categorization of impacts and activities to the maritime field cannot help in identifying the increasingly complex impact of globalization in a Postmodern context—especially so in the light of its overt recognition of interdependence, plurality, and cross-disciplinarity. This section will begin to look at its application across a number of related fields before we concentrate upon the maritime sector in some depth.

Ross (1958, p. 10) provided an indication of the variety of applications of Postmodernism that have been made. This is far from exhaustive as the concept refers to a complete societal aura, a mélange of all things human. However, the breadth and scope of what is suggested here is almost breathtaking in its extent. At the same time this also suggests that its application to the governance of policy-making is an inevitable result. We have seen earlier how Hebdige (1988, pp. 118–119) provided a selection of contexts, tendencies, and objects to which Postmodern refers and this diverse, cultural currency defies attempts to purify the concept and reflects the very nature of what has been going on—a complete societal shift that requires a similar shift in how society is governed and governs itself—through measures of governance and the policies that result. Hassan (1980, p. 122, 1985) followed this up by questioning whether Postmodernism is only a literary tendency—where much of its earliest recognition took place—or is it a 'cultural phenomenon'? He saw that the latter was clearly the case but then was troubled by how the 'various aspects of this phenomenon' could be brought together, identifying the psychological, philosophical, economical, and political as paramount.

He concluded:

> We ask about Postmodernism. But men die, women suffer, children starve in the dust, torturers heed no screams, technicians don their leaded aprons, octogenarians reach for an empty cup, and physicists discover traces of the original moment of the universe. I suspect that these and countless other events, are not entirely alien to the 'question of Postmodernism'. We are, after all, engaged in a task of reflection that seeks to encompass the reality of our time. The pressure of the unspeakable, the unnamable, the raw necessities of the human condition, threatens always to disrupt our discourse even as we try to give it dignity and shape. Yet that pressure and that disruption also save us from our categorical selves. In some recoverable sense, no question we ask is innocent of pain and our mortality.

8.1 Postmodernism and Organizations

The effect of Postmodernism in the maritime sector has to have part of its roots in the relationship of the movement to change in organizations. In this section we attempt to address this issue.

Postmodernism and its relationship to organizations has received some considerable attention and one that has been of substance. Rouleau and Clegg (1992), Morgan (1997), Alvesson and Deetz (1996), Flax (1990), and Hassard and Parker (1993) provided selected reviews. Boje and Dennehy (1993, p. 32) attempted to show how Postmodern management had emerged from the Premodern and the Modern; the Premodern concentrated upon planning and influencing with little firm direction; the Modern was a response to this and was focused upon control and organization; the Postmodern is different again in that it is characterized by leadership, influence, and organization, in a sense combining its predecessors. Kreiner (1992, pp. 38–39) traced the change from a Modernist to a Postmodernist organization suggesting that there was a noticeable trend away from organizations as 'machines' characterized by 'efficiency', 'predictability', and 'conformity'. Benton (1990, p. 8) was particularly sarcastic:

8.1 Postmodernism and Organizations

In the 1950s, most people worked in structured hierarchies, assigned to particular tasks, governed by the discipline of the Corporation and the orders of top managers who reserved to themselves the strategic decisions, and those that balanced the competing requirements of the separate specialities. Those organizations resembled vast clockwork constructions, with a myriad of cogs turning on their spindles, mechanical, each oblivious of the machinery beyond its own sprockets.

Kreiner (1992, p. 39) went on to describe his vision of the Postmodern organization:

> the machine has been dismantled into 'locally rational' parts (Simon 1947; Cyert and March 1963); into political 'factions' (Simon and March 1957; Crozier 1964); into loosely coupled and decoupled processes (Cohen et al. 1972; March and Olsen 1976; Weick 1976). In other contributions, the tangible machine has evaporated into, for example, cultural and ideological systems (Pondy et al. 1983; Alvesson 1987).

Kreiner saw these theories disputing the existing and fixed ideas about what an organization was. They did not imply coherence, solely united by their resistance to what had gone. Together they suggested that the 'classical conception' of an organization had been overthrown.

Featherstone (1987, p. 69) saw the Postmodern organization as taking a distinctly new form:

> traditional distinctions and hierarchies are collapsed, polyculturalism is acknowledged... kitch (*sic*), the popular and difference are celebrated. Their cultural innovation proclaiming a *beyond* is really a *within*, a new move within the cultural game which takes into account the circumstances of production of cultural goods, which will itself in turn be greeted as eminently marketable by the cultural intermediaries.

Perhaps his ideas about cultural goods can be extended to commodities, services, and institutions?

Clegg (1990, p. 181) offered his view on the Postmodern organization. He saw Modernist organizations as rigid, focused on mass, premised on technological determinism, differentiated, demarcated, and de-skilled. Postmodern organizations are characterized by niches, technological choice, dedifferentiation, dedemarcation, and multi skills.

> If organizations were to mirror art... Williams (1987, p. 52) would have us rediscover organizational 'community' in the neglected, alternative tradition of the past century. While we would see in these neglected traditions a democratic imperative, it is by no means clear that this should be so. Communitarian conceptions of organizations have left no locational monopoly within the imagination of reformers of a 'left' persuasion. As we have seen in the appeal of post Confucianism, the familiar image of an imagined organic past can as readily illuminate the contemporary reformers of the 'right'.

Cooper and Burrell (1988, p. 92) summarized the organizational debate:

> Modernism with its belief in the essential capacity of humanity to perfect itself through the power of rational thought and Postmodernism with its critical questioning, and often outright rejection, of the ethnocentric rationalism championed by Modernism. Apart from the radical revaluation of the whole process of modernization which this dialogue evokes, there are significant implications for how we understand the role and nature of organizations in the modern world. Not least of these is the shift away from the prevailing

Table 8.1 The new organizational paradigm

Modernist model	Context	Postmodern model
Hierarchy	Organization	Network
Self-sufficiency	Structure	Interdependents
Security	Work expectation	Personal growth
Autocratic	Leadership	Inspirational
Homogenous	Work Force	Culturally diverse
By individuals	Work	By teams
Domestic	Markets	Global
Cost	Advantage	Time
Profits	Focus	Customers
Capital	Resources	Information
Board of directors	Governance	Variety of constituents
What is affordable	Quality	No compromises
Doing less with more	Efficiency	Doing more with less

Source derived from Byrne (1992, p. 1)

definition of organization as a circumscribed administrative-economic function... to its formative role in the production of systems of rationality. This is clearly a return to the grand concerns that Weber introduced into the study of modern social systems, in which bureaucratic organization had created the 'iron cage' of the modern economic order, and whose other significant effect had been to purge the world of its auratic and magical. In other words, Weber made us see modern organization as a process which emblemized the rationalization and objectification of social life, and it is to this process that the current debate returns us, but with a fresh twist which directs our attention to the concept of *discourse* and its place in institutional structures.

Byrne (1992) provided a succinct comparison of the Modern and Postmodern organization (Table 8.1). He extensively discussed what he saw as the new Postmodern organization—possibly representing what we might expect to see fronting up policy-making institutions in the maritime sector. Much of what he suggested was acceptable but has little relevance to our discussion as it focused on workforce characteristics, employer/employee relationships, skills and loyalty, but despite this some issues were directly relevant. These included moves toward expanding service provision; internationalization; increased stakeholder empowerment, involvement and ownership; lowering of confidence in institutions and management; shunning incremental change and stimulating radical and dramatic changes; organization focused on smaller units and upon processes rather than outcomes; the creation of networks of relationships. Most importantly he suggested that the focus must be on networks rather than hierarchies; global over domestic markets; empowerment at the local level; continuous reinvention ('change is not an event'). He stressed the value of reengineering companies and institutions which implied starting with a completely clean sheet of paper and reinventing an organization from scratch—something that the framework of maritime institutional policy-making has never even begun to address.

8.1 Postmodernism and Organizations

Despite suggesting that 'many new management ideas are yesterday's theories warmed over and disguised under a sauce of new buzzwords', Byrne's comments are valuable and may suggest a way ahead which involves re—(or even de)—institutionalizing the maritime policy sector, creating new frameworks which reflect the globalized, Postmodern world. However, as Parker (1992, p. 10) suggested, it is important to remember that there is no single Postmodern organizational type. To quote Power (1990, p. 121):

> the Postmodernist perspective flows from a denial that there is any single, ultimate or deep language game that is uniquely determinative of organizational stability. The organization theorist must be sensitive to the diversity and fluidity of the 'life' organization and no one model will suffice to orientate research.

Mayntz (1976), quoted in Cooper and Burrell (1988) brought together many of the Postmodern organizational issues, suggesting that the existing dominant models of organizational decision-making are 'normatively rational' in that 'goals are set by the organization and this step is followed by a search for the best solution from competing alternatives' (Mayntz 1976, p. 119). Action is therefore 'touched off by preconceived goals or purposes'. Mayntz, supported by Jeffcutt (1994), suggested that the Postmodern organization should be characterized by reactions to local perturbations:

> organizational activity in general and policy-making in particular is primarily triggered by situational factors which constitute a pressure to act, rather than being generated by deliberations on how certain abstract values can be achieved.

This last point may be of particular relevance to existing maritime policy-making processes which appear to be driven by normatively rational organizations striving to achieve abstract values which may not reflect those of the market to which they are directed. Cooper and Burrell (1988, p. 103) continued:

> in contrast to the normative-rational model, decisions are rarely, if ever, taken by individuals but are invariable embedded in an active network of people within a division of labour.

Hirschman and Lindblom (1962) noted much earlier that policy-making should be triggered by situations rather than by attempts to achieve predefined values citing examples in economics, technology, and development—all relevant to the maritime situation. Cooper and Burrell (1988, p. 103) suggested that:

> there is no perfect theoretical solution to problems which we can prepare in advance and that decisions have to be made as remedial moves in situations marked by uncertainty, disorder and imbalance. In these analyses, practice usurps theory, and organization, far from being a structure of calculated, deliberate actions, is in reality the automatic response to an impending threat. The analyses thus suggest that rational control subserves a more fundamental process which acts autonomously in much the same way that Varela (1979) suggests that referentiality subserves self-referentiality. In short the process of organization is self-originating and automatic.

8.2 Postmodernism, Transport, and the Maritime Sector

The impact of Postmodernism is well represented in the transport and maritime sectors and it is to these specific applications that we now turn with the objective of illustrating the relevance and significance of the Postmodern movement to our interests in maritime governance and the failures that are apparent in maritime policy-making. The discussion will cover not just policy-making and its related developments where we have shown already that although a Postmodern appreciation of policy-making is necessary, it is one that has yet to be fully realized, but we shall also attempt to reveal a sample of evidence from the full societal maritime spectrum. Hence we shall touch upon, everything from vessel technology and architecture to arts and culture, from housing and urban planning to port renaissance, and colonization. However, we shall start with the broader discipline of transport. Table 8.2 summarizes many examples of the Postmodern effect on the sector.

Hansen referred to Urry's (2000) concept of 'instantaneous time' replacing 'scheduled time' as technology increases speed to the point when time is compressed into insignificance. This Postmodern feature of globalized trading is clearly identified in the maritime sector through the movement of information concerning ownership, finance, and the like even if commodities themselves still take an appreciable time to reach destinations—albeit commonly less than they used to. We shall return to the issue of speed, and space/time compression, its significance for the Postmodern and its relationship to maritime governance in our conclusion but the work in particular of Virilio needs to be noted here (1977).

Finally Hansen also reflected upon what he termed the third Postmodern characteristic of transport which relates to flows, and the changing relations that have occurred between individuals, organizations, business, etc., and widely discussed by Castells (1989, 1996, 1997, 2000). These ideas emerged from the developments in technology and speed/time compression that have occurred along with and as a consequence of globalization—and are an essential part of it.

These characteristics of Postmodern transport are neatly summarized in the construction of the Oresund Link providing a road and rail crossing between Denmark and Sweden and built in the context of a rapidly changing and highly demanding Postmodern transport market (Fig. 8.1). Its replacement of regulated, conventional ferry links by a more variable and diverse road and rail crossing and its effective dissolution of state borders are both highly Postmodern in their effects.

Meanwhile Hansen concluded by reflecting on the significance of the Postmodern for transport in general and it is useful for us to consider how this also applied more specifically to the maritime sector—and hence to its governance:

> These three characteristics or driving forces of the Postmodern society function as preconditions for the analytic coupling between general societal changes and corresponding changes within logistics and transport. This coupling should not be perceived as a simple cause-effect relationship, where changes on the societal level directly results in changes on the more specific level of firms, organizations and individuals. The relationships should

Table 8.2 Postmodernism, transport, and the maritime sector

Transport and maritime category	The impact
Transport planning	Infrastructure
	Instantaneous time; speed-time compression
	Flows
	Roads as social zones
	Planning
Gender	Gender blind transport
	Women in shipping
	Physical features of transport and the relationship to women
The third world	Colonialism and transport
	Decentralizing and the abandonment of grand theories
	Impact of cruise liners and cheap air tourism
The waterfront	Ports
	Urban waterfronts. Docklands
	Architecture in ports
	Varied development—tourism, recreation, culture, heritage, etc
	New port uses
	JIT, logistics, ports and reconfiguration
	Flexible labour and ports
	Relationship between ports and cities
Architecture	Public buildings; commercial buildings; container housing
	Building design as imagery
	HSBC Hong Kong; Lloyd's London;
	Guggenheim, Bilbao; Maritime Hotel, New York
Blogs, pods, and e-mails.	Increase, cheapness, and diversity of communications in the maritime sector
	Individual communication and expression of opinion. Facebook, Twitter
E-booking and internet cargo management	Container shipping internet-based cargo booking
Media, interest groups, and NGOs	Increase in the role of the media. You-Tube. Camera phones
	Rise of non-governmental organizations, interest groups, etc
Shipping technology	Skysails
Labour	Labor deregulation
	Flexibility. The EU ports policy
Green shipping and ports	Rise of environmental concerns
	Shipping and ports use of 'green'
Port infrastructure, organization, and Technology	Single Point Moorings. Inland Container Terminals. The remote port. The virtual port
Logistics	JIT. Lean supply chain management
Shipping organization	Landlocked Flags of Convenience
Marketing	'Pink Lady'
	EasyCruise
Ownership	Privatization. Globalized ownership of ports. Diversification

(continued)

Table 8.2 (continued)

Transport and maritime category	The impact
Arts	Sex and violence in maritime literature. Ballard's 'Crash'; Kerouac's 'On the Road'
	Films, tourism, and transport
	Arks as art
	Postmodern Stalinist representation
	Malcolm Morley
Naval strategy	Naval strategy
	Razzle Dazzle ships
Ocean governance	Moves from abuse to holistic view of the value of the ocean to man

Source Author

Fig. 8.1 The Oresund link. Postmodern transport infrastructure. *Source* NASA

instead be seen as mutually affecting each other and thereby treating changes in transport and material flows as capable of affecting the spatial, time and relational structures within society (Hansen 2003, p. 4).

Sharma (2009, p. 2) contrasted what he described as the Modernist approach to transport planning—based on enforcement, education, and engineering whereby traffic rules are enforced, the public educated and roads upgraded—with a Postmodern approach. The Modernist approach to transport planning applied categorization and segregation throughout the twentieth century typified by the work of

8.2 Postmodernism, Transport, and the Maritime Sector

Eugene Henard (Wolf 1969), Holroyd Smith (Lay 1992, p. 187), Arthur Tuttle, and Edward Holmes and subsequently actively pursued in Buchanan's 'Traffic in Towns' (UK Ministry of Transport 1963).

The Postmodern approach was different treating roads as social zones, integrating car, and pedestrian movements including that of children's play, taking the best from the Netherlands 'Woonerfs' developed by de Boer and Vahl (De Boer 2005). Applied in the Danish city of Christiansfield, traffic signs were withdrawn and road signs removed. No mode (including pedestrians) had priority and eye contact was the only means of negotiating junctions. The numbers killed or seriously injured over the next 3 years fell to nil.

Dear and Flusty (1998, p. 54) quoting Relph (1987, pp. 242–250) provided a checklist of what they saw as the contrasting Modernist and Postmodernist approaches to city planning. The Modernist city is characterized by:

- Mega-structural bigness (few street entrances to buildings; little architectural detailing).
- Straight space/prairie space (city centre canyons and endless suburban vistas).
- Rational order and flexibility (the landscapes of total order verging on boredom).
- Hardness and opacity (including freeways and the displacement of nature).
- Discontinuous serial vision (deriving from the dominance of the automobile).

Conversely the Postmodern city is more 'detailed, handcrafted and intricate', celebrating difference, polyculturalism, variety and stylishness' (Dear and Flusty 1998, p. 54):

- Quaint-space (a deliberate cuteness).
- Textured facades (for pedestrian absorption, rich in detail).
- Style (appealing to the fashionable, chic, and affluent).
- Reconnection with the local (deliberate historical and geographical reconstruction).
- Pedestrian-automobile split (to redress the Modernist bias toward the car).

Meanwhile Newman (1995, pp. 258–260) also considered the relationship between urban planning, transport, and Postmodernism and provided a useful guide to the issues which characterized the sector in the late twentieth century, which may be of value when considering the need to adapt governance for the maritime sector. His analysis was somewhat vague but the trend was clear when assessing a Postmodern approach to a transport related issue. The most significant issues were determined as:

- Values and communities: Postmodern planning is not value free but the values needed are those which represent the individual. This clearly suggests a need for specific, comprehensive, and directed stakeholder involvement and a recognition of diverse communities.
- Sustainability and the environment: were seen as core issues whatever the planning focus.

- Social justice: was also seen to be central. However, in the commercial world of global shipping this is an area under constant pressure.
- Heritage: Postmodernism, unlike Modernism demands an understanding and appreciation of the past.
- The public realm: planning takes place and has impact in public spaces. Hence it must recognize this and cannot plan social interchange out of existence. The Modernist route was to overcome social niceties by organization and efficiency.
- Economic efficiency: remains a significant part of the Postmodern planned environment. Just because Postmodernism looks for variety, flexibility, and is less focused on organization does not mean it can neglect the economic menu.
- Diversity: the most significant of all the Postmodern values expressed by Newman, this featured all manner of diversity from transport to culture and also stressed the need to understand and overcome boundaries—particularly relevant in the new world of Postmodern nations and their implication for state territory.

To loosely quote Newman with a contribution and apologies from this author (1995, p. 266):

> One of the key reasons for the age of uncertainty (*or maritime policy failure in our case*) is the collapse of the (Modernist institutional regime) as a paradigm for planning. (*section in italics added*)

8.2.1 Gender, Postmodernism, Transport, and Shipping

Law (1999, p. 572) provided a critical review of the research that stemmed from feminist critique of gender blind transport research and planning suggesting that the study of gender and transport has been 'affected by the shifting winds of fashion in feminist theory… as the increasing prominence of poststructuralist perspectives directed attention from structural constraints to discursive constraints'. McDowell (1993a, b) saw this as a move from 'rationalist feminist empiricism' (concerned primarily with social relations) to an antirationalist/standpoint perspective (concerned primarily with gender symbolism) and subsequently to a postrational or Postmodern feminism (concerned primarily with the construction of gender identities)'.

The issue of gender also features elsewhere in the Postmodern maritime agenda. WISTA is the Women in Shipping and Trade Association that has grown in leaps and bounds since its formation and represents the gender diversity that is apparent in new Postmodern maritime activity. The IMO first produced a strategy for the training of women in 1988 and since then there have been considerable efforts at least by women to raise the profile of shipping and to recognize the contribution made and potential to be made by women (Lloyd's List 2008; Tradewinds 2010c, d). This even extended to the Secretary-General of the IMO urging more women to enter the seafaring professions representing an 'undeveloped resource'.

Law, along with Hanson and Pratt (1988), Matless (1995), Wolff (1990), McDowell (1996) and Cresswell (1997) also suggested that metaphors of mobility 'abound in recent works of social theory' especially in poststructuralist (and by implication Postmodern) writing. Writers cited include Minh-ha (Foss 1999; Hutnyk 2004), Ferguson (Ball 1987), Spivak (Eagleton 1999), Mohanty (Alexander and Mohanty 1996), Grosz (O'Sullivan 2008), Deleuze and Guattari (Deleuze and Guattari 1984), Grossberg (Grossberg et al. 1989), and De Certau (De Certau 1984; Highmore 2006). Law in particular went on to discuss the relationship between gender, mobility, and the physical features of the built environment, linking this with cultural studies of urbanism and the lived experience of Modernity and Postmodernity (Jameson 1985; Jukes 1991).

8.2.2 Postmodernism, the Third World, and Transport

Simon (1996) meanwhile reflected upon the change in approach to understanding the relationship between transport and development in the Third World with particular reference to the abandonment of earlier theories based around colonialism, neo-colonialism, and the idealization of precolonial models. He suggested that the Postmodern model of transport and economic development in Third World countries is especially appropriate as it recognized the rejection of the Modernist ideal centred around the nation-state and the idea that there is a single best way with a monopoly of the truth. Third World countries have a particular resistance to the nation-state principle which was imposed upon them almost without exception by European imperialism. Simon (1996, p. 55) suggested that the application of Postmodern ideals to transport and Third World development involved:

- The decentring (i.e., downgrading, replacement) of universalizing theories and methodologies:
- Concern instead with a multiplicity of more locally relevant and sensitive perspectives, insights, and theories: in other words, various forms of pluralism;
- Downgrading of the primary role accorded in theory and political practice to the nation-state, in favor of more local and responsive forms of social and political organization;
- Replacement of monolithic styles in art, architecture, and the like with diversity and mixtures ('pastiche') to reflect the different historical periods, the diverse cultures and social groups within almost all societies today.

This antistate and anarchic approach to understanding society can be extremely relevant to transport in the Third World (and by extension to elsewhere) where the state is seen as a colonial imposition by many and a majority of societies, especially in the southern hemisphere, have never experienced a Modernist phase and thus the understanding of Postmodern as 'anti-modern' has a particular place. Simon points out the implications of the cruise liner and the wide-bodied jet for Third World tourism and travel, representing an 'unreal' phenomenon that is inherently globalized and Postmodern in structure, promoting a real experience in sanitized communities for Western tourists.

8.2.3 The Postmodern Waterfront

Norcliffe Bassett and Hoare (1996) provided one of the earliest attempts to provide a Postmodern understanding of ports and more specifically the emergence of Postmodernism on urban waterfronts. In looking at the evolutionary sequence of change in port-city relations they suggested that

> this sequence corresponds with some of the features of the succession from a Fordist phase of industrial capitalism to a succeeding phase in which Postmodern values are more predominant. (Norcliffe et al. 1996, p. 124)

They began by looking at a Fordist interpretation of the development of economies of scale in ports during the early and mid-twentieth century. Fordism achieved its 'apotheosis in the standardization of the moving assembly line' and similar features can be identified throughout society. This included architecture—see, for example, the functional reinforced concrete of Mies van der Rohe and Le Corbusier (Relph 1987; Cooke 1990)—but more significantly here, what Norcliffe et al. termed 'public vandalism' in the inner city.

These developments were constructed to serve the port and central business district of cities using a selection of public buildings, expressways, and other transport facilities designed to be efficient and functional (Berman 1982). This Norcliffe et al. (1996, p. 127) described as monotheism—the 'pursuit of a functional, unadorned uniformity of style, irrespective of geographical context'. This is certainly something easily identifiable in the port waterfront during the twentieth century with much evidence still in existence today.

Norcliffe et al. went on to stress that port operations also displayed similar Fordist characteristics with a highly demarcated and structured, unionized work force (Doeringer and Piore 1971); operations formed integrated sequences of loading and unloading, warehousing, and shipping; and the ports themselves began to move downstream but still retaining their original central city locations, creating extensive 'port-industrial spaces' much of which soon became superfluous. To summarize:

> modern ports... characterized by increased scale of operations, specialization in sequential tasks in handling particular commodities within a port, specialization too in those commodities handled between ports, and embeddedness within dominant, spatially-integrated economic systems functioning on a global scale. (Norcliffe et al. 1996, p. 128)

By the 1970s this Fordist model for ports (and industry in general) seemed to be anachronistic. In particular, the notion of flexible specialization was beginning to have a serious effect (Harvey 1988; Scott 1988; Harvey and Scott 1989). Despite some with the opposite view (for example Amin and Robins 1990), the process in particular of deregulation had a dramatic effect. Ports found the trading environment considerably more competitive and the whole process more entrepreneurial. Speed and cost became the prime issues. Privatization was widespread and spare land was sold off in sizeable amounts for residential property development (Harrison 1994).

The operating environment changed too with the trend toward reductions in batch size, consumer segmentation, and shorter 'shelf-lives' encouraged competition from air freight and enhanced the process of containerization. Workforces were reduced substantially and demarcation lines largely eliminated. Turnaround times for ships reduced markedly.

This reversal in Fordist fortunes has led into the Postmodernism of ports which we see today. Harvey (1990) recognized its links to the production side of economies and as such ports are an excellent indicator. This is evident not only in the ways that ports are managed and operated but also in their visual style, their range of activity, and their urban profile. To quote Norcliffe et al. (1996, p. 129):

> Not only does the crenellated topology which maximized quay-side length translate into a miscellany of intimate niche-like environments for contemporary, non-port activities, but their long and important histories offer rich veins of opportunities to mine for space and time-specific styles of contemporary redevelopment... the centrality of many abandoned city waterfronts provides them with a highly visible set of locations from which to proclaim the conscious individualism of Postmodernism... Finally the scale and speed of waterfront abandonment of many such prime potential sites have encouraged public-sector intervention to facilitate revitalization.

These processes freed up space for a 'new round of capital accumulation including housing, hotels, heritage, sports, recreation, tourism and local commerce' (Desfor, Goldrick and Merrens 1988) something found more specifically in for example Toronto Harbour (Desfor 1993). Thus:

> the Postmodern renaissance of the urban waterfront is, in a fuller historical perspective, just another round in the accumulation of capital. (Norcliffe et al. 1996, p. 130)

Norcliffe et al. (1996) identified five main expressions of Postmodernism in the waterfront—employment, housing, recreation, hospitality, and culture and heritage. Each in turn can be summarized:

- *Employment*: Postmodern evidence is characterized by Canary Wharf and the development of London's Docklands, but also Battery Park in Lower Manhattan (New York)—both developments by the same architect and both justified on the grounds of the increase in City activities created by the defragmentation and privatization of economies—a Postmodern phenomenon. Other similar employment generators come from the development of cafes and restaurants, recreation pursuits, heritage activities, etc.
- *Housing*: Postmodern development here is substantial with both the gentrification of older stock and the creation of large quantities of new apartments and town houses in converted port areas, often with waterfront views, and always

with close access. Evidence comes from all over the world—London, Plymouth, Swansea, Cardiff, Southampton, New York, Montreal, Baltimore, Toronto, etc.
- *Recreation*: typified by the marina replacing the old wharves and docks, these Postmodern manifestations on the waterfront are perhaps the most obvious expression of the new era. They are accompanied by a proliferation of water sports—sailing, wind-surfing, etc.
- *Hospitality*: Postmodernism has brought new hotels as well as conversions of old port buildings. These are inevitably accompanied by a cluster of restaurants, cafes, bars, and clubs.
- *Culture and Heritage*: this includes not only the remains of older, port industrial activity—cranes, locks, docks, chains, etc.—but also a variety of museums, art galleries, souvenir opportunities, and so on.

The new waterfront exemplifies maritime Postmodernism and reflects all the characteristics of change in the Postmodern era that we have noted already. It is material, social, economic, and cultural evidence of a fundamental change that has taken place in the maritime sector that we can extrapolate to the difficulties faced in maritime policy-making in general. Much of the development noted above not only has occurred worldwide, but also has been financed and organized by international interests whose focus may commonly be over a wide range of non-maritime related activities. Competition between these Postmodern developments is similarly non-national, fragmentary, and diverse. At the same time globalization has been the very impetus for their emergence creating an homogenized market for the Postmodern waterfront.

Daamen (2007, pp. 8–10) came to similar conclusions. He saw that the relationship between ports and their cities was one that has changed along with that of industrial change and capitalist transition. The 'Modern to Postmodern transition of the western world, made sense of the movement of ports out of cities'. He emphasized Harvey's (1973, 1990) contribution to this debate, identifying a notable break in the early 1970s evidenced by:

> major structural shifts in time (just-in time), space (global segmentation and regionalization of production), function (demand-driven), and organizational forms (lean management, small and medium sized enterprizes, and vertical disintegration) of capitalist production systems.

Ports therefore reacted producing the waterfronts we see today. The Postmodern waterfront commoditizes traditional spaces of production, particularly those with waterfront qualities, mirroring the sociocultural trends of the city, and its wider society.

Finally, Danyluk and Ley (2007, p. 2195) examined the relationship between gentrification and the journey to work in Canada placing developments in a Postmodern context and reflecting on the appropriateness of the Postmodern to society in general:

> Nearby, waterfront boardwalk and a cluster of public parks provide space for cycling and jogging, perfect for the 'active person who loves urban life' (Concord Pacific 1999). The

downtown commercial district is a close 10–15 min walk away. The street experience is cosmopolitan, presenting Postmodern urban vitality and sensuality." (Hutton 2004, p. 1976)

8.2.4 Postmodernism and Maritime Architecture

We have already noted the close relationship that exists between the Postmodern and architecture and there are many examples of where they have coincided in a maritime context. Albert Ledner provides an introduction to the emergence of Postmodern maritime architecture from the modern with his series of designs for the National Maritime Union in the USA. These included buildings in San Francisco, San Juan, New Orleans, and most famously, the Curran/O'Toole Building in Greenwich Village, New York constructed in 1964, and currently threatened with demolition. Similarly The Maritime Hotel in New York was also a Ledner design of 1966 for the Union which once accommodated sailors and subsequently has had a variety of other uses. It is now a hotel and leisure complex.

The examples of the intrinsically maritime HSBC building in Hong Kong (Fig. 8.2) and the Lloyd's Building in London are both characteristic of the phase of transition between the Modern and the Postmodern.

Meanwhile, moving wholly to the Postmodern, Michael Graves is an established international architect with a reputation for a number of projects reflecting maritime and riparian themes that suggest the breadth of the influence that the Postmodern has had in maritime (and wider) society. Examples with a maritime flavor include a Beach House in Malibu, a development for the Delaware Port Authority; the Riverbend Music Center in Cincinatti, Ohio; Washington Ballpark Redevelopment; and the Walt Disney World Dolphin Hotel at Lake Buena Vista, Florida.

More influential and familiar is the Guggenheim Museum in Bilbao, Spain designed by Frank Gehry, located alongside the Nervion River, and opened in 1997. Considered a classic of *Deconstructivism* it is essentially Postmodern. Resembling a ship but with hints of fish-like forms, it is radically sculpted and essentially organic.

Thoroughly both Postmodern and maritime are the increasing number of container-architecture developments that are emerging. One such is Container City at Trinity Buoy Wharf in London's Docklands, consisting of 5-storeys of ocean-going containers making up a variety of living spaces and workshops. Designed by Nicholas Lacey and Buro Happold, the scale of the development is large and its flexibility increased by adapting the containers rather than just bolting together single units.

Meanwhile containers have been used elsewhere in Postmodern architecture. Lot-Ek's prefabricated container houses and their collaboration with fashion retailer Uniqlo to produce mobile retail outlets are notable. Cargo Town is another

Fig. 8.2 The HSBC building, Hong Kong. *Source* Ian Lambot

Postmodern residential development made up of used sea containers located in the Port of Seattle.

Other Postmodern maritime container architectural developments include De Maria's Redondo Beach Shipping Container House in California made up of eight containers (http://www.youtube.com/watch?v=UvcUe_yPHdg), further examples at Venice Beach also in California, the Port-a-Bach folding steel container system from New Zealand, Cove Park Artists' Retreat in Scotland, New Jersey architect Adam Kalkin's Quik House and what is claimed to be the largeest container housing development in the world of student accommodation at Keetwoonen, Amsterdam in the Netherlands.

8.2.5 Maritime Postmodernity: Blogs, Pods, and E-mails

Essentially Postmodern in that the emergence of consumer sites that allow access to previously opaque institutions, concerns and issues, the role of e-mails, blogs, and pods has been notable. Perhaps typified by the formality of Facebook and Twitter and contrasted by the individual expression provided by e-mail addresses

8.2 Postmodernism, Transport, and the Maritime Sector

for the (worldwide) general public to the IMO, EU, national ministries, and so on, these technological developments in instant and free communication have changed the face of modern maritime policy-making allowing anyone to have their say (and perhaps be ignored), and more importantly to feel as if they have a right to be heard and their opinion considered. The result is that policy-makers have no choice but at a minimum, to pay lip-service to current public opinion and to recognize that through file sharing facilities such as You-Tube, the maritime sector is exposed in its entirety with little if any control by those who once upon a time legislated policy for, and almost without consideration of others.

There are numerous examples:

- Facebook pages for: 'Shipping'; 'Maritime Shipping'; 'Anver Maritime Relations'; 'Merseyside Maritime Museum'; 'Florida Maritime Leadership Coalition', etc.
- You-Tube files including 'shipping' (548,000 listings); 'maritime' (132,000 listings); 'port' (705,000 listings); 'seafarer' (1420 listings); 'shipping and environment' (3,090 listings); 'oil spill' (59,300 listings) (October 27th, 2011).
- Direct public e-mail addresses for individual and institutional opinion on policy proposals from the EU, numerous national government agencies and ministries around the world, pressure groups, etc.
- Blogs including for example the ubiquitous Twitter plus individual and informal sites, such as sea-fever (http://sea-fever.org); gcaptain (http://gcaptain.com/maritime/blog; the Black Pig informal shipping, and logistics blog (www.handyshippingblog.com); and more formal locations such as those for the Maritime Union of New Zealand, the European Union,
- The example of the highly traditional UK Chamber of Shipping introducing a Twitter site [@britishshipping] (Lloyd's List 2011).

Nothing escapes inspection, comment, and exposure and this in turn has significant impact upon policies that are acceptable and ones that can be implemented and enforced. Modern policy-making structures and institutions have begun to understand the need to recognize the effect of the revolution in communication that has taken place—but have yet to change what they do and how they do it, whom they should include in policy making, and how this inclusion is structured.

The likes of maritime blogs and Twitter have even featured in Lloyd's List (2010a) where it was noted how outdated the IMO website was, located in a 'technocratic fog' and an 'impenetrable digital thicket' compared with the likes of the European Union, The Baltic Exchange, IMarEST, and the Sailors' Society. This was followed up by further reports of the IMO's 'communication inertia' and comment on the unlikelihood of the IMO in the near future accommodating Twitter or blogging and failing to understand the power of 'Linked-in' (Lloyd's List 2010c). The maritime industry is:

> no longer… a romantic picture of a sea-captain spending six months taking cargoes to Tahiti and other tropical paradises… today it is an international business linked with political decisions, multi-billion pound contracts and sophisticated technology.

Tradewinds (2010f) suggested many more blogging sites where the Postmodern phenomenon of Twittering was an important communications mechanism for the industry, both serious, and not so. Easily dismissed as irrelevant to the generation and implementation of effective policy, their presence, and impact are substantial and growing.

8.2.6 Screen Trading, E-booking, and Internet Cargo Management

Traditional methods of cargo management and booking are beginning to change as well reflecting the rise of new technology, the development of globalization, and the new Postmodern environment in which the industry must work. Although screen trading was still in its infancy in October 2011 (Tradewinds 2011b) it did represent about 5 % of all forward-freight agreements conducted through what was described as a 'plethora of screens', including SSY, GFI, Baltex, and Cleartrade Exchange. Two further examples of internet cargo management reflect the growth in this type of Postmodern commercial activity that is now in the marketplace.

GoCargo has created what its inventors call an 'e-commerce platform' aimed at the container shipping industry allowing shippers and service providers to trade online using what they term is a 'reverse auction format'. Shippers obtain bids from a large number of ocean carriers and freight forwarders anywhere in the world while service providers can reduce sales costs and optimize their cargo carrying capacity. Spot and long-term contracts can be accommodated along with all manner of cargo types—FCL, refrigerated, breakbulk, etc. Its Postmodern credentials of flexibility and time and space insensitivity are reflected in GoCargo's 2009 publicity:

> GoCargo.com's unique technology is secure, and highly scalable, and provides users with a suite of productivity tools, such as automatic alerts from the exchange, personalized folders and templates, rating of trading partners, and advanced cargo search capabilities. (www.GoCargo.com).

Of course, GoCargo is not alone and independent competitors include Maritime Direct and Level Seas. Maersk has also introduced an online bunker adjustment factor calculator with the aim of showing how fair it is in the surcharges it is charging for rising fuel prices (Tradewinds 2008b).

Resistance in the traditional shipping forum remained high for some time. However, following trends in the low cost airline industry, container shipping is now beginning to find that it is no longer possible to restrict dialog to long-term relationships, conducted with single providers, where comparisons are impossible. The result has been a plethora of e-commerce tools represented by two consortia—INTTRA, backed by MSC, Maersk Sealand, CGA CGM, P&O Nedlloyd and Hamburg Sud. and Global Transportation Network (GTN) backed by APL, ANZDL, Canada Maritime, Cast, Contship Containerlines, Lykes Lines, TMM

Lines, Hanjin, Hyundai, K Line, Mitsui OSK Lines, Senator Lines, Yang Ming, and Zim Israel Navigation Company (Tradewinds 2008c).

Meanwhile Maersk's own e-booking 'module' allows shippers to provide real-time information of container location to their client. Bookings can be made at any time of day or night providing enormous flexibility to a system that once was dictated by the opening hours of booking offices across the world. Youship.com is not an online booking system but provides data on all-in freight rates, initially for a limited range of services from Hong Kong, Rotterdam, and Antwerp to destinations in the UK, Netherlands, Morocco, Turkey, China, Japan, and Australia (Tradewinds 2008a). Benefits include lowers rates, instant confirmation, and space guarantees. Like budget airlines, rates can be high but at the same time also very low with some offers as little as US$1 for last minute bookings where demand is low plus marketing strategies such as free flowers by post. Running alongside existing booking portals CargoSmart, GT Nexus, and INTTRA, the ultimate aim is to merge Youship with Maersk's main web site. However, following the economic downturn from 2009 Youship was temporarily suspended awaiting better times when space on vessels was more constrained and in greater demand.

Many further developments in Postmodern market platforms have occurred since then not least the intense competition arising over trading screens to rival the Baltic Exchange system (Tradewinds 2010a) and exchange-based screen trading platforms for forward-freight agreements (FFAs) (Tradewinds 2010b). Meanwhile work continues on developing an e-bill of lading linked to a web-accessed database (Tradewinds 2010g) and on an electronic derivatives platform through Cleartrade Exchange (Tradewinds 2011a).

8.2.7 The Media, Interest Groups, and NGOs

The rise of the significance of the media in the maritime sector is paralleled only by its rise across all other global activities. The ability of individuals to record images and sequences by camera phone and then through e-mail to communicate them to the media worldwide has revolutionized news capture for the maritime sector. No longer is evidence of ship spills, near misses, and collisions hidden by distance and time and the consequential rise of image as a serious contender within financial, engineering, and technical disciplines is wholly Postmodern. Shipping giants now employ media consultants and lawyers to ensure that the best image is put forward to the world media and the global media industry is now a central part of the policy-making process whether desired or not. Simply consider the impact of the deliberate grounding of the *MSC Napoli* off Branscombe in Dorset, UK in 2007 where the news story was played out across YouTube, various blogs, and on TV screens with images gathered by individuals on shore by mobile phone (Independent 2007).

Not everyone is enamored by the transparency that now increasingly characterizes some parts of the shipping industry. Claims have been made that shipping is in fact more transparent than the aviation sector and the information now available could actually compromise the industry's attempts at image building (Lloyd's List 2006). Quite.

Similarly Postmodern has been the enormous rise in number and influence of maritime interest groups and other non-governmental organizations (NGOs). These include both what might be termed traditional maritime interest groups, a selection of which are given in Table 8.3, and those of a more indirect but diverse and increasingly significant nature, such as Greenpeace, Friends of the Earth, the UK Institution of Civil Engineers—Maritime Special Interest Group. Their role in policy-making may commonly remain unofficial but their presence is both ubiquitous and widely recognized as long-term (Pallis 2005, p. 6, 2006; Pallis and Tsiotis 2006, 2008). Most policy-making institutions now accept a role for these types of organizations in policy making—take the recent acceptance of the UK-based Nautical Institute by the IMO at some of its meetings—but their impact and relationship has yet to be fully integrated. Their variety, scope, and tendency toward being ephemeral are essentially Postmodern and contrast significantly with the formalized and grand structures of the modernist IMO, WTO, EU, etc.

8.2.8 Port Infrastructure, Organization, and Technology

It is not just the port waterfront that has seen Postmodern changes as there has been considerable activity in terms of how ports work, their technology, and organization which are characteristic of the Postmodern. Chlomoudis et al. (2000, p. 2) emphasized the extent of the changes which have taken place moving from the post-war mass production paradigm to one which has had to accommodate the increased flexibility and diversity that Postmodern economic activity demands.

A notable development has been the emergence of the port which is not located on land but out to sea—commonly known as a single buoy mooring (SBM) or single point mooring (SPM). These are loading buoys located offshore—often at considerable distance—acting as a connection point for tankers for loading and offloading fluids or gas. Because of their location—and the reason commonly for their existence—they can be positioned to accommodate any size of ship and as a result provide opportunities for cost savings and flexibility that would otherwise be impossible. In addition, they have security advantages in that they are located away from land-based activity.

Their Postmodernism comes from their near infinite, spatial flexibility in that they can be positioned almost at will within territorial waters thus opening up a very large range of possibilities for port operators, shipping companies, and associated players. This detachment from conventional port spatiality, manifesting itself in three dimensions releases the activity from almost all port constraints that normally exist (or at least normally must be taken account of). Conventional consideration and analysis of ports and their spatial characteristics can thus be forgotten with the port no longer active within its own boundaries.

8.2 Postmodernism, Transport, and the Maritime Sector

Table 8.3 Traditional EU policy-making maritime interest groups

Interest group	Type	Member states represented	Members
AMSI (association of marine scientific industries)	Business association	4	17 EU 1 non EU
CEMT (confederation of European maritime technology societies)	Business association	8	9 EU
CESA (community of European shipyards association)	Business association	12	12 EU 2 non EU
CLECAT (liaison committee of European forwarders)	Business association	14	21 EU 5 non EU
EBA (European boating association)	Business association	14	22 EU 4 non EU
EBU (European barge union)	Business association	7	9 EU
ECASBA (EC association of ship brokers and agents)	Business association	20	21 EU 1 non EU
ECSA (European community shipowners association)	Business association	15	15 EU 1 non EU
EIA (European intermodal association)	Business association	20	81 EU 4 non EU
EMEC (European marine equipment council)	Business association	8	10 EU 2 non EU
ESC (European shippers council)	Business association	9	9 EU 3 non EU
ESPO (European sea port organization)	Business association	20	20 EU 3 non EU
EUDA (European dredging association)	Business association	9	11 EU 9 non EU
EUROGIF (European oil and gas innovation forum)	Business association	5	10 EU 5 non EU
FEPORT (federation of European ports private operators)	Business association	13	13 EU
CRPM (conference of peripheral maritime regions)	Regional interest	20	15 EU 6 non EU
HELCOM (baltic marine environment protection committee)	Regional interest	8	8 EU 1 non EU
EHMC (European harbourmasters committee)	Trades union	19	94 EU 15 non EU
ETF (European transport workers federation)	Trades union	25	215 EU 15 non EU
ELAA (European liner affairs association)	Business association	24	24 companies

(continued)

Table 8.3 (continued)

Interest group	Type	Member states represented	Members
OCEAN (Organization of European community ship suppliers)	Business association	12	12 EU 1 non EU
FEMAS (federation of European maritime associations of surveyors and consultants)	Business association	8	9 EU
INE (inland navigation Europe)	Business association	6	6 EU
EFIP (European federation of inland ports)	Business association	11	17 EU 8 non EU
EUROPIA (European petroleum industry association)	Business association	0	20 multi-nationals
FEAP (federation of European aquaculture producers)	Business association	20	19 EU 2 non EU
EURMIG (European recreational marine industry group)	Business association	16	Sub-committee of international association
EURACS (European association of classification societies)	Business association	3	Sub-committee of international association
IIMS(EG) (International institute of marine surveyors)	Business association	Na	Sub-committee of international association
EMPA (European maritime pilots association)	Trades union	Na	Na
EMF (European Metalworkers Federation)	Trades union	11	65 members
ETA (European tug operators)	Business association	Na	Na
ETTC (European towing tank community)	Business association	Na	Na
EAFPO (European association of fish producer organizations)	Business association	Na	Na
EUROPECHE (Association of national organizations of fishing enterprizes)	Business association	Na	Na
IAMI (EU) (Organization of Maritime Institutes of the EU)	Business association	Na	Na

Source After Pallis (2005–6)

Similarly, the manifestation of Inland Container Terminals (ICTs) across the world is another Postmodern characteristic of the maritime marketplace allowing a traditional port function to disassociate itself from the waterfront and consequently opening up an almost infinite number of locations (Steenken, Voss and Stahlbock 2004; IBI Group 2006; Steenken and Voss 2008). This spatial liberation is inherently both global and Postmodern. Whilst the transshipment at ICTs is commonly between train and truck and clearly does not involve a ship (although it may involve canal transport), the activities taking place are those which also normally occur at maritime terminals and in some cases (for example near Trieste in Italy), the ICT has been developed solely to relieve pressure in the traditional port located in constrained and historic surroundings. Similar cases, although not quite as traditional, can be found in Hong Kong and Pusan (Korea) where congestion by the sea has become intense. Modern technology and communications have made ICT's that much more viable.

In a similar fashion the recent proliferation of dedicated terminals in ports is another Postmodern trend as the traditional universal characteristic of the port as open to all becomes fragmented into specialized and sole user locations. Both bulk and container shipping have seen this take place around the world and the continued process is one that reflects the individualization that Postmodernism demands.

8.2.9 Shipping and Technology

The Postmodern also commonly mixes developments and design across time with little respect for traditional views on progress, change, and modernity. The whole process of containerization and the more broadly defined intermodalism has been seen by some as a Postmodern phenomenon (Bonacich 2003, p. 3). This is also the case with the latest ideas on commercial cargo ships partially powered by sails (Guardian 2008; Daily Telegraph 2008; Tradewinds 2011c). Claiming to be the world's first, the *MS Beluga Skysails* completed its maiden voyage from Bremerhaven (Germany) to Guanta (Venezuela) using a sail system designed in Germany. The sail system is designed for use on medium-sized cargo ships, cruise liners, and trawlers, and is claimed to reduce annual fuel costs by between 10 and 35%. The 'large towing kite resembles a paraglider and is shaped like an aircraft wing' (BBC 2008). Operating at 100–300 m above sea level the kite can be used in winds of 7–40 knots. Further debate continued into 2010 with details emerging of 'wind-powered hybrid ships', 50 of which might be built in the UK to carry 30 million tonnes a year of woodchips, wood pellets, and biocoal pellets from Scandinavia (Tradewinds 2010e). The combination of old and new technology is characteristically Postmodern in its fusion of the traditional and the new.

Another manifestation of the Postmodern maritime sector is the unmanned ship—originally emerging in the 1970s and 1980s it was based around the idea of an unmanned engine room, and rather than an unmanned ship this would still have engineers elsewhere in the vessel monitoring engine performance remotely. Later discussion suggested how many officers and other posts would be reduced and eventually might disappear completely as technology kicked in. The ideas reemerged in 2010 centring on EMSA's SafeSeaNet, further extension of their e-maritime principles and the e-navigation concepts of the IMO and focusing upon a 'NASA-like control room' (Lloyd's List 2010b). Meanwhile further evidence of substantive change in the maritime sector comes from advances in satellite communications and software systems that encourage moves toward paperless commerce and operations through extended use of e-mail, bidirectional e-fax, text and data transmission, and voice telephony (Lloyd's List 2010d). The most obvious move is that paper-based charts may now be abandoned replaced by electronic chart display and information systems mandated by the IMO's Maritime Safety Committee in 2009.

8.2.10 Logistics, JIT, and Lean Supply Chain Management (SCM)

The significance of logistics in commerce and the inexorable rise of the concepts of Just-in-Time (JIT) and lean supply chain management are a product of globalization and a reflection of a new Postmodern era. The paper by Boje and Dennehy (2008) is one of many emphasizing the significance of JIT to modern industry and how its role has been driven by the technical and organizational impacts of globalization. The ability to communicate almost instantaneously and to assemble and share vast quantities of data have made JIT not just a concept but an essential reality in manufacturing if companies are to remain competitive and the fragmentation of both the physical and information aspects of business that results is substantial. This fragmentation is essentially Postmodern driving industry away from economies of scale, storage systems and bulk orders to variety, flexibility, and responsiveness to a degree previously unimaginable.

Notteboom and Winkelmans (2001, p. 73) related the Postmodern trends to outsourcing in particular enabling 'a producer to make fixed costs variable and to liberate internal resources for investments in the core activities' and generating economies of scope and flexibility. In their discussion of the successful port authority their central hypothesis was that adaptation to the new (Postmodern) rules of the maritime sector, had to be made.

Lean supply chain management (and its other guises as value chains and demand pipelines for example) takes this further and can be applied to both manufacturing and non-manufacturing sectors, looking for the elimination of waste, and non-value added activities. A clear product of Postmodern management and technology. New and Ramsay (1997) were some of the earliest commentators on the identification of waste in terms of time, costs, and inventory—although all three are clearly one and much of the same thing. They and others have identified areas where opportunities exist to reduce costs because of Postmodern innovations in technology, information systems and management, and concentrating on 'strong and effective relationships' and 'operational integration'. Together they create commitment and trust through the use of fewer suppliers thus reducing transaction costs (Williamson 1975; 1979) and emphasize the holist view of a business. The areas where lean supply chain management has impacted include procurement (focusing on reducing points of contact for suppliers, introducing internet based approaches, and streamlining invoicing/payment); manufacturing (focusing on routing in production, reducing material wastage, and refining equipment); warehousing (reducing inventory to reduce space and finance excesses); and transportation (minimizing transport suppliers, outsourcing, and route improvements). Lean supply chain management also has Postmodern characteristics in that it relies upon a combination of focusing upon the elements of the supply chain whilst also looking at its overall implications. It is multifacetted, comprehensive, and relies on a flexible and dynamic approach to problems and concerns. It contrasts significantly with its Modernist predecessor characterized by large-scale manufacturing and storage, economies of scale and limited consumer choice, market inflexibility and single, optimum solutions. The result is a Postmodern 'lean enterprise' represented as:

> a group of individuals, functions, and legally separate but operationally synchronized companies. The group's mission is collectively to analyze and focus a value stream so that it does everything involved in supplying a good or service in such a way that provides maximum value to the customer. (Womack and Jones 1994)

8.2.11 Marketing and the Postmodern Maritime Sector

Examples of the Postmodern in marketing of the maritime sector can be found in modern ship appearances with a classic example coming from the Contship 'Pink Lady', completed in 1992 and named and painted as a token to Cecilia Bratistella, then one of the owners. The vessel was repainted in more conventional colors after the company was sold to Canadian Pacific in 1997.

Meanwhile ship's interior décor has also been described (albeit anonymously) as Postmodern with examples on one vessel of walls covered in Greek Gods, Ancient Egyptians, Art Nouveau, Napoleon, medieval castles, bingo parlours, totem poles, Impressionists, and scenes from Shanghai bars (Anonymous 2007).

Although sold to Hellenic Seaways in 2009 and subsequently lost to the marketplace, EasyCruise was an example of a Postmodern approach to a long-standing and traditional industry. The orange livery of the vessels and cabin interiors both contrasted with the traditional cruise styles and booking/ticketing was concentrated on the internet in a manner similar to modern airlines. In addition to the website, EasyCruise contrasted markedly in style with traditional cruise ship companies in approach and design. The emphasis was on the contemporary and on the difference that there was between EasyCruise and others in the market. Cruise schedules and destinations were also designed around a young rather than mature market, allowing hopping on and off rather than fixed holiday periods of 7, 10, or 14 days for example. This Postmodern fragmentation contrasted clearly with the traditional (Modernist) cruise organization.

8.2.12 Transport, The Arts, and Postmodernity

Transport, literature, and Postmodernity are also linked in a way that indicates the all-embracing notion of the Postmodern and how it is a societal trend rather than simply one that refers to culture or the arts (Schulz 1992). Hence its application to industry, business, globalization, and enterprise is both realistic and rational. Hassan (1971) provided a detailed background to the relationship that exists between the Postmodern and literature generally whilst two books stand out as classic Postmodern literature with transport as their central theme—albeit being used as a metaphor for much else in society. Ballard's *Crash* has been described as an hallucinatory novel where the car provides a 'Hellish tableau in which Vaughan, a TV scientist, experiments with erotic atrocities among crash victims, each more sinister than the last':

> Ultimately he craves a union of blood, semen and engine coolant in a head-on collision....

Ballard (1973, p. 6) described *Crash* as an:

> extreme metaphor for an extreme situation, a kit of desperate measures only for use in an extreme crisis... (concerned) with a pandemic cataclysm that kills hundreds of thousands of people each year and injures millions. Do we see, in the car crash, a sinister portent of a nightmare marriage between sex and technology? Will modern technology provide us with hitherto undreamed-of means for tapping our own psychopathologies? Is this harnessing of our innate perversity conceivably of benefit to us? Is there some deviant logic unfolding more powerful than that provided by reason?...the ultimate role of 'Crash' is cautionary, a warning against that brutal, erotic and overlit realm that beckons more and more persuasively to us from the margins of the technological landscape.

Jack Kerouac's *On the Road* was published in 1957 (Kerouac 1957). Cresswell (1993, pp. 254–255) suggested that it is an:

> exuberant resistance to the hegemonic ideals of home and family, (filled with characters) happy to be nowhere and everywhere.

McDowell (1996, p. 414) noted that homeless bums, dropouts, junkies and drunks are key figures in the text.

> In travelling, the characters find a spiritual experience that apparently eludes them in cities and at home.

Cresswell suggested that Kerouac valorizes male friendships and heterosexual relationships, emphasizing the 'tired dualism of male/female and public/private'. However, underneath this is a contradictory concept of home as both a place that provides security—'home, mom and apple pie'—and one that needs to be left behind by travelling and taking to the road, images that reflect in their divergence the antagonism of masculinity and femininity. The social metaphors are substantial and the interrelationship of the Postmodern with transport and the widest of social considerations are clear.

In addition, Alessandro Baricco presented a Postmodern interpretation of psychological, existential, and erotic human malady in his book 'Ocean Sea' using the sea as a means of deliverance. At a remote shoreline inn, an artist uses seawater to paint a portrait of the ocean, a scientist pens love letters to someone he is yet to meet, an adulteress searches for relief from her proclivity for sex and a young girl seeks medical relief. Their lives become entwined with destiny and desire to create a Postmodern tale wholly related to the sea (Baricco 2000).

Meanwhile, Postmodernism, music, and the maritime world are themes that have interacted through the Canadian Music Centre where in 2005, the Upstream Guerilla Orchestra performed improvised music at the Museum of the Atlantic in Halifax, Nova Scotia, that were aimed to suggest submarine warfare, Morse code, nausea, storms, frozen water, and bad food (http://dofundodomar.blogspot.com).

Mazierska (2002) provided a similar integration of Postmodernism, transport, and social change in her discussion of tourism, relocation, and the films of Eric Rohmer suggesting that his discourse on tourism (for example *A Summer's Tale*, 1996; *Claire's Knee*, 1970; and *The Green Ray*, 1986) had borrowed much from the Postmodern work of Bauman (1997) and Urry (1990, 1995). Both discussed tourism as an 'actual experience and as a metaphor of the Modern/Postmodern condition' (Mazierska 2002, p. 224). In particular, she singled out Bauman who suggested that initially the position of a tourist appeared attractive signifying 'freedom and choice' but in 'time it can become painful'.

> Postmodern tourists often change into vagabonds, who travel because they have no other choice. (Bauman 1997, p. 94)

Urry (1995, pp. 148–151) took a much broader view of the Postmodern and tourism and the travel implications of those involved emphasizing the changes in the growth of the Postmodern from 'authentic' sights of mountains and seas, to the 'simulacrum' of theme parks. This mirrors changes throughout society as a virtual world of entertainment, society, trade, and work emerges in a multitude of forms including that of the maritime sector, globalization, and international trade.

Meanwhile Postmodernism is represeneted in maritime installation art with Mark Bradford's post-Katrina 'Ark' in New Orleans. Twenty-two feet high and 64 feet long, it is constructed from plywood fencing salvaged from local construction sites and has a core of two ocean containers. However, this was not the first of its kind. Kea Tawana also built an Ark in Newark, New Jersey in 1987 as a protest against the neglect of the city following riots which took place 20 years earlier. She was ordered to destroy it as it had no planning permission and its ephemerality and temporality was classically Postmodern.

Meanwhile to emphasize the breadth of social influence that the Postmodern era has had even in the restricted maritime area one can turn to the use of maritime artistic images to promote the profile of Stalin both in his lifetime and even today. Brent (2009) stressed the Postmodern characteristics of the modern conception of Stalin that is currently being put forward in Russia which includes the use of the ship as a vehicle to represent his leadership qualities.

Possibly the most iconic of many Postmodern images is that of Malcolm Morley's "Ocean Liner" (Fig. 8.3). Morley's huge oil painting is of a postcard of the 'SS Amsterdam in Front of Rotterdam (1966)'. Morley inverted a real

Malcolm Morley. *S.S. Amsterdam in Front of Rotterdam*, 1966
Liquitex on canvas. 64 x 84 inches (157,5 x 213,4 cm)
Norman and Irma Braman Collection. Courtesy Sperone Westwater, New York

Fig. 8.3 SS Amsterdam in Front of Rotterdam, 1966

postcard, projected it on a screen, covered the projection with a grid, and copied it, upside down. Termed by Crowther (1993, p. 187) super realism, its aim is to question expectations of art as what Butler calls 'high culture' creating art which 'internalizes and displays the problematic of its own socio-cultural status'. Crowther, cited in Olivier (2007), saw Morley's painting as key to the transition from Modern to Postmodern, providing fundamental questioning and skepticism of the form of art. Think what you like about all this, the maritime theme at its heart is clear, if fortuitous. Morley has also produced a substantial number of other Postmodern maritime themed paintings in the Postmodern genre (Francis 1996).

We can also identify a cross-over between architecture and visual art in the work of architect Stanley Tigerman and his collage, *The Titanic* which features a relatively well-known building—Mies Van De Rohe's School of Architecture at the Illinois Institute of Technology–depicted as the SS Titanic sinking. The objective of the collage was to distance himself from established Modernist architecture of Van De Rohe and to begin to place himself within the Postmodern.

Finally in our consideration of art, Postmodernism and maritime influence there is the ship 'Regentag'. Originally the 'San Guiseppe T', it was acquired by the

Fig. 8.4 Friedensreich Hundertwasser. 'Kolumbus Regentag' Screenprint. *Source* Hundertwasser Non Profit Foundation

Austrian painter and architect Friedensreich Hundertwasser in 1967 and refurbished over the following 7 years. Acting as his home, country, and headquarters, he sailed it around the world until 1999 when the hull was coated with ferrocement and the waterline decorated with tiles. An example of Hundertwasser's depiction of the ship is given in Fig. 8.4. It currently lies on the Danube at the Minoritenkloster Abbey in Tulin, Austria.

8.2.13 Naval Strategy and the Postmodern

It is not only in art that the modern and the Postmodern have been significant in the maritime sector. For example, Modernism as the foundation to Postmodernism has had a close influence upon naval strategy in a tradition that is unexpected as much as it is entertaining. The Postmodern impact and its relation to globalization we will consider in a moment, but initially we can examine the UK Royal Navy's Modernist roots. Quoted on a number of blog sites including Tosh (2007):

> War has inspired many great artistic moments (*sic*) but how often have artists returned the favour?

Tosh suggests only once. The British navy during the First World War was hemorrhaging vessels and conventional camouflage appeared to be incapable of providing protection against both sea and sky backgrounds (Roskam 1987). Norman Wilkinson, a British naval officer and painter suggested 'Razzle Dazzle', whereby a ship's silhouette would be broken up with brightly contrasting geometric designs which would confuse U-boats. Edward Wadsworth (1918) depicted a number of ships in Razzle-Dazzle camouflage. These overtly Modernist designs were both surprising and hinted at Postmodern approaches to tradition and convention that were to follow later in the twentieth century. In addition, these designs were also applied to passenger liners with the HMS Mauritania inspired by the work of Pagliacci.

The approach was followed by a marked reduction in losses but the convoy system was introduced at the same time and undoubtedly was more effective. To quote Tosh:

> The only way these designs would have confused the U-boat captains was if the sky and sea had been created by Mondian and Duchamp.

Till (2007a, b) took a Postmodern view on naval strategy and its relationship to globalization. Navies were seen to be driven on by the search for free markets and by individuals' demands for free information and open access. This created a society formed around nation-states that was incapable of taking independent action and which he suggested are 'thoroughly sea-based'. A world which 'picks out continents, trade routes and capital flows but ignores nations'. Till also stressed, however, the vulnerability of this sea-based society where goods were moved over vast distances, where industry demands a JIT attitude, where containers are plotted at quarter hour intervals, and where security may well be at its lowest ebb (Till 2007a, p. 570). In this context modern navies take on a Postmodern view:

> An internationalist, collaborative and almost collective world outlook. They see their role as defending the system, directly at sea and indirectly from the sea. (Till 2007a, p. 571)

8.2.14 Labor

Fragmentation and the destruction of the grand, single narrative are each central parts of the Postmodern and this is also the case in the maritime sector. Two examples from the ports sector will clarify.

The UK National Docks Labor Scheme (NDLS) was introduced in 1947 to provide job security for casual dock labor although the latter continued to exist until the Scheme was amended in 1967 (Richardson 2008). It was abolished in July 1989 to free port employers from its constraints and to facilitate the privatization of UK ports which followed (Finney 1999; Rayner 1999; Stoney 1999; McNamara and Tarver 1999). This trend toward fragmentation of labor relations which was paralleled in the UK and across the world in many industries, not least the railways, is truly Postmodern as it reflects moves away from national trades unions, national regulation and pay bargaining and toward individuality and multiple narratives in the sector (Evans 1999). Modernist ideals were centerd on the values of central administration, standardized rules and structures, clear labor demarcation, and predictability and regulation. Postmodern labor relations reflected individual choice, flexibility in duties, and the recognition of the marketplace as the main driver (Barton and Turnball 2002).

Meanwhile the European Union has been pursuing a ports policy for over 20 years but with little success (Pallis 1997; Farrell 2001; Perez-Labajos and Blanco 2004; Psaraftis 2005; Verhoeven 2009). Reflecting the EU ideals of competition, the successive attempts by the EU have focused in particular upon introducing compulsory competition between service providers within ports, whereby the larger ports would have to ensure that through a competitive bidding process, alternatives were available to port users. This would cover all major services, including stevedoring, roll-on/roll-off ramps, warehousing, pilotage, etc. and also provision for shipowners to insist on using their own on-board facilities and manning to load and unload vessels. Despite repeated failure to introduce these provisions by 2009 (see Chap. 1) their flexible and diverse nature make them suggestive of a Postmodern approach to ports that characterized the moves away from the UK NDLS and similar restrictive practices elsewhere. The monolithic, state-owned, state organized and regulated port, dominated by restrictive labor practices and little choice is being slowly succeeded by a Postmodern arrangement that reflects option, and flexibility.

8.2.15 Green Shipping

Shipping has long been viewed as a fine example of clean transport compared with most other modes (see for example Mair 1995) but in recent years its environmental credentials have been increasingly and intensely scrutinized and this has been noted with some concern by the industry itself (Turtiainen 2005). Johnson et al. (2005, p. 6) suggested that the primary drivers behind green shipping initiatives were a heady mix of cost savings, compliance, and public relations, which in themselves represent a Postmodern view on environmental policy, particularly the emphasis on profile. To quote:

> leading edge companies will attract ethical business: environmental efficiency is strongly linked with safety: a relationship with environmental regulators is important and corporate sustainability reporting relies on good news and continual progress.

Quite. Green shipping is not so much about the environment but more about image, style, and relationships. Meanwhile, the conclusions of the report were fairly pessimistic in that public perception of the industry remained skeptical—suggesting a distinct dichotomy between the real and the perceived, something again suggestive of a Postmodern world. The report went on to stress that:

there is broad agreement that public opinion, largely expressed through the media, drives the regulatory agenda. Recognizing the power of positive press is, however, more subtle and elusive... cruise companies are convinced of the marketing value of environmentally friendly good (*sic*) and services but public indifference to maritime matters, other than high profile casualties, has been a concern for the industry for some time.

In an editorial commentary in 2002, John Lyras, President of the Union of Greek Shipowners, maintained the need for a collective stance through the 'informed, directed and widespread advocacy of quality shipping'. He called for major shipping organizations to work together to present the case for quality shipping to governments and publicize the benefits associated with quality shipping." (Johnson et al. 2005, p. 15).

8.2.16 *Shipping Organization*

The past 30 years has also seen a marked change in the regulation (or deregulation) of the maritime sector which can be associated with the fragmentation and flexibility inherent within Postmodernism. Bonacich (2003, p. 4) placed these organizational and regulatory changes in the light of the US Ocean Shipping and Reform Act (OSRA) of 1998 but plenty of other examples exist including the EU's moves to eliminate liner conferences, their attempts at exposing EU ports to more competition, and a variety of other impacts worldwide often associated with the growth of privatization.

Bonacich (2003, p. 5) also saw Postmodern trends in warehousing—where modern warehouses look to have goods continually moving in and out rather than sitting around—and she emphasized the impact of cross-docking as an essentially Postmodern feature of ports with ships arriving one side of the warehouse and trucks taking the goods away from the other. Postmodern warehouses are more about process than form.

Similarly, the moves that have taken place throughout the logistics and supply chain management industries toward outsourcing, flexible production, global logistics, lean manufacturing, and the like are all organizational manifestations of Postmodernism in the maritime (and related) sectors. In particular the whole concept of supply (and increasingly demand) chain management has occurred as a result of the end of the Fordist (read Modernist) movement in ports and shipping and the growth of flexibility. Product proliferation, designed to specific demands, production only when demand is registered, minimization of storage, and the development of flexible production relationships provides ample evidence of a maritime industry that is working in a Postmodern environment (Bonacich 2003, pp. 2–9).

There are many other examples of organizational change in the Postmodern—the proliferation of landlocked flag states is one (Mongolia is a classic example) and also those in remote and largely unknown locations (for example the Kerguelen Islands), unimaginable at the time that the international and supranational

Table 8.4 The stage-based model of ocean governance

Societies	Phases	Duration	Triggering factors
Modern	Take-off	1760–1880s	First industrial revolution
	Maturity	1880–1970s	Second industrial revolution
Postmodern	Take-off	1970–1990s	New international division of labor
			Rise of the environmental question
	Maturity	1990s and on	Globalization

Source Vallega (2001)

organizations that dominate the maritime sector were designed but now something almost commonplace.

8.2.17 Postmodern Maritime Governance

Vallega (2001) viewed the emergence of Postmodernism from an ocean governance viewpoint suggesting that between about 1972 and 1992, the transition from a modern to a Postmodern approach to the ocean took place. This coincided with the UN Conference on the Human Environment and the UN Conference on Environment and Development. He questioned whether 'the present ocean governance patterns are consistent with the Postmodern spirit', something that sounds particularly relevant to our consideration of maritime policy-making and governance in a new Postmodern era. He provided a useful if simplistic stage-based model of the Modern-Postmodern transition in ocean governance (Table 8.4). Vallega (2001, p. 400) went on to suggest that:

> Modern society essentially perceived the ocean as an enormous reservoir from which to supply human communities with food, energy and mineral resources. Post-modern society has adopted a much wider viewpoint also including the representation of the ocean as a space rich in cultural heritage and essential for maintaining the Earth's ecological balance. Not only is the post-modern perspective more articulated than its predecessor, in the sense that a larger number of roles are attributed to the ocean but it also epistemologically recalls the idea of a complex system, in that the ocean is perceived as a system which requires consideration as a whole. The awareness that the ocean is a holon... must be the most influencing factor for both our future cultural and governance patterns.

Vallega's view of the Postmodern interpretation of ocean governance is something we will return to when we examine a way of moving forward in maritime policy-making but for the time being it is useful to see how he viewed oceans should be considered and the foundations upon which their governance should be based. The need to incorporate more flexibility, more uncertainty, more opinion, and to be more willing to adapt, change, alter and migrate with the varying pressures that exist is apparent. These are all Postmodern ideals that any governance system should reflect in a Postmodern society.

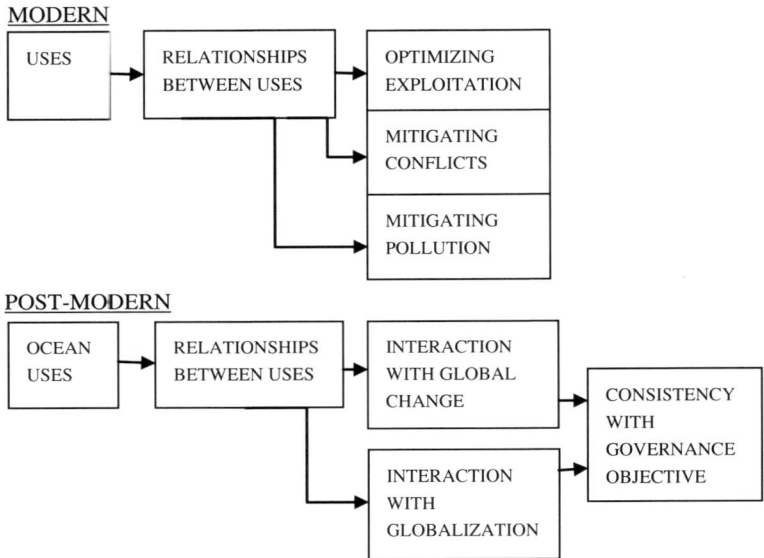

Fig. 8.5 The Changing Approach to Ocean Governance. *Source* Vallega (2001, p. 407)

Vallega went on to explain that he saw future Postmodern ocean usage as dividing into two scenarios: that relating to current usage and that for the future. Current usage includes its use as a resource reservoir (fish, oil, salt, biomass, etc.): navigation and transportation; communications (submarine cables etc.); a geopolitical space for sovereignty; and defence purposes. Future use will include settlement (on the sea's surface and floor); cultural heritage; and a primary component of the earth's ecosystem governance. He continued to visualize his theory of ocean governance as shown in Fig. 8.5 and Table 8.5.

As Vallega suggested, governance (of the maritime sector):

> should be supported by a clear definition of goals and targets, it should be operated in geographical spaces wholly embracing one of more contiguous (systems) and should embrace (systems) and social contexts in integrated ways.

Currently maritime governance does none of these things adequately—and some not at all. Table 8.6 summarizes some of the significant maritime interpretations of Postmodernism and in so-doing indicates some of the changes that have taken place and those that still need to take place. Issues of innovation, extended stakeholder involvement, changes in command and control, the manipulation of traditional jurisdictions, and the extension of flexibility and adaptability are all clear. In the final chapter, we shall examine proposals for change which derive from the move from the traditional Modernist order toward a new Postmodern maritime governance.

8.2 Postmodernism, Transport, and the Maritime Sector

Table 8.5 Disjunctive versus conjunctive logic; a postmodern interpretation applied to the ocean

Disjunctive logic	Conjunctive logic
Cartesian thought	*Theory of the general system*
Principles	Pertinence
Evidence	Describing those elements of the ocean which are perceived as essential to management
Describing only those elements of the ocean area which are clear	
Reduction	*Holism*
Disaggregating the components of the ocean and then describing them separately	Describing the ocean as a unique system with its external environment
	Considering a detailed knowledge of the ocean as not essential
Causality	*Teleology*
Moving from the simplest elements of the ocean toward the most complicated ones	Leaving out the existence of a cause-effect relationship between the elements of the ocean
Supposing that the role of the elements is regulated by a cause-effect relationship	Focusing on feedback and on circular relationships
	Considering the evolution of the ocean with reference to its project i.e., the goals which are pursued through its organization
Exhaustiveness	*Aggregation*
Assessing all the elements of the ocean in detail	Selecting those elements that pertain to ocean management and leaving out the rest
Ensuring that nothing has been left out	
Postulating that, only where these conditions are met, the subsequent knowledge is objective	Being aware that knowledge is relativist and partisan per se

Source After Vallega (2001)

Table 8.6 A maritime interpretation of postmodern governance

Postmodern governance	In the maritime sector
A move away from hierarchy and competition as alternative models for delivering services	New institutional frameworks for policy-makers based on networks and cross-hierarchies. Redefinition of the role (or even the existence) of the IMO, WTO, UNCTAD, national ministries, etc
A recognition of the blurring of boundaries and responsibilities for tackling social and economic issues	Moving beyond traditional political decision-making and responsibility. More power to individuals, interest groups, e.g Greenpeace; the shipping media; individual responsibility; private/public partnerships; the family
The recognition and incorporation of policy networks into the process of governing	Networks of cross-jurisdictional individuals and organizations. Moves away from traditional maritime hierarchies of IMO/EU/state Ministry for example. Mixed jurisdictional governance and decision-making. Emphasis on regional and local collaboration with individuals, pressure groups, and state bodies

(continued)

Table 8.6 (continued)

Postmodern governance	In the maritime sector
The replacement of traditional models of command and control by governing at a distance	Moral and ethical maritime responsibilities. Governance by discussion and interaction. The rise of blogs and e-mails
The development of more reflective and responsive policy tools	Less legislation and rules and more publicity, persuasion, and open-ness in admitting failure
The role of government shifting to a focus on providing leadership, building partnerships, steering and coordinating, and providing system-wide integration and regulation	Distancing of government from the industry where possible. Reformed governance of states and international organizations. Less intervention and more persuasion. Port and shipping company privatization. Delegation of ship inspections to the private sector. Loosening of flag ties. Development of offshore registers
The emergence of negotiated self-governance in communities, cities, and regions based on new practices of coordinating activities through networks, and partnerships	Increased role of self-regulation in the maritime sector. Increased significance of shipping associations, labour unions, professional bodies, etc. setting standards
The opening up of decision making to greater participation by the public	More openness; less secrecy; more true democratic procedures; blogs; e-mails; You-Tube, etc
Innovations in democratic practice as a response to the problem of the complexity and fragmentation of authority and the challenges this presents to traditional democratic models	More meaningful stakeholder involvement in policy-making
A broadening of focus by government beyond institutional concerns to encompass the involvement of civil society in the process of governance	Widening of stakeholder definition in the maritime sector—to include interest groups, media, political parties, women, etc

Source Derived from Newman (2001, p. 24)

References

Alexander, M. J., & Mohanty, C. T. (Eds.). (1996). *Feminist genealogies colonial legacies, democratic futures (thinking gender)*. London: Routledge.

Alvesson, M. (1987). *Organizational theory and technocratic consciousness: Rationality, ideology and quality of work*. Berlin: de Gruyter.

Alvesson, M., & Deetz, S. (1996). Critical theory and postmodernism approaches to the study of organization. In S. Clegg, C. Hardy, & W. Nord (Eds.), *Handbook of organization studies* (pp. 191–217). Thousand Oaks CA: Sage.

Amin, A., & Robins, K. (1990). The re-emergence of regional economics? The mythical geography of flexible accumulation. *Environment and Planning D, 8*, 7–34.

Anonymous, (2007). Bad ship's décor, postmodernism and "a night at the museum". http://www.this-side-of-glory.com

Arthur, W. B. (1994). *Increasing returns and path dependence in the economy*. Ann Arbor MI: University of Michigan Press.

Ball, T. (1987). *Idioms of enquiry*. NY: SUNY Press.

Ballard, J. G. (1973). *Crash*. London: Jonathan Cape.
Baricco, A. (2000). *Ocean sea*. London: Hamish Hamilton.
Barton, H., & Turnball, P. (2002). Labour regulation and competitive performance in the port transport industry: The changing fortunes of three major European seaports. *European Journal of Industrial Relations, 8*(2), 133–156.
Bauman, Z. (1997). *Postmodernity and its discontents*. Cambridge: Polity Press.
BBC, (2008). Gone with the wind on 'kite ship, January 23rd. http://news.bbc.co.uk/1/hi/world/europe/7205217.stm
Benton, P. (1990). *Riding the whirlwind Benton on managing turbulence*. Oxford: Blackwell.
Berman, M. (1982). *All that's solid melts into air: The experience of modernity*. Harmondsworth: Penguin.
Boje, D. M., & Dennehy, R. F. (1993). *Managing in the postmodern world; America's revolution against exploitation*. Kendall/Hunt: Dubuque IA.
Boje, D. M., & Dennehy, R. F. (2008). *Managing in the postmodern world; America's revolution against exploitation information age* (2nd ed.). Dubuque IA: Kendall Hunt.
Bonacich, E. (2003). Pulling the plug: labor and the global supply chain. *New Labor Forum, Summer, 12*(2), 4–6.
Brent, J. (2009). Postmodern Stalinism, the chronicle of higher education, September 21st
Byrne, J. A. (1992). Paradigms for postmodern managers. *Business Week, 23*, 62–63.
Castells, M. (1989). *The informational city: Information technology, economic restructuring and the urban regional process*. Oxford: Blackwell.
Castells, M. (1996). *The rise of the network society*. Cambridge, MA: Blackwell.
Castells, M. (1997). *The information age*. Oxford: Blackwell.
Castells, M. (2000). *The rise of the network society* (2nd ed.). Cambridge: Blackwell.
Chlomoudis, C. I., Karalis, A. V., Pallis, A. A. (2000). *Transition to a New Reality: Theorising the Organizational Restructuring of Ports*, Infomare, International Workshop, Special Interest Group on Maritime Transport and Ports, Genoa, June 8–10.
Clegg, S. R. (1990). *Modern organizations organization. Studies in the postmodern world*. London: Sage.
Cohen, A. P., March, J. G., & Olsen, J. P. (1972). A garbage can model of organizational choice. *Administrative Science Quarterly, 17*(1), 1–25.
Concord Pacific, (1999). *Where is Concord Pacific Place?* Concord Pacific Place. http://www.concordpacific.com/location.html.
Cooke, P. (1990). *Back to the future. Modernity, postmodernity and locality*. London: Unwin Hyman.
Cooper, R., & Burrell, G. (1988). Modernism, postmodernism and organizational analysis. *Organization Studies, 9*(1), 91–112.
Cresswell, T. (1993). Mobility as resistance: a geographical reading of Kerouac's 'On the Road.'. *Transactions of the Institute of British Geographers, 18*(2), 249–262.
Cresswell, T. (1997). Imagining the nomad: mobility and the postmodern primitive. In G. Benko & U. Strohmayer (Eds.), *Space and social theory, interpreting modernity and postmodernity* (pp. 360–382). Oxford: Blackwell.
Crowther, P. (1993). *Critical aesthetics and postmodernism*. Oxford: Clarendon Press.
Crozier, M. (1964). *The bureaucratic phenomenon*. Chicago IL: University of Chicago Press.
Cyert, R. M., & March, J. G. (1963). *A behavioral theory of the firm*. Englewood Cliffs NJ: Prentice Hall.
Daamen, T. (2007). *Sustainable development of the port-city interface*. Rotterdam: Sustainable Urban Areas Conference.
Daily Telegraph, (2008). *Kite-powered ship sets sail for greener future*, January 20th
Danyluk, M., & Ley, D. (2007). Modalities of the new middle class: ideology and behaviour in the journey to work from gentrified neighbourhoods in Canada. *Urban Studies, 44*(11), 2195–2210.
De Boer, N. A. (2005). *De stad van Niek de Boer*. Delft: Publicatieburo Faculteit Bouwkunde TU-Delft.

De Certeau, M. (1984). *The Practice of Everyday Life (S. Rendall, Trans.)*. Berkeley CA: University of California Press.
Dear, M. J., & Flusty, S. (1998). Postmodern urbanism. *Annals of the Association of American Geographers, 88*(1), 50–72.
Deleuze, G., & Guattari, F. (1984). *Anti-Oedipus: capitalism and schizophrenia*. Minneapolis MN: University of Minnesota Press.
Desfor, G. (1993). Restructuring the Toronto Harbour Commission. *Journal of Transport Geography, 1*(3), 167–181.
Desfor, G., Goldrick, M., & Merrens, H. R. (1988). Development of the North American water frontier: the case of Toronto. In B. S. Hoyle, D. A. Pinder, & M. S. Hussain (Eds.), *Revitalizing the waterfront* (pp. 92–113). London: Belhaven Press.
Doeringer, P. B., & Piore, M. J. (1971). *Internal labour marketsand manpower analysis*. Lexington MA: Heath.
Eagelton, T. (1986). *Against the grain: Essays 1975–1985*. London: Verso Press.
Eagleton, T. (1999). In the gaudy supermarket, *London review of books*, May 15th
Evans, R. (1999). The Australian waterfront: Breaking monopoly power. *Economic Affairs, 19*(2), 31–35.
Farrell, S. (2001). If it ain't bust, don't fix it: The proposed EU directive on market access to port services. *Maritime Policy and Management, 28*, 307–313.
Featherstone, M. (1987). Lifestyle and consumer culture. *Theory, Culture and Society, 4*(1), 55–70.
Finney, N. (1999). Editorial: A decade of relative peace and productivity. *Economic Affairs, 19*(2), 1–4.
Flax, J. (1990). *Thinking fragments*. Berkeley CA: University of California Press.
Foss, K. A., Foss, S. K., & Griffin, C. L. (Eds.). (1999). *Feminist rhetorical theories*. Thousand Oaks CA: Sage Publications.
Francis, R. (1996). *Malcolm Morley* (p. 55). Spring: Bombsite.
Grossberg, L., Fry, T., Curthoys, A., & Patton, P. (1989). *It's a sin: Essays on postmodernism politics and culture*. Sydney: Power Publications.
Guardian, (2008). *Giant sail technology could make shipping greener*, January 2nd
Hansen, L. G. (2003). *Impacts of fixed links on firms' organization of logistics and transport*. Nectar Conference 7, European Transport Systems Between Efficiency, Equity and Sustainability, Umea, Sweden
Hanson, S., & Pratt, G. (1988). Spatial dimensions of the gender division of labor in a local labor market. *Urban Geography, 9*, 367–378.
Harrison, B. (1994). *Lean and mean: The changing landscape of corporate power in the age of flexibility*. NY: Basic Books.
Harvey, D. (1973). *Social justice and the city*. Baltimore MD: The John Hopkins University Press.
Harvey, D. (1988). Voodoo cities. *New Statesman and Society, 1*(17), 33–35.
Harvey, D. (1990). *The condition of postmodernity*. Cambridge MA: Blackwell.
Harvey, D., & Scott, A. (1989). The practice of human geography: theory and empirical specificity in the transition from Fordism to flexible accumulation. In B. Macmillan (Ed.), *Remodelling geography* (pp. 217–229). Oxford: Blackwell.
Hassan, I. (1971). POSTmodernISM: A paracritical bibliography. *New Literary History, 3*(1), 5–30.
Hassan, I. (1980). The question of postmodernism. *Bucknell Review, 26*(2), 117–126.
Hassan, I. (1985). The culture of postmodernism. *Theory, Culture and Society, 2*(3), 119–131.
Hassard, J., & Parker, M. (1993). *Postmodernism and organizations*. London: Sage.
Hebdige, D. (1988). *Hiding in the light: On images and things*. London: Routledge.
Highmore, B. (2006). *Michel de Certeau: Analysing culture*. London: Continuum.
Hirschman, A. O., & Lindblom, C. E. (1962). Economic development, research and development, policy making. Some emergent views. *Behavioral Science, 7*, 211–222.

Hutnyk, J. (2004). Introduction to Trinh T. Minh-ha. Not you/like you: post-colonial women and the interlocking questions of identity and difference. In K. A. Foss, S. K. Foss, & C. L. Griffin (Eds.), *Readings in feminist rhetorical theory*. Thousand Oaks CA: Sage Publications.

Hutton, T. A. (2004). Post-industrialism, post-modernism and the reproduction of Vancouver's central area: re-theorising the 21st century city. *Urban Studies, 41*(10), 1953–1982.

IBI Group. (2006). *Inland Container Terminal Analysis*. Vancouver: Final report, IBI Group.

Independent, (2007). *Grounded container ship is refloated*, July 9th

Jameson, F. (1985). Postmodernism and consumer society. In H. Foster (Ed.), *Postmodern culture* (pp. 111–125). London: Pluto Press.

Jeffcutt, P. (1994). From interpretation to representation in organizational analysis: Postmodernism, ethnography and organizational symbolism. *Organization Studies, 15*(2), 241–274.

Johnson, D., Pike, K. and Walmsley, S. (2005) *Global green shipping initiative: audit and overview*. World Wildlife Fund to the Issue Group on Sustainable Shipping, North Sea Ministerial Meeting, Hamburg, 1–2 March

Jukes, P. (1991). *A shout in the street: An excursion into the modern city*. Berkeley CA: University of California Press.

Kerouac, J. (1957). *On the road*. NY: Viking Press.

Kreiner, K. (1992). The postmodern epoch of organization theory. *International Studies of Management and Organization, 22*(2), 37–52.

Law, R. (1999). Beyond 'women and transport': Towards new geographies of gender and daily mobility. *Progress in Human Geography, 23*(4), 567–588.

Lay, M. G. (1992). *Ways of the world: A history of the world's roads and the vehicles that used them*. Chapel Hill NC: Rutgers University Press.

Lloyd's List (2006) *Transparency 'could damage shipping's image'*, March 23rd

Lloyd's List (2008) *Wista redefines maritime role*, October 13th

Lloyd's List (2010a) *Me, you and Mitropoulos: giving the IMO a voice*, June 9th

Lloyd's List (2010b) *Making ship navigation attractive to today's youth*, June 24th

Lloyd's List (2010c) *Maritime industry needs a new image*, July 14th

Lloyd's List (2010d) *The issues raised by a paperless planet*, January 28th

Lloyd's List (2011) *UK Chamber joins Twitter*, May 17th

Mair, H. (1995). 'Green' shipping. *Marine Pollution Bulletin, 30*(6), 360.

March, J. G., & Olsen, J. P. (1976). *Ambiguity and choice in organizations*. Bergen: Universitetsforlaget.

Matless, D. (1995). Culture run riot? Work in social and cultural geography, 1994. *Progress in Human Geography, 19*, 395–403.

Mayntz, R. (1976). Conceptual models of organizational decision-making and their application to the policy process. In G. Hofstede & S. Kassem (Eds.), *European contribution to organization theory* (pp. 114–125). Amsterdam: Van Gorcum.

Mazierska, E. (2002). Road to authenticity and stability: Representation of holidays, relocation and movement in the films of Eric Rohmer. *Tourist Studies, 2*(3), 223–246.

McDowell, L. (1993a). Space, place and gender relations, part 1. Feminist empiricism and the geography of social relations. *Progress in Human Geography, 17*, 157–179.

McDowell, L. (1993b). Space, place and gender relations, part 2, identity, difference, feminist geometries and geographies. *Progress in Human Geography, 17*, 305–318.

McDowell, L. (1996). Off the road: alternative views of rebellion, resistance and 'the beats'. *Transactions of the Institute of British Geographers, 21*(2), 412–419.

McNamara, M. J., & Tarver, S. (1999). The strengths and weaknesses of dock labour reform—Ten years on. *Economic Affairs, 19*(2), 12–17.

Morgan, G. (1997). *Images of organization*. Thousand Oaks CA: Sage.

New, S., & Ramsay, J. (1997). A critical appraisal of aspects of the lean chain approach. *European Journal of Purchasing and Supply Chain Management, 3*(2), 93–102.

Newman, J. (2001). *Modernising governance*. London: Sage.

Newman, P. (1995). Sustainability and the post-modern city: some guidelines for urban planning and transport practice in an age of uncertainty. *The Environmentalist, 15*, 257–266.

Norcliffe, G., Bassett, K., & Hoare, T. (1996). The emergence of postmodernism on the urban waterfront. *Journal of Transport Geography, 4*(2), 123–134.

Notteboom, T. E., & Winkelmans, W. (2001). Structural changes in logistics: how will port authorities face the challenge? *Maritime Policy and Management, 28*(1), 71–89.

Olivier, B. (2007). Beauty, ugliness, the sublime and truth in art. *South African Journal of Art History, 22*(3), 1–16.

O'Sullivan, S. (2008). *Art encounters Deleuze and Guattari: Thought beyond representation.* Palgrave Macmillan: Basingstoke.

Pallis, A. A. (1997). Towards a common ports policy? EU proposals and the ports industry's perceptions. *Maritime Policy and Management, 24*(4), 365–380.

Pallis, A. A. (2005–2006). Maritime interest representation in the EU. *European Political Economy Review, 3*(2), 6–28.

Pallis, A. A. (2006). Institutional dynamism in EU policy-making: evolution of the EU maritime safety policy. *Journal of European Integration, 28*(2), 137–157.

Pallis, A. A. and Tsiotsis, S. G. P. (2006). *Inside EU-level maritime interest groups: structures, lobbying practices and governance,* Interim Results of a Research Project at the Jean Monnet Centre in European Port Policy Department of Shipping Trade and Transport, University of the Aegean, Chios

Pallis, A. A., & Tsiotsis, S. G. P. (2008). Maritime interests and the EU port services directive. *European Transport/Trasporti Europei, 38*, 17–31.

Parker, M. (1992). Post-modern organizations or postmodern organization theory? *Organization Studies, 13*(1), 1–17.

Perez-Labajos, C., & Blanco, B. (2004). Competitive policies for commercial sea ports in the EU. *Marine Policy, 28*, 553–556.

Pondy, L. R., Frost, P. J., Morgan, G., & Dandridge, T. C. (Eds.). (1983). *Organizational symbolism.* Greenwich CT: JAI Press.

Power, M. (1990). Modernism postmodernism and organization. In J. Hassard & D. Pym (Eds.), *The theory and philosophy of organizations* (pp. 109–124). London: Routledge and Kegan Paul.

Psaraftis, H. N. (2005). EU ports policy: Where do we go from here? *Maritime Economics and Logistics, 7*, 73–82.

Rayner, J. (1999). Raising the portcullis: Repeal of the National Dock Labour Scheme and the employment relationship in the docks industry. *Economic Affairs, 19*(2), 5–11.

Relph, E. (1987). *The modern urban landscape.* Baltimore MD: John Hopkins University Press.

Richardson, M. (2008). *State intervention and the abolition of the National Dock Labour Scheme.* The Bristol Experience, Centre for Employment Studies Research CESR Review (April), University of Bristol: Bristol

Roskam, A. (1987). *Dazzle painting, Kunst als Camouflage, Camouflage als Kunst.* Uitgeverij Van Spijk: Stichting kunstprojecten (Museum Catolgue).

Ross, N. S. (1958). Organized labour and management: The UK. In E. M. Hugh-Jones (Ed.), *Human relations and modern management.* North Holland: Elsevier.

Rouleau, L., & Clegg, S. R. (1992). Postmodernism and postmodernity in organization analysis. *Journal of Organizational Change, 5*(1), 8–25.

Schulz, M. (1992). Postmodern pictures of culture. *International Studies of Management and Organization, 22*(2), 15–35.

Scott, A. J. (1988). *New industrial spaces: Flexible production, organization and regional development in North America and Western Europe.* London: Pion.

Sharma, S. (2009). *Post-modernism and transport planning,* April 19th. http://cityrenewal.blogspot.com/2009/04/post-modernism-transport-planning.html

Simon, D. (1996). *Transport and development in the third world.* London: Routledge.

Simon, H. A. (1947). *Administrative behavior. A study of decision-making processes in administrative organizations.* NY: Macmillan.

Simon, H. A., & March, J. G. (1957). *Organizations.* NY: Wiley.

Soja, E. W. (1987). The postmodernization of geography: A review. *Annals of the American Association of Geographers, 77*(2), 289–294.

Steenken, D., & Voss, S. (2008). Operations research at container terminals—A literature update. *OR Spectrum, 30,* 1–52.

Steenken, D., Voss, S., & Stahlbock, R. (2004). Container terminal operation and operations research—a classification and literature review. *OR Spectrum, 26,* 3–49.

Stoney, P. (1999). The abolition of the National Dock Labour Scheme and the revival of the port of Liverpool. *Economic Affairs, 19*(2), 18–22.

Thrift, N. J. (1999). The place of complexity. *Theory, Culture and Society, 16,* 31–69.

Till, G. (2007a). *Maritime strategy in a globalizing world* (pp. 569–575). Fall: Orbis.

Till, G. (2007b). *Globalization: Implications of and for the modern/post-modern navies of the Asia-Pacific,* S. Rajaratnam School of International Studies Paper No. 140, Singapore.

Tosh, (2007). *Toshism: Modernist Art in Camouflage.* http://toshism.com/main/2007/11/14/modernist-art.

Tradewinds, (2008a). *Maersk in fresh online initiative,* January 25th.

Tradewinds, (2008b). *New online calculator fires storm,* February 1st.

Tradewinds, (2008c). *'One stop' system,* February 22nd.

Tradewinds, (2010a). *Baltic faces challenge in screen battle,* September 17th.

Tradewinds, (2010b). *'Slow-motion war' over FFA screen trading,* September 17th.

Tradewinds, (2010c). *A breath of fresh air from the girls,* October 1st.

Tradewinds, (2010d). *Shipping would benefit from more women—IMO chief,* October 8th.

Tradewinds, (2010e). *Big ambitions for biomass transport,* October 8th.

Tradewinds, (2010f). *'Tweets' bring out a fresh dimension,* October 15th.

Tradewinds, (2010g). *Fronting an electronic revolution,* June 11th.

Tradewinds, (2011a). *Screen team remains confident,* October 14th.

Tradewinds, (2011b). *FFA trading is yet to pass screen test,* October 28th.

Turtiainen, M. (2005). Green shipping. *ABB Review, 3,* 54–57.

Tradewinds (2011c) Kite a flyer for Anbros and Cargill, Dec 9th.

UK Ministry of Transport. (1963). *Traffic in towns, (the Buchanan Report).* London: HMSO.

Urry, J. (1990). *The tourist gaze.* London: Sage.

Urry, J. (1995). *Consuming places.* London: Routledge.

Urry, J. (2000). *Sociology beyond societies: Mobilities for the twenty-first century.* London: Routledge.

Vallega, A. (2001). Ocean governance in a post-modern society—A geographical perspective. *Marine Policy, 25,* 399–414.

Varela, F. J. (1979). *Principles of biological autonomy.* NY: North Holland.

Verhoeven, P. (2009). European ports policy: Meeting contemporary governance challenges. *Maritime Policy and Management, 36*(1), 79–101.

Virilio, P. (1977). *Speed and politics, an essay on dromology.* NY: Semiotext(e).

Waldrop, M. M. (1993). *Complexity.* NY: Viking.

Weick, K. E. (1976). Educational organizations in loosely coupled systems. *Administrative Science Quarterly, 21*(1), 1–19.

Williams, D. (1987). *The specialized agencies, the united nations: The system in crisis.* London: C. Hurst and Company.

Williamson, O. E. (1975). *Markets and hierarchies: Analysis and antitrust implications.* NY: The Free Press.

Williamson, O. E. (1979). Transaction cost economics: The governance of contractual relations. *Journal of Law and Economics, 22,* 233–261.

Wolf, P. M. (1969). *Eugene Henard and the beginning of urbanism in Paris, 1900–1914.* The Hague: International Federation for Housing and Planning.

Wolff, J. (1990). The invisible flaneuse; women and the literature of modernity. In J. Wolff (Ed.), *Feminine sentences, essays on women and culture.* Cambridge: Polity Press.

Womack, J. P., & Jones, D. T. (1994). From lean production to the lean enterprize. *Harvard Business Review, 72*(2), 93–103. (March–April).

Chapter 9
So, What Next?

> We playwrights who have to cram a whole human life or an entire historical era into a two-hour play, can scarcely understand the rapidity (of change) ourselves. And if it gives us trouble, think of the trouble it must give political scientists who have less experience with the realm of the improbable. Vaclav Havel, speaking to the US Congress, February 1990.

> …for now the commercial dealings with the Newts will also be crowned with statesmanlike foresight, which will take steps to see that the machinery of the New Age will not collapse in ruins. In London a group of Maritime States is holding a conference to work out and approve of an international convention for the Salamanders. The High Contracting Parties bind themselves as between each other not to send their own Newts into the territorial waters of the others; that they will not permit their Newts, in any way whatsoever, to violate the territorial sovereignty or recognized sphere of interest of any other State; that in no way will they interfere with the Newt affairs of another Sea Power; that in the case of dispute between their own and Salamanders belonging to a Foreign Power, they will submit to arbitration by the Hague Tribunal; that they will not arm their Newts with any weapons whose caliber exceeds that of the standard water pistol for use against sharks… and that they will not allow their Newts to enter into close relations of any kind with the Salamanders subject to another Power; they will not with the aid of the Newts construct new continents or enlarge their own territories without the previous approval of the Permanent Sea Commission at Geneva, and so on. Karel Capek, *Válka s mloky (War With The Newts)* (1936), p. 243.

The discussion so far has ranged over a multitude of issues which have been stimulated by the clear inadequacy of modern shipping policy to address the failures of the maritime sector. Seas continue to be polluted, administrations continue to argue, regulations are unenforced, security continues to be breached, and people continue to die. All these occur at an alarming rate and despite the vast quantity of energy and effort undoubtedly put into maritime policy-making by politicians, regulators, civil servants, pressure groups, and the like. To continue with the same will bring no more success than we have seen in the past and it is an option only for those who either do not really care or who actually benefit from the situation as it stands. It is to this last group that we turn our attention in the first instance.

9.1 The Attitude of the Shipping Industry

Shipping is not ambivalent to the failures in policy. It publically espouses the advantages of clean, secure, and safe seas through its representatives in the EU, the IMO, through national Ministries and ship-owning associations, through its own interest groups, and professional associations. Yet it remains an industry out of control, situated within a governance framework that has no ability to affect its organization and practice in a meaningful way unless as an industry, it wants to adapt.

The causes of this loss of control have been discussed in the earlier chapters. They center on the inadequate and inappropriate jurisdictional framework that remains central to maritime governance; a framework that emerged through the twentieth century, based upon Modernist principles of control and command, rooted deeply in the nation-state regardless of jurisdiction. The new, globalized world, a product of the major capitalist shift to Postmodernism with its profile of flexibility and diversity and characterized by Harvey's vision of time-space compression, has little in common with the traditional nation-state and the jurisdictional structures that are the feature of maritime governance—the IMO, the EU Commission, national Ministries, and regional and local governments effectively controlled by national budgets and priorities. Maritime legislation and policy derived from such a nation-state-dominated framework clashes with the globalized characteristics of the maritime sector generating a whole series of policy failures and allowing the shipping industry to take commercial advantage of the structural failure within the sector.

This may not be a popular message either within the industry or for policy makers, but few will deny the significant changes in society that have taken place from the second half of the twentieth century. Few can deny that maritime policy-making is fraught with disappointment and rather short of success. And few could genuinely argue that the nation-state remains the best vehicle for maritime policy-making as technological and social change in the new Postmodern world generates the need for a new view on maritime institutions, governance frameworks, stakeholder involvement, media presence, and the like.

Central to all this is the shipping industry itself. Understandably it wants to sustain a commercial environment that maximizes profit. Not all shipowners or participants in this market are willing to sacrifice anything to achieve this but the opportunity that current governance failure provides to abuse laudable policy aims to create additional profit is a major factor in the inadequacies of current policy making. There are many operators and owners, ancillary actors, and interested parties that are willing to take advantage of the governance failures that exist and who will only adopt acceptable policies if made to do so. The current governance framework, despite good intentions, produces policies that are commonly inadequate, often inappropriate, frequently ineffective, and widely abused.

The success of the nation-state in retaining its influence in maritime policy-making bodies at all jurisdictions is remarkable in the light of the Postmodern

9.1 The Attitude of the Shipping Industry

revolution and especially for an industry where the process of time-space compression has been so influential. Mangat (2001, p. 9) suggested that this is even more impressive because globalization has created a post-Westphalian world where nation-states have lost the majority of their sovereignty and as a consequence have neither the 'capacity nor the willingness to develop and enforce policy'. This contradiction between a nation-state's apparent impotence and the substantial policy role it retains is a reflection of governance inertia which in turn is the greatest cause of the inadequacies in maritime policy.

Alderton and Winchester (2002, p. 39) went further in discussing flags of convenience and the failure of policy makers to provide an adequate response to their deficiencies. International regulations are enacted on a state-by-state basis, reflecting the substantial residual power that the state has even in a Postmodern, globalized world. Quoting from the International Labour Organization's (2001) *Characteristics of International Labour Standards*, 'transforming these universally accepted goals and rules into a binding legal obligation is each State's sovereign privilege'. The state is privileged in that it can not only choose to act, but also not to act—and many do when it comes to policy for flags of convenience (and many other areas as well). As Alderton and Winchester (2002, p. 39) went on to say:

> Where the nation-state is the bulwark of international regulation, sovereignty is for sale in the context of ship registration, and the state enjoys privilege.

Further, Alderton and Winchester (2002, p. 40) added that there is clear evidence of:

> the deliberate attempt of the state to restrict the possibility of exerting its authority as an explicit economic decision. (Flags of convenience) are formed in such a way that a path between the flag state and the ship owner is, at best, obscure and minimal.

They went on to suggest that the increased standards applied to Flag States has made the failure of policy in this area worse as they have encouraged unscrupulous states to develop registries to accommodate those vessels which would otherwise be unregisterable. Thus the nation-state remains significant not only in policy generation and approval but also in a beneficiary of policy evasion.

> In the context where international regulation is enacted upon a nation by nation basis then it is no wonder that this situation occurs. Where legislation still relies on a state as the analytical model, yet the context itself is irredeemably global, there is always a remainder, a remainder that, due to its sovereign privilege may create an unregulated environment where capital is free to act as it pleases. (Alderton and Winchester 2002, p. 42)

This was reinforced by Bauman (2000, p. 192) citing Hobsbawm (1998, pp. 4–5):

> What we have today is in effect a dual system, the official one of the 'national economy' of states, and the real but largely unofficial one of transnational units and institutions… [U]nlike the state with its territory and power, other elements of the 'nation' can be and easily are overridden by the globalization of the economy. Ethnicity and language are the two obvious ones. Take away state power and coercive force, and their relative insignificance is clear.

And if this insignificance is so clear, why on earth do nation-states retain such a powerful force in maritime policy-making?

Putnam (1988, pp. 433–434) contributed to this debate in citing Walton and McKersie's (1965) 'behavioral theory' of social negotiations. They saw much policy-making—particularly that which involves international issues (and this is epitomized by shipping)—as a two-level game. At the national level, domestic groups (for example shipowners) pressurize governments to adopt favourable policies and domestic 'politicians seek power by constructing coalitions amongst those groups'. At the international level, national governments seek to 'maximize their own ability to satisfy domestic pressures, while minimizing the adverse consequences of foreign developments'. Each nation state appears at both game boards.

> The unusual complexity of the two-level game is that moves that are rational for one player at one board may be impolitic for that same player at the other board. The political complexities for the players... are staggering. Any key player at the international table who is dissatisfied with the outcome may upset the game board and conversely, any leader who fails to satisfy his fellow players at the domestic table risks being evicted from his seat. On occasion however, clever players will spot a move on one board that will trigger realignments on other boards, enabling them to achieve otherwise unattainable objectives.

This view of policy making across jurisdictions as a two-level game was supported by Druckman (1978), Axelrod (1987), and Snyder and Diesing (1977) among many others. Shipping interests have been extremely adept (and fortunate) in playing this game, using the pressure that inevitably falls on nation-states in international negotiations to ensure global fortune and domestic bliss.

The inadequate relationship between the state and shipping has been recognized for some time and even before the wider implications of the Postmodern epoch had started to emerge. Strange (1976, p. 358) had no compunction in suggesting that the 'authority of states over the operators and the market is generally rather weak' and contrasted this to other markets (for example the airline industry) where she saw considerably more authority. The significance of this is not whether the state should or should not be a central authority in maritime policy-making—this is a doctrinal argument of which we have no part—but whether it is realistic to sustain a maritime governance framework centered on the state if the relationship between it and the industry is so inadequate. Shipping is an industry, as we have seen again and again, where controlling its operation and administration is made more difficult perhaps than any other because it takes place so much outside national, territorial control and international law has recognized this situation in 'pronouncing the freedom of the high seas and the rule that at sea, the master exercises sole authority over his ship'. How can this be reconciled with national authority, national policy-making, and national ambitions?

This very weakness of state-shipping relationships despite the nation-state's continued domination of all jurisdictions and consequently all policy-making encourages the highly mobile shipping industry to take advantage of this inherent governance failure. Clear examples exist in the playing off of nation-states, the EU, and the IMO in the creation of tonnage taxes (providing privileges to a

9.1 The Attitude of the Shipping Industry

specific industry unlike those available in any other), the rise of Open Registries and Second Registries, the continued flaunting of international regulations, and the sustained protectionism that the shipping industry enjoys, heavily subsidized even by the leader of free-market ideals, the USA. Strange (1976) saw the attitude of the USA to shipping governance as central to the failure of national and international regulation and policy making and this remains the case today with the USA Merchant Marine Act of 1920 (otherwise known as the Jones Act), a prime example of state protectionism retained (with some revision) at the expense of international policy optimization, and one welcomed (of course) by the shipping interests themselves. Strange's comment that

> Considering the multitude of international problems which the transnational operations of ship-operators create, the impact of international law and organization on shipping is still relatively weak. (Strange 1976, p. 361)

…remains as valid today as when written and also reflects the shipping industry's ability to take advantage of the failure in governance not only to extract help from nation-state governments under threat of flag or administrative emigration, but also the impact their inherent ability to move national allegiance and confuse national location through the multiplicity of national associations they can command, can have on international regulation. The shipping industry has the unique ability to choose to associate with whichever nation it wishes, to change this at will, and to mix national associations effortlessly. In doing so, it can avoid what it wishes to avoid, ignore what it wishes to ignore and to become nation-less or nation-orientated whichever is most attractive. Shipping is today ungovernable.

Strange went on to emphasize in particular how weak the IMO (then IMCO) was and little has changed. Regulations are slow to emerge, routinely ignored by the nation-states that agreed them, and its concentration on safety, security, and the environment along with the exclusion of trading issues makes its activities divorced from the real world. nation-states make up the IMO, but the relationship between these same states and the shipping companies make international regulation much more ineffective. Strange, in a later publication (1996, p. 14) followed this up:

> the fundamental responsibilities of the state in a market economy… are not now being discharged by anyone. At the heart of the international political economy, there is a vacuum, a vacuum not adequately filled by inter-governmental institutions or by a hegemonic power exercising leadership in the common interest.

In fact the vacuum is a convenience for the international shipping industry, generated by a nation-state-focussed governance framework that has become anachronistic in the light of the Postmodern change exemplified by globalization—made even more ironic in that the shipping industry has benefitted perhaps more than any other from the process of globalization itself. Thus, shipping delights in the new globalized world while taking advantage of the lax governance that this new world permits. In the meantime policy-making frameworks, processes, and institutions take their time to catch up. To quote Strange again (1996, p. 14):

> The diffusion of authority away from national governments has left a yawning hole of non-authority, ungovernance it might be called.

And this was 1996. So little has changed since. The result is that the shipping industry is headed for 'increased inequity and continued instability' (Strange 1976, p. 364). The:

> need for control increases every year that shipping expands its role in the world economy and with every technical change that brings new problems. But compared with a hundred years ago, the general impression is not of increased but rather of diminished political authority, from whatever source. (Strange 1976, p. 364)

Strange is not alone in her comments. Cerny (1995, pp. 620–621), while not referring directly to the industry, said much that is pertinent to the governance difficulties that characterize the shipping sector. He cited Andrews (1994) who saw the state as an agent for the 'commodification of the collective, situated in a wider, market-dominated playing field' (Cerny 1995, p. 620). The nation-state then can be seen increasingly to be free riding.

The state has found it increasingly difficult to balance its societal and commercial interests—something clearly apparent in shipping where the effective application of policy in areas of societal interest (for example safety, security, and the environment) is compromised through commercial desires to retain a national shipping interest (through for example the introduction of a tonnage tax) with the shipping industry happily encouraging such compromises either directly (through flag desertion) or indirectly (through threat of flag desertion).

Cerny went on to emphasize the governance problems that are faced by a globalized world, where the nation-state remains supreme. He suggested that a globalized world remains anarchic but the structural composition of that anarchy changes. Whereas before globalization took off in the new Postmodern milieu nation-state relationships were paramount in governance, now it is the relationships between the functional spheres of economic activity (for example shipping) and the new institutional structures which have filled the nation-state power void that really matter. This change of relationship has not been recognized in maritime governance and policy failure is the result. To quote Cerny (1995, pp. 620–621):

> Different economic activities... increasingly need to be regulated through distinct sets of institutions at different levels organized at different optimal scales. Such institutions, of course, overlap and interact in complex ways but they no longer sufficiently coincide on a single optimal scale in such a way that they could be efficiently integrated into a multitask hierarchy like the nation-state.

Thus, the nation-state is no longer capable of being an effective vehicle for policy making in a Postmodern, globalized industry. We have entered what Cerny terms a 'complex, world-wide evolutionary process of institutional selection' to find a way ahead for meaningful policy making—but a process that the maritime sector has yet to publically recognize as it continues to exercise its global strengths to manipulate nation-state-dominated maritime governance.

9.1 The Attitude of the Shipping Industry

Picciotto (1991, p. 46) summed it up neatly, and using the neoclassical argument of Rugman (1982) and the Marxist arguments of Brett (1985), and Jenkins (1988) suggested that internationalized ownership of capital, exemplified by shipping, emerged, through international corporate groups facilitated by the existence of nation-state protectionist regulation–tariffs, national procurement policies, national financial protection measures. Foreign-owned capital is offered minimal national treatment (for example through tonnage taxation regimes which avoid the normal national taxation rules applied to others), but then becomes the staunchest of all nation-state defenders. Shipping thus obtains a significant competitive advantage by exploiting national differences (at the EU, IMO, WTO and so on) both politically and economically. Such internationalization resulting from globalization is highly contradictory, reflecting both homogenization and differentiation, and makes effective policy-making impossible.

Chowdhury (2006, p. 141) also saw a close relationship between capital (in this case shipping), globalization and the role of the nation-state and how one relies upon the other but at the same time takes clear advantage of it—to the detriment of its organization, management, control, and governance. He quoted Wood (1997):

> It is the essential nature of capitalism that appropriation will always be separate from, and yet require enforcement by, legal, political, and military instruments external to the 'economy', as well as support from extra-economic social institutions. The economic reach of capital will always and increasingly, exceed the grasp of the extra-economic means required to reproduce and reinforce it. However, global the economy becomes, it will continue to rely on spatially limited constituent units with a political, and even an economic logic of their own.

Or to put it another way, globalization can occur only in conjunction with nation-states (or something resembling a nation-state) as capital (i.e., shipping) uses their political space to generate wealth. This might well (and does) involve abuse of the nation-state, trading off one against another, and placing true allegiance to a national flag against commercially generated loyalty. Fine—but the outcome is a policy-making process that is only allowed to be effective if it permits capital's excesses to continue.

Lambert (1991, p. 14) saw the confusion caused by the development of the EU and its relationships with member nation-states as ideal for big business, typified not only by the shipping sector but also applicable in others. Capital has become increasingly footloose and none so much as shipping capital which has adapted to the new, Postmodern, globalized environment to further its interests. Lambert suggested that 'national prestige industries (he cited automobiles, aeronautics and computers but shipping has been equally as active) milked the nation-states for massive subsidies only to link up with their rivals in other countries when competitive conditions dictated'. In shipping's case they even abandon ship and emigrate to those with the best offer. Major firms with plans for investment could be seen to be playing off nation-states against each other to obtain 'maximum fiscal concessions and subsidies, only to move on when it suited them'.

As Berg (1993, p. 495) suggested, reality and perception thus differ. Kant's *dualism* of reality (*noumena* or *things in themselves*) and the individual's concepts of reality (*phenomena* or *what we can know of things*) can be seen to represent the process of globalization—the reality—and the perception of reality in the fictional significance that the nation-state retains. The two are not the same, and an effective governance framework will need to be framed in reality rather than perception if it is to work.

Denning LJ, in his judgement upon *HL Bolton (Engineering) Co Ltd v TJ Graham and Sons Ltd* [1957] 1 QB 159, although not once referring to shipping, passed comment on the modern company and its function in society, said at 172 with some relevance to the discussion here:

> A company may in many ways be likened to a human body. It has a brain and nerve centre which controls what it does. It also has hands which hold the tools and act in accordance with directions from the centre. Some of the people in the company are mere servants and agents who are nothing more than hands to do the work and cannot be said to represent the mind or will. Others are directors and managers who represent the directing mind and will of a company, and control what it does. The state of mind of these managers is the state of mind of the company and is treated by the law as such.

These same directors and managers are those who will also exploit the failure of shipping policy and the anachronistic structure that frames its origins and designs to their own purposes, using the hard to define, metaphysical characteristics of 'the company'. The example of the Open Registry ranks with that of tonnage tax as clear abuse of weak governance and inadequate policy-making. Examples abound but one clear policy failure that exemplifies abuse of good governance by the shipping sector can be found on a Panamanian website advertising the advantages of incorporation in Panama and clearly directed toward the shipping sector (Bowman-Gilfillan 2004):

- No reporting requirements for taxes.
- No piercing of the corporate veil.
- Anonymous ownership.
- No capital requirements.
- Nominee directors (This is accompanied by "When we appoint nominee directors for the entities that we establish for our clients, we also provide our clients with pre-signed, undated letters of resignation from the directors so that our clients can replace those directors at any time". Need we say more?).

Meanwhile, Zacher and Sutton (1996, p. 64) stressed the significance of the Open Registry failure in that the majority of nation-states view it favorably, willing to overlook the inadequacies of the system so that national shipping companies might benefit, and combining this with state subsidies through tonnage taxation and other priveliges to maintain a national flagged fleet. Policies to control and regulate Open Registries are encouraged by the same nation-states through global jurisdictional authorities while at the same time the enforcement of such regulations are commonly given less than full attention as domestic priorities predominate. And through all this, the shipping industry understandably takes advantage—and will continue to do so until the governance of the sector is revamped.

We can conclude by considering developments that have taken place through 2009–2010 in Greece, where the Ministry of Mercantile Marine was absorbed by the Ministry of Economic Development, Competitiveness, and Shipping following the election of a new government. One immediate impact supported by the London-based Greek Shipping Co-operation Committee (GSCC) was that the number of ships registered under the Greek flag dropped by around 20 % within 3 months. The result—reinstatement of a stand-alone Greek shipping ministry in September 2009 accompanied by a Piraeus-based Minister and Chios originating Deputy (Tradewinds 2010). While the strength of the Greek shipping community is apparent, this contrasts with the denationalization of shipping and the absence of any allegiance to any flag when it proves to be inconvenient, regardless of nation or tradition.

> Observers might see in this volte face, which was clearly forced on the... administration, yet more evidence that shipping cannot be treated like any other branch of the economy. It moves to a different beat, where cycles and sentiment, gut feeling and statistics meld. (Tradewinds 2010)

9.2 The Problems of Implementing Governance and Policy Change in the Maritime Sector

Neither governance nor policy change is ever easy and yet it is change that is required for the globalized maritime industry. The governance of policy making and implementation is fundamental to its success and it is how policy is created and how it is put into effect that is far more significant than the policies selected. These will have much more success once the structural problems that face the industry are corrected.

Crosby (1996, pp. 1404–1405) addressed what he called the 'unpredictable nature of policy change' and how it is closely related to the task of successful implementation. Several features make implementation difficult:

- Commonly, the need to change policy comes from an outside stimulus as international bodies and other nation-states encourage, insist on direct policy renewal (Nelson 1989; Gordon 1994). This makes change often unwelcome and subject to indirect influences as change in one policy may be closely linked to changes in others (Haggard 1990; Callaghy 1990).
- Changing policy will also change relationships between stakeholders and the distribution of power and benefits. This will always be unpopular with some actors and generate resistance (Haggard 1995; Lindenberg and Ramirez 1989).
- While politicians negotiate, it is normally what Crosby terms 'technocrats' who implement. However, their criteria and standards may be very different. Thus,

politicians negotiate at the IMO and the European Commission (EC) while it is the industry itself that has to implement the policies agreed. This dichotomy is far from helpful. Technocrats may not be amenable to political tradeoffs; politicians unresponsive to technical/operational niceties.
- The hierarchical nature of policy making is very unresponsive. Kahler (1989) described it as a 'top-down, non-participative process confined to a narrow set of decision-makers' with little sense of ownership of the policy decisions.
- Those involved with substantial policy change are also likely to be new to the process or certainly peripheral to the 'establishment', making their proposals unpopular regardless of content, stimulus, or impact.
- New policy arrangements and proposals are never cheap. Costs are always incurred in implementing new arrangements and even if these can be argued away as there are savings elsewhere, the balance of expenditures and those incurring them would have changed. This makes for suspicion and resistance.
- There is always resistance to change from those entrenched regardless of the benefits that may follow. The established maritime governance is made up of a range of institutions and individuals with long-standing arrangements, procedures, relationships, and rewards. New governance structures and policy making and implementation processes will interrupt the existing 'comfortable' situation.

Crosby cited the theoretical 'linear model' of policy formulation, selection, and implementation to which current maritime policy-making subscribes but from which it deviates as and when it is convenient and significant stakeholders require. In practice, policy making is 'fragmented and open'. In one way policy implementers are commonly excluded from the process of policy formulation and selection—carried out by politicians advised by the industry—and the latter may feel it has little ownership of what is agreed (Ayee 1994; Mazmanian and Sabatier 1989; White 1990). One result is failure of those policies, with numerous examples from the maritime sector including EU policies for ports, liner shipping, places of refuge, a European flag, and various fiscal and financial measures, not least tonnage tax. The other is a tendency for the implementers (the shipping industry) to be selective in policy implementation, and active in policy evasion. Hence, the need for new governance, new stakeholder involvement, and new institutions—the latter emphasized by Sletmo (2002, p. 8) in his discussion of maritime policy and his claim that 'no policy can be better than the institution implementing it'. De Vivero et al. (2009, p. 623) bring us right up-to-date in their discussion of maritime policy and institutions. They suggest that maritime policy is growing into a significant feature of governance and consequently the state adjusts its politico-administrative structure; but as such it needs new institutions to accommodate ever widening objectives:

> Piecing together the maritime territory puzzle, connecting and linking geographical areas on different scales and with different features, sharing powers and competences between states and, in the states, between the various tiers of the administration, does seem to require an institutional complex in which the state, despite playing a key role, is not the sole actor, and neither is sufficient on its own for maritime space to be effectively governed.

9.2 The Problems of Implementing Governance and Policy Change

They consider that a reflexive or adaptive governance orientation (Folke et al. 2005) would seem more appropriate characterized by very limited technical control and a predominance of uncertainty and risk. Experiment in extended public participation and stakeholder responsibility would be 'opportune'; and a long-term, systemic approach should be adopted.

Stoker (1998, p. 21) viewed it as all about achieving legitimacy of power, without which governance must fail. He quoted Beetham (1991, p. 19):

> For power to be fully legitimate... three conditions are required; its conformity to established rules: the justification of the rules by reference to shared beliefs; and the express consent of the subordinate, or the most significant among them, to the particular relations of power.

Clearly, the rules of power can be made more or less legitimate and those that currently apply to maritime policy making tend more to the less than the more. A system can be designed to decrease or increase legitimacy and the current system is in Stoker's terms a 'myth'. What is needed is an effective move toward enhancing legitimacy.

Jordan (2001, p. 198) is less than convinced that this can ever be achieved, seeing the realism of domestic politics reducing international policy implementation at home to a 'transmission belt', translating 'international imperatives into state politics' (Evans et al. 1993). Hoffmann's (1995, p. 5) view was that domestic politics always lie at the base of international or supranational policies with states 'ferrying between the national and international sphere of policy-making' always keeping in mind their domestic agendas. While the nation-state retains control of global, international, and supranational policy making—as it does in the maritime sector—then the priorities of the nation-state will always come first and policy making will remain vulnerable to abuse by those trading off one state against another.

This was something that Puchala (1972, p. 275) had pointed out many years before in explaining the 'realist' case, supported by a number of others (Aron 1966; Hoffmann 1964; Morgenthau 1967). International policy making is seen as a 'process of mutual exploitation wherein governments attempt to mobilize and accumulate the resources of neighbouring states in the interest of enhancing their own power'. International organizations are created and tolerated simply to achieve international autonomy, military security, diplomatic influence, and heightened prestige for the nation-state. They are created solely to be used by national governments for their own self-interest. 'They are made at the convergent whims of those (national) governments and flounder or fossilize as their usefulness as instruments of foreign policy comes into question' (Hoffmann 1964, pp. 179, 219–230).

> What we are observing 'out there' and calling international integration are really international marriages of convenience, comfortable for all partners as long as self-interests are satisfied, but destined for divorce the moment any partner's interests are seriously frustrated. Hence, international integration drives not toward federalism or nationalism or functionalism, but toward disintegration. It never gets beyond the nation-state. (Puchala 1972, p. 276)

Unfortunately, so convinced are many that no alternatives exist, or perhaps they do not wish to look for any, that the question is never seriously asked whether 'international actors other than national governments may independently influence the allocation of international rewards' (Puchala 1972, p. 276). Nowhere is this more true than in shipping and nowhere is more advantage taken of the inequities and inefficiencies in policy that result. And to draw the whole discussion back to the debate of the Modernism epoch and its transmogrification into Postmodernism, Cooper and Burrell (1988, p. 92) remind us that Modernism was associated with the power of rational thought and the essential capacity of humanity to perfect itself with implications for the 'organization' (think policy maker) as a circumscribed, administrative-economic function. This contrasts with the Postmodern organization with its focus on discourse, and questioning and rejection of ethnographic rationalism. New institutions are a requirement of the new Postmodern epoch that can accommodate the changed environment in which shipping works, one no longer centred on the nation-state but truly globalized, with massively changed dynamics, ideals, and measures of success and failure. New organizations, new institutions, new policies, and a new governance to match the new world in which shipping operates. However, 'our analytic capabilities are rooted in methodological territorialism' (Scholte 2000, pp. 66–67) made less appealing with the 'ever-growing porosity of domestic–foreign boundaries' (Rosenau 2000, p. 1). Shipping provides an excellent example.

Brenner (1998, p. 463), citing Harvey (1982, p. 423) insisted that any improvement in governance would have to understand the multi-scalar notion of the spatial-fix, whereby jurisdictions, although commonly hierarchically modeled and forming the basis of current policy-making institutionalism, are actually not only artificial but also inappropriate. Policy-making institutions based upon jurisdictional hierarchies ignore the multiple and overlapping scales and the complex intermeshing of 'transnational corporations, monetary regimes, legal codes, interurban networks and state regulatory institutions' that actually exist in the maritime sector. Consequently policy is inadequate. The circulation of capital reflected in the shipping industry constantly creates tensions between 'concentration and dispersal, between local commitment and global concerns'. The intensification of globalization is the latest phase in capital's reconfiguration of itself, creating a Postmodern world where the notion of fixed jurisdictional hierarchies centered on the nation-state is nonsensical (Harvey 2000, p. 58).

9.3 So What Is to be Done?

Now the difficult bit. It is commonly easier to see that there is a need for change rather than how to change. However, in the case of maritime policy-making and governance things may be different in that there appears to be a marked resistance not to the changes themselves so much as to the need to make changes at all. The jurisdictional hierarchy centered upon the state, and dominated by traditional interests remains a comfortable and convenient structure for those central to it as well as those at whom policies are mainly directed—the shipping industry. To instigate and sustain a debate on governance change in these circumstances is difficult if not impossible.

Despite this there are a number of proposals that can be made which may help to direct change for the future, when and if this is allowed to occur. Remember, maritime policy is not only about commercial success and competition, but also about supporting society, inhibiting terrorism, providing for clean seas, and restricting illegal immigration, illegal drugs, money laundering, and death and injury. Surely we owe it to ourselves to ensure that policy making is as effective as it can be regardless of the wrench this might require in revamping the governance of the sector?

So what can be proposed?

9.3.1 Institutional Change

This is going to be unpopular. There are considerable vested interests in the governance of shipping remaining largely as it is now—take for example many shipowners who as we have seen already benefit from the chaos and confusion that currently characterizes shipping policy-making. Also, many shipping ancillary activities who through tradition or self-interest see no need for change or fear that change will threaten their comfortable existence (brokers, P and I Clubs, the legal and banking industries etc.); governments of nation-states who dominate the policy-making process—traditional maritime nations, Open Registry states; locations of major maritime institutions—New York, London, Paris, Geneva, Copenhagen, etc.; and of course those interests focussed upon the nation-state in more general terms who see a move toward a different institutional structure with a reconfigured jurisdictional emphasis as a threat. Lined up against change is all that encompasses the capitalist establishment—politicians, governments, business, and so on.

However, just because a structure for policy making has become institutionalized does not mean that fundamental change cannot nor should not occur. Remember that before the traditional, Modernist, maritime jurisdictional framework focussing upon the state was established centering on institutions such as the IMO, WTO, OECD, EU Commission, and the national ministry, there were other patterns of policy making. These may well have been less adequate but simply that change occurred to develop the modern-day policy-making structure suggests that change can occur again. Postmodern, globalized society is here; the twentieth century Modernist society has moved on and the need to revise the institutional structure for maritime governance to match Postmodern trends is clear.

Institutional change is supported from many directions. Peters and Pierre (2001, p. 133) were clear in their desire for change, linking policy change, and institutional change in both directions:

> Policy changes trigger or necessitate institutional changes and similarly, institutional changes frequently entail some degree of policy change.

They see the dual process of political integration (exemplified by the growth of size and influence of the EU) and decentralization to the local and regional levels (classically Postmodern), stimulated by globalization and these twin impacts as requiring institutions to change not only in structure and interrelationships from the formalized hierarchy of present but also to widen their dialog to include a much greater range of stakeholders.

Zurn (2003, p. 354) emphasized the need to develop institutions to reflect the moves toward governance beyond the nation-state as society becomes 'denationalized'. Although it is hard to disagree with his stress that 'the shape of more recent inter, trans and supranational institutions is hardly compatible with the traditional notion of state sovereignty in the national constellation' it is possible, in fact desirable, to go further and to suggest that perhaps governance based on jurisdictional institutions is irrelevant in the Postmodern, globalized world where a new configuration of more complex but relevant policy-making relationships needs to be found reflecting similar relationships in society.

Storper (1997, pp. 188–190) made many similar points relating in particular to territorially bound institutions—which of course characterize the shipping sector. Describing territorialization as declining or just simply low (for example with increasing globalization), 'locational substitution' becomes increasingly possible (again shipping but also many other activities with globalization). There occurs what he terms 'a race for the bottom'; effectively a competitive bidding war for economic activity—a frenetic activity of 'states and localities'.

Because major global business organizations (for example shipping companies) interact across many territorial economies and not just a single nation-state, there exists 'little harmony between the rules by which such firms intend to relate to these environments and the relational assets already built up in those places'. Policy-making institutional change becomes a necessity to reflect these new relationships.

Institutional change is necessitated by the changes that have occurred in the globalized, Postmodern world which makes policy-making institutions structured around formalized nation-state principles an anachronism. Consequently, not only might the established institutions of the UN, WTO, EC, and the like have to change the way they work, the relationship they have between jurisdictional levels and the formulation and exercise of their power, they might even have to contemplate the unimaginable—giving up maritime governance for a new mechanism entirely. This new structure remains unclear but this is no reason to avoid debate, especially as effective maritime governance remains elusive under the current system.

9.3.2 Stakeholders

The relationship between effective governance and policy making is at the same time both clear and confused—clear because it is seemingly obvious that to create effective policies you need to include those most affected by them in drawing them up. Only then can ownership be achieved and if something is not owned then it

9.3 So What Is to be Done?

is not uncommon that is also unloved. Confusing because deciding who is a stakeholder, what role they play, and whom to include is never straightforward. While shipowners, freight forwarders, and port operators are clearly maritime stakeholders, what about 3PL organizers, maritime consultancies, and the owners of Lloyd's List? And then there is always the general public, environmental pressure groups, and the broadcast media; politicians, banks, lawyers, and virtually every industry in the world.

The literature on stakeholders is vast and we shall make little attempt here to be comprehensive. Instead we shall concentrate upon recent attempts in the maritime sector to incorporate more stakeholder interests in policy making before going on to see how current maritime governance has adapted to the increased recognition of the significance of the stakeholder in the generation of policies.

Examples of the stakeholder literature abound across all disciplines and even includes the maritime sector (for example Notteboom 2002). Freeman (1984) related stakeholders to effective strategic management, Goodpaster (1991) assessed the relevance of stakeholder analysis to business ethics, while Gordenker and Weiss (1995) indicated the significance of NGOs as stakeholders at the UN particularly in the context of international relations. Donaldson and Preston (1995) examined corporations and the importance of stakeholders, Dicken et al. (2001, p. 91) suggested that it was vital to try and include all stakeholders when analyzing the global economy, and Hemmati et al. (2002) in discussing governance generally, and Hess (2008, p. 455) considering value chains and networks, both included stakeholders as central features of their analysis.

Theoretical studies relating to stakeholders are even more extensive. Examples include Charan and Freeman (1979) examining the issues of stakeholder negotiation; Savage (1981) who assessed strategies for the management of organizational stakeholders; Hill and Jones (1992) who looked at stakeholder agency theory; Altman and Petkus Jr. (1994, pp. 38–40) who developed a network-based stakeholder approach to public policy-making; Clarkson (1995) and social performance; Mitchell et al. (1997, pp. 885–862, 874) who provided an extensive discussion of the identification of stakeholders, the historical development of stakeholder theory and a typology of stakeholders; Harrison and Freeman (1999) who considered stakeholders in the context of social responsibility; Gioia (1999) who assessed the problems of stakeholder theorizing; Fletcher et al. (2003) who suggested strategies for stakeholder mapping; and Donaldson (1999) and Friedman and Miles (2002) who each provided extensive theoretical backgrounds. Other commentaries come from Charan and Freeman (1979) discussing stakeholder negotiations; Castells (2000a, pp. 12–13) examining interactions with policy making and the media; and Borzel (2007, p. 5) who looked at the role of 'actors' in governance. This latter issue was also taken up by Bennett (2000, p. 879):

> There is increasing acknowledgement that 'governance has been viewed primarily as intergovernmental relationships, but it must now be understood as also involving non-governmental organizations (NGOs), citizen's movements, multinational corporations, and the global capital market'. (Commission on Global Governance, Commission on Global Governance 1995, pp. 2–3)

The significance of stakeholders in our discussion of governance in the maritime sector and the central role of the changing nation-state is emphasized by the existing work on the relationships between these areas. Examples come from Picciotto (1998, p. 3) who suggested like many others that the state had become significantly more fragmented representing what he called a move from government to governance. The consequences have been a shift from 'direct government economic intervention and management' to 'public semi-autonomous bodies' operating as market regulators. These new regulators now formed the new public sphere composed of the stakeholders who constitute each respective sector—banks, police, social services, utilities, and the like—and in the maritime sector such bodies as shipbrokers, port authorities, P and I clubs, lawyers, freight forwarders, and logisticians. Thus, stakeholders have tended to replace politicians and government employees as the process of governance has developed and refined.

MaCleod and Goodwin (1999, pp. 506–507) confirmed this trend suggesting that the drift from government to governance was clear and unstoppable (Wolch 1989) and was reflected in the state's relative decline in 'direct management and sponsorship of social and economic projects' and what they called an 'analogous engagement of quasi and non-state actors in a range of public–private partnerships and networks'. Stone (1989, p. 3) agreed:

> What makes governance… effective is not the formal machinery of government but rather the informal partnership between City Hall and the downtown business elite. This informal partnership and the way it operates constitutes the city's regime; it is the means through which major policy decisions are made.

Jessop (1997, p. 574) also confirmed this process of moving from government to governance with evidence of a decline in state sponsorship and political hegemony to partnerships between 'governmental, para-governmental and non-governmental organizations in which the state apparatus is often only first amongst equals'. Governments have always relied on outside agencies to realize state objectives but that this reliance has been 'reordered and increased'. The state of course can also benefit through this by enhancing 'their capacity to project state power and achieve state objectives by mobilizing knowledge and power resources from influential non-governmental partners or stakeholders' (Jessop 1997, p. 575).

The role of stakeholders in governance has been taken up in earnest by the EU which has recognized at least in principle, how important they are to the creation of effective policy. The Commission's 2001 White Paper on Governance (Commission of the European Communities 2001, pp. 14–15), further developed through their 2009 White Paper (Commission of the European Communities 2009, pp. 3 and 6), emphasized the importance of societal contribution to policy making and the involvement of those most directly affected while their 2008 White Paper on Maritime Policy stressed the need for a 'stakeholder consultation structure' feeding opinion and expertise into the development of maritime policy and 'allowing the exchange of best practices' (Commission of the European Communities 2008, p. 6).

9.3 So What Is to be Done?

An appreciation of the significance of stakeholders in maritime sector policy-making has grown in recent years reflecting the need to incorporate them more in the process of governance (see for example Aspinwall 1995 and Bennett 2000, p. 893). Sutherland and Nichols (2006, p. 6) stressed that governance of marine spaces was actually the 'management of stakeholder relationships' and that good governance would only come with the 'recognition of the interests of all stakeholders'. Central to this was stakeholder identification, engagement, and management of inputs. Pomeroy and Douvere (2008) discussed in some detail the application of stakeholder theory to marine spatial planning including the value of stakeholders to the policy-making process, how to define stakeholders and their relative significance, and how to empower them. However, in most cases the definition of a maritime stakeholder remains conservative and this should stimulate a reappraisal of what is meant by the term, whom else should be included (and whom not) and in what way.

Pallis concentrated upon the role of Maritime Interest Groups, their representation, structures, and influence on policy making. Pallis (2005–2006) and Pallis and Tsiotsis (2006) identified 37 maritime interest groups made up of business associations (82 %), trades unions (10 %), and regional interests (8 %). They considered the input of port interest groups in the European Union when the Port Services Directive was proposed in 2005–2006 identifying nine groups with direct port interests. This was not claimed to be comprehensive but represented those who had expressed interest during the consultation phase. Once again only those with direct (rather than indirect interests) were represented.

Notable is the large number of interest groups that exist in the maritime sector—and possibly the significance of their input to maritime policy-making as a result; and that here we are keeping the definition of a stakeholder very narrow—those with direct, close, and well-defined interests. Current policy making in the EU and at other jurisdictional levels already consults with these organizations and it is these who tend to respond to calls expressing interest when maritime policies are being discussed. However, it excludes all the indirect stakeholders upon which maritime policy impinges to a greater or lesser degree—interest groups, other sectors of the economy, individuals, politicians, media, and so on.

Notteboom and Winkelmans (2002) had earlier considered the significance of stakeholders to ports policy, recognizing the importance they should have in the policy-making process. Here, they did not confine their definition to only those with clear and direct interests but extended the discussion somewhat further to include community groups and environmentalists, economic and contractual stakeholders, those related to public policy, and those representing the community. Stakeholders were expected to vary according to the issue, the port, the purpose of the policy making, and much more. In all cases ports were viewed as characterized by multiple stakeholders. This was further taken up by Wang and Slack (2002), Wang et al. (2004, p. 238), Dooms et al. (2004), and Coeck and Dooms (2007).

De Langen (2007, pp. 459–460), along with Brett and Roe (2010), outlined the importance of stakeholders to port clusters including rather wider interests than those indicated by Pallis. These covered importers and exporters, transport service

Table 9.1 Influential actors in the globalization of regulation

Community	Examples of key actors
Organizations of states	IMO, EU
States	UK, US, Japan
International business organizations	IACS, ICS, ICC
National business organizations	None
Corporations	Lloyd's register of shipping, classification societies, insurers
International NGOs	IMO, friends of the earth, international confederation of free trade unions
National NGOs	None
Mass publics	Yes—catalyzed by the titanic, torrey canyon, and herald of free enterprise disasters
Individuals	Edward Lloyd, Justinian
Epistemic communities of actors	Strong

Source Braithwaite and Drahos (2000, pp. 476–477)

providers, environmental groups, regional government, manufacturing industry, and port labor. However, many more might also have been included if maritime governance was to take full account of stakeholder interests. Brooks and Pallis (2008, pp. 413 and 419) took up this message citing De Langen (2007) as well as Wang and Slack (2002) to suggest that port governance required consideration of three axes—spatial-jurisdictional context, logistical capabilities, and not least the stakeholder community which they saw as including much more than the primary customers of the port.

As part of a much wider discussion, Braithwaite and Drahos (2000, pp. 476–477) outlined what they saw as the significant stakeholders in the globalization of maritime policy-making, divided between various definitions of the community (Table 9.1).

Furger (1997, pp. 446, 449, and 453) meanwhile saw stakeholders as intermediaries in the policy-making process and noted that 'institutionalized distinction of regulators and regulatees makes government agencies blind to a large number of institutions which cannot be equated to regulators or regulatees but which have nevertheless an impact on safety and environmental protection as significant as traditional regulations'. Or in plain English, policy makers are trapped by inertia into considering only the 'regular' stakeholders they always have. New stakeholders are considered 'an avoidable nuisance rather than a social resource' and this is despite the fact that the maritime sector is largely shaped by them. Sletmo (2002, p. 3) meanwhile recognized the range of stakeholders in shipping policy suggesting this had to encompass 'national and international agencies as well as national and global shipping corporations, their customers, their owners and labour'. He went on to emphasize that 'certain aspects of policy are strongly promoted by or supported by groups far removed from the business of shipping and maritime transport'.

Sletmo's view is far from all embracing and in many ways reflects that adopted by the current range of policy makers including the EU. The contention here is that meaningful maritime policy-making will need to be far more comprehensive encompassing a much wider range of those affected if it is to start to avoid the failures in policy that we have identified.

An example of what can be done is provided by Hosseus and Pal (1997) who analyzed policies and policy instruments for the Canadian shipping industry and in so doing implied a much wider range of stakeholders who were involved. Some 473 shipping topics were derived, most of which could be associated with identifiable stakeholders. These were subsequently consolidated into more limited areas shown in Table 9.2 which again reflected a very wide range of those influenced and affected by maritime policy-making—far wider than conventionally consulted. The unconventional expression of some of the topics was related to the content analysis approach that was utilized but despite this, the implications for stakeholder choice were clear.

Stakeholder issues are topical in the current debate about governance and much has been said about widening and deepening participation. However, the truth is that whilst recognition of the need to involve stakeholders in the maritime policy-making process would help to reduce the severity and extent of maritime governance failure, it is not enough. Far more needs to be done to involve those both directly and indirectly affected by maritime activities—the media, politicians, electorates of all sorts and jurisdictions, individuals, interest groups—rather than the lip service paid to stakeholder involvement by current processes of consultation with the traditional institutions and disciplines.

9.3.3 Speed

Karl Marx was convinced;

> Thus the creation of the physical conditions of exchange – of the means of communication and transport – the annihilation of space by time – becomes an extraordinary necessity for it. (Marx 1992, pp. 349 and 524).

Santos (1995, p. 71) viewed accelerations as 'culminating moments in history' which 'concentrate powers that explode to create something new' and he cited Michelet (1833) in stressing the significance of time and acceleration in the course of history. He suggested that the first reaction to acceleration is to adore the 'underlying velocity' and noted the emergence of railways and steam boats to support this. We may not be so interested in the emotional reaction to speed and acceleration, but its importance more generally is clear. Meanwhile Nielsen and Oldrup (2001) stressed the strong relationship between time, mobility and speed in the transport sector through their editing of a series of related papers, which included in particular the contribution by Nielsen and Jespersen (2001).

Table 9.2 Shipping policy topics

Abandoning of	Department	Industries	Passenger lists	Shipping
Act	Deregulation	Information	Periodicals	Shuttles
Aeronautics	Developing countries	Information system	Pilots	Signals and signaling
Aids to navigation	Disadvantaged	Insulation	Pneumatic equipment	Simulators
Allowance	Division	Interior	Services	Sound-proofing
And pier	Domain	Iron and steel	Price	Speed
Astronautics	Earth stations	Legislation	Producing	Sponsored
Automation	Education	Liability	Productivity	State aid
Automotive	Efficiency	Lighthouses	Navigation	Storage and moving
Berth	Electronic equipment	Load line	Radar in navigation	Stranding of ships
Builders	Emergency employee	Losses	Rail transport	Studies
Buoys	Engineering	Maintenance and repair	Rates	Study and teaching
Careers	Environment	Maneuvrability	Rat-proof construction	Subsidies
Cathodic Protection	Facilities	Manning	Reform	Surveys
Characteristics	Fenders	Maneuvring	Registration	Survivability
Classification	Financing	Maritime	Regulation	Taxation
Coastwise shipping	Fisheries navigation	Mathematical models	Repair facilities	Techniques
Collisions at sea	Fouling	Medicine	Replacement	Tele communications
Common Carriers	Fuel	Models	Routes	Test centers—US
Conferences	Fumigation	Movement	Routing and scheduling	To transit
Consolidation	Government aid	Nautical paraphernalia	Classification rules	Transfers to foreigners
Containerization	Guides	Network	Rural transportation	Transit time Economics
Contracts, maritime	Hazards	Of disabled persons	Safety	Transportation
Convoys	Highway	Officials	Sailing cards	Trucking
Corporations	Hulls	Operation	Satellites	Underwater navigation
Costs	Hydrography	Operators	Scrapping	Unitized cargo
Data Processing	Hydro statics	Optimum ship routing	Sector	User fees
Decks	Impact forces	Painting	Ship handling	Vehicles

Source Hosseus and Pal (1997, p. 411)

9.3 So What Is to be Done?

Much has been written in the past 30 years about the notion of speed and its relationship to society. This has extended to debate about acceleration and its role with respect to flows, which are considered in the following section. An appreciation of how speed, acceleration, flows and governance interact centers upon the evidence of a new societal era that we have conveniently termed Postmodern, the emergence of an intensely globalized world community, and how the characteristics of the maritime sector and the policies that it brings are largely determined by the relationship between these stimuli.

One of the earliest commentators on the significance of speed was Porter (1968, pp. 5–6) who suggested that the second industrial revolution had occurred based on what Michael (1962) termed the 'silent conquest of cybernation'. This Postmodern cybernated society had been accompanied by enormous changes both in the means and speed of travel. Although our consideration of the Postmodern places it somewhat later than Michael, this may just reflect our current knowledge now of the substantial changes that have occurred, whereas Michael had the vision to see them as they began. Certainly changes in speed—in their effect and more specifically acceleration—have been central to the new era that has emerged and which places such stress upon maritime governance.

Stalder (2006, p. 155) focused upon the relationship between Modernist time—formal, mechanized, predictable and globally defined—and that of the Postmodern era where suddenly the characteristics of time have been altered so that virtual instantaneity has replaced the traditional temporal concepts that applied to Modernist markets. He defined multiple times dependent upon the context, user, place and so on for which hierarchies are very inappropriate.

> Rigid hierarchies are too inflexible to deal with multiple temporalities; their historical rise to dominance, from the seventeenth century onward, was connected with the imposition of a dominant temporality, clock time, on everyone. Flexible (Postmodern) networks, by their very flexibility, are not capable of doing this, and nor do they require it. (Stalder 2006, p. 158)

Crogan contributed significantly to the debate about speed, society and the new era and along with Brugger (1999) commented on the work of Virilio (for example 1995, 1999) who has been foremost in emphasizing the relationships that existed.

> all aspects of human experience are increasingly determined by the dominance of what (Virilio) terms 'logistics' over politics, culture and society – a total reorientation of economic activity and social/political organization which has occurred in conjunction with the development of mass and total war in the last two hundred years, culminating in the post-war period. (Crogan 1999a, p. 142)

Whilst we may not agree with the time period—seeing the Postmodern era emerging from the Modern in the twentieth century, and the reference to war is a distraction (although its significance and accuracy may well be undeniable)—the drift of Crogan's (and Virilio's) argument is clear. Crogan emphasized Virilio's identification of three types of speed in modern society which we can suggest underlies the Postmodern, globalized society—nomadic or revolutionary speed (riot and guerrilla warfare); state appropriated and regulated speed (management of public ways); and global, total warfare, planetary over-armament speed (nuclear strategy) (Crogan 1999a, p. 142; Deleuze and Guattari 1988, pp. 137–138). Speed was seen as a fundamental component of logistics in which a 'nation's potential is transferred to its armed forces' resulting in the decline of the significance of the nation. Crogan (1999b, p. 164) went on quoting Derrida and his consideration of the continuing growth in significance that speed to the new society at that time was just becoming apparent:

> Are we having today, another, a different experience of speed? Is our relation to time and to motion qualitatively different? Or must we speak prudently of an extraordinary—although qualitatively homogenous—acceleration of the same experience? (Derrida 1984, p. 20)

Agnew (1994, p. 72) suggested that the continued acceleration of society could be interpreted as stimulating the 'disappearance of space' with wealth untied to territory and nation-states consequently undermined or at least their pecking order rejigged to reflect changed status. Beckmann (1999a, b, 2004, p. 98) summarized much of the debate on speed and in more detail discussed its Postmodern relationships (2004, p. 88) commenting on Bauman's (2000, p. 11) view of mobility as a function of 'exitability'. Individuals' tendency toward increasing 'escape velocity' suggests that speed and mobility 'enables absenteeism just as much as it permits proximity'. Increasing velocity reflects escape, slippage, elision and avoidance. This 'capacity to disengage, withdraw and move away is the privilege of the motile hybrid'. Think Postmodern shipping.

Hassan (1999, pp. 1–2) noted that the present day institutions are unsuited to the new 'neoliberalised, high-speed network society' situated within a 'hierarchy of speed' where time is seen as social and not absolute, rejecting the assumptions of Newtonian physics that space and time were containers, and abstract and absolute forces of nature. Time forms an essential part of societal change the most significant recently being that of globalization. The new society is 'predicated upon acceleration (and) the flexibility of economy and society' (Hassan 2009, p. 11).

Armitage and Graham (2001, pp. 113–116) emphasized the links between the globalized society, speed, mobility and wealth. Economic growth in contemporary capitalism depends upon increasing rates of the processes of production which includes the movement of materials and the exchange of information, or in their words, 'trade is dependent on the overproduction of speed' (115). Similarly, 'what is required, above all, is recognition of the centrality of speed in contemporary societies' (121). Shipping is no exception. Capitalism demands ever-increasing production (what Armitage and Graham termed 'over-production') and this requires 'ever-more efficient use of fractured, punctuated and rigidly organized social time—seconds, hours, days, months or years'. Today's globalized economy is centred upon space and time and especially the 'increased efficiencies of time, acceleration, (and) increased rates of increasing speed. 'In trade, acceleration is sought to reduce production, consumption and circulation time' (Armitage and Graham 2001, p. 116).

9.3 So What Is to be Done?

In recent years the relationship between speed, acceleration, globalization and society has centred around the work of Virilio and his numerous contributions to the debate (in particular Virilio 1977, 1995, 1999). Commentators on Virilio abound (see for example Benko 1997, p. 24; Dickens and Ormrod 2006, pp. 61, 89–90; Armitage 1999a, b, c, 2000a, b). There has been a substantial number of other publications widely ranging in opinion but throughout there has been a degree of consistency in accepting that the Postmodern era has a close relationship to the increased emphasis on speed and acceleration in society. In the case of the maritime sector this is clearly manifested through the sector's relationship with globalization, and the need to understand all that it brings to the policy-making process and the structure of governance. Examples of related work include Der Derian (1990, 1999), Conley (1999), Crogan (1999a), Cubitt (1999), Gane (1999), Kellner (1999), Leach (1999), McQuire (1999), and Zurbrugg (1999) all of which featured in a special edition of *Theory, Culture and Society* which focussed on Virilio's ideas. Additional commentary can be found in Der Derian (1992, 1999), Gilfedder (1994), and Wark (1988).

There is no doubt that Virilio has many supporters in his views of the importance of speed to the Postmodern era. Take Der Derian (1999, p. 215):

> A single Virilio sentence, full of concatenated clauses and asyndetic phrases, can collapse a century of political thought, dismantle a foundation of scientific absolutes. His take on a deterritorialized, accelerated, hyper-mediated world redefines *outlandish*. Nonetheless, when shit happens—events that defy conventional language, fit no familiar pattern, follow no conception of causality—I reach for Virilio's conceptual cosmology.

And as in Der Derian's words, shit has happened in the maritime sector in the form of a new Postmodern era that has changed all the rules, regulations, and relationships upon which maritime governance is founded. Der Derian went on to provide a serious commentary on the contribution Virilio has made to the debate on speed, time, acceleration and the Postmodern. Speed is seen as shrinking the globe and the Postmodern effect is one derived from acceleration, causing mental confusion of 'near and far, present and future, real and unreal' which affects all society. Governance and policy-making can no longer hold on to the historical tradition of the known, the fixed, the close and the well-tried but must now in a sense become more 'unreal' and recognize that the acceleration of speed has generated a totally new scenario where time and space are one, where the meaning of domesticity and the nation-state is unclear, and where there is little that is fixed, immovable, inflexible or guaranteed. Where the individual can be as influential as the corporation; where communication is not reserved to those within a sector but is extended to influences far beyond; where the media can both create reality and substitute one for another almost at will. Virilio saw all these trends in modern society and we can see how they affect policy-making and governance in all sectors, not least shipping.

> Real time now prevails above both real space and the geosphere. The primacy of real time, of immediacy, over and above space and surface is a fait accompli and has inaugural value (ushers a new epoch). (Virilio 1995, p. 1)

That new epoch is the Postmodern and it requires a new epoch for maritime governance.

Despite this Virilio is not wholly convinced that the Postmodern represents anything different from the Modern. Armitage (1999a, pp. 6, 8, 11) suggested that Virilio contributes significantly to critical cultural and social theory by concentrating on the relationship between military space, territorial organization, dromology ,and the esthetics of disappearance and dismissing what he termed the sterile debate over Modernism and Postmodernism. Virilio is certainly not a self-defined Postmodernist, something claimed for him by Harvey (1989, p. 351), Waite (1996, p. 116), Sokal and Bricmont (1998, pp. 159–166), and Gibbins and Reimer (1999, p. 143). Virilio in fact consistently refers to Modernist writers such as Kafka and Aldous Huxley and Modernist artists such as Marinetti and Duchamp. He wrote with optimism of key Modernist features:

> it's the global dimensions of the twentieth century that interest me—both the absolute speed and power of the twentieth century's telecommunications, nuclear energy and so on, and at the same time the absolute catastrophe of this same energy! (Virilio 1998, p. 2)

Far from making his work of less relevance his preoccupation with speed and acceleration remains highly pertinent and reflects the changes that have taken place in society and which impact upon maritime governance.

The change that we have seen occurring in Postmodernism and the manifestation of the next era of capitalism is something Virilio has long associated with speed and its position at the 'heart of the organization and transformation of the contemporary world' (Armitage 1999a, p. 1). He identified 'productive interruptions', 'jumps', and 'creative dynamics'.

Jumping to our final section, Dickens and Ormrod (2006, p. 11) made the link between time-space compression and the next phase of capitalism, beyond the Postmodern. This sees the expansion of capitalism to new and even more distant territories, facilitated by the ever-increasing power of speed over distance, territory, and space. While currently new spatial fixes as envisaged by Harvey (2006) are being made in Japan, Brazil, Russia, and China, encouraged by new technology, this same new technology is also beginning to open up extraterritory, outside the established and traditional global marketplace in the form of space-based infrastructure (satellites, shuttles etc.) and undoubtedly, eventually capital investment on other planets and moons. At this point (soon to occur) capitalism's new territorial boundaries are almost unlimited—perhaps only by the human imagination—and the role of speed and its accumulation of space is central. More of this in the final section after a consideration of the importance of 'flows'.

9.3.4 Flows

Luke (1991, pp. 320–321) was an early advocate of the notion that society was becoming less spatially fixated and increasingly more process and even more

9.3 So What Is to be Done?

specifically flow orientated. A central feature of the Postmodern era has been the move away from territory and space (characterized by the Westphalian notions of the nation-state) defining what is possible and toward a refocus toward the movement of information, data, capital, people, and products over increasing distances and in shorter periods of time. The process of globalization has required and in fact only been possible because of space–time compression is itself a notion that is centered upon process rather than object. Flow rather than space.

Luke suggested that:

> Moving from place to flow, spaces to streams, introduces non-perspectival, anti-hierarchical and dis-organizational elements into traditional spatial/industrial/national notions of sovereignty.

This makes the traditional Westphalian state outdated in the new globalized society and the characteristics of maritime governance—hierarchy, organization, nation-state directed—of increasingly less relevance. Modern governance needs to be adapted to accept movement, flow, stream, and change rather than a 'one size fits all', rigid structure to policy making that cannot accommodate the 'globalized world. To quote Luke again:

> The ethnographic settings of self-rule defined by the classical Westphalian universe of borders, shorelines and airspaces in spatially construed grids of/for sovereignty increasingly collide in the transnational universe of technoregions generated out of global monetary transactions, commodity exchanges, technical commerce, telecommunication links, and media markets.

He saw the unconstrained access by policy makers to flows, rather than the existing closed domination of place, as a crucial attribute. The 'reality of place, expressed in terms of a sociocultural context of spatial location, gradually is being resituated within the hyperreality of flow'. Cooke (1990, p. 141) agreed stressing that globalized society has no center, rather it is a decentered space of flows, while Blatter (2001, pp. 176, 178) also emphasized how the traditional Westphalian notion of nation-state policy making is being eroded by flows of information, capital, services, goods, and people.

The space of flows is a widely debated concept whose origins can be found largely in the work of Castells (1989, pp. 348–353, 1996, pp. 378–478, 2000a, pp. 13–14) who has long been the main proponent of what he terms the 'network society'. We shall not debate the concept here but note its widespread acceptance as a characteristic of the new, globalized (and what we have termed Postmodern) society. For further discussion see for example Stalder (2006, pp. 46, 152–154) who considered the historical development of Castells' concepts; Friedman (2000, p. 113), who debated the relationship of flows to governance; Hassan (2009, p. 11) who saw the rise of digital capitalism creating a new societal morphology that has no relationship to space; Allen (2003, pp. 60–64) who placed the flow of power into a geographical context; Dicken et al. (2001); Yeung (2000, p. 201) who examined the development of network flows in Hong Kong business; Webster (2002, pp. 97–123) and Soja (2000, pp. 212–216) who provided extensive analyses; Watts (1991); and Taylor (2000c, p. 161) who examined the relationship between flows, states, and cities.

Castells also stressed the significance of a wide range of interactive stakeholders (2000a, p. 12), the role of the media (2000a, p. 13), the importance of networks in replacing hierarchies (2000a, p. 12, 19), and how all this relates to continued globalization. The rise of flows and networks is not a short-term or temporary thing. Its relevance to the debate on the new maritime governance is clear.

Taylor (2005, p. 705), along with Stalder (2006, p. 10), highlighted the link between globalization and Castells' notions of flows suggesting that nation-states increasingly are being undermined and replaced by networks of cities. However, rather than seeing space and flow as alternatives he emphasized that the two work together with nation-state responsibilities exercised through processes and flows with consequential impact upon their effectiveness, power, and relationships. Earlier Castells (2000b, p. 1111) had stressed how our global conceptions were dominated by the 'mosaic of states' forming the 'global political map through which we 'view world spaces'. As (Lewis and Wigen 1997, p. 9) noted, it is a 'key geography… through which people order their knowledge of the world'.Taylor (2000b, p. 1111) stressed that these uncriticized and unexamined meta-geographies were taken for granted and consequently 'ripe for radical reconsideration at a time of global transformation'. As Arrighi (1994, p. 81) stated 'deficiencies in our perceptual habits—causing non-territorial spaces-of-flows—have gone unnoticed alongside the national spaces-of-places throughout the history of the modern world system'.

Meanwhile, Storper (1997, p. 170) agreed globalization was making 'contemporary economies… placeless, mere flows of resources'. He also saw no future for the nation-state within these flows with the:

> locus of control over important dimensions of the economic development process… passing from territorialized institutions such as states to deterritorialized institutions such as intrafirm, international corporate hierarchies or international markets that no know no bounds.

Storper (1997, pp. 177–178, 182) went on to discuss the issue of flows at some depth. The globalized supply chain would see resources flowing within and between companies and markets with no dependence on any particular place creating what he called a 'flow economy'. This is where a 'location offer (*sic*) only those factors of production that could potentially be substituted by a large number of other locations'. Such conditions which he described as 'non-specific, locationally substitutable and perfectly elastic' are increasingly close to reality. Shipping presents one very specific case which exhibits these features more than any other. The result is a globalized sector, characterized more by flows rather than spaces, which remains caught in a governance time-warp, with policy making allowed to remain dominated by a Modernist, state-centric, hierarchical institutional framework that was designed for territorialized economic activity and wholly inappropriate for the purpose. Table 9.3 is one interpretation of the flow/territory dichotomy. In particular, Cell 1 is represented by firms with high territoriality and high international flows—something reminiscent of the shipping

9.3 So What Is to be Done?

Table 9.3 Flows and territories

	HIGH	LOW
HIGH	1 Intrafirm trade with asset specificities. Intermediate inputs of FDI. International markets served from territorial cores. Industrial districts. Inter-firm and inter-industry trade.	2 International divisions of labor (e.g in routinized manufacturing). International markets (e.g in consumer services) Interfirm and interindustry trade without territorial core
LOW	3 Locally serving production to specialized tastes with low international competition.	4 Local commerce in basic services not delivered via big-firm hierarchies.

Source Storper (1997, p. 192)

industry before the onset of serious globalization and supported by the institutional policy-making framework we have today. Cell 2 meanwhile is represented by low levels of territorialization yet high international flows—very much the modern shipping industry—but which lacks a policy-making structure to support it.

Friedmann (2000, p. 113), quoting Castells (1997, p. 349) took up this theme in relating Castells' notion of flows and the need to accommodate this into governance.

> The space of flows... dominates the space of places of people's cultures. Timeless time as the social tendency toward the annihilation of time by technology supersedes the clocktime logic of the industrial era (*Postmodern governance supersedes Modern governance*). Capital circulates, power rules and electronic communication swirls through flows of exchanges between selected, distant locales, while fragmented experience remains confined to spaces. Technology compresses time to a few, randomized instants, thus de-sequencing society and de-historicizing history. By secluding power in the space of flows, allowing capital to escape from time, and dissolving history in the culture of the ephemeral, the network society disembodies social relationships, introducing the culture of real virtuality. (*section in italics added*)

Modern governance needs to accept that capital is constantly attempting to 'escape from time' and that globalization increasingly allows this to happen. As a central feature of globalization, shipping governance should be designed accordingly. Taylor (2005, p. 706) agreed but saw it as a complex interrelationship of three levels of flows which represent three different networks generated by three different agents. The three networks are what might be termed the suprastate network, the interstate network, and the ultra-state network:

> The first is a network of flows between states; it represents the contemporary operation of the Westphalia process through cities. The second is a network of flows above states: it represents an interpretation of globalization processes as an increasing geographical scale of operation. The third is a network of flows across and beyond states: it represents an interpretation of globalization processes as transcending states and their boundaries.

Dickens and Ormrod (2007, pp. 105–107) also stressed the network significance of the move from an emphasis of place to one of flow while assessing the moves by capital to continue expansion beyond the conventional terrestrial limits of globalization to that of the cosmos.

The Postmodern credentials of the space of flows are well documented. Castells (1989, pp. 16–17) was direct about the close relationship seeing flexibility as a key part of the move toward flows, processes and change in production, consumption, and management. Under the Postmodern arrangements, the advantages of economies of scale and depth of organizational power are retained while divesting the rigidity and difficulties of adapting to the new environment that comes with traditional Modernist models. Thus large-scale shipping activities can be pursued in a framework of flexibility with few constraints that allows it to take advantage of the Postmodern market and the laxity of the existing policy-making structures. Waterman (1999, p. 358) concurred seeing a close relationship between Castells' ideas, globalization, and the Postmodern evidenced throughout much of the literature (for example Beck 1992; Giddens 1990; Hall et al. 1992; Harvey 1989, 1996; Melucci 1989; Poster 1984, 1990; and De Sousa Santos 1995).

9.4 Cosmic Capitalism and the Outer Spatial Fix: The Face of Shipping Future?

So where does capitalism go from here? It is beginning to exhaust the resources of this planet with few locations left to exploit. Shipping has done its bit in opening almost everywhere to everything and along with modern communications the process of globalization has ensured that the capitalist structure has the opportunities to generate and dispose of capital in almost endless ways. Shipping has been fundamental to this and is even active in exposing the last outposts of the capitalist desert to the desires and addictions of modern society—the Antarctic and the Arctic Oceans and the shipping routes north of Russia and Canada as classic examples. Where next is the cosmos.

Far-fetched as this might sound it is far from it and the debate on extraterrestial change agents is intensifying (see for example Pelton (2004) and Pelton et al. (2004)). Although it is beyond the scope of this book the expansion of capitalism into the limitless space that surrounds this planet is going on already—and in the process capital can be generated through the exploitation of new markets, territory, and resources. Maybe even the exploitation of new labor if some time in the future a colony of Martians (or Venusians... etc.) who are happy to undercut the Filipino, Vietnamese, or even North Korean seafarer, is discovered. You can bet that the experienced shipowner from the planet Earth will not miss the chance to employ them at lower rates, in poorer conditions, and less well trained if the opportunity affords itself. And think of flagging? Offshore registers might become off-planet registers with names such as Mars, Venus, Jupiter, and beyond becoming ever more common. With the poverty of rules, regulations, and legislation to control the Earth-like activities, the complete absence of the nation-state, and the perpetuation of state-structured Earthly extraterrestrial governance institutions (e.g the UMO), the potential of the cosmos for the maritime sector is mind-blowing.

9.4 Cosmic Capitalism and the Outer Spatial Fix: The Face of Shipping Future?

Examples of those already commenting on the capitalist market in the cosmos center around Dickens and Ormrod (2006, 2007), Dickens (2009), Parker (2009), and Parker and Bell (2009a, b). In particular, Dickens and Ormrod (2007, pp. 49–67) emphasized how Harvey's (1989, 1990) conception of the crisis of capitalism—in needing a constant fix of new sources of capital and its disposal and hence the rush toward globalization—can be extended to territory beyond the planet Earth and what they termed the 'humanization of the universe'. they called this the 'outer spatial fix' whereby capitalism searches for new markets and sources of resources (labor, land, materials, etc.) extraterrestrially. While shipping may find itself a little constrained by the need for oceans, it is certainly not beyond imagining the use of extraterrestrial land forming the administrative basis for new flags, legal regimes, and methods of avoiding terrestrial regulations and constraints—something that might well appeal to many a shipowner. Thus, cosmic capitalism might well center around the use of outer space to facilitate the capitalist dream on the Earth. And the consequence—is the need for cosmic governance to administer the generation and application of cosmic legislation and policy that affects the Earthly activity. Given that we at present have ineffective terrestrial governance for the shipping industry, the prognosis is not good. But perhaps that is for the next book.

9.5 Some Final Thoughts

We have come a long way from Van Loon's 'little mound' that we contemplated at the very outset. Our travels through globalization, governance, Modernism, and the Postmodern have reflected the enormous complexity of policy making especially in a sector that has such close interrelationships with other sectors, across jurisdictions, continents, disciplines, and social networks. Shipping is a highly global, highly political, human activity that makes attempts to control and maneuvre its activities by nation-states, governments, institutions often impractical and always difficult. The current maritime policy-making and governance system is inadequate for this purpose despite its lofty ambitions, extensive good will, and the determination and dedication of many to succeed. New approaches are needed.

These new approaches will have to involve substantial change. The current maritime governance situation has remained largely unaltered since the early twentieth century and rests upon the established series of institutions that may have been added to (for example the emergence of the EU) but have remained the same in form, operation, jurisdiction, power, and influence. New institutions may be needed, certainly new ways of organizing them and in particular the communications and power that flows between them. New ways of looking at how the maritime sector can be organized and controlled are undoubtedly needed to ameliorate the problems of flag-hopping, the advantages taken by the shipping industry of the nation-state-dominated structure at present, and to absorb and accommodate the entirely globalized environment within which shipping now operates.

What is perhaps most important, however, is not so much that the specific changes which have been outlined here are carried out, not even that the need for change is accepted, but that the question is asked whether the current maritime governance system is optimal, appropriate for its purpose, and could not be improved? Despite the continued clear manifestation of maritime policy failure there is little (if any) debate about whether things might need to change.

We have spent considerable time emphasizing the substantial, profound changes that take place in society generally—through Modernism and Postmodernism over the past century or so— which in turn have major ramifications for governance and policy making in all sectors. Globalization and the altered role of the nation-state are significant reflections of these changes and shipping is a significant global player yet working in a governance framework designed around obsolete societal structures. The result is that the role and characteristics of maritime governance need at least to recognize that societal change has occurred and that in turn implies the need for reflection and consideration of governance change as well. A Postmodern society (or whatever else it might be called, but clearly a changed society) requires Postmodern governance if policies are to be respected, effective, and relevant. Clearly, that is far from the case at the moment.

References

Agnew, J. (1994). The territorial trap: The geographical assumptions of international relations theory. *Review of International Political Economy, 1*(1), 53–80.

Alderton, T., & Winchester, N. (2002). Globalization and de-regulation in the maritime industry. *Marine Policy, 26*, 35–43.

Allen, J. (2003). *Lost geographies of power*. Oxford: Blackwell.

Altman, J. A., & Petkus, E., Jr. (1994). Towards a stakeholder-based policy process: an application of the social marketing perspective to environmental policy development. *Policy Sciences, 27*, 37–51.

Andrews, D. M. (1994). Mobility and state autonomy: toward a structural theory of international monetary relations. *International Studies Quarterly, 38*(2), 193–218.

Armitage, J. (1999a). Paul Virilio: An introduction. *Theory, Culture and Society, 16*(5–6), 1–23.

Armitage, J. (1999b). From modernism to hypermodernism and beyond: An interview with Paul Virilio. *Theory, Culture and Society, 16*(5–6), 25–55.

Armitage, J. (1999c). Paul Virilio: A select bibliography. *Theory, Culture and Society, 16*(5–6), 229–240.

Armitage, J. (2000a). *Beyond postmodernism? Paul Virilio's hypermodern cultural theory*, ctheory.net. www.ctheory.net/articles.aspx?id=133.

Armitage, J. (Ed.). (2000b). *From modernism to hypermodernism and beyond*. London: Sage.

Armitage, J., & Graham, P. (2001). Dromoeconomics: Towards a political economy of speed. *Parallax, 7*(1), 111–123.

Aron, R. (1966). *Peace and war; A theory of international relations*. Garden City, NY: Doubleday.

Arrighi, G. (1994). *The long twentieth century*. London: Allen and Unwin.

Aspinwall, M. (1995). *Moveable feast*. Avebury: Aldershot.

Axelrod, R. (1987). *The gamma paradigm for studying the domestic influence on foreign policy*. Washington, DC: Annual Meeting of the International Studies Association.

References

Ayee, J. R. A. (1994). *An anatomy of public policy implementation: The case of decentralization policies in Ghana*. Avebury: Aldershot.
Bauman, Z. (2000). *Liquid modernity*. Cambridge: Polity Press.
Beck, U. (1992). *Risk society*. London: Sage.
Beckmann, J. (Ed.) (1999a). *SPEED—A workshop on space, time and mobility*, Transportradets notatserie, nr. 99–05. Transportradet, Kobenhavn.
Beckmann, J. (1999b). Introduction—Risks, benefits and tools of speed. In J. Beckmann (Ed.), *SPEED—A workshop on space, time and mobility* (pp. 7–10), Transportradets notatserie, nr. 99–05 Transportradet, Kobenhavn.
Beckmann, J. (2004). Mobility and safety. *Theory Culture and Society, 21*(4/5), 81–100.
Beetham, D. (1991). *Legitimation*. London: Macmillan.
Benko, G. (1997). Introduction: Modernity, postmodernity and the social sciences. In G. Benko & U. Strohmayer (Eds.), *Geography, history and social science* (pp. 1–48). Dordrecht: Kluwer.
Bennett, P. (2000). Environmental governance and private actors: enrolling insurers in international maritime regulation. *Political Geography, 19*, 875–899.
Berg, L. (1993). Between modernism and postmodernism. *Progress in Human Geography, 17*(4), 490–507.
Blatter, J. K. (2001). Debordering the world of states: Towards a multi-level system in Europe and a multi-polity system in North America? Insights from border regions. *European Journal of International Relations, 7*(2), 175–209.
Borzel, T. A. (2007). *European governance—Negotiation and competition in the shadow of hierarchy*. Montreal, Canada: European Studies Association Meeting.
Bowman-Gilfillan. (2004). *The use or abuse of the corporate veil in shipping: Is legislation needed to unveil it, if so, for what reasons?* http://www.bowman.co.za.
Braithwaite, J., & Drahos, P. (2000). *Global business regulation*. Cambridge: Cambridge University Press.
Brenner, N. (1998). Between fixity and motion: Accumulation, territorial organization and the historical geography of spatial scales. *Environment and Planning D, 16*(4), 459–481.
Brett, E. A. (1985). *The World economy since the war—The politics of uneven development*. Basingstoke: Macmillan.
Brett, V., & Roe, M. S. (2010). The potential for the clustering of the maritime transport sector in the Greater Dublin region. *Maritime Policy and Management, 37*(1), 1–16.
Brooks, M. R., & Pallis, A. A. (2008). Assessing port governance models: process and performance components. *Maritime Policy and Management, 35*(4), 411–432.
Brugger, N. (1999). A critical introduction to the works of Paul Virilio. In J. Beckmann (Ed.), *SPEED—a workshop on space, time and mobility* (pp. 11–21), Transportradets notatserie, nr. 99–05. Kobenhavn: Transportradet.
Callaghy, T. (1990). Lost between state and market: The politics of economic adjustment in Ghana, Zambia and Nigeria. In J. Nelson (Ed.), *Economic crisis and policy choice: The politics of adjustment in the third world* (pp. 257–321). Princeton, NJ: Princeton University Press.
Čapek, K. (1936). *Válka s mloky (War with the newts)*. London: George Allen and Unwin.
Castells, M. (1989). *The informational city: Information technology, economic restructuring and the urban regional process*. Oxford: Blackwell.
Castells, M. (1996). *The rise of the network society*. Cambridge, MA: Blackwell.
Castells, M. (1997). *The information age*. Oxford: Blackwell.
Castells, M. (2000a). Materials for an exploratory theory of the network society. *British Journal of Sociology, 51*(1), 5–24.
Castells, M. (2000b). *The rise of the network society* (2nd ed.). Cambridge: Blackwell.
Cerny, P. G. (1995). Globalization and the changing logic of collective action. *International Organization, 49*(4), 595–625.
Charan, R., & Freeman, R. E. (1979). Building bridges with corporate constituents. *Management Review, 68*(11), 8–13.

Chowdhury, K. (2006). Interrogating "newness". Globalization and postcolonial theory in the age of endless war. *Cultural Critique, 62*, 126–161.

Clarkson, M. (1995). A stakeholder framework for analyzing and evaluating corporate social performance. *Academy of Management Review, 20*(1), 92–117.

Coeck, C., & Dooms, M. (2007). *The practical application of stakeholder management in ports: A cross-case analysis of best practices in world ports*. Annual Conference of the International Association of Maritime Economists (IAME), Athens, Greece.

Commission of the European Communities. (2001). *European governance. A white paper*, COM (2001) 428, Brussels.

Commission of the European Communities. (2008). *An integrated maritime policy for the European Union*, Brussels.

Commission of the European Communities. (2009). *Towards an integrated maritime policy for better governance in the Mediterranean*, Communication from the Commission to the Council and the European Parliament, COM (2009) 466 Final, Brussels.

Commission on Global Governance. (1995). *Our global neighbourhood*. Oxford: Oxford University Press.

Conley, V. A. (1999). The passenger: Paul Virilio and feminism. *Theory, Culture and Society, 16*(5–6), 201–214.

Cooke, P. (1990). *Back to the future. Modernity, postmodernity and locality*. London: Unwin Hyman.

Cooper, R., & Burrell, G. (1988). Modernism, postmodernism and organizational analysis. *Organization Studies, 9*(1), 91–112.

Crogan, P. (1999a). Theory of the state. Deleuze, Guattari and Virilio on the state, technology and speed, *Angelaki. Journal of the Theoretical Humanities, 4*(2), 137–147.

Crogan, P. (1999b). The tendency, the accident and the untimely: Paul Virilio's engagement with the future. *Theory, Culture and Society, 16*(5–6), 161–176.

Crosby, B. L. (1996). Policy implementation: The organizational challenge. *World Development, 24*(9), 1403–1415.

Cubitt, S. (1999). Virilio and new media. *Theory, Culture and Society, 16*(5–6), 127–142.

De Langen, P. W. (2007). Stakeholders, conflicting interests and governance in port clusters. In M. R. Brooks & K. P. Cullinane (Eds.), *Devolution, port governance and port performance* (pp. 457–477). London: Elsevier.

De Sousa Santos, B. (1995). Globalization, nation-states and the legal field, from legal diaspora to legal ecumene? In B. de Sousa Santos (Ed.), *Towards a new common sense: Law, science and politics in the paradigmatic transition* (pp. 250–377). New York: Routledge.

De Vivero, J. L. S., Mateos, J. C. R., & del Corral, D. F. (2009). Geopolitical factors of maritime policies and marine spatial planning: state, regions and geographical planning scope. *Marine Policy, 13*, 624–634.

Deleuze, G., & Guattari, F. (1988). *A thousand plateaus: Capitalism and schizophrenia*. London: The Athlone Press.

Der Derian, J. (1990). The (s)pace of international relations: Simulation, surveillance, and speed. *International Studies Quarterly, 34*(3), 295–310.

Der Derian, J. (1992). *Antidiplomacy: Spies, terror, speed and war*. Oxford: Blackwell.

Der Derian, J. (1999). The conceptual cosmology of Paul Virilio. *Theory, Culture and Society, 16*(5–6), 215–227.

Derrida, J. (1984). No apocalypse, not now (full speed ahead, seven missiles, seven missives). *Diacritics, 14*, 20–31.

Dicken, P., Kelly, P. F., Olds, K., & Yeung, W. (2001). Chains and networks, territories and scales: towards a relational framework for analysing the global economy. *Global Networks, 1*(2), 89–112.

Dickens, P. (2009). The cosmos as capitalism's outside. *Sociological Review, 57*(1), 66–82.

Dickens, P., & Ormrod, J. S. (2006). *The outer spatial fix. critical approaches to outer space panel*. BISA Annual Conference, Cork, Ireland.

Dickens, P., & Ormrod, J. S. (2007). *Cosmic society. Towards a sociology of the universe.* London: Routledge.
Donaldson, T. (1999). Making stakeholder theory whole. *Academy of Management Review, 24*(2), 237–241.
Donaldson, T., & Preston, L. E. (1995). The stakeholder theory of the corporation: Concepts, evidence and implications. *Academy of Management Review, 20*(1), 65–91.
Dooms, M., Macharis, C., & Verbeke, A. (2004, August 25–29). *Proactive stakeholder management in the port planning process: Empirical evidence from the port of Brussels.* European Regional Science Association 44th European Congress, Porto, Portugal.
Druckman, D. (1978). Boundary role conflict: Negotiations as dual responsiveness. In J. W. Zartman (Ed.), *The negotiation process: Theories and applications* (pp. 100–101). Beverly Hills: Sage. 109.
Evans, P., Jacobson, H., & Putnam, R. (Eds.). (1993). *Double edged diplomacy.* Berkeley, CA: University of California Press.
Fletcher, A., Guthrie, J., Steane, P., Roos, G., & Pike, S. (2003). Mapping stakeholder perceptions for a third sector organization. *Journal of Intellectual Capital, 4*(4), 505–527.
Freeman, R. E. (1984). *Strategic management: A stakeholder approach.* London: Pitman.
Friedman, A. L., & Miles, S. (2002). Developing stakeholder theory. *Journal of Management Studies, 39*(1), 2–21.
Friedmann, J. (2000). Reading Castells: Zeitdiagnose and social theory. *Environment and Planning D, 18*, 111–120.
Furger, F. (1997). Accountability and systems of self-governance: The case of the maritime industry. *Law and Policy, 19*(4), 445–476.
Gane, M. (1999). Paul Virilio's bunker theorizing. *Theory, Culture and Society, 16*(5–6), 85–102.
Gibbins, J. R., & Reimer, B. (1999). *The politics of postmodernity: An introduction to contemporary politics and culture.* London: Sage.
Giddens, A. (1990). *The consequences of modernity.* Stanford, CA: Stanford University Press.
Gilfedder, D. (1994). Virilio: The cars that ate Paris. *Transition, 43*, 36–43.
Gioia, D. A. (1999). Practicability, paradigms, and problems in stakeholder theorizing. *Academy of Management Review, 24*(2), 228–232.
Goodpaster, K. E. (1991). Business ethics and stakeholder analysis. *Business Ethics Quarterly, 1*, 53–73.
Gordenker, L., & Weiss, T. G. (1995). Pluralising global governance: Analytical approaches and dimensions. *Third World Quarterly, 16*(3), 357–387.
Gordon, D. (1994). *Sustaining economic reform in Sub-Saharan Africa: Issues and implications for USAID, Implementing policy change project* (Working Paper No. 6). Washington, DC: US Agency for International Development.
Haggard, S. (1990). The political economy of the Philippine debt crisis. In J. Nelson (Ed.), *Economic crisis and policy choice: The politics of adjustment in the third world* (pp. 215–256). Princeton, NJ: Princeton University Press.
Haggard, S. (1995). *The reform of the state in Latin America.* Unpublished Draft Manuscript, Graduate School of International Relations and Pacific Studies, University of California, San Diego, CA.
Hall, S., Held, D., & McGrew, T. (Eds.). (1992). *Modernity and its futures.* Cambridge: Polity Press.
Harrison, J. S., & Freeman, R. E. (1999). Stakeholders, social responsibility, and performance: Empirical evidence and theoretical perspectives. *Academy of Management Journal, 42*(5), 479–485.
Harvey, D. (1982). *The limits to capital.* Chicago, IL: University of Chicago Press.
Harvey, D. (1989). *The urban experience.* Oxford: Blackwell.
Harvey, D. (1990). *The condition of postmodernity.* Cambridge MA: Blackwell.
Harvey, D. (1996). *Justice, nature and the geography of difference.* Oxford: Blackwell.
Harvey, D. (2000). *Spaces of hope.* Edinburgh: Edinburgh University Press.

Harvey, D. (2006). *Spaces of global capitalism. Towards a theory of uneven geographical development.* London: Verso.

Hassan, I. (1999). Globalism and its discontents: Notes of a wandering scholar. *Profession*, 59–67.

Hassan, R. (2009). Crisis time: Networks, acceleration and politics within late capitalism, ctheory.net, www.ctheory.net/articles.aspx?id=618.

Hemmati, M., Dodds, F., & Enayati, J. (2002). *Multi-stakeholder processes for governance and sustainability: Beyond.* London: Earthscan.

Hess, M. (2008). Governance, value chains and networks: an afterword. *Economy and Society*, 37(3), 452–459

Hill, C. W. L., & Jones, T. M. (1992). Stakeholder-agency theory. *Journal of Management Studies*, 29(2), 131–154.

Hobsbawm, E. (1998). The nation and globalization. *Constellations*, 5(1), 1–9.

Hoffmann, S. (1964). Europe's identity crisis: Between past and America. *Daedalus*, 93(4), 12–44.

Hoffmann, S. (1995). *The European Sisyphus.* Boulder, CO: Westview Press.

Hosseus, D., & Pal. L. A. (1997). Anatomy of a policy area: The case of shipping. *Canadian Public Policy*, XXIII(4), 399–415.

Jenkins, R. (1988). *Transnational corporations and uneven development.* London: Routledge.

Jessop, B. (1997). Capitalism and its future: Remarks on regulation, government and governance. *Review of International Political Economy*, 4(3), 561–581.

Jordan, A. G. (2001). The European Union: An evolving system of multi-level governance…or government? *Policy and Politics*, 29(2), 193–208.

Kahler, M. (1989). International financial institutions and the politics of adjustment. In J. Nelson (Ed.), *Fragile Coalitions: The politics of economic adjustment* (pp. 139–159). Washington, DC: Overseas Development Council.

Kellner, D. (1999). Virilio, war and technology: Some critical reflections. *Theory, Culture and Society*, 16(5–6), 103–125.

Lambert, J. (1991). Europe: The nation–state dies hard. *Capital and Class, Spring*, 43, 9–24.

Leach, N. (1999). Virilio and architecture. *Theory, Culture and Society*, 16(5–6), 71–84.

Lewis, M. W., & Wigen, K. E. (1997). *The myth of continents.* Berkeley, CA: University of California Press.

Lindenberg, M. M., & Ramirez, N. (1989). *Managing adjustment in developing countries.* San Francisco, CA: International Center for Economic Growth.

Luke, T. W. (1991). The discipline of security studies and the codes of containment: Learning from Kuwait. *Alternatives*, 16, 315–344.

MacLeod, G., & Goodwin, M. (1999). Space, scale and state strategy: Rethinking urban and regional governance. *Progress in Human Geography*, 23(4), 503–527.

Mangat, R. (2001). The death of distance? Globalization of international relations, *E-merge. Student Journal of International Affairs*, 2, February.

Marx, K. (1992). *Das kapital* (Vol. 1). London: Penguin.

Mazmanian, D. A., & Sabatier, P. A. (1989). *Implementation and public policy.* Lanham, MD: University Press of America.

McQuire, S. (1999). Blinded by the (speed of) light. *Theory, Culture and Society*, 16(5–6), 143–159.

Melucci, A. (1989). *Nomads of the present: Social movements and individual needs in contemporary society.* London: Hutchinson.

Michael, D. N. (1962). *Cybernation: The silent conquest.* Santa Barbara, California: Center for the Study of Democratic Institutions.

Michelet, J. (1833). *Histoire de la France* (Vol. 1). Paris: A Lacroix and Co.

Mitchell, R. E., Agle, B. R., & Wood, D. J. (1997). Toward a theory of stakeholder identification and salience: Defining the principle of who and what really counts. *Academy of Management Review*, 22(4), 853–886.

Morgenthau, H. J. (1967). *Politics among nations: The Struggle for power and peace.* New York: Alfred Knopf.

Nelson, J. M. (Ed.). (1989). *Fragile coalitions: The politics of economic adjustment.* Washington, DC: Overseas Development Council.

Nielsen, L. D., & Jespersen, P. H. (2001). Time and space in freight transport. In L. D. Nielsen & H. H. Oldrup (Eds.), *Mobility and transport* (pp. 63–72). Denmark: Transportradet, Aalborg University, Department of Development and Planning.

Nielsen, L. D., & Oldrup, H. H. (Eds.). (2001). *Mobility and transport*. Denmark: Transportradet, Aalborg University, Department of Development and Planning.

Notteboom, T. (2002). *The quest for sustainable port development: Managing stakeholder relations in a highly competitive environment*. Proceedings of the 8th IACP Conference on Port Cities and World Trade, Dalian, China: Urban Strategies and Industrial Dynamics, International Association of Cities and Ports.

Notteboom, T. E., & Winkelmans, W. (2002). *Stakeholder relations management in ports: Dealing with the interplay of forces among stakeholders in a changing economic environment*, Proceedings of the International Association of Maritime Economists (IAME) Conference, Panama City.

Pallis, A. A. (2005–2006). Maritime interest representation in the EU. *European Political Economy Review, 3*(2), 6–28.

Pallis, A. A., & Tsiotsis, S. G. P. (2006). *Inside EU-Level maritime interest groups: Structures, lobbying practices and governance, Interim results of a research project at the Jean Monnet Centre in European Port Policy*. Department of Shipping Trade and Transport, University of the Aegean, Chios.

Parker, M. (2009). Capitalists in space. *Sociological Review, 57*(1), 83–97.

Parker, M., & Bell, D. (Eds.). (2009a). *Space travel and culture: From Apollo to space tourism*. Malden, MA: Blackwell.

Parker, M., & Bell, D. (2009b). Introduction: Making space. *Sociological Review, 57*(1), 1–5.

Pelton, J. N. (2004). Satellites as worldwide change agents. In J. N. Pelton, R. J. Oslund, & P. Marshall (Eds.), *Communications Satellites. Global change agents*. Mahwah, NJ: Lawrence Erlbaum Associates.

Pelton, J. N., Oslund, R. J., & Marshall, P. (Eds.). (2004). *Communications satellites. Global change agents*. Mahwah, NJ: Lawrence Erlbaum Associates.

Peters, B. G., & Pierre, J. (2001). Developments in intergovernmental relations: Towards a multi-level governance. *Policy and Politics, 29*(2), 131–135.

Picciotto, S. (1991). The internationalization of the state, capital and class. *Spring, 43*, 43–63.

Picciotto, S. (1998, April 16). *Globalization, liberalization, regulation*. Conference on Globalization, The Nation–state and Violence, University of Sussex.

Pomeroy, R., & Douvere, F. (2008). The engagement of stakeholders in the marine spatial planning process. *Marine Policy, 32*, 816–822.

Porter, J. (1968). The future of upward mobility. *American Sociological Review, 33*(1), 5–19.

Poster, M. (1984). *Foucault, Marxism and history: Mode of production versus mode of information*. Cambridge: Polity Press.

Poster, M. (1990). *The mode of information: Poststructuralism and social context*. Cambridge: Polity Press.

Puchala, D. J. (1972). Of blind men, elephants and international integration. *Journal of Common Market Studies, 10*(3), 267–284.

Putnam, R. D. (1988). Diplomacy and domestic politics: the logic of two-level games. *International Organization, 42*(3), 427–460.

Rosenau, J. N. (2000). *The Governance of fragmegration: Neither a world republic nor a global interstate system*. Quebec: Congress of the International Political Sciences Association.

Rugman, A. (Ed.). (1982). *New theories of the multinational enterprize*. London: Croom Helm.

Santos, M. (1995). Contemporary acceleration: World-time and world-space. In G. Benko & U. Strohmayer (Eds.), *Geography, history and social sciences* (pp. 171–176). Dordrecht: Kluwer.

Savage, G. T. (1981). The *Significance of interruptions in same sex dyadic converstaions*. Invited Panel Verbal and Non-Verbal Investigations in Dyadic Conversations. Non-verbal Communication Interest Group, Pittsburgh, PA.

Scholte, J. A. (2000). *Globalization*. London: Palgrave.

Sletmo, G. K. (2002, November 13–15). *National shipping policy and global markets: A retrospective for the future.* International Association of Maritime Economists Annual Conference (IAME), Panama City.

Snyder, G. H., & Diesing, P. (1977). *Conflict among nations: Bargaining, decision-making and systems structure in international crises.* Princeton, NJ: Princeton University Press.

Soja, E. W. (2000). *Postmetropolis: Critical studies of cities and regions.* Oxford: Blackwell.

Sokal, A., & Bricmont, J. (1998). *Intellectual impostures.* London: Profile Books.

Stalder, F. (2006). *Manuel Castells.* Cambridge: Polity Press.

Stoker, G. (1998). Governance as theory: Five propositions. *International Social Science Journal, 50*(155), 17–28.

Stone, C. (1989). *Regime politics: Governing atlanta 1946–1988.* Lawrence, KS: University Press of Kansas.

Storper, M. (1997). *The regional world. Territorial development in a global economy.* New York: The Guilford Press.

Strange, S. (1976). Who runs world shipping? *International Affairs, 52*(3), 346–367.

Strange, S. (1996). *The retreat of the state. The diffusion of power in the world economy.* Cambridge: Cambridge University Press.

Sutherland, M., & Nichols, S. (2006). Issues in the governance of marine spaces. In M. Sutherland (Ed.), *Administering marine spaces: International issues* (pp. 6–20). Copenhagen: International Federation of Surveyors.

Taylor, P. J. (2000a). World cities and territorial states under conditions of contemporary globalization. *Political Geography, 19,* 5–32.

Taylor, P. J. (2000b). Worlds of large cities; pondering Castells' space of flows. *Third World Planning Review, 21*(3), 3–10.

Taylor, P. J. (2000c). World cities and territorial states under conditions of contemporary globalization II: Looking forward, looking ahead. *GeoJournal, 52,* 157–162.

Taylor, P. J. (2005). New political geographies: global civil society and global governance through world city networks. *Political Geography, 24,* 703–730.

Tradewinds. (2010). *Shipowners turn Athens screw,* September 10.

Virilio, P. (1977). *Speed and politics. An essay on dromology.* New York: Semiotext(e).

Virilio, P. (1995). *Speed and information: Cyberspace alarm!* ctheory.net, www.ctheory.net/articles.aspx?id=72.

Virilio, P. (1998). Military space. In J. Der Derian (Ed.), *The Virilio reader.* Oxford: Blackwell.

Virilio, P. (1999). Indirect light: Extracted from polar inertia. *Theory, Culture and Society, 16*(5–6), 57–70.

Waite, G. (1996). *Nietzsche's corps/e: Aesthetics, politics, prophecy or, the spectacular technoculture of everyday life.* Durham, NC: Duke University Press.

Walton, R. E., & McKersie, R. B. (1965). *A behavioral theory of labor negotiations: An analysis of a social interaction system.* New York NY: McGraw-Hill.

Wang, J. J., Slack, B. (2002). *Port governance in China: A case study of Shanghai.* Occasional Paper Series, (Paper No. 9). The Centre for China Urban and Regional Studies: Hong Kong Baptist University, Hong Kong.

Wang, J. J., Ng, A. K., & Olivier, D. (2004). Port governance in China: A review of policies in an era of internationalizing port management practices. *Transport Policy, 11,* 237–250.

Wark, M. (1988). On technological time: cruising Virilio's over-exposed city. *Arena, 83,* 82–100.

Waterman, P. (1999). The brave new world of Manuel Castells: What on earth (or in the ether) is going on? *Development and Change, 30,* 357–380.

Watts, M. J. (1991). Mapping meaning: Denoting difference, imagining density; Dialectical images and postmodern geographies. *Geografiska Annaler, 73*(B), 7–16.

Webster, F. (2002). *Theories of the information society.* London: Routledge.

White, L. G. (1990). *Implementing policy reforms in LDCs: A strategy for designing and effecting change.* Boulder, CO: Lynne Reinner Publishers.

Wolch, J. (1989). The shadow state: Transformations in the voluntary sector. In J. Wolch & M. Dear (Eds.), *The power of geography: How territory shapes social life* (pp. 197–221). London: Unwin Hyman.

Wood, E.M. (1997) Modernity, postmodernity or capitalism? *Review of International Political Economy, 4*(3), 539–560.

Yeung, H. W. (2000). Embedding foreign affiliates in transnational business networks: The case of Hong Kong firms in south-east Asia. *Environment and Planning A, 32,* 201–222.

Zacher, M. W., & Sutton, B. A. (1996). *Governing global networks.* Cambridge: Cambridge University Press.

Zurbrugg, N. (1999). Virilio, Stelarc and terminal technoculture. *Theory, Culture and Society, 16*(5–6), 177–199.

Zurn, M. (2003). Globalization and global governance; from societal to political denationalization. *European Review, 11*(3), 341–364.

Author Biography

Professor Michael Roe currently holds the Chair of Maritime and Logistics Policy at the Centre for Maritime Logistics, Economics and Finance, Plymouth Business School, Plymouth University. He has previously worked with the Greater London Council, West Midlands County Council (Birmingham, UK), and the Universities of Aston, Coventry, London Guildhall and City. A graduate in Geography, with postgraduate qualifications in Transport Planning and Engineering and a Doctorate in Transport Welfare Economics, he is the author of over 60 refereed journal papers, 12 books and a large number of other publications. His research interests focus upon the problems of maritime governance. His wife, Liz, provides moral and intellectual support while his two children, Joe and Siân, provide entertainment and expenses. He has active interests in traveling, modern European literature, the historical development of Soviet Europe, restoring aging VW Beetles, the work of Patti Smith and most importantly, the exploits of Charlton Athletic FC.

Index

4056/86, 3

A
Acceleration, 123, 324, 243, 413, 415–418
Accounting, 200, 201, 299, 303
Architectural, 271, 272, 274, 279, 361, 368
Architecture, 53, 82, 88, 119, 147, 255, 264, 265, 268, 271, 272, 274, 279, 281, 296, 298–301, 303, 325, 336, 358, 363, 364, 367, 380
Art, 208, 230, 243, 262, 265, 268, 272, 281, 325, 342, 355, 363, 366, 377, 379, 380, 381
Arts, 255, 265, 275, 296, 300–303, 305, 306, 358, 378
Association of Southeast Asian Nations (ASEAN), 112

B
Ballast water, 21, 87
BIMCO, 19, 30, 32
Blogs, 368, 369
Blue Paper, 29, 303, 33, 5
Blue Policy Paper, 160
Borders, 120, 142, 144, 147, 148, 150, 152, 154–157, 161, 162, 170, 173, 177–179, 182, 199–206, 219, 239, 240, 262, 323, 324, 334, 340, 358, 419, 80, 85
Bulk carriers, 31
Bunkering, 21
Boundaries, 41, 43, 44, 46, 63, 68, 71, 77, 79, 83, 91, 112, 123, 125, 129, 137, 138, 141, 144, 145, 147-149, 151, 152, 155, 156, 162, 164–167, 176–180, 182, 197, 200, 205, 206, 219, 224, 227, 228, 235, 236, 239, 240, 263, 269, 274, 306, 323, 324–325, 329, 330, 337, 362, 372, 406, 418, 422

C
Cabotage, 5, 30
Capital accumulation, 119, 123, 124, 143, 151, 152, 156, 168, 177, 201, 202, 205–207, 213–216, 220, 221, 228–232, 235, 239, 245, 255, 271, 288, 301, 312, 330, 365
Capitalism, 64, 118, 122, 124, 125, 137, 144–146, 154, 156, 158, 163, 173, 182, 198, 200, 202, 206, 208–211, 213–219, 221, 222, 224–234, 236, 237, 241, 242, 245, 255, 256, 258, 262–264, 266, 268–271, 274–276, 279, 281, 288, 298, 302–304, 308, 309, 312, 313, 329–332, 334, 336, 338, 342, 364, 401, 416, 418, 420, 423
Capitalist, 74, 138, 139, 143, 144, 151, 165, 168, 177, 201, 205–211, 213–217, 220, 222, 224, 227–231, 233, 237, 255–257, 262, 265, 266, 269, 275, 278, 279, 281, 299, 306, 307, 312, 313, 325, 329, 330, 332, 338, 343, 366, 407, 422, 423
Chronopolitics, 119, 324
Cluster, 366, 55, 68, 92
Clusters, 68, 122, 411
CO_2, 21–23
Colonialism, 153, 207, 208, 298, 363
Colonization, 208
Commission, 3–15, 17, 18, 20, 22–25, 28, 29, 30, 32, 42, 44, 49, 82, 89, 396, 409, 410

C (cont.)

Complexity theory, 353
Container, 9, 20, 159, 161, 162, 167, 177, 213, 236, 367, 368, 370, 371, 375
Containers, 79, 159, 219, 235, 367, 368, 379, 382, 416
Corporations, 123, 124, 129, 155, 157, 158, 160, 173, 199, 203, 204, 206, 232, 236, 267, 308, 331, 406, 409, 54, 84, 85
Cosmic Capitalism, 422
Cosmos, 210, 220, 422, 423
Criminalization of seafarers, 18
Crises, 124, 163, 200, 207, 209–211, 214–216, 221, 229, 230, 233, 237, 275, 308, 313, 330
Crisis, 117, 152, 154, 163, 182, 210, 211, 214, 216, 219, 221, 225, 229, 231, 237, 281, 30, 305, 306, 308, 311, 330, 331, 332, 333, 336, 378, 423, 78
Crises of capitalism
CRPM, 28
Cruise liner, 363
Culture, 53, 94, 148, 149, 162, 166, 202, 204, 231, 238, 241, 242, 263, 264, 268, 277–279, 299–302, 305–307, 314, 329, 335, 343, 358, 362, 365, 378, 380, 415, 421
Cultures, 53, 177, 224, 263, 276, 335, 363, 421

D

Democracy, 52, 54, 58, 79, 123, 142, 173, 200, 212, 238, 275, 278, 311
Deregulation, 10, 205, 206, 238, 365, 384
Dialectic, 206, 220, 257, 259, 262
Dialectical, 221, 237, 257, 262
Disorganized capitalism, 200, 222, 228, 232, 234, 245, 255, 270, 275, 281, 313, 329, 330
Double-hulled tankers, 6, 87
Dromology, 210, 418

E

E-booking, 370
E-commerce, 370
European Community Shipowners' Association (ECSA), 20, 24, 25, 29, 30, 33
EEDI, 26
Elites, 61, 71, 73, 139, 158, 235, 257, 313
Email, 180, 210, 368, 369, 371, 376
Emissions, 21–26
Engels, 198, 212, 221
Environment, 8, 25, 26, 30, 53, 70, 75, 84, 87, 116, 119, 127, 145, 161, 170, 183, 184, 197, 202, 224, 242, 267, 269, 281, 297, 311, 325, 327, 328, 33, 334, 339, 361–363, 365, 369, 370, 383, 384, 396, 397, 399, 400, 401, 406, 422, 424
Environmental, 6, 13, 16, 18, 20, 21, 25, 32, 80, 85, 89, 123, 156, 163, 172, 175, 203, 313, 383, 409, 412
Epoch, 47, 163, 178, 221, 225, 231, 241, 259, 298, 311, 314, 326, 333, 335, 336, 342, 343, 398, 406, 417, 418
European Commission (EC), 3, 4, 8–11, 15, 16, 18, 19, 21, 22, 25, 27, 29, 404, 408
European Community Shipowners Association, 5, 18
European Conference of Peripheral Maritime Regions, 28
European Maritime Safety Agency (EMSA), 20, 29, 375
European Union
Energy Efficiency Design Index, 22, 25
EU, 2–26, 28, 31, 32, 45, 46, 48, 53, 64, 68, 70, 71, 75, 82, 85, 87–90, 93, 112, 115, 118, 121, 125, 128, 130, 137, 144, 146, 150, 155, 158, 159, 163, 164, 166, 170, 176, 179, 180, 182, 201, 215, 234, 238, 242, 258, 265, 288, 297, 303, 322, 326, 369, 372, 383, 384, 396, 398, 401, 407, 408, 410, 411, 423
EU governance, 82, 89
EU ports policy, 9, 10
EUROS, 30, 4–6

F

Facebook, 368, 369
Films, 296, 379
Failure, 2, 3, 5–7, 9, 10, 13, 14, 16, 18–20, 25, 28, 29, 31, 32, 46, 41, 52, 55, 63, 78–80, 82, 85, 87, 88, 91, 93, 94, 112, 116, 117, 138, 146, 151, 153, 154, 161, 167, 174, 180, 183, 209, 231, 236, 237, 239, 241, 255, 258, 271, 274, 277–280, 296, 299, 302, 304, 311, 314, 338, 339, 342, 353, 362, 395–400, 402, 404, 406, 413, 424
Fixation of space, 208
Flag states, 1, 7, 14, 17, 84, 91, 206, 384
Flags of convenience, 4, 91, 397
Flexible accumulation, 64, 129, 216, 218, 229, 241–243, 288, 308, 323
Flexible production, 288, 384
Flexible specialization, 224, 228, 308, 364

Index 437

Flow, 47, 79, 129, 143, 145, 155, 165, 199, 257, 258, 267, 202, 207, 224, 241, 242, 309, 325, 328, 419, 420, 422
Flows, 60, 62, 73, 77, 78, 92, 144, 145, 148, 149, 151, 152, 164, 165, 170, 176, 202, 211, 219, 224, 227, 240, 242, 301, 323–325, 328, 340, 357, 358, 360, 382, 415, 419–422, 424
Fordism, 121, 224, 225, 227, 233, 238, 241, 243, 269, 273, 274, 288, 307, 308, 310, 313, 315, 326, 364
Fordist, 78, 85, 121, 122, 201, 224, 225, 232, 233, 241, 265, 268, 273, 275, 288, 307, 310, 364, 365, 384
Fragmegrating, 259
Fragmegration, 121
Futurism, 244, 265
Futurist, 243, 265

G

GATS, 155
Gender, 300, 362, 363
Global governance, 147, 153, 76
Global value chains, 74
Globalization and
  The Nation-State, 203
Globalization, 1, 30, 41, 43, 47, 51, 53, 60, 63, 65, 66, 68, 71, 77, 78–81, 83–85, 89, 93, 94, 113, 115, 116, 119–124, 127, 130, 137, 141–157, 159, 160–162, 164–183, 197–210, 213, 215–222, 224–229, 231, 232, 235–245, 255, 258, 262, 274, 278, 280, 288, 304, 307, 309, 311, 317, 321, 323–326, 329, 333, 337, 338, 340, 342, 353, 358, 366, 370, 376, 378, 379, 381, 382, 397, 399–402, 406, 408, 412, 416, 417, 419–423
Glocalization, 148
Good governance, 68, 80, 89, 93, 402, 411
Governance, 2, 3, 6, 11, 13, 14, 17, 19, 23, 25, 33, 41–57, 59, 60–66, 68–94, 111–130, 137, 140–148, 150–161, 165, 167, 170–172, 174–176, 178–184, 197, 202–206, 209, 210, 215–220, 224–228, 230–245, 255, 258, 259, 261, 262, 266–269, 271, 274, 277, 278, 280, 281, 288, 296, 300–304, 306, 307, 309, 310, 311, 313–315, 317, 320, 322–328, 330, 333, 334, 336–340, 343, 353, 354, 358, 361, 385, 386, 396–409, 410–413, 415, 417–419, 420–424

Governance Failure, 76
Governance models, 55, 72
Green Paper, 5, 9, 28–30, 383

H

Hegel, 207, 208, 217, 237, 257
Hegemony, 140, 49, 212–214, 269, 410
Hierarchical, 6, 8, 12, 13, 28, 29, 43, 44, 50, 53, 57, 60–63, 65, 69, 70, 72, 74, 75, 77, 78, 81–87, 89, 90, 111–130, 137, 140, 151–153, 157, 159, 161, 170, 172, 174–177, 201, 203, 205, 228, 232, 238, 242, 244, 257, 263, 267, 269, 270, 278, 280, 281, 308, 323, 327, 238, 337, 339, 404, 419, 421
Hierarchy, 5, 6, 9, 11, 12, 32, 43–45, 47, 51, 54, 57, 60, 61, 63, 69, 70, 73, 89, 94, 111, 113, 115, 117–125, 128–130, 176, 177, 262, 266, 270, 201, 203, 230, 240, 315, 323, 327, 328, 330, 400, 416, 419
Hollowing out, 63, 65, 158, 179, 180
Hundertwasser, 381

I

International Association of Classificiation Societies (IACS), 32, 33
International Labour Organisation (ILO), 13, 31, 32, 88
International Monetary Fund (IMF), 183, 310
International Maritime Organisation (IMO), 6, 11–19, 21–26, 29, 31, 45, 60, 74, 80, 84, 86–89, 91, 112, 144, 146, 150, 170, 173, 180, 182, 183, 205, 215, 234, 242, 265, 277, 297, 308, 310, 317, 339, 362, 369, 372, 375, 396, 398, 399, 401, 404, 407
Imperialism, 145, 207, 208, 213, 216, 304, 363
Implementing Policy Change, 403
Inland container terminals (ICT), 374
Installation, 379
Institutional Change, 407
Institutional, 17, 41, 45, 53, 54, 56, 63, 70, 77, 80, 88, 115, 117, 119, 120, 122, 123, 125, 126, 138, 146, 147, 160, 163, 168, 171, 180, 182, 183, 200, 205, 215, 240, 255, 268–271, 277, 278, 295, 301–303, 307, 309, 311, 313, 322–324, 333, 334, 337, 340, 356, 362, 369, 400, 404, 407, 408
Interest groups, 23, 53, 55, 60, 62, 67, 68, 73, 76, 78, 81, 84, 159, 166, 372, 396, 411, 413

I (*cont.*)
International Association of Classification Societies, 32
International Chamber of Shipping (ICS), 11, 18, 19, 22, 32
International Monetary Fund (IMF), 288, 310
International relations, 53, 65, 117, 141, 162, 172, 173, 176, 277, 304, 409
International trade, 122, 123, 47, 379
Iron cage, 269, 271, 333, 356
Iron triangle, 58–60
Iron triangles, 58, 60
IT, 308

J
Just-in-time (JIT), 270, 376
Jurisdiction, 2, 3, 8, 21, 24, 18, 30, 67, 85, 92, 111, 112, 115, 121, 125, 138, 142, 144, 150, 159, 162, 17, 179, 183, 184, 200, 205, 396, 423
Jurisdictional, 2, 5, 11, 17, 19, 23, 27, 32, 33, 41, 49, 65, 72, 76, 83, 84, 89–91, 93, 94, 112, 113, 115, 124–127, 129, 130, 280, 145–148, 152, 174, 176, 180, 182, 205, 239, 242, 322, 323, 396, 402, 406–408, 411, 412
Just-in-time, 241

K
Kant, 116, 198, 265, 402

L
Labour, 13, 32, 71, 382, 397
Lean manufacturing, 384
Liner conferences, 3, 4, 384
Liner shipping, 3, 20, 22, 404
Literature, 41, 43, 52, 53, 55, 57, 59, 71, 82, 138, 197–199, 204, 265, 296, 306, 326, 342, 378, 409, 422
Logistics, 28, 85, 94, 210, 225, 241, 242, 122, 270, 299, 325, 358, 369, 376, 384, 415, 416

M
Management, 21, 22, 26, 42, 70, 75, 77, 84, 94, 155, 221, 224, 225, 265, 267–269, 273, 274, 281, 295, 299, 303, 308, 313, 317, 354, 356, 357, 366, 370, 376, 384, 401, 409–411, 415, 422

Marine Environment Protection Committee (MEPC), 21, 22
Maritime Governance, 240, 385
Maritime policy, 2, 4, 8, 9, 11, 12, 17, 18, 28, 30, 32, 54, 61, 72, 83, 84, 91, 130, 137, 163, 174, 217, 245, 263, 295, 301, 302, 310, 311, 315, 326, 327, 353, 395–397, 404, 410, 411, 423
Market, 1, 8–10, 24, 26, 30, 43, 44, 53, 57, 60, 61, 69, 74, 78, 80, 86, 91, 94, 118, 123, 126, 138, 155, 158, 172, 182–184, 198, 204, 209, 211, 214, 215, 227, 230, 231, 237, 242, 256, 266, 275, 279, 301, 303, 304, 306–309, 314, 324, 327, 336, 357, 358, 366, 377, 396, 398–400, 409, 410, 418, 422, 423
Markets, 57, 262
Marxist, 113, 161, 175, 198, 216, 219, 231, 232, 303, 401
Media, 48, 5, 13, 29, 164, 255, 305, 306, 324, 332, 371, 384, 396, 409, 411, 413, 417, 419, 420
Member states, 11–20, 23–25, 28–30, 33, 4–9, 45, 80, 81, 83, 87, 88, 118, 125, 126, 322
MEPC, 21, 24–26
Mobius web, 73
Modernism, 200, 214, 238, 243, 255–258, 262–276, 278–281, 288, 295, 297, 298, 299, 302, 306–308, 311–313, 317, 320, 329, 330, 343, 355, 362, 381, 406, 418, 423, 424
Modernist, 80, 85, 86, 94, 140, 154, 162, 205, 225, 228, 231, 233, 241, 244, 245, 255, 256, 263–265, 267–280, 288, 295, 296, 301–315, 317, 323, 327–329, 331332, 334, 336, 338, 339, 341, 343, 353, 354, 355, 360, 361–363, 377, 380–384, 386, 396, 407, 415, 418, 421, 422
Multi-level governance, 82, 128, 179
Music, 265, 281, 325, 379

N
NAFTA, 112, 201
National shipping policy, 83
Nation-state decline, 153
Nation-state, 14, 52, 77, 79, 80, 83, 112, 116, 118, 119, 137, 141, 142, 144, 148–157, 159–165, 168, 170–177, 179–183, 200, 203–205, 208, 209, 211–213, 215, 217, 227, 232, 235, 238, 239, 244, 258, 262, 277, 304, 310, 328, 331, 363, 396–398, 400, 401, 405, 406, 419

Index 439

Naval strategy, 381, 382
Neo-functionalism, 176
Network, 43, 54, 55, 57–66, 68–72, 74, 81, 88, 91–93, 115, 119, 120, 122, 126, 130, 145, 147, 158, 159, 161, 170, 201, 210, 214, 240, 262, 267, 304, 308, 322, 327, 328, 339, 340, 357, 409, 416, 419, 421, 422
Networks, 43–46, 54, 55, 57–66, 68–72, 74–76, 91, 117, 119, 120, 124–126, 129, 146–149, 159–161, 163, 170, 179, 181, 202–204, 206, 262, 307, 315, 326–328, 340, 356, 406, 409, 410, 415, 42, 420, 421, 423
NGO, 78
Northern, 25, 271

O

Organisation for Economic Cooperationand Developemnt (OECD), 10, 13, 60, 74, 144, 150, 160, 180, 200, 215, 265, 288, 310, 407
Open method of coordination (OMC), 126
Oresund Link, 358
Organizational, 44, 52, 62, 63, 65, 66, 71, 72, 151, 157, 170, 181, 220, 221, 233, 237, 241, 297, 299, 303, 313, 315, 323, 327, 331, 336, 355, 357, 366, 376, 384, 409, 419, 422
Organization, 6, 8, 12, 16, 19, 32, 52, 58, 60, 66, 69, 70, 74, 75, 81, 88, 89, 90, 91, 112, 116, 118–121, 123, 124, 129, 139–141, 146, 151, 152, 159, 162, 175, 178, 182, 183, 202, 206, 213, 215, 220–222, 226, 228, 230–232, 234, 237, 241, 256, 257, 257, 268, 269, 270, 265, 266, 268, 269, 274, 275, 281, 288, 299, 300, 302, 303, 305, 306, 309, 311–313, 317, 320, 321, 323, 324, 325, 327, 331–334, 336, 338, 342, 354–357, 362, 363, 372, 377, 396, 399, 401, 406, 415, 418, 419
Organizations, 9, 14, 15, 23, 31, 32, 43, 45, 49, 52, 60, 62, 64, 65, 66, 68, 69, 72, 77–81, 88, 119, 120, 123, 126, 142, 144, 150, 154, 155, 158–160, 176, 183, 204, 214, 215, 227, 235, 238, 257, 268, 271, 278, 288, 297, 299, 306, 311, 313, 314, 317, 325, 327, 328, 330, 333, 336, 354, 355, 357, 358, 372, 384, 385, 405, 406, 408, 409–411
Organizational theory, 62
Outer spatial fix, 422
Outsourcing, 85, 241, 327, 376, 377, 384
Ownership, 88, 92–94, 111, 112, 145, 202, 203, 209, 227, 356, 358, 401, 402, 404, 408

P

Piracy, 16, 84, 305
Places of refuge, 19, 404
Pluralism, 48, 50–55, 71, 117, 257, 258, 165, 305, 332, 336, 363
Pluralist, 50, 51, 52, 53, 54, 55, 58, 322, 334, 336, 336
Pods, 180, 368
Policy, 1–8, 10–15, 17–24, 26–33, 41, 43, 45–49, 51–65, 68, 70–73, 75, 76, 78–80, 82–91, 83, 93, 94, 111, 112, 114–119, 124, 125, 127–130, 137, 146, 147, 150, 152, 154, 157–161, 163, 166, 167, 170–177, 179–182, 184, 197, 203–205, 209, 215–219, 221, 224, 226–235, 237–245, 255, 258, 259, 261–263, 267–271, 274, 277, 278, 281, 288, 295, 296, 301, 302, 304, 306, 307, 309, 311–315, 317, 320–328, 330–334, 336–340, 343, 353, 354, 356–358, 362, 366, 369–372, 383, 385, 395, 396–413, 417, 419, 421–424
Policy networks, 49, 55, 58, 62, 65, 68
Policy-makers, 28, 48, 71, 85, 271, 174, 335, 338, 369
Political, 3, 6, 12, 13, 19, 21, 23, 25, 26, 41, 44–46, 49, 52–55, 58, 59, 61–65, 68, 72, 74, 76, 80–84, 88, 92, 94, 112, 113, 115, 116, 119–122, 124, 125, 137–141, 143, 144, 146, 147, 150–156, 158–173, 175–180, 182, 184, 197–204, 206, 211–214, 216–218, 220, 225, 227–230, 233, 235, 236, 239, 241–244, 255, 257, 258, 265, 266, 268, 270, 275, 277, 278, 296, 301, 303–305, 308, 309, 311–313, 323, 325, 325, 331, 332, 334, 336, 337, 339, 354, 355, 363, 369, 395, 398–401, 404, 408, 410, 415, 417, 420, 423
Politics, 22, 23, 47, 53, 55, 59, 61, 63, 75, 80, 81, 126, 130, 150, 155, 156, 158, 160, 162, 166, 167, 171, 175, 178, 179, 184, 199, 200, 212, 218, 228, 231, 232, 237, 241, 243, 255, 274, 278, 296, 300, 301, 303, 304, 311, 312, 324, 326, 331, 332, 339, 342, 343, 405, 415
Pollution, 12, 13, 18, 19, 21, 25, 84, 204, 313
Port Policy, 9
Port services, 9, 10

P (*cont.*)
Port, 9, 10, 84, 85, 91–94, 270, 273–275, 358, 364–366, 369, 372, 374–376, 382, 383, 409–411
Ports, 5, 9, 10, 28–31, 84, 85, 92–94, 210, 275, 364–366, 372, 375, 382–384, 404, 411
Post-fordist, 85, 122, 224, 225, 232, 307–309
Postmodern, 47, 53, 54, 78, 80, 85, 86, 91, 122, 127, 129, 145, 149, 150, 154, 161–164, 173, 180, 182, 197, 202, 205, 214, 216, 218, 221, 224, 225, 228, 232, 241, 243, 244, 257, 258, 261, 262, 268, 270, 271, 274, 275, 278–280, 288, 295–309, 311–315, 317, 320–343, 353–358, 360, 361–368, 370–372, 374–386, 396–401, 406–408, 415–419, 421–424
Postmodernism, 53, 256–259, 275–281, 164, 214, 232, 233, 238, 243, 244, 288, 295, 296–308, 310–313, 320, 322, 324–333, 335–343, 353–355, 358, 361–367, 372, 375, 379–381, 384–386, 406, 418, 424
Postmodernist, 228, 231, 241, 245, 255, 271, 278–280, 296, 308, 311, 312, 314, 315, 317, 329, 354, 357, 361, 418
Power, 14, 18, 23, 29, 42, 44–47, 52–54, 57, 62, 65, 67, 70–72, 78, 81, 82, 92, 115, 116, 118, 122, 129, 130, 137–140, 143–145, 148, 151–153, 157, 158, 160–162, 164–168, 170, 173, 173, 175–178, 180, 181, 183, 199–202, 205, 206, 211–215, 217, 219, 220, 226, 230, 233–236, 239, 241, 243, 256, 258, 262–266, 270, 275, 277, 298, 299, 301, 305, 323, 324, 328, 331, 332, 334, 338, 340, 343, 355, 369, 384, 397–400, 403, 405, 406, 408, 410, 418, 420–423
Premodern, 243, 315, 345
Premodernism, 281
Pressure groups, 13, 29, 45, 369, 395, 409
Prestige, 6, 87
Public administration, 125, 303
Punctuated equilibria, 221

R
Razzle dazzle, 382
Regional policy, 27
Risk society, 155, 156, 158, 175
Roads, 360, 361
Rubbish, 220, 225, 226

S
Sails, 375
Scales, 79, 113, 115, 116, 119, 122–124, 129, 146, 148, 168–171, 217, 221, 236, 238, 259, 400, 404, 406
Scrapping, 13
Security, 6, 8, 12, 20, 30, 31, 48, 80, 126, 154, 158, 160, 172, 279, 202, 212, 234, 243, 303, 313, 323, 372, 378, 382, 395, 399, 400, 405
Ship Efficiency Management Plan (SEMP), 26
Shipbuilding, 31
Shipping, 2–8, 11, 13, 16, 18–20, 22–25, 30–33, 53, 80, 83–85, 87, 91, 94, 119, 122, 130, 143–145, 150, 154, 157–160, 163, 170–173, 178, 180, 183, 184, 202, 204, 205, 209, 210, 215, 219, 226, 23, 232, 236, 236, 239, 242–245, 262, 270, 271, 275, 304, 305, 320, 321, 324, 327, 328, 330, 338, 340, 362, 364, 369–372, 375, 383, 384, 395, 396, 398–404, 406–408, 413, 416, 417, 421–424
Single buoy, 372
Single point mooring (SPM), 372
Soft law, 10
Sovereignty, 16, 23, 43, 87, 91, 118, 138–144, 146, 147, 150, 153, 154, 156–161, 164, 166, 172, 173, 175, 183, 202, 205, 216, 232, 233, 235, 244, 266, 277, 304, 322–325, 334, 340, 397, 408, 419
Space, 30, 45, 55, 63, 64, 79, 84, 94, 113, 119, 122, 123, 126, 140, 142–153, 156, 160, 162, 164, 167–170, 175, 177, 179, 181, 182, 199, 202, 204, 207–210, 215–219, 220–222, 224, 225, 230, 231, 233, 235, 237, 239–244, 272, 274, 277, 299, 312, 322–326, 328, 329, 334–336, 338, 358, 361, 365, 366, 370, 371, 377, 385, 401, 404, 413, 416–423
Spaces, 41, 66, 71, 72, 82, 116, 118, 129, 149, 164, 166, 202, 220, 240, 242, 267, 275, 278, 311, 323, 362, 364, 366, 367, 386, 411, 419–421
Space of flows, 421
Space-time compression, 221, 338, 419
Spatial, 64, 70, 72, 78, 79, 86, 112, 115, 116, 118, 119, 122–124, 129, 143–148, 162, 164, 166–169, 175, 177, 202–204, 206–211, 214–222, 228–233, 235–244, 266, 271, 304, 305, 323, 325, 325, 326,

Index 441

334, 338, 360, 372, 374, 406, 411, 412, 418, 419, 423
Spatial fix, 123, 208, 210, 215, 217, 218, 231, 232, 237, 241–243, 271
Spatial fixation, 217, 219
Spatial scales, 116, 148
Speed, 13, 23, 71, 142, 181, 210, 219, 220, 225, 243, 244, 271, 298, 323–325, 334, 358, 365, 413, 415–418
Stakeholders, 2, 4–6, 10, 23, 29–33, 49, 53, 65, 66, 71, 81, 86, 89–91, 115, 119, 123, 126, 128, 152, 205, 234, 270, 311, 325, 327, 337, 339, 340, 403, 404, 408–413, 420
State, 1, 10–16, 18–20, 23, 29, 30, 33, 112, 113, 115, 117–120, 122–126, 128–130, 256–259, 262, 264, 266, 270, 272, 277, 278, 280, 288, 137–184, 197–200, 202–205, 207–209, 211–215, 217, 219–222, 226–233, 235–240, 244, 298, 301, 304, 305, 308, 309, 315, 320–324, 326, 328, 330, 331, 333, 334, 336–340, 358
State-centred, 53, 159
State-centricism, 259, 277
State-centric, 79, 112, 114, 128, 144, 151, 168, 170, 177, 182
State decline, 154
STCW, 16
Styles, 279, 305, 333, 336, 363, 377
Subsidiarity, 49, 89, 118, 128
Sulfur, 21, 25, 26, 213
Supply chain management, 122, 225, 270, 376, 384
Supranational, 2, 5, 6, 8, 11, 19, 22, 30, 41, 47, 66, 71, 82, 89–91, 112, 119, 123, 125, 126, 128, 129, 144, 145, 152, 155, 158–160, 164, 166, 167, 169, 176, 180, 183, 204, 215, 235, 238, 259, 322, 341, 384, 405, 408
Surplus capital, 220, 329
Survival, 161

T
Tankers, 6, 7, 13, 31, 87, 372
Territorial, 18, 63, 79, 113, 118, 119, 121, 123, 124, 126, 127, 139–145, 147, 148, 151, 152, 155, 156, 158, 159, 162, 164–169, 171, 173, 176–178, 181, 182, 206, 208, 215–217, 221, 232, 233, 236–239, 259, 288, 334, 372, 398, 408, 418, 420
Territoriality, 79, 122, 140, 143–145, 147, 150, 164, 165, 168, 169, 169, 172, 175–177, 181, 213, 218, 233–239, 244, 326, 421

Territorialization, 140, 151, 167, 169, 181, 205, 220, 221, 237, 259, 408, 421
Territories, 139–142, 144, 145, 147–149, 170, 206, 208, 216, 219, 236, 325, 418
Territory, 77, 113, 129, 138–141, 143–150, 153, 156, 161, 162, 166–168, 171–173, 177, 204, 205, 208, 210, 212, 213, 219, 224, 233, 235, 236, 240, 244, 270, 295, 322–326, 335, 362, 397, 404, 416, 418, 419, 421–423
The North American Free Trade Agreement (NAFTA), 112
Third world, 201, 363
Time, 15, 19, 20, 113, 118, 124, 125, 130, 140, 140, 143–145, 148, 150, 153, 159, 160, 162, 164, 165, 170, 171, 175, 175, 181, 181, 182, 202, 202, 204, 206, 207–211, 215–222, 225, 229–231, 233, 235–237
Time-space, 113, 124, 143, 182, 206, 219, 241, 311, 312, 335, 396, 397, 418
Tonnage tax, 7, 30, 154, 204, 400, 402
Tonnage taxation, 7
Total quality management (TQM), 268, 303
Tourism, 365, 379
Tourists, 225, 334, 364, 379
Tramp shipping, 4
Trans-Atlantic Conference Agreement (TACA), 3
Transport, 14, 20, 28, 42, 54, 127, 149, 183, 201, 207, 210, 215, 219, 225, 236, 242, 274, 305, 324, 358, 360–364, 375, 377–379, 383, 411, 413
Transportation, 168, 183, 206, 210, 340, 377
Treaty of Rome, 3, 7, 8, 30, 340
Treaty of Westphalia, 139, 181, 199, 237, 340
Twitter, 368, 369

U
UNCLOS, 17, 18, 87
UN Framework Convention on Climate Change (UNFCCC), 16, 22–26
Unbundling, 154, 164, 322, 326
UNCTAD, 180
United Nations (UN), 22, 80, 142, 183, 277, 288, 408, 409
Unmanned ship, 375
Urban planning, 274, 281, 358, 361

V
Violence, 2, 42, 65, 140, 160, 169, 211–215, 244, 264, 298, 304

W

War, 17, 80, 85, 117, 157, 180, 211–213, 215, 233, 261, 268, 270, 271, 278, 305, 307, 311, 330, 340, 372, 408, 415, 415

Warehousing, 275, 364, 377, 383, 384

Waterfront, 275, 304, 364–366, 372, 374

Weber, 112, 120, 212, 213, 269, 295, 315, 333, 340, 356

Westphalia, 140, 141, 153, 172, 322, 421

Westphalian, 141, 153, 154, 177, 181, 182, 205, 266, 288, 321, 322, 324, 326, 397, 419

WISTA, 362

World Bank (WB), 43, 80, 183, 310, 288

World Shipping Council (WSC), 20, 22, 33

World Trade Organisation (WTO), 31, 60, 74, 80, 142, 150, 155, 180, 181, 200, 265, 310, 372, 401, 407

Y

You-Tube, 369

Printed by Printforce, the Netherlands